Water Resources Engineering

Water Resources Engineering

Edited by **Herbert Lotus**

SYRAWOOD
PUBLISHING HOUSE
New York

Published by Syrawood Publishing House,
750 Third Avenue, 9th Floor,
New York, NY 10017, USA
www.syrawoodpublishinghouse.com

Water Resources Engineering
Edited by Herbert Lotus

International Standard Book Number: 978-1-68286-320-6 (Hardback)

Printed in China.

Contents

Preface

The world is advancing at a fast pace like never before. Therefore, the need is to keep up with the latest developments. This book was an idea that came to fruition when the specialists in the area realized the need to coordinate together and document essential themes in the subject. That's when I was requested to be the editor. Editing this book has been an honour as it brings together diverse authors researching on different streams of the field. The book collates essential materials contributed by veterans in the area which can be utilized by students and researchers alike.

Water resource engineering is an emerging field of study that aims to analyse the distribution and quality of diverse water resources. The main aim of this field is to evaluate and prevent the contamination of water resources and ensure supply of clean water. This book covers in detail some prominent concepts and topics revolving around water resource engineering such as waste water treatment, environmental engineering, climate change, analysis of water quality, etc. From theories to research to practical applications, case studies related to all contemporary topics of relevance to this field have been included in this book. It will prove immensely beneficial to professionals and students involved in this area at various levels.

Each chapter is a sole-standing publication that reflects each author´s interpretation. Thus, the book displays a multi-facetted picture of our current understanding of applications and diverse aspects of the field. I would like to thank the contributors of this book and my family for their endless support.

Editor

Contamination of boreholes water by 76 pesticides molecules in the cotton zone of Kérou

ELEGBEDE Bernadin[1,3]*, EDORH A. Patrick[2,3], CLEDJO Placide[4], KOUMOLOU Luc[3], BOKO Michel[4], Messanvi GBEASSOR[5], E. E. CREPPY[6]

[1]Direction Départementale de l'Hydraulique, Ouémé-Plateau, 04 BP 1412, Bénin.
[2]Département de Biochimie et de Biologie Cellulaire, Université d'Abomey-Calavi (UAC), 01 BP 526 Cotonou, Bénin.
[3]Centre Interfacultaire de Formation et de Recherche en Environnement pour le Développement Durable (CIFRED), Université d'Abomey-Calavi (UAC), 03 BP 1463, Jéricho, Cotonou, Bénin.
[4]Laboratoire Pierre Pagney : Climat, Eau, Ecosystème et Développement (LACEEDE), Faculté des Lettres, Arts et Sciences Humaines, Université d'Abomey-Calavi, 03 BP 1122 Cotonou, Bénin.
[5]Département de Physiologie/Pharmacologie, Faculté des Sciences, Université de Lomé (UL), BP 1515 Lomé, Togo.
[6]Laboratory of Toxicology and Applied Hygiene/UFR of Pharmaceutical Sciences, 146, Street Léo Saignat, 33076 Bordeaux Cedex France.

Analysis campaign of underground water was done in 2010 on five boreholes water samples situated in agricultural cotton zone. The obtained results showed the presence of various active matters of at least 76 residues of pesticides, especially insecticides, herbicides and fungicides, with accumulated contents which could average 0.350 or 350 µg/L per borehole. Indeed, all the prospected boreholes were contaminated. Every molecules of pesticides analyzed were present with at least 1 µg/L, some molecules concentration such as: aldrine (7 ± 0.7 µg/L), dieldrin (2 ± 0.7 µg/Ls), phorate (7 ± 2 µg/L) and terbufos (7 ± 1 µg/L), highly exceeded maximum residue limits. The most affected zones of the contamination of the underground water by pesticides were those areas where the cotton faming is practiced.

Keywords: Kérou, cotton farming, water of boreholes, pesticides.

INTRODUCTION

Human consumption, health and all activities depend on sufficient availability of quality water. Ground waters which are the main source of drinking water are supplied by the rainwater which penetrates with gravity in the pores and cracks of rocks. Ground waters are formed in the impermeable layers; the tablecloths (Bennabidate, 2000). One could access this subsoil through wells and drillings. In the study area there are 102 boreholes and 30 wells (Elegbede, 2007). This work proposes to focus on boreholes for certain reasons. First of all, the greater part of water consumption comes from boreholes in this big cotton area. The chemicals used for the protection of the cotton farming largely contributed to increase and regularize the agricultural outputs.

However, the agrochemical products were incriminated to be the cause of the degradation of the quality of surface and ground waters (Hani et al., 1997). Thus, insecticide residues of weed killers and fungicides were already detected in water reserves, for example, lakes, lagoons and even in the food (Calvert, 1992; Traore et al., 2003). The contamination of surface water was the consequence of pollution of the grounds by the plant health treatment (Bedding et al., 1983; Cabrindec, 1988). The infiltration of the polluted water into ground waters constitutes one of important subject nowadays. Especially when the plant health products were suspected to have undesirable side effects for human health and the environ-

*Corresponding author. E-mail: Elegbedebern@yahoo.fr.

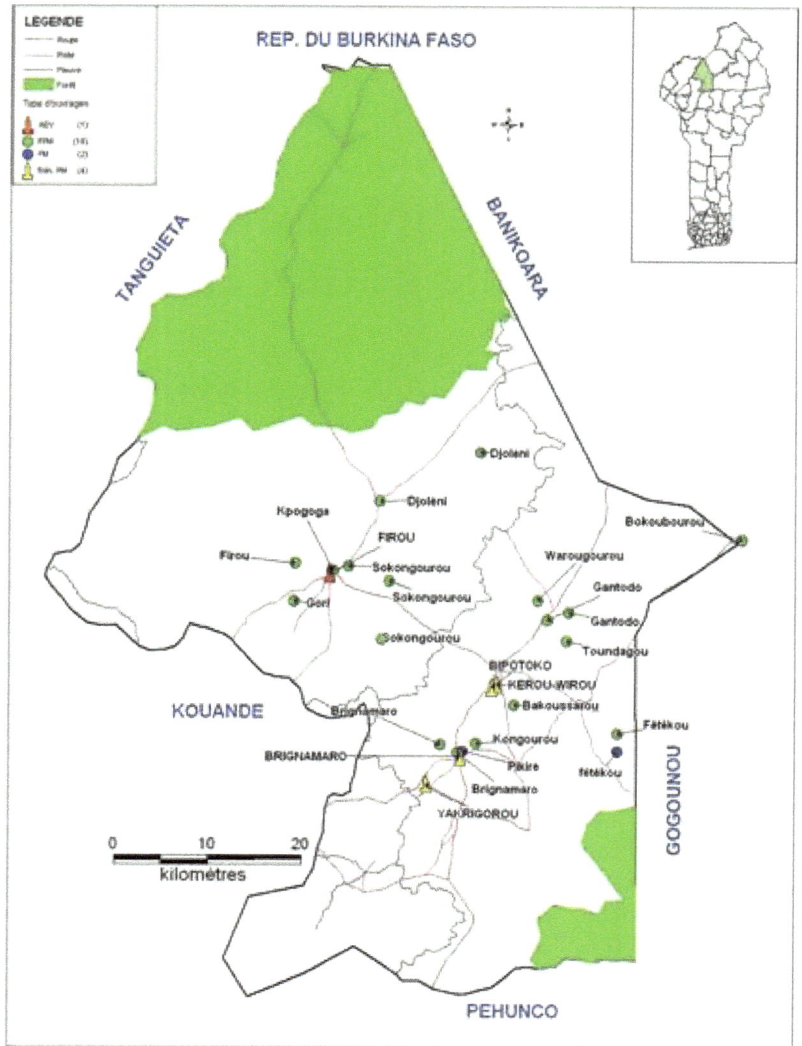

Figure 1. Location Map of the Study Area Showing Sampled boreholes.

ment (Gibons et al., 1987; Fournier et al., 2000). To evaluate the level of contamination of ground waters, a sampling campaign of borehole water was carried out last December 2010 in Kérou's Commune. The objective of this work is based on the identification of the residues of pesticides and their impacts on the quality of water.

MATERIAL AND METHODS

Study area

Figure 1 depicts the study area. Kerou lies on latitude 10°49′29″ North of the Equator and on longitude 2°06′30″ east of the Greenwich median. Kerou stands on the attitude of about 326 m above the sea level. Kerou municipality is limited at North- West by the Municipality of Tanguieta with which it shares the Pendjari Park and at North-East by the Municipality of Banikoara which is the largest producer of cotton at the national level. In the South, it shares its borders with the cotton belt of Pehonco, and Kouande Gogounou. It is divided into four districts that are Brignamaro, Kerou, Centre Kaobagou and Firou. The climate of the region is governed by the North Sudanese regime characterized by two distinct seasons: a dry one from October to April and a rainy one from May to September. The study area Kerou is the second largest producer of cotton at the national level; it is located in the water-shed of the Niger which is controlled by the stream tributaries of Pendjari such as Sota and Mekrou Rivers. Annual rainfall is around 900 to 1000 mm.

Its hydrogeology is dominated by Precambrian granitic-gneiss covered with a thick layer of clay-altered lateritic. Infiltration of rainwater is very intense due to the high density of the fracture network and the nature of the soil generally clay lateritic with high permeability. The crystalline basement consists of igneous and metamorphic rocks and is waterproof, facilitating runoff chemicals substances in the beds of rivers. These geological formations explain the nature and properties of soils which are extremely fragile. The area is dominated by cotton cultivation with uncontrolled use of pesticides and fertilizers. They are sometimes undeniable traces of metals found in the river and in groundwater.

Sampling and analysis

For economic reasons, water from four (4) boreholes and Mekrou River were chosen for the study and their waters were collected and analyzed.

Table 1. Distribution of drillings (F) in the districts of Kérou.

Borehole	District	Populations (PDC, 2006)	Area
F1	Firou	10.204	Gone Firou
F10	Kérou	27.889	Konigourou Centers
F14	Brignamaro	12.687	Ouinra-center
F16			Banbaba
F19			Mékrou River

The river Mékrou (F19) was chosen as control because the exchanges between the rivers and the underground water were permanent. Moreover the water of the Mékrou River is largely consumed by the cotton farmers because of difficult access to drinking water in spite of the efforts of different Government authorities of Benin.

These water samples were collected in December which was dry season in order to have a better assessment of the increment of the concentration of the molecules of pesticides. The water samples were collected with a sampler of glass and were transferred in ambers bottles washed thoroughly and decontaminated (Rodier, 1976; Keith, 1990; Saliot et al., 1992). An aluminum foil was placed on the neck in order to prevent every contact between the samples and the plastic stoppers. Samples were preserved in the ice during their movement and then preserved at 4°C till the time of analyses.

Measurements of temperature, pH, total hardness and electrical conductivity were performed in situ. The pH and the temperature were measured with the help of WTWLF-340 multi-parameter by submersing the sensor in the water. The conductivity was measured by a conductivity meter (WTWLF-340). Nearly, all the values of temperature, pH and conductivity recorded in the boreholes compared favorably with WHO standards for drinking water.

Analyses of residues of pesticides has been carried out by means of chromatography (gaseous phase) and spectrometry (CPG-5M and CPG- ECD). In order to detect several residues, CPG-SM has been added. A total of 76 molecules were found.

The water samples were collected with a glass sampler and were transferred in amber bottles which were washed and sterilized beforehand (Rodier, 1976; Keith, 1990; Saliot et al., 1992). An aluminum foil is placed on the neck in order to prevent any contact between the sample and the stopper of plastic. These samples were stored in ice during their transport and then preserved at 4°C until their analyses.

RESULTS

Table 1 illustrates the distribution of drillings (F1, F10, F14, F16 and F19) in the districts of Kerou.

The results of proportioning were presented in Table 2 and on Figure 2 (contents of different pesticides in water samples of boreholes), and Figure 3 showed cumulated contents of pesticides in water of boreholes. The analysis of the water samples of 5 drillings in the cotton zone of Kérou showed that 76 residues of pesticides were detected in approximately the totality of the drillings sampled close to the cotton plantations. As a whole, according to Table 2 and Figure 2, all the five prospected drillings were contaminated by pesticides. Different pesticides were detected at the same time in the water samples, which were main source of drinking water. The detected residues of pesticides (76) including: insecticides, weed killers and fungicides, with cumulated contents reaching a mean concentration of 0.345 mg/L per drilling, that is, 345 µg/L. The totality of prospected drillings is contaminated.

DISCUSSION

All the molecules of pesticides analyzed were present and the lowest concentration recorded was 1 µg/L. For some pesticides molecules, the concentrations largely exceed the permissive limit which is for aldrin (7 ± 0.7 µg/L), dieldrin (2 ± 0.7 µg/L), phorate (7 ± 2 µg/L) and terbufos (7 ± 1 µg/L).

Additionally, other insecticides such as DDT and its metabolites were also found in the analyzed samples. The weed killers and certain fungicides were detected with at trace levels. However, higher concentrations of these chemicals reaching 9 µg/L were found in some samples; this could be undoubtedly connected to the level of contamination of the grounds. The concentrations recorded in the drillings were relatively in the range of (345 µg/L); they were largely above the maximum permissive limits (2 µg/L). Average contents of HCH, one of the isomers of lindan, which was detected in all drillings were higher than the standard limit (0.1 µg/L). The endosulfan was detected in all contaminated samples of water. It was presented in its both isomer forms: α endosulfan and β endosulfan. The maximum concentrations measured in all the samples were respectively 1 and 4 µg/L. With regard to endosulfan sulphate, the concentrations were 7 µg/L in the water of Mékrou and 6 µg/L in drillings. Their presence in all drillings lets predict of a generalized use of these products despite the restrictions of their use. In addition, the persistence and the remanence of certain pesticides cannot alone explain the values recorded even if several concentrations do not exceed the standard for drinking water. According to Figure 3, the comparison of the concentrations recorded between zones makes it possible to identify the commune of Kérou with its 27,889 inhabitants as the most contaminated commune since it hosts the most contaminated drillings (363 µg/L). This concentration is 726 times higher than the permissive value which is 0. 5 µg/L. The other found residues of pesticides were sometimes specific to certain zones: dieldrin was especially recorded in Mékrou while aldrin was everywhere in the zone. Dieldrin concentrations were respectively 9 µg/L in the river and

Table 2. Results of proportioning of the water samples in pesticides.

Pesticide	Result on the level of drillings (mg/l)					Result on the level of drillings (mg/l)					Pesticide
	F10	F16	F1	F14	F19	F10	F16	F1	F14	F19	
2,4-DDE	0.005	0.003	0,007	0.004	0.003	0.008	0.006	0.009	0.009	0.009	Fensulfothion
2,4-DDE	0.002	0.009	0.001	0.006	0.006	0.007	0.008	0.004	0.004	0.004	Fenthion
2,4-DDT	0.003	0.004	0.004	0.002	0.002	0.001	0.001	0.001	0.001	0.001	Fonofos
4,4-DDE	0.009	0.006	0.009	0.006	0,006	0,005	0,004	0.006	0.006	0.006	Formothion
4,4-DDE	0.004	0.006	0.008	0.001	0.001	0.004	0.006	0.008	0.008	0.008	HCH alpha
4,4-DDT	0.006	0.001	0.004	0.004	0.005	0.004	0.002	0.002	0.002	0.002	HCH beta
Aldrin	0.006	0.008	0.007	0.009	0.009	0.008	0.006	0.004	0.004	0.004	Heptachlor
Azinphos-ethyl	0.001	0.002	0.001	0.001	0.001	0.005	0.001	0.002	0.002	0.002	Epoxy Heptachlor has
Azinphos-methyl	0.008	0.009	0.006	0.005	0.005	0.006	0.004	0.006	0.006	0.006	Heptachlor epoxy B
Bromophos	0.002	0.004	0.004	0.004	0.004	0.001	0.009	0.006	0.006	0.006	Hexachlorobenzene (HCB)
Bromophos-methyl	0.009	0.005	0.008	0.008	0.008	0.005	0.001	0.008	0.008	0.008	Iodofenphos
Carbophenothion	0.004	0.006	0.001	0.001	0.001	0.004	0.005	0.001	0.001	0.005	Isofenphos
Chlorbenside	0.005	0.004	0.006	0.006	0.006	0.006	0.004	0.005	0.005	0.005	Lindane (gamma HCH)
Chlorfenvinphos	0.006	0.008	0.004	0.004	0.004	0.007	0.008	0.004	0.004	0.004	Malathion
Chlormephos	0.004	0.002	0.003	0.003	0.003	0.009	0.001	0.004	0.004	0.004	Mecarban
hlorpyriphos ethyl	0.008	0.005	0.005	0.005	0.005	0.001	0.006	0.001	0.001	0.001	Methidation
Chlorpyriphos-methyl	0.002	0.004	0.001	0.001	0.001	0.006	0.004	0.005	0.005	0.005	Mevinphos
Chlorthiophos	0.005	0.006	0.006	0.006	0.006	0.008	0.003	0.004	0.004	0.004	Omethoate
Demeton-s-methyl-sulfone	0.004	0.007	0.009	0.009	0.009	0.002	0.005	0.001	0.001	0.001	Parathion
Dialiphos	0.006	0.009	0.002	0.002	0.002	0.006	0.001	0.006	0.006	0.006	Parathion-methyl
Diazinon	0.007	0.002	0.009	0.009	0.009	0.004	0.006	0.009	0.009	0.009	Pentachloroaniline
Dichlofenthion	0.009	0.004	0.001	0.001	0.001	0.009	0.009	0.006	0.006	0.006	Phorate
Dichlorvos	0.002	0.001	0.004	0.004	0.004	0.005	0.002	0.005	0.005	0.005	Phosalone
Dicofol	0.004	0.006	0.004	0.004	0.004	0.008	0.009	0.008	0.008	0.008	Phosmet
Dieldrin	0.001	0.009	0.001	0.001	0.001	0.008	0.001	0.006	0.006	0.006	Phosphamidon
Diethion Dimethoate	0.006	0.004	0.006	0.006	0.006	0.001	0.004	0.001	0.001	0.001	Profenofos
Dosimfptp.	0.009	0.006	0.009	0.009	0.009	0.004	0004	0.004	0.004	0.004	Propetamphos
Dotamoùfps	0.004	0.001	0.004	0.004	0.004	0.004	0.001	0.009	0.009	0.009	Pyrazophos
Edifenphos	0.006	0.004	0.006	0.002	0.002	0.008	0006	0.001	0.001	0.001	Pyridafenthion
Endosulfan alpha	0.001	0.001	0.001	0.001	0.001	0.004	0009	0.001	0.001	0.001	Pyrimiphos ethyl
Endosulfan beta	0.004	0.004	0.004	0.004	0.004	0.006	0.001	0.009	0.009	0.009	Pyrimiphos methyl
Endosulfan sulphates	0.001	0.002	0.001	0.001	0.001	0.007	0.006	0.001	0.001	0.001	Quinalphos
Endrin	0.004	0.001	0.008	0.008	0.008	0.006	0.004	0.005	0.005	0.005	Quintozene (PCNB)
Ethoprophos	0.002	0.008	0.002	0.002	0.002	0.001	0.003	0.005	0.005	0.008	Sulfotep
Etrimphos	0.001	0.002	0.005	0.005	0.005	0.004	0.005	0.004	0.004	0.004	Tecnazene
Fenamiphos	0.004	0.004	0.004	0.004	0.004	0.005	0.001	0.008	0.008	0.008	Terbufos
Fenchlorphos	0.001	0.002	0.003	0.003	0.003	0.001	0.006	0.006	0.006	0.006	Thiometon

1 µg/L in the districts. No residue was detected with concentrations lower than 0. 1 µg/L enforced by international organizations (standard for only one residue). Certain organophosphorous pesticides present the average concentrations highest in average sample, all zones together taken, with in particular the malathion (7 to 9 µg/L). The levels of these concentrations, could find their explanations in the abusive frequent use of these active molecules for the treatment of the cultures (Quinio, 1981, Diallo, 1993). Their detection in the medium, despite their capacity of being degraded quickly in the environment (Tomlin, 1997) represented not only their strong use but also the proximity of the plantations and sometimes the weak protection of the drillings. Organochlorine pesticides were also present in all contaminated water. The mean concentrations in the sample were 5 µg/L for lindane, 6 to 7

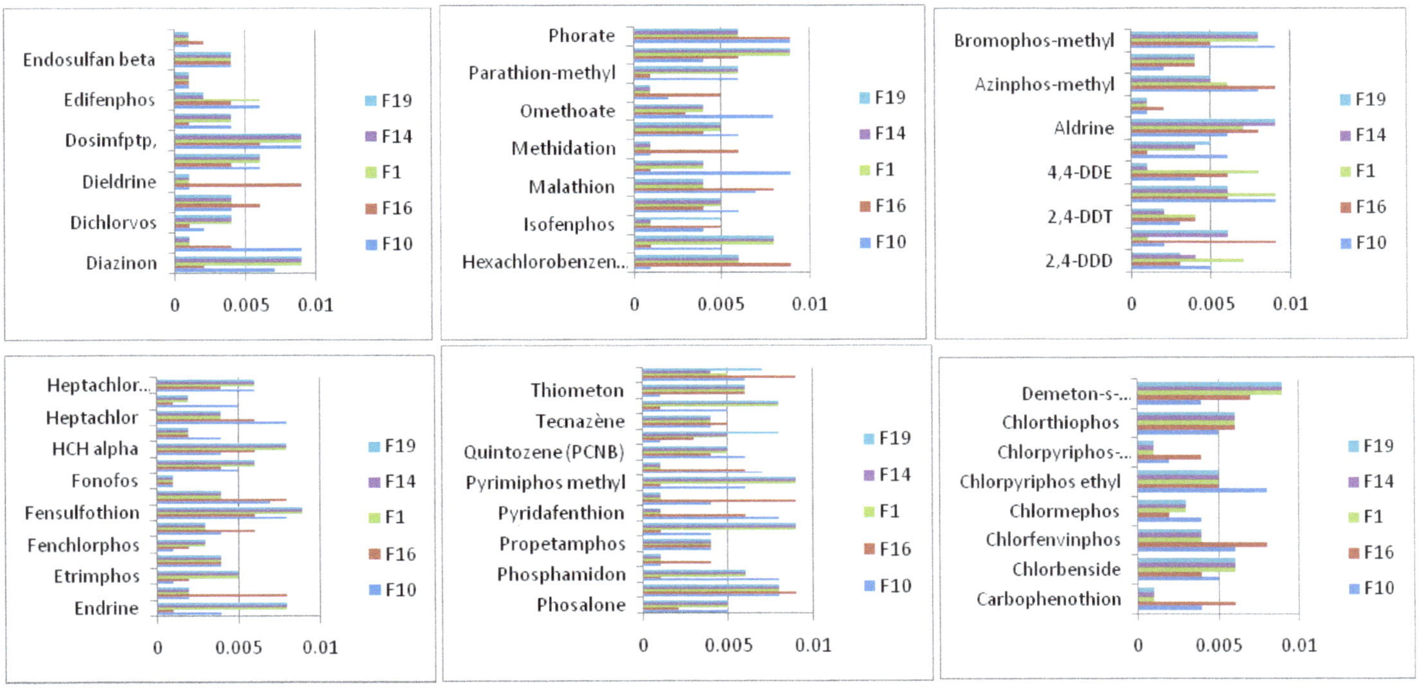

Figure 2. Contents of pesticides in water samples of boreholes.

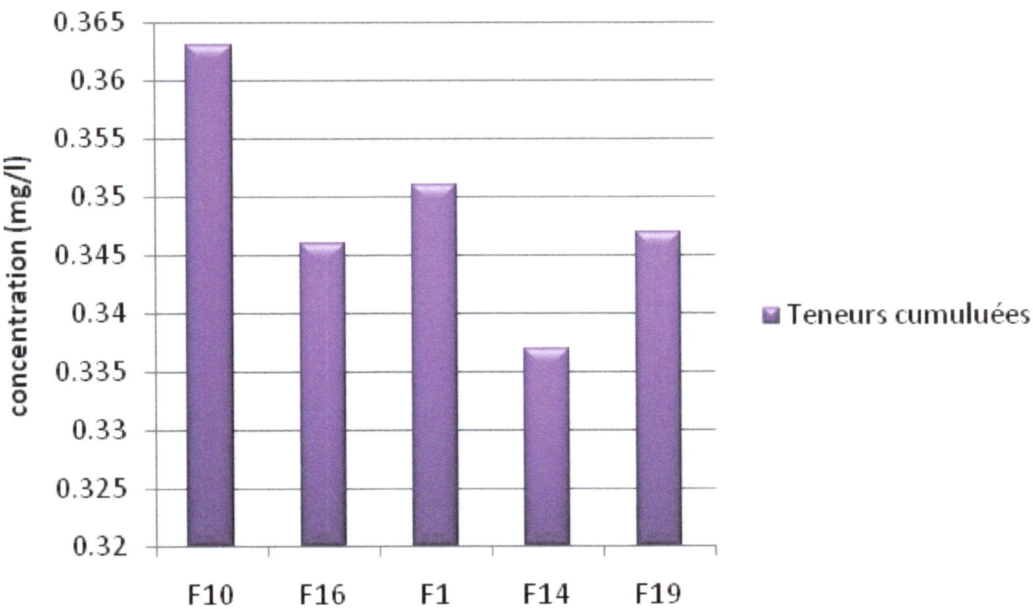

Figure 3. Cumulated contents of pesticides in water of boreholes.

µg/L for dosulfan, 4 to 8 µg/L for heptachlore and dieldrin rising up to 9 µg/L in the drilling of Brignamaro (F16). Their presence nowadays in spite of various bans on their use (Ramade, 1992; Tomlin, 1997) could be due to their persistence in the environment (Vennetier, 1988) or to a fraudulent use. The residues of pesticides are omnipresent in the majority of the sampled tablecloths. Approximately, some of the detected pesticides are prohibited since years (Vennetier, 1988) which indicates that they are no biodegradable and still persist in the ground and that they infiltrate slowly in ground waters or that they are still used fraudulently nowadays especially due to the leniency of the borders with Nigeria. In fact the case of organochlorine (endosulfan, lindane and heptachlor) pre-

prevails in water of the cotton zones. Elsewhere, the most remarkable pesticides are organophosphorous (profenofos and malathion) and others finally are insecticides. This shows that in the zone, the cultures are very diverse (Vennetier, 1988) and occupy of large surfaces. That testifies to the many plant health problems that the cultures meet. Also they receive very intensive chemical treatments, in particular with the nematicides and fungicides, the weed killers. The results of our analyses show that the subsoil water was of good quality contrary to the traditional belief which supposes that the ground removes from water all its contaminants at the time of its infiltration towards the underground tablecloth.

The moisture-holding capacity of the micropolluants by the ground was related to its chemical nature, with its lithographic structure, its high content in organic matter and the depth of the underground tablecloth. These results were alarming in comparison with the recommendations on the quality of the water intended for human consumption (WHO, 1994).

These values of residues of pesticides in water obtained are higher than those recorded in Oueme River by Pazou et al. (2010) in Benin, aldrin (7 ± 0,7 µg/L), dieldrin (1 ± 0,7 µg/L), phorate (5 ± 2 µg/L) and terbufos (3 ± 1 µg/L).

The value for aldrin (7 ± 0.7 µg/L) obtained by N'Dong in the cotton zone of river Senegal is similar to that of Pazou et al. (2010).

However the results in the present investigation are by far lower than those of Adam et al. 2010, who recorded a mean value of 4 µg/L for dieldrin.

Conclusion

The analyses carried out on boreholes water showed a contamination of water by the pesticides with low disparities of contamination according to the zone of culture considered, even if some boreholes were contaminated than others. Concentrations sometimes are higher than the maximum residue limits relative to the drinking water. This report constitutes one of the main public health problems in the study area, because of the cumulative effects of these products on health. It is a growing concern, knowing that water is consumed in great quantity. There is thus a real risk to health. And this risk is higher as the molecules present were used more and more in significant quantity in the zone of studies and also stable in water. Added to this, the risk related to the vulgarizing made of the use of pesticides. It is advisable to recall that the most effective means to reduce the health risk remains the decline of the use of chemicals.

REFERENCES

Adam Soulé , Edorh PA, Totin H, Koumolou L, Amoussou E, Aklikokou K, Boko M (2010). Pesticides et métaux lourds dans l'eau de boisson les sols et les sédiments de la ceinture cotonnière de Gogounou Kandi et Banikoara (Bénin). Int. J. Biol. Chem. Sci. 4(4): 105-114

Bedding N, Mcintyre A, Perry R, Lester J (1983). Organic contaminants in the aquatic environment II- Sci. Total Environ. 26 255-312.

Bennabidate L (2000). Caractérisation hydrologique du bassin versant de Sebou : hydrogéologie, qualité des eaux et géochimie des eaux thermales. Thèse Doc. Es-sc., F.S.T. Fès, pp. 250 .

Cabrindec R (1988). Comportement des substances chimiques en milieu aquatique en fonction de leurs propriétés physicochimiques. Océanie, 14 735-49.

Calvert R (1992). Considérations générales sur le transfert des produits phytosanitaires dans l'eau. Un point sur la protection de plantes. Paris : INRA,

Diallo S (1993). L'agriculture urbaine en Afrique de l'Ouest: revue et perspective de recherche. Cities Feeding People Report 5, IDRC, Ottawa, Canada.

Elegbede MB (2007). Pollution de l'eau des puits et de la nappe phreatique par les pratiques humaines et son impact sur la sante des populations dans la commune de kerou, premier producteur du coton dans les departements de l'atacora/donga. DEA, pp.111.

Fournier MP, Brousseau, Bernier J, Cyr D, Pillet S, Rooney A, (2000).Les perturbateurs endocriniens, Université du Québec, INRS, Institut Armand-Frappier – Santé humaine, pp. 171.

Gibons SM, Hoss VG, Laseter LJ, Rea WJ (1987). Physician's Clinical Guide - A clinical guide to toxic chemicales; enviro-health information center, Richardson, Texas, pp.87-89.

Hani K, Narcisse A, Traoré SK, Koné M, Dembelé A, Houenou P, (1997). Recueil des travaux scientifiques des 3èmes Journées annuelles de la SOACHIM, Lomé.

Keith LH (1990). Environmental sampling: a summary. Environ. Sci. Technol. 24 610-617.

Quinio J (1981). Pollution des eaux superficielles par les pesticides organochlorés. Etude de niveau et des facteurs de contamination. Conséquences au niveau de la potabilisation des eaux. Mémoire section ingénieurs. ENITRTS, Strasbourg.

Ramade F (1992). Précis d'écotoxicologie. Masson, Paris, pp. 300.

Rodier J (1976). L'analyse de l'eau : eaux naturelles, eaux résiduaires, eau de marée. 5e éd. T. II, Bordas, Paris, pp. 364.

Saliot A, Bouloubassi I, Lipiatou E (1992). Stratégies d'échantillonnage et d'analyse appliquées à l'étude des polluants organiques dans les eaux côtières. Hydroecol. Appl. 4: 9-12.

Tomlin CDS (dir.) (1997). The Pesticide Manual: a World Compendium, 11e éd., The British Crop Protection Council, Farnham, Surrey, UK, pp. 1606.

Traoré SK, Koné M, Dembelé A, Lafrance P, Houenou P (2003). Etude comparative du niveau de résidus de pesticides organochlorés chez trois espèces de poissons du lac de Buyo (sud-ouest de la Côte d'Ivoire) et estimation du potentiel de risques pour la santé humaine, J. Soc. Ouest – Afr. Chim. 016 : 137-152.

Vennetier P (1988). Urbanisation, production agricole et autosuffisance alimentaire: réflexions sur le cas africain, *Cahiers d'outre-mer*, 41 : 163.

WHO (1994). Water quality guidelines. Geneva, OMS 1.

Contribution of some water bodies and the role of soils in the physicochemical enrichment of the Douala-Edea mangrove ecosystem

Aaron Suh Tening[1], George Bindeh Chuyong[2], Godswill Azinwie Asongwe[3], Beatrice Ambo Fonge[2], Lydia Likowo Lifongo[1], Antoine David Mvondo-Ze[4], Vivian Bih Che[3] and Cheo Emmanuel Suh[3]

[1]Department of Chemistry, University of Buea, P.O Box 63, Buea, Cameroon.
[2]Department of Botany and Plant Physiology, University of Buea, P.O Box 63, Buea, Cameroon.
[3]Department of Geology and Environmental Science, University of Buea, P.O Box 63, Buea, Cameroon.
[4]Department of Soil Science, Faculty of Agronomy and Agricultural Sciences, University of Dschang, BP 67 Dschang, Cameroon.

The effect of enrichment of water bodies could be of serious crises to the mangrove ecosystem. Changes in physicochemical properties of some water bodies in the Douala-Edea mangrove ecosystem was investigated alongside the potential role of soils in controlling these parameters. Water and soil samples within the Douala industrial zones were collected in February 2010 and analysed using standard methods. The concentrations of cations and chlorides (Cl^-) in the rivers increased from upstream to downstream and with depth. These parameters were not distinct with other anions which showed higher fluctuations around confluences. Many anomalies were obtained in streams and wells at vicinity of the industries. Mean Cl^- concentrations in streams and wells around River Wouri (135.1 and 57.9 mg/l, respectively) were higher than those around River Dibamba (59.3 and 38,2 mg/l, respectively). A low retention capacity of the soils was observed by the non significant ($P > 0.05$) relationship between the clay fraction and cation exchange capacity (CEC). This makes the mangrove ecosystem vulnerable to the increase nutrient from anthropogenic activities as indicated by the occurrence of Nypa Palms (*Nypa fructicans*) and Water Hyacinths (*Echhornia cassipes*). It is therefore imminent that the Douala-Edea Mangrove Ecosystem is being degraded.

Key words: Soils, water, physicochemical properties, mangrove ecosystem.

INTRODUCTION

Rivers constitute a major source of water. Their usage constitutes a major criterion towards sustainable growth and development of a region and its economy (Katte et al., 2003; Gleitsmann et al., 2007). Over the world, a few rivers are unaffected by upstream transportation of their fresh water inflow (Boynton et al., 1995). Unprecedented increase in population and rapid growth of urbanization has seen remarkable impact of man on the environment (Morris et al., 1994). This has tended to a wide scale contamination of soil, water from industrial effluents and domestic sewage discharges (Morris et al., 1994; Lener, 1995). Untreated wastes are often channelled or piped from the industries and/or households onto and/or into the ground or into surface water in coastal environments lined by mangroves. This could change the physicochemical properties of such water bodies rendering these waters undesirable to the mangrove ecosystem.

Soils are natural components of the ecosystem which interact with other components by moving and cycling nutrients and energy (Yerima and Van Ranst, 2005). Soils serve many important functions in our society (Zhao et al., 2009). They serve as a vital "ecological crossroad" of the environment linking the atmosphere, lithosphere and hydrosphere (Rowel, 1994). Most plants survive by absorbing nutrients (essential elements) from the soil. The availability of these nutrients depends on the physic-cochemical properties of the soil, which vary considerably in tropical soils (Yerima and Van Ranst, 2005). Soils rich in organic matter and clays such as montmorillonite hold nutrients and water better than sandy soils, which are deficient in absorptive sites.

Public attention is increasingly focused on environmental pollution and its effects on man and other entities of the ecosystem such as the mangroves. Among the major pollutants are excessive nutrients originating from waste from an urbanized or industrialized society. The soil is one of the primary recipients of these wastes. Once these materials enter the soil, some find themselves in water and become part of the cycle which could eventually affect the mangrove ecosystem.

The mangrove ecosystem is one of the unique features of intertidal zones throughout tropical and subtropical regions. It is located between latitudes 25°N and 25°S of the Equator (Choudbury, 2002) and covering an area of approximately 15 million hectares worldwide (FAO, 2005). The mangrove forest has been traditionally used by indigenous people of the tropics for a variety of purposes: fisheries, fuel wood, scenic beauty, cultural and inspirational well being, medicinal plants and the provision of food (Suratman, 2008). One of the most important aspects of the mangrove ecosystem is its role in the sequestration of carbon, which contributes to the global carbon cycle (Twilley et al., 1992). This role helps to maintain a healthy coastal environment by modifying the climate (UNFCCC, 1998). Despite the ecological importance of mangroves, they are exposed to effluents (potential pollutants in soils, water and vegetation) from households, industries and agricultural establishments (Chiras, 2000). Changes in the physicochemical properties of water have the potential to influence the chemistry of mangrove soils which is not desirable for the mangrove ecosystem. Excess of nutrient inputs (nitrogen) can shift the mangrove ecosystem from carbon sinks to carbon sources and vice-versa (Aurela et al., 2002). The nutrients are harmful to the pneumatophores and propagules (Lafleur et al., 2003; Roulet et al., 2007). Nutrient characterization equally influences regeneration (McGuinness, 1997), forest structure (Ball, 1988), the presence of invasive species such as the Nypa palms (Spalding et al., 2010), and increase in primary productivity (Wickramasinghea et al., 2009). Almost 225000 metric tons of carbon sequestration potentials is lost each year resulting from disturbed mangrove as mangroves take up about 1.5 metric tones of carbon per hectare per year from the atmosphere (Mangrove Action Project, 2010). Disturbed mangrove soils release greater than 11 million metric tones of carbon annually (Mangrove Action Project, 2010). When disturbed, carbon and other polluting substances from the mangrove soils are released back to the atmosphere, contributing to carbon emissions (Mangrove Action Project, 2010). Mangroves may be one of our last defences against the perils of climate change and global warming. The mangrove last stand may be humanity last stand.

Douala, the most industrialized municipality in Cameroon, has witnessed within the last decades a rapid increase in the growth of factories mainly in the Bassa and Bonaberi industrial zones. These factories include: food processing, breweries, metal works, cement production, oil processing and paper processing. Wastes from these industries and households are often channelled or piped onto/or into the soil or into the surface water courses that empty into Rivers Wouri and Dibamba, lined by the mangrove forest. According to Ajonina and Usongo (2001), the Douala-Edea mangrove forest has been declining at a rate of 0.33% under pressure mainly by human activities. Independent studies of water samples had been done by Ekane and Oben (2001) revealing high nitrate levels not desirable for fishes, while Asaah et al. (2006) studied soils of the Bassa Industrial Zone for trace metals and found out that they were contaminated.

Owing to soil-water-plant interaction, monitoring these compartments as a block remains a better tool towards understanding their relationships. This will enhance sustainable development within the societal, economic, conservational and climate contextual need.

This study was therefore designed to (i) examine the possible effects of the water chemistry of Rivers Wouri and Dibamba on the Douala-Edea mangrove ecosystem and (ii) assess the potential role of soils of this area in controlling the transportation of cations and anions into the different water bodies.

MATERIALS AND METHODS

Area of study

The area lies between latitudes 3° 14' and 4° 93' N and longitudes 9° 30' and 10° 05' E (Figure 1). It harbours the Bassa and the Bonaberi Industrial Zones of Douala, a city with about 1.8 million inhabitants (MINEPAT, 2010).

The two industrial zones account for the bulk of industrial activities in the country, but depict contrasting features in terms of physical landscape. The Bonaberi Industrial Zone evolved almost entirely on the aquatic terrain located on lagoon marginal depressions. This necessitated extensive land reclamation to obtain space on which industrial activities had to be built. The Bassa Industrial Zone terminates into estuarine creek formation of the Dibamba River to the East of the city. The Bonaberi Industrial Zone complex

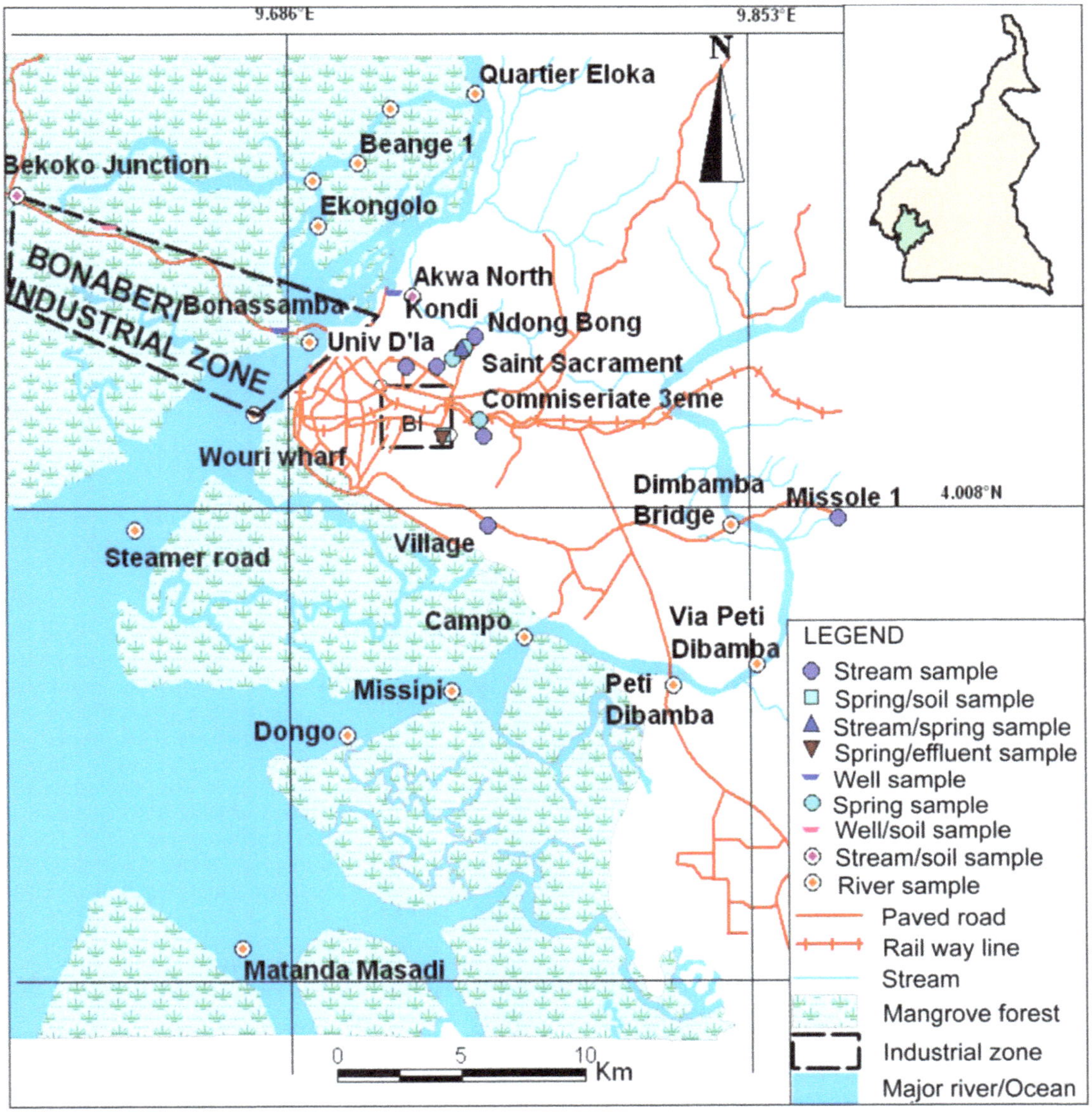

Figure 1. Map of study area showing the sampling sites (Tening et al., 2011). B.I. = Bassa Industrial Zone ; Univ. D'la = University of Douala.

has encroached into the Wouri River and this most likely provokes increased discharge of effluents into it.

Geology and soils

The zone is located within the Douala Sub-Basin of the Douala Basin, a sedimentary basin of Cretaceous to Tertiary age having a total surface area of 7000 km^2 and of maximum width of 60 km (Dumort, 1968; Regnoult, 1986). The origin and structure of this basin is associated with the opening of the South Atlantic Ocean, genetically related to the break-up of Gondwanaland. The stratigraphy of the basin consists of the Cretaceous Mungo River Formation, overlain by the Tertiary Mpundu Formation. The Mungo River Formation consists mainly of sandstone with a few intercalations of limestone and shale. The Mpundu Formation consists of poorly consolidated grits and sandstones that occasionally display bedding. However, the Douala metropolis lies on the Wouri member of the Mpundu Formation, which is dominantly made up of gravely sandstone with clay matrix. The soils of this area are alluvial resulting from the decomposition of sedimentary rocks. They are essentially sandy and commonly rich in quartz and clay minerals. These soils are very permeable in some areas. However, in other areas, the soils are very rich in clays to the extent that percolation of rainwater takes quite a long time.

Drainage

The Wouri and the Dibamba rivers constitute the drainage system in the entire Douala zone and flow all year round. The Wouri flows within the Akwa, Bonaberi and part of the Bassa while the Dibamba flows in the Eastern outskirt of the town. In the Bassa and Bonaberi Industrial Zones, there are many intermittent streams that are normally loaded with most of the solid and liquid waste channelled from industries, households as well as from waste dumps. These streams feed larger streams which empty into the Rivers Wouri and Dibamba.

Table 1. Major activities in the Douala metropolis adapted from Horan (1990) and Montgomery (1992).

Activity	Company	Product	Chemical entity
Agro-industry and food processing	Guinness Cameroon SA. Brasseries, La PASTA, ISEBERG, SIC CACAO.	Drinks/processed foods	NO_3^- NO_2^-, PO_4^{3-}, organic substances
Chemical/pulp	CIMENCAM, CEP, SAPCAM, UNALOR, SOPARCA, SIPCA, CCC, PILCAM, PLASTICAM. SOCAME, SCTB	Cement, oils, paints, detergents, vanish, soap, butter, plastics, matches, batteries, fertilizers	NO_3^-, acids, Hg, Cu, Pb, PO_4^{3-}, SO_4^{2-}, NO_2^-
Textiles	CICAM, SACC.	Cloths	Acids, Hg, organic compounds
Petroleum	SCDP, TEXACO	Aviation fuels, petrol, diesel fuel, wax.	Hydrocarbons, Pb, NO_3^-, PO_4^{3-}
Metallurgy alloys	ALLUCAM, SOCAAFERE,	Metals of various types.	As, Be, Bi, Cd, Cu, Pb, Hg, Ni, Zn.
Glass	SOCAVERE	Bottles	As, Be, Bi, Cd, Cu, Pb, Hg, Ni, Zn.
Domestic activities		Human waste and organic substances	PO_4^{3-}, SO_4^{2-}, NO_3^-

Climate

The climate is humid. The coast is covered by mangroves and dominated by Rhizophora and Avicenia species (Neba, 1987). Mean monthly temperature usually goes above 31°C. January and February are the hottest months with mean monthly temperature of 32 and 33°C, respectively (CWCS, 1998). The dry season runs from November to March accompanied by very high temperatures. Mount Cameroon forms a great barrier to the rain bearing winds causing the air to rise with moisture it carries and falling back as relief rain in Douala. Daily rainfall is very heavy and occurs throughout the year though heavy rains set in from April and subside in October.

Human activities

The metropolitan area is heavily loaded with diverse forms of human activities ranging from domestic, primary, secondary to tertiary industries (Table 1).

Sampling

Sampling was carried out within the Bassa and Bonaberi Industrial Zones and along the Wouri and Dibamba Rivers in February 2010. Sampling sites were established using a 12-channel Garmin etrex Global Positioning System (GPS). Sampling sites in rivers were areas of stream inflow, high coastal activities or near villages. Sample collection in the rivers was accomplished by the use of an engine boat. Land sampling included areas of potential influence by industrial activities.

Water samples

Surface water samples were collected from Rivers Wouri and Dibamba, and some streams, wells and springs feeding these rivers by dipping the containers in the water and filling them to the brim. Depth water samples (one, three and five metres) were collected at two points (upstream and downstream) in each river using a 1-L depth sampler (The Science Source: WIFFLE BALL KING, Reg. No. 1149044, USA). A total of 50 water samples were collected. In all cases, 0.5 L plastic containers were used. Before the samples were collected, the bottles were rinsed several times with the sample to be collected. The pH and electrical conductivity (EC) were measured *in situ* using a Tracer PockeTester™ field conductivity meter, model pH/TDS/salts. They were then transferred immediately into a cooler containing ice blocks. Before chemical analysis was carried out, the samples were filtered using Whatman 40 filter papers. Determinations of Bray II phosphates, (PO_4^{3-}) by spectrometry, nitrates (NO_3^-) and ammonium (NH_4^+) by distillation, chlorides (Cl^-) by titration with silver nitrate, sulphates (SO_4^{2-}) by gravimetry, calcium (Ca^{2+}) and magnesium (Mg^{2+}) by compleximetry, potassium (K^+) and sodium (Na^+) by flame spectrophotometry, organic carbon (Org. C) by Walkley and Black method were carried out by methods largely described by Pauwels et al. (1992). In all the parameters determined, triplicate samples were analysed and their mean values were calculated.

Soil samples

Eight top soil (0 to 15 cm) samples were collected proxy to some of the water collection points to cut across all water bodies (streams, wells, rivers, springs and mangroves) with particular avoidance of points of disturbance. The samples were collected into black polyethylene bags using a hand trowel. They were air dried and passed through a 2-mm sieve. The samples were analysed for CEC by distillation using a Kjeldhal set. The pH in water and in 1 N KCl was determined by 1:10 soil: solution ratio. Particles size analysis was carried out by the hydrometer method as largely described by Pauwels et al. (1992).

Experimental data was analysed with the statistical package SPSS11.0 and EXCEL 2007 for Windows. Univariate statistics, factor analysis, descriptive and correlation analyses were performed on the various data to evaluate and trace the sources of the different chemical entities into the mangrove ecosystem.

RESULTS AND DISCUSSION

The composition of cations and anions vary at different sites of both Rivers Wouri and Dibamba (Tables 2 and 3). Variations in the physicochemical properties from upstream to downstream may be caused by changes in quantity and/or composition of freshwater discharges. This can

Table 2. Physicochemical properties of water samples from River Wouri.

Site	Location	pH	EC (µS/cm)	Organic C (%)	NH_4^+	Ca^{2+}	Mg^{2+}	K^+	Na^+ (mg/l)	NO_3^-	PO_4^{3-}	SO_4^{2-}	Cl^-
Quartier Eloka	040914.2N 009 4516.3E	7.6	43	0.09	23.80	0.16	0.8	13.7	0.8	23.8	0.15	29.60	1.1
Beange	04 08 54.8N 009 43 24.9E	7.3	244	0.06	29.40	1.12	0.48	13.7	4.6	15.4	0.16	28.00	63.9
Beange1	04 07 44.7N 009 42 40.7E	7.2	801	0.08	22.40	7.84	75.36	25.1	17.4	23.8	0.14	23.00	85.2
Todo Njabale	04 07 21.9N 009 41 41.9E	7.2	1127	0.07	18.20	5.92	62.24	30.1	24.8	30.8	0.27	24.70	92.3
Ekongolo	04 06 25.1N 009 41 49.0E	7.3	1710	2.74	35.00	11.04	33.76	65.6	61.3	30.8	0.15	28.00	142.0
Wouri Bridge	04 03 59.9N 009 41 36.3E	7.3	6960	0.11	39.20	24.8	108.32	226.0	229.7	22.4	0.15	26.30	873.3
Wouri Wharf	04 02 27.8N 009 40 23.9E	7.4	9600	0.11	21.00	28.0	119.52	220.6	217.6	29.4	0.13	11.50	582.2
Steamer Road	04 00 01.5N 009 37 48.4E	7.5	15000	0.14	33.60	47.52	252.96	469.5	505.4	25.2	0.16	36.20	28.4
Mean		7.4	4436	0.43	27.83	15.80	81.68	133.0	132.7	25.2	0.16	25.91	233.6

alter the chemistry of the environment, such as hydrogeology and temperature not suitable for the mangrove ecosystem. Nutrients from fresh water inflow do not only affect water quality but equally species diversity (Doering, 1996). Average pH was 7.4 in River Wouri and 7.5 in River Dibamba indicating that the rivers were neutral with values within the permissible range advocated by WHO (2008). The pH decreased and increased from upstream to downstream in both rivers (Figure 2a). The decrease in pH could be associated to nearness of such areas to the industrial zones with probable releases and deposition of acidic gases. The EC of River Wouri were generally lower than those of River Dibamba with mean values of 4436 and 12496 µS/cm, respectively. Mean concentrations of the cations stood at $Na^+ > K^+ > Mg^{2+} > NH_4^+ > Ca^{2+}$ and $Na^+ > NH_4^+ > K^+ > Mg^{2+} > Ca^{2+}$ for Rivers Wouri and Dibamba respectively. Anions followed a trend of $Cl^- > NO_3^- > SO_4^{2-} > PO_4^{3-}$ for both rivers. There was a significant difference ($P<0.01$) between the means of both rivers. This could be an indication that from River Wouri the sampling scenario had no effect on the statistical detection of temporal trends in physicochemical properties of water samples implying that the higher values of River Dibamba could be accounted for by the nature of activities of the area. In both rivers, PO_4^{3-} concentrations were low. They ranged from 0.14 to 0.20 mg/l in River Wouri and 0.13 to 0.27 mg/l in River Dibamba.These current values seem to present no threat to the mangrove environment when compared with the 50 mg/l maximum advocated by the former Ministry of Mines and Energy in Cameroon. No consistent trend was observed in this parameter from upstream to downstream (Figure 2b). This inconsistency of values could be allied to addition by anthropogenic activities through stream flow which are continually increasing as a result of increase in population and urbanisation. However, sample number 4 located between Ekongolo and Beange 1 (Figure 1) had the highest PO_4^{3-} which could be ascribed to its location at a confluence with anthropogenic inputs from both stream flows. Nitrate concentration ranged from 15.4 to 30.8 mg/l (Table 2) and 42.0 to 341.6 mg/l (Table 3) for Rivers Wouri and Dibamba, respectively. These values were very high in some sites in Dibamba when compared with the Russian (40 mg/l) maximum permissible values for fisheries and other aquatic plants. Increase NO_3^- concentration as a consequence of Industrialisation puts a considerable burden on forest,

Table 3. Physicochemical properties of water samples from River Dibamba.

Site	Location	pH	EC (μS/cm)	Organic C (%)	NH_4^+	Ca^{2+}	Mg^{2+}	K^+	Na^+	NO_3^-	PO_4^{3-}	SO_4^{2-}	Cl^-
					(mg/l)								
Dibamba Bridge	04 00 6.5N 009 50 46.9E	7.8	2210	0.10	497.0	11.04	67.36	84.8	82.0	63.0	0.14	29.60	255.6
Via Peti Dibamba	03 57 10.4N 009 51 19.6E	7.2	6250	0.11	74.2	18.72	108.00	168.9	242.2	56.0	0.22	32.90	284.0
Peti Dibamba	03 56 43.7N 009 46 30.6E	7.3	9810	0.13	86.8	35.04	195.36	384.0	452.0	58.8	0.20	29.60	773.9
Campo	03 57 44.7N 009 46 16.6E	7.4	14300	0.15	78.4	40.80	239.84	458.5	505.4	60.2	0.21	32.90	1107.6
Missipi	03 56 37.6N 009 44 41.8E	7.4	16500	0.15	78.4	60.00	216.80	651.9	771.2	341.6	0.18	36.20	582.2
Dongo	03 55 41.8N 009 42 23.6E	7.5	18500	0.15	63.0	53.28	298.72	506.0	662.3	60.2	0.18	29.60	1192.8
Matanda Masadi	03 51 09.0N 009 40 04.8E	7.6	19900	0.15	632.8	63.84	341.28	530.5	864.3	42.0	0.14	24.70	1249.6
Mean		7.5	12496	0.13	215.8	40.39	209.62	397.8	511.3	97.4	0.18	30.79	778.0

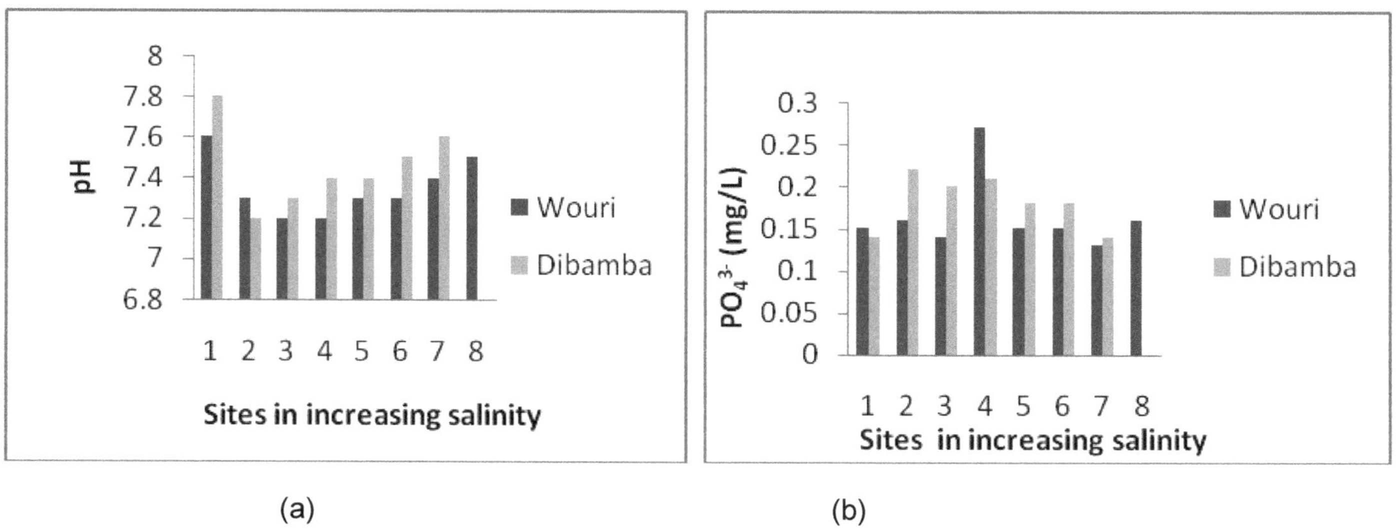

Figure 2. Variation of (a) pH and (b) PO_4^{3-} with salinity gradient.

grassland and aquatic plants thereby causing undesirable changes in biodiversity and acidification of soils and water (Stevens et al., 2004). Calcium concentration ranged from 0.16 to 47.52 mg/l with a mean value of 15.80 mg/l in River Wouri (Table 2) and 11.04 to 63.84 mg/l with a mean value of 40.39 mg/l in River Dibamba (Table 3). These values are below the 700 mg/l guideline set by WHO (2008) for domestic and industrial use. Mean Mg^{2+} concentration stood at 81.68 mg/l in River Wouri and 209.62 mg/l in River Dibamba. Major sources of magnesium in water include amphiboles, olivine, pyroxene, dolomites and clay minerals (Todd, 1980). Todd (1980) observed that dolomites have high concentration of Mg^{2+} while clay minerals have low concentration. If the concentration of this element were from clay mineral alone, concentrations would have been low. Thus, the high concentration could be from the combination of these sources and anthropogenic activities. Potassium had an average concentration of 133.0 mg/l in River Wouri and 397.8 mg/l in River Dibamba. Major sources of potassium in water include feldspers (orthoclase and microclin) and clay minerals (Todd, 1980). The concentration of potassium should be 10 mg/l in natural water. The high concentrations of potassium in these rivers could partly be from industrial sources such as the salt producing industry (Sel Camerounaise (SELCAM)) in this area.

Streams (Tables 4 and 5) revealed concentrations that vary considerably with the characteristics of the areas in which they are situated. Sulphates ranged from 0.00 to 26.30 mg/l in streams feeding River Dibamba and 23.00 to 32.90 mg/l in streams feeding River Wouri. The concentrations were lower than the 250 mg/l advocated by WHO (2008). These values varied proportionately with sites of varying activities, implying that their main source could be anthropogenic activities. Sample number 4 (Table 4) had a pH value of 11.2 with an EC of 1800 µS/cm. Similar high values were obtained from sample number 6 (Table 5). The EC value and Na^+ ion concentration of sample number 6 were of values 5540 µS/cm and 242 mg/l, respectively. These values were higher than the 1000 µS/cm and 200 mg/l maximum allowable values set by WHO (2008) for EC and Na^+, respectively. Major sources of Na^+ include: feldspar, clay mineral, evaporates such as halites (NaCl), and industrial waste. Long term discharge of these ions which function as dispersants into the mangrove environment can change the structural properties of the swamp land leading to stress events. Mean Ca^{2+} and Mg^{2+} ions concentrations of 2.24 and 1.65 mg/l, respectively, in River Dibamba and 3.59 and 1.58 mg/l, respectively in River Wouri were lower than the 700 mg/l and 250 mg/l, respectively, set by WHO (2008). Sample number 1 (Table 5) had the lowest EC value of 9 µS/cm and the highest NH_4^+ ion concentration of value 660.8 mg/l as compared to other streams in the area. This could be attributed to the decay of luxuriant natural vegetation in the area. In all the streams, NH_4^+ ion concentrations were higher than the 0.5 mg/l maximum level stipulated by WHO (2008). The mean Cl^- concentration of 135.1 mg/l in streams feeding River Wouri was higher than the mean value of 59.3 mg/l in streams feeding River Dibamba. The mean Cl^- concentration in streams feeding River Wouri was highly influenced by samples numbers 3 and 4 (Table 4). Probable sources of Cl^- here include sea spray from Atlantic Ocean, and industrial waste which could produce point source. The mean NO_3^- concentrations of the streams feeding both rivers were generally higher than the recommended standards of 40 mg/l by WHO (2008). These NO_3^- concentrations might have been a consequence of agricultural activities (excess application of inorganic nitrogenous fertilizers), waste water disposal, oxidation of nitrogenous waste products in human and animal excreta, septic tanks and deposition from industrial establishments. High NO_3^- concentrations promote vegetative growth in trees such as mangroves but equally increase vulnerability to lodging. High NO_3^- values are undesirable in water bodies as they can result in eutrophication and incidences of metamoglobinemia (Kross et al., 1993). Mean K^+ concentrations stood at 46.2 and 20.0 mg/l in Rivers Wouri and Dibamba, respectively. These values are higher than the maximum permissible values advocated by WHO (2008). The variation in physicochemical properties in streams influenced by human activities could be some of the major sources on changes in properties along the rivers.

Physicochemical properties of wells of Wouri and Dibamba (Tables 6 and 7) were generally lower than those of the streams. This is an indication that ground water contamination is minimal and thus human activities are responsible for the high levels in streams. Wells feeding both Rivers Wouri and Dibamba were acidic with an average pH of 5.8 and 6.3, respectively. Sample number 1 (Table 6) had the highest EC value (450 µS/cm). This is probably due to the fact that it is an open well and located near a cement block industry. Sample number 4 (Table 6) had the least pH value of 4.3 for wells and springs feeding River Wouri. This could have been enhanced by its proximity to the Bassa Industrial zone with probable deposition of acidic gases. Sample number 3 (Table 6) had the highest ammonium ion concentration (365.40 mg/l). This could be associated to its proximity to la Pasta, the food processing industry and breweries. Sample number 3 (Table 7), was the most acidic with a pH value 4.7 for wells and springs feeding River Dibamba. Average Cl^- concentration of wells feeding River Wouri (57.9 mg/l) was higher than those feeding River Dibamba (38.2 mg/l). Though these values are lower than the 250 mg/l set by WHO (2008), higher Cl^- concentrations of wells feeding River Wouri could represent a good example of human disturbance (Gove et al., 2001).

Cation concentration increased with depth in both rivers (Figures 3). This could be an indication that one of the

Table 4. Physicochemical properties of streams feeding River Wouri.

Site	Location	pH	EC (µS/cm)	Organic C (%)	NH_4^+	Ca^{2+}	Mg^{2+}	K^+	Na^+	NO_3^-	PO_4^{3-}	SO_4^{2-}	Cl^-
					(mg/l)								
Akwa North	04 04 56.5N 009 43 51.8E	7.5	404	0.08	44.80	2.08	1.44	58.3	20.9	33.6	1.83	24.70	71.0
Bekoko Junction.	04 07 06.1N 009 35 16.1E	7.5	10	0.07	30.80	2.40	0.16	4.2	0.2	19.6	0.13	23.00	1.2
CamTel	04 03 28.7N 009 43 43.0E	7.0	361	0.07	32.20	11.68	5.28	105.2	8.7	37.8	2.15	32.90	376.3
Univ. Douala	04 03 28.9N 009 44 24.1E	11.2	1800	0.07	44.80	2.72	1.44	65.6	89.5	46.2	4.64	31.30	284.0
Ndog Bong 1	04 03 50.6N 009 44 58.9E	6.7	170	0.07	44.80	1.76	1.44	32.7	1.7	53.2	0.44	28.00	71.0
Ndog Bong 2	04 03 53.3N 009 44 56.8E	6.5	156	0.08	42.00	1.76	0.48	24.7	1.7	40.6	0.37	24.70	71.0
Kondi	04 04 06.3N 009 45 13.5E	7.3	283	0.08	67.20	2.72	0.80	32.7	6.5	50.4	0.45	26.30	71.0
Mean		7.7	455	0.07	43.80	3.59	1.58	46.2	18.5	40.2	1.43	27.27	135.1

Table 5. Physicochemical properties of streams feeding River Dibamba.

Site	Location	pH	EC (µS/cm)	Organic C (%)	NH_4^+	Ca^{2+}	Mg^{2+}	K^+	Na^+	NO_3^-	PO_4^{3-}	SO_4^{2-}	Cl^-
					(mg/l)								
Stream Missole	04 00 16.0N 009 53 07.2E	7.6	9	0.09	660.8	0.80	2.40	4.6	0.2	25.2	0.13	0.00	0.7
Stream Village	04 00 07.1N 009 45 29.6E	7.6	246	0.09	120.4	2.40	1.12	31.2	4.6	43.4	0.23	26.30	42.6
Stream Nkongi	04 02 01.2N 009 45 23.7E	5.9	69	0.08	54.6	2.08	1.76	9.1	0.8	204.4	0.20	18.10	28.4
Spring and Stream CCC	04 02 02.2N 009 44 32.3E	6.7	223	0.09	47.6	3.04	1.44	24.3	4.6	37.8	0.72	24.70	56.8
Stream CCC	04 02 02.2N 009 44 32.3E	6.8	187	0.09	46.2	3.68	1.12	29.9	6.5	103.6	0.78	21.40	28.4
Stream and Effluent CCC	04 01 57.3N 009 44 31.0E	11.8	5540	0.22	81.2	1.44	2.08	24.3	242.0	98.0	10.03	26.30	198.8
Mean		7.7	1046	0.11	168.5	2.24	1.65	20.6	43.1	85.4	2.02	19.47	59.3

Table 6. Physicochemical properties of wells and springs feeding River Wouri.

Site	Location	pH	EC (µS/cm)	Organic C (%)	NH_4^+	Ca^{2+}	Mg^{2+}	K^+	Na^+	NO_3^-	PO_4^{3-}	SO_4^{2-}	Cl^-
					(mg/l)								
Well Akwa North	04 05 02.7N 009 43 29.8E	7.5	450	0.09	92.40	20.32	40.48	112.5	4.6	89.6	0.19	24.70	56.8
Well Bonassama	04 04 15.2N 009 41 00.6E	7.5	382	0.09	53.20	12.32	34.72	66.8	4.6	75.6	0.18	23.00	85.2
Well C. Mutzig	04 06 25.9N 009 37 17.6E	6.4	75	0.08	365.40	1.44	1.12	12.6	0.8	67.2	0.20	24.70	2.2
Spring Saint Sacrament	04 03 39.7N 009 44 45.2E	4.3	143	0.09	32.20	18.08	1.26	31.1	1.7	44.8	0.26	36.20	142.0
Spring Ndog Bong	04 03 53.7N 009 44 59.8E	4.8	92	0.09	44.80	2.40	0.48	12.6	0.8	47.6	0.16	21.40	56.8
Spring Commesarait	04 02 21.8N 009 45 19.0E	4.5	61	0.22	92.40	1.44	0.80	10.4	0.2	75.6	0.19	26.30	4.3
Mean		5.8	201	0.11	113.40	9.33	13.14	41.0	2.1	66.7	0.20	26.05	57.9

Table 7. Physicochemical properties of wells and springs feeding River Dibamba.

Site	Location	pH	EC (µS/cm)	Organic C (%)	NH_4^+	Ca^{2+}	Mg^{2+}	K^+	Na^+	NO_3^-	PO_4^{3-}	SO_4^{2-}	Cl^-
					(mg/l)								
Well Missole	03 59 46.4N 009 53 41.5E	7.6	159	0.09	44.8	0.8	1.76	11.5	0.2	53.2	0.16	28	56.8
Well Camdem	03 59 48.7N 009 50 29.2E	7.6	208	0.1	39.2	2.4	0.48	2.4	0.2	49	0.14	29.6	0.9
Well Nkongi	04 02 02.0N 009 45 20.4E	4.7	92	0.1	37.8	0.8	2.72	13.9	1.7	50.4	0.18	26.3	52.4
Spring CCC	04 02 02.2N 009 44 32.3E	5.2	199	0.07	67.2	2.7	1.44	13.9	4.6	36.4	0.13	24.7	42.6
Mean		6.3	164	0.09	47.3	1.7	1.6	10.4	1.7	50.9	0.16	27.97	38.2

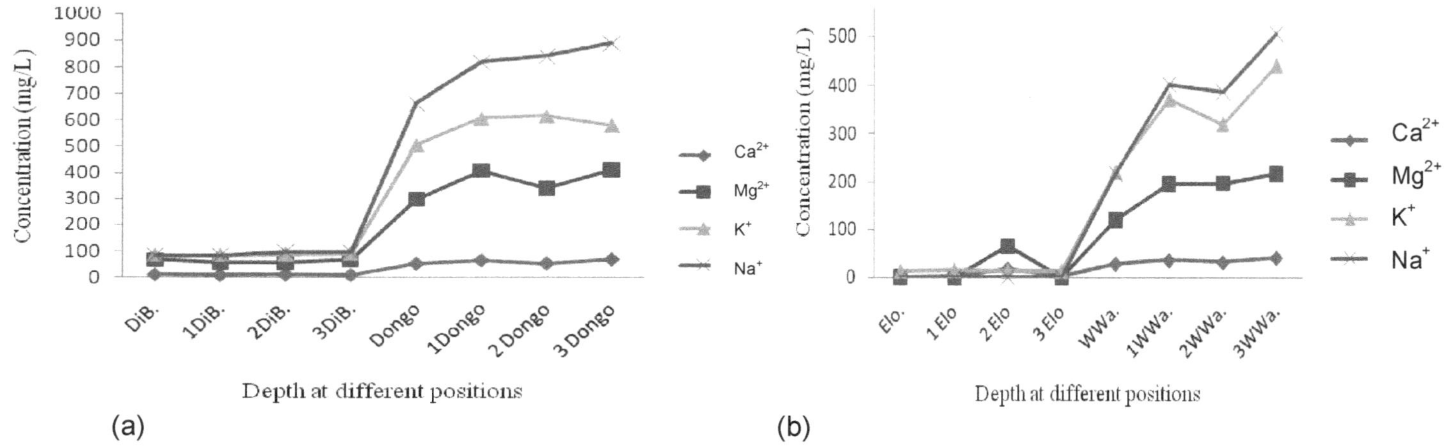

Figure 3. Cation concentration with depth in (a) River Dibamba (b) River Wouri. DiB. = Dibamba Bridge 1, 2 and 3= 1, 3 and 5 m depths, respectively. Elo = Quartier Eloka, WWa = Wouri 1, 2, and 3 = 1,3 and 5 m, respectively.

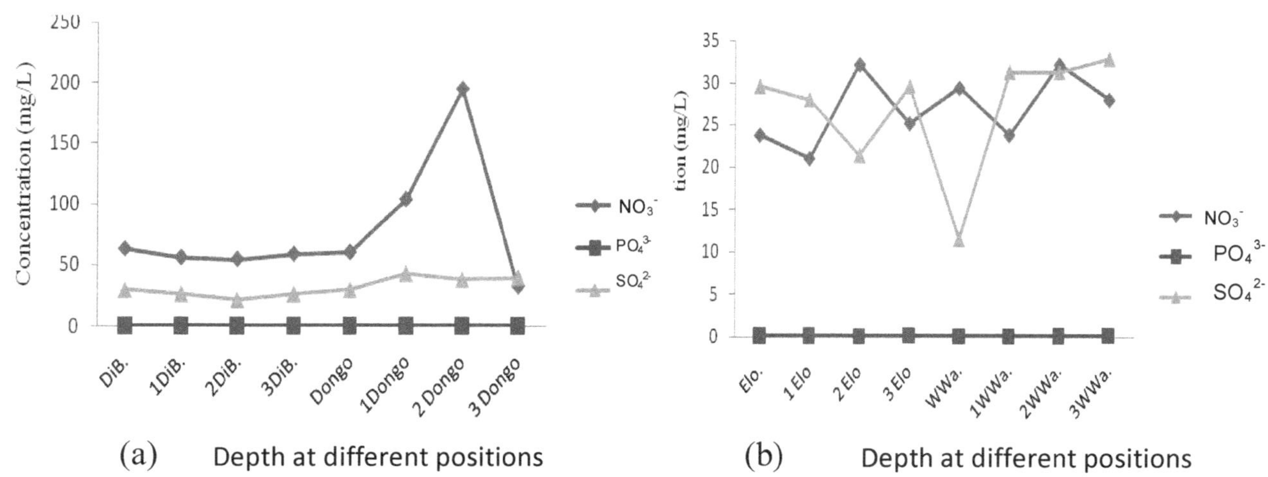

Figure 4. Variation of anions with depth in (a) River Dibamba (b) River Wouri. DiB. = Dibamba Bridge, 1, 2 and 3 = 1, 3 and 5 m, respectively. Elo. = Quartier Eloka, WWa = Wouri Wharf ,1, 2, and 3 = 1, 3 and 5 m, respectively.

major sources of the cations is the breakdown of bedrock materials. There were positive relationships ($P<0.01$) within the cations when correlation analysis was carried out indicating that these cations were coming from the same source. On the other hand, anions did not show an increase with depth in both rivers except chlorides (Figure 4). Salinity profiling revealed perturbations at sampling sites, a probable indication of anthropogenic origin with severity in River Wouri. Correlation analysis showed that anions from both rivers were significantly correlated ($P<0.01$) with each other except the chlorides. Therefore, the major sources of cations and anions are not the same.

From factor analysis for River Wouri (Tables 8 and 9), three factors accounting for 97.4% of the total variance, explained how the various sites could be associated. Factor 1 consists of all the sites of River Wouri and contributed 87.2% of the total variance. The grouping of these sites could be allied to anthropogenic activities as phosphates and nitrates were within a close range. Factor 2 contributed 8.2% of the total variance and was made of sample number 1 (Table 9) alone. This could be partly due to domestic discharges, which are rich in phosphates and nitrates in the remote areas with the presence of water hyacinths in areas of enhanced nutrients. Factor 3 was a single factor and made up of sample number 5 (Table 9). It contributed 2.0% of the total variance. This could be linked to food processing activities that generate high organic carbon content.

For River Dibamba (Tables 10 and 11), three factors accounting for 99.8% of the total variance, also explained site groupings. Factor 1 accounted for 98.4% of the total variance and was composed of all sites (Table 11). The association of these sites could be due to natural activities which could be from the weathering of bedrock material and the decomposition of natural vegetation. Fac-

Table 8. Total variance, cumulative Eigen values and percentage contribution from various sites of water samples from River Wouri.

Site	Initial Eigen values (%)			Extraction sums of squared loadings (%)		
	Total	Variance	Cumulative	Total	Variance	Cumulative
Quartier Eloka	6.972	87.210	87.210	6.972	87.210	87.210
Beange	0.654	8.174	95.384	0.654	8.174	95.384
Beange 1	0.158	1.980	97.364	0.158	1.980	97.364
Todo Njebale	0.141	1.761	99.125			
Ekongolo	0.055	0.682	99.807			
Wouri Bridge	0.009	0.116	99.924			
Wouri Wharf	0.004	0.050	99.973			
Steamer Road	0.002	0.027	100.000			

Extraction method: Principal component analysis.

Table 9. Three component matrix (a) from various sites of water samples from River Wouri.

Site	Component		
	1	2	3
Quartier Eloka	0.757	0.631	-0.058
Beange	0.906	0.334	0.166
Beange 1	0.983	-0.048	-0.015
Todo Njebale	0.984	-0.032	-0.129
Ekongolo	0.930	-0.199	0.311
Wouri Bridge	0.977	-0.187	0.004
Wouri Wharf	0.944	-0.228	-0.002
Steamer Road	0.974	-0.119	-0.130

Extraction method: Principal component analysis.

Table 10. Total variance explained from various sites of water samples from River Dibamaba.

Site	Initial Eigen values (%)			Extraction sums of squared loadings (%)		
	Total	Variance	Cumulative	Total	Variance	Cumulative
Dibamba Bridge	6.890	98.435	98.435	6.890	98.435	98.435
Via Peti Dibamba	0.073	1.036	99.471	0.073	1.036	99.471
Peti Dibamba	0.026	0.376	99.847	0.026	0.376	99.847
Campo	0.008	0.116	99.963			
Missipi	0.002	0.032	99994			
Dongo	0.000	0.004	99.999			
Matanda Masadi	0.000	0.004	100.000			

Extraction method: Principal component analysis.

Table 11. Three component matrix (a) from various sites of water samples from River Dibamba.

Site	Component		
	1	2	3
Dibamba Bridge	0.974	0.225	0.032
Via Peti Dibamba	0.998	0.005	0.041
Peti Dibamba	0.999	-0.042	-0.013
Campo	0.997	-0.064	-0.023
Missipi	0.991	-0.071	0.102
Dongo	0.995	-0.095	-0.030
Matanda Masadi	0.992	0.047	-0.107

Extraction method: Principal component analysis.

Table 12. Some physicochemical properties of soils in the Wouri and Dibamba area.

Site	Location	pH (Water)	pH (1 N KCl)	CEC (cmol/kg)	Sand (%)	Silt (%)	Clay (%)
Akwa North	04 05 02.7N 009 43 29.8E	5.5	4.8	13.2	64.4	30.8	4.8
Bekoko Junction	04 07 11.8N 009 35 17.6E	5.7	4.3	7.98	78.2	17.8	4.0
Commesariate 3eme	04 02 21.8N 009 45 19.0E	4.3	3.7	22.2	52.2	25.4	22.4
Saint Sacrament	04 03 39,7N 009 44 45.2E	5.3	4.6	5.0	74.0	14.8	11.2
Ndog Bong	04 03 50.6N 009 44 58.9E	4.5	3.7	5.9	79.0	14.2	6.8
Carrefour Mutzig	04 0625.9N 009 3717.6E	4.3	3.9	2.16	72.8	14.0	13.2
Missole	03 59 46.4N 009 53 41.5E	5.7	4.5	11.7	20.2	72.0	7.8
Dibamba	03 59 54.8N 009 50 25.4E	5.5	4.3	11.3	37.2	13.8	49.0

Figure 5. Nypa palms along the banks or River Dibamba(a) and floating water hyacinths in River Wouri (b).

tor 2 accounted for 1.0% of the total variance. It was made up of sample number 1 (Table 11) alone with high NH_4^+ concentration which might have resulted from remains of living tissues used by humans in the nearby Dibamba village. Factor 3, contributing 0.4% of the total variance, was not dominated by any site. This could be an indication of the effects of distant activities.

All soils of the area were acidic (Table 12). The soil pH in water ranged from 4.3 to 5.7 in water and 3.7 to 4.8 in 1N KCl. Soil pH plays a great role in the occurrence of some mangrove species such as the Nypa palms. Generally, soils of Wouri area are high in sand while those of Dibamba area are rich in clay. The high clay values of Dibamba soils may be as a result of variation in the stratigraphy within the Douala basin (Ntamak-Nida et al., 2010). The soils also showed varying CEC values. There was a positive correlation ($P<0.05$) between Wouri clays and CEC. On the other hand, correlation between Dibamaba clays and CEC showed a highly significant negative correlation ($P<0.01$). This implies that Dibamba clays have mostly positively charged colloids, which can retain anions but have low cation retentive capacity. Cations released into these bodies will easily find themselves into water bodies. On the other hand, the clays of Wouri have negatively charged colloids and will therefore retain cations and will not retain anions (PO_4^{3-} and NO_3^-) sent into the environment. These anions will easily find themselves into water bodies exposing them to threats such as eutrophication. The presence of these substances within the mangrove environment changes the hydrology resulting in the emergence of less desirable species such as the Nypa palms and Water hyacinths (Figure 5). Though beautiful plants, water hyacinths are regarded as pests in tropical and subtropical regions, they reproduce very rapidly in water environments with enhanced nutrient concentrations and are therefore abundant in eutrophicated water bodies. The presence of these plants in Rivers Wouri and Dibamba is an evidence of changes in physicochemical properties of these water bodies. Though NO_3^- and PO_4^{3-} were lower in River Wouri than River Dibamba, the presence of water hyacinths, which are high consumers of these entities, might have been responsible for their low levels in River Wouri. Similar appearances were identified in the Kafue River in Zambia (Alsterhag and Petersson, 2004). Changes in hydrology of the mangrove environment due to influx of cations and anions can equally convert the mangrove environment from carbon sinks to carbon sources (Roulet et al., 2007).

Conclusions

A greater proportion of cations in the Douala-Edea mangrove ecosystem came from weathering of soils and rocks while anions came from anthropogenic activities. With variability in the retentive capability of soils in the area, there is the need for an elaborate characterization of wastewaters and water bodies, and evaluation of treatment facilities of municipal and industrial waste to assess the status and control of water pollution. Clay mineralogical studies should be conducted to ascertain the retention capacity of the soils and used like a tool in the demarcation of industrial areas or land for waste disposal.

ACKNOWLEDGEMENTS

The authors gratefully acknowledge the financial and material support from the International Atomic Energy Agency (IAEA), Vienna, Austria, the University of Buea, Cameroon and the Flemish inter-University Commission (VLIR), Belgium towards the completion of this work. The chemical analysis was carried out in the Laboratory of the Bristol Biogeochemical Research Centre, School of Earth Sciences, University of Bristol, UK and Soil and Water Laboratory, Department of Soil Science, Faculty of Agronomy and Agricultural Sciences, University of Dschang, Cameroun.

REFERENCES

Ajonina GN, Usongo L (2001). Preliminary quantitative impact assessment of wood extraction on mangroves of the Douala-Edea Forest Reserve Cameroon. Trop. Biodivers. 7: 137-149.

Alsterhag E, Petersson L (2004). Nutrient loading in the Kafue River between Mazabuka and Kafue Town, Zambia. Uppsala University.

Asaah HA, Ambimbola AF, Suh CE (2006). Heavy metal concentration in surface soils of the Bassa Industrial Zone I, Douala Cameroon. Arabian J. Sci. Eng. 31: 147-158.

Aurela M, Laurila T, Tuovinen J (2002). Annual CO_2 balance of sub-artic fen in Northern Europe: Importance of winter efflux. J. Geophys. Res. 107: 4607-4619.

Ball PW (1988). Nutrient inputs to estuaries from nine Scottish East Coast Rivers: Influence of estuarine processes on inputs to the North Sea. Estuar. Coast Shelf Sci. 39: 329-352.

Boynton WR, Garber JH, Summers RK (1995). Inputs, transformations and transport of nitrogen and phosphorus in Chesapeake Bay and selected tributaries. Estuaries 18: 285-314.

Chiras DD (2000). Environmental Science: Creating a sustainable future. Jonesan Bartlet Publishers, Inc. Sudbury p.728

Choudbury JK (2002). Sustainable management of coastal mangroves, forest development and social needs. FAO Publication, Rome. 265-285.

CWCS (Cameroon Wildlife Conservation Society) 1998. Cameroon Mangrove Coverage Yaounde, Cameroon.

Doering PH (1996). Temporal variability of water quality in St. Lucie Estuary Southern Florida. Water Resour. Bull. 32: 1293-1306.

Dumort JP (1968). Notice explicative sur la feuille de Douala-Ouest. Direction Mines et de la Geologie du Cameroun, Yaounde.

Ekane C, Oben PM (2001). Biomedical indicators of marine pollution in the Douala and Limbe estuary. In: Lambi CM, (Ed.), Environment issues, problems and prospects (Unique Printers, Bamenda, Cameroon. pp.119-132.

FAO (Food and Agricultural Organisation) Report (2005). Global Forest Resources Assessment.

Gleitsmann BA, Kroma MM, Steenhuis T (2007). Analysis of rural water supply project in three communities in Mali. Participation and sustainability. Nat. Resour. Forum 31:142-150.

Gove NE, Edwards RT, Conquest LL (2001). Effects of Scale on land use and water quality relationships: A longitudinal basin-wide perspective. J. Am. Water Resour. Assoc. 37: 1721-1734.

Horan NJ (1990). Biological waste water treatment systems: Theory and operation. John Wiley and Sons, New York, USA. p.215.

Katte VY, Fonteh MF, Guemuh GN (2003). Domestic water quality in urban centres in Cameroon: A case study of Dschang in West Province. Afr. Water J. 25: 15-56.

Kross BC, Halberg GR, Bruner R, Cherryholmes K, Hohnson KJ (1993). The nitrate contamination of private well water in Lowa. Am. J. Public Health 83: 270-272.

Lafleur PM, Roulette NT, Bubier JI, Frakings S, Moore TR (2003). International variability in the peatland atmosphere carbon dioxide exchange at an ombrotrophic bog. Global Biogeochem. Cycles 17: 1036-1050.

Lener DN (1995). Chlorinated solvent pollution of an industrialized urban area. In: Nash H, Mc, Call GJH (Eds.), Groundwater Quality, Chapman and Hall, London. pp.185-190.

Mangrove Action Project, 2010. Climate change. http://mangrogeactionproject.org/issues/climage-change downloaded in February 2010.

McGuinness KA (1997). Dispersal, establishment and survival of ceriops tagal propagules in a North Australian mangrove forest. Ecologia 109: 80-87.

MINEPAT (2010). National Population and Household Census. Ministry of Employment, Planning and Territorial Administration, Yaounde.

Montgomery CW (1992). Environmental Geology. Brown Publishers, New York. p.465

Morris BL, Lawrence AR, Stuart ME (1994). Impact of Urbanization on Groundwater Quality BGS Project WC/94/56.

Neba SA (1987). Modern Geography of the Republic of Cameroon. Neba Publishers, Bamenda. p.244

Ntamak-Nida MJS, Bourquin J, Makong F, Baudin JE, Mpesse CI, Ngouem PB, Komguem PB, Abolo GM (2010). Sedimentology and sequence stratigraphy from outcrops of the Kribi-Campo sub-basin: Lower Mundeck Formation (Lower Cretaceous, southern Cameroon). J. Afr. Earth Sci. 58: 1–18.

Pauwels JM, Van Ranst E, Verloo M, Mvondo-Ze AD (1992). Manuel de Laboratoire de pedologie: methods d'Analyses de sols et de plantes, Equiptment, Gestion de stocks de verrerie et de produit chimique. Publication Agricoles, Dschang, Cameroun.

Regnoult JM (1986). Synthese Geologique du Cameroun. Ministere des Mines et des la Energie, Yaounde Cameroun.

Roulet NT, Lafleur PM, Richard PJH, More TR, Humpherys ER, Bubier JL (2007). Contemporary carbon balance and late Halocene Carbon accumulation in Northern Peatland. Glob. Change Biol. 13: 397-411.

Rowel DL (1994). Soil Science: Methods and Application. Prentice hall, Hockley. p.350.

Spalding M, Kainuma M, Collins L (2010). World Atlas of mangroves, UK.

Stevens CJ, Dise NB, Mountford JO, Gowing DJ (2004). Impact of nitrogen deposition on the specific richness of grassland. Science 3003: 1876-1879.

Suratman MN (2008). Carbon sequestration potential in Southeast Asia. Managing Forest Ecosystem 17: 297-315.
Tening AS, Hornibrook E, Chuyong GB, Lifongo LL (2011). The influence of seasonal changes in hydrology and nutrient cycling in West African tropical mangroves. Technical Report, Research Project No. RC-14039, International Atomic Energy Agency (IAEA), Vienna, Austria. pp. 15.
Todd DK (1980). Groundwater Hydrology. John Willey and Sons, New York, USA. p.535.
Twilley RR, Chen RH, Hargis T (1992). Carbon sinks in mangroves and their implications to carbon budget in tropical coastal ecosystems. Water Air Soil Pollut. 64: 265-288.
UNFCCC (1998). Report of the United Nations Framework Convention on Climate Change: A follow-up of mitigation actions of the Kyoto Protocol. United Nations, Buenos Ares.
WHO (World Health Organisation) (2008). Guideline for Drinking-water Quality, WHO, Geneva.
Wickramasinghea S, Borin M, Kotagama SW, Cochard R, Anceno AJ, Shipin OV (2009). Multi-functional pollution mitigation in a rehabilitated mangrove conservation area. Ecol. Eng. 35: 898-907.
Yerima BPK, Van Ranst E (2005). Introduction to Soil Science, Soils of the Tropics. TRAFFORD Publishers, Victoria. p.397.
Zhao D, Huo X, Dong S (2009). Investigation of the transport and fate of Pb, Cd, Cr (VI) and As (V) in soil zones derived from moderately contaminated farm lands in North East China. J. Hazard. Mater 170: 570-577.

Nitrate concentration in drinking water supplies in selected communities of Ibadan Southeast local government, Ibadan, Nigeria

Elizabeth O. Oloruntoba[1] , Mynepalli K. C. Sridhar[1], Ramon T. Alabi[2] and Stephen A. Adebowale[3]

[1]Department of Environmental Health Sciences, Faculty of Public Health, College of Medicine, University of Ibadan, Ibadan, Nigeria.
[2]International Institute of Tropical Agriculture, Ibadan, Nigeria.
[3]Department of Epidemiology and Medical Statistics, Faculty of Public Health, College of Medicine, University of Ibadan, Ibadan, Nigeria.

Elevated levels of nitrate in drinking water have been associated with adverse health effects. Most susceptible to nitrate toxicity are infants under six months of age and pregnant women. This study assesses the nitrate concentration of 48 randomly selected wells in an urban-slum setting in Ibadan South East Local Government Area (IBSELGA), Nigeria. The coordinates of the wells were mapped with a hand-held Global Positioning System (GPS). The nitrate concentration ranged between 0.00 and 42.80 mgL^{-1} with a mean of log_{10} of 0.735 and a geometric mean of 5.43 mgL^{-1}; and 0.00 and 93.30 mgL^{-1} with a mean of log_{10} of 0.696 and a geometric mean of 4.97 mg L^{-1} for the wet and dry seasons, respectively. During the wet season, the levels of nitrate in all the wells were within the WHO permissible limit of 45 mg L^{-1} NO_3. However, during the dry season, few of the wells; six (12.5%) dried up completely. Out of the remaining 42 wells, six (14.3%) had nitrate concentration which exceeded the permissible limit. This same set of wells lacked sanitary features such as lining and cover. Out of the 48 wells, 30 (62.5%) and 24 (50%) were located at ≤10 m from the septic tank/pit latrine and refuse dumps, respectively. Nitrate maps were developed using Georeferenced data for the water points. The nitrate exceedence map produced showed water sources within the study area with nitrate concentration exceeding the recommended limit and therefore requiring urgent intervention. The results of this study will serve as indicator for determining risk areas with respect to nitrate concentration in drinking water and therefore help in decision making activities.

Key words: Global positioning system, Ibadan, nitrate, nitrate exceedence, wells, urban slum.

INTRODUCTION

The general practice in many developing countries in the past was that a large proportion of people in the urban areas had access to piped water supply to the neglect of those in the rural and peri-urban areas; but now, the situation has changed. For instance, in Ibadan, Nigeria, the Water Corporation of Oyo State supplies 107 million liters of water/day for a population of 3.6 million through the Eleyele and Asejire dams. This quantity is adequate for only 20% of the population (Itama et al., 2006). The outcome of this inadequacy is that most people in the city resort to water from other sources such as boreholes (for institutions and privileged few), dug wells, springs (both protected and unprotected), ponds and streams. Study by Ochieng et al. (2011) assessed the sanitary features, prevailing pollution and quality of water in some communities in IBSELGA of Ibadan metropolis. They concluded that 85% of the respondents were dependent on wells for drinking, while the remaining 15% depends on other drinking water sources such as public tap, pond and water from commercial vendors. The same water sources were also used for cooking, washing, toileting and other household works. The study area, Ibadan South East Local Government Area (IBSELGA) is one of the five urban local councils in Ibadan metropolis. It is a densely populated area and fit the description of an urban slum.

In an interview conducted by Fourchard in 2002, the Chairman of IBSELGA defined slums as "areas which concentrate low income earners, low cost houses, possibly mud houses, no layout and poor inhabitants" (Fourchard, 2003). Fourchard (2003) identified three types of slums in Ibadan based on age, location and size. These include:

i) The oldest and largest slum is the core area of the city, which covers the entire pre-colonial town. A large part of the ancient walled city can be seen as a slum, even if the inhabitants do not agree that they live in a slum for historical reasons.
ii) A few small-scale slums, on land occupied illegally by squatters, can be found at the margins of the planned city.
iii) Numerous slums, generally occupied by tenants on legal lands, are found at the outskirts of the city along major roads or close to local labour markets. Their size, history, socio-economic and cultural features differ from one slum to another.

This study restricted itself to the first type of urban slum, the core area of the city. Many communities in the study area belong to the category. The inner city area is the oldest, has the lowest quality residence and the highest population density in the city. According to Asiyanbola (2012), the traditional core, high density or indigenous areas of Ibadan correspond roughly with Mabogunje's core and older suburbs and Ayeni's high density residential areas. Also, in terms of ethnic status, traditional core areas are relatively homogeneous in the sense that majority of the residents are indigenes of Ibadan. Nitrate represents the most oxidized chemical form of nitrogen found in natural systems. Nitrate concentrations of less than one (1) mgL^{-1} as nitrate-N in natural groundwater sources is not a major source of exposure. However, excessive amounts can pass through the soil and contaminate groundwater (Bernard et al., 1998). Elevated levels of nitrate in groundwater may result from human activities such as overuse of chemical fertilizers and improper disposal of human and animal wastes (Nugent et al., 1993). Municipal and industrial wastewaters, animal pens, runoff or leachate from refuse dumps, manured or fertilized agricultural lands, urban drainage and on-site sanitary systems (pit latrines/septic tanks) are the major sources in fast growing urban centres. Power plants and automobiles may also contribute through emissions containing nitrogen compounds, which are carried through rainfall.

The assessment of nitrate concentration in drinking water sources in Pakistan by Tahir and Rasheed (2008) revealed that 19% of the total samples (747) collected from a wide range of irrigated or non-irrigated regions in 16 cities had nitrate concentration greater than the permissible safe limit of 10 mg/L falling in the concentration range of 11 to 160 mg/L nitrate. Also, study by Mondal et al. (2008) to determine the extent of nitrate and the possible sources of nitrate pollutants in groundwater in selected regions of north and south Krishna Delta, India revealed that nitrate concentration ranged from nitrate from 10 to 135 mg/L with about 39% of 79 groundwater samples showing high nitrate contents which is more than the permissible limit of 50 mg/L in drinking water. Nitrate pollution level was found more in dug wells compared to hand pumps/bore wells. The study identified the possible sources of high nitrate level in groundwater to be excessive utilization of nitrogenous fertilizers for agricultural purposes. Contamination of drinking water by nitrate is an evolving public health concern globally. Nitrate in groundwater is of concern not only because of its toxic potential, but also because it may indicate serious groundwater pollution (Nugent et al., 1993). Ingested nitrate from dietary sources and drinking water can be converted to nitrite and ultimately to N-nitroso compounds, many of which are known carcinogens (Sandor et al., 2001; Weyer et al., 2006). When present in high concentrations, they may give rise to potential health risks particularly in bottle-fed infants (methemoglobinemia) and pregnant women (Jennings and Sneed, 1996; Kelter et al., 1997; Kempster et al., 1997; NRDC, 2003). Nitrate toxicity also can cause recurrent acute diarrhoea (Gupta et al., 2001). However, a study of methemoglobin levels in infants receiving low dietary nitrates in Eastern Nigeria conducted by Mbanugo et al. (1990) concluded that there was no significant association between high methemoglobin levels observed in the blood of infants and levels of nitrate ingested.

Ward et al. (2010) investigated the association of nitrate intake from public water supplies and diet with the risk of thyroid cancer and self-reported hypothyroidism and hyperthyroidism in a cohort of 21,977 older women in Iowa who were enrolled in 1986 and who had used the same water supply for >10 years. Outcome of the study revealed an increased risk of thyroid cancer with higher average nitrate levels in public water supplies and with longer consumption of water exceeding 5 mg/L nitrate-N [for ≤5 years at >5 mg/L, relative risk (RR) = 2.6 (95% confidence interval (CI) = 1.1-6.2)]. Even though there was no observed association with prevalence of hypothyroidism or hyperthyroidism; increasing intake of dietary nitrate was associated with an increased risk of thyroid cancer [highest vs. lowest quartile, RR = 2.9 (1.0 to 8.1); *P* for trend = 0.046] and with the prevalence of hypothyroidism [odds ratio = 1.2 (95% CI = 1.1 to 1.4)], but not hyperthyroidism. Khademikia et al. (2013) also investigated the relationship between the amounts of nitrate, nitrite, and total organic carbon (TOC) in two drinking water sources used for drinking, industrial and agricultural consumption in Iran and their relationship with some gastrointestinal diseases. Even though the study did not document significant association of nitrite, nitrate, and TOC content of water with gastrointestinal diseases, it concluded that such health hazards may develop over time, and the quality of water content should be controlled to prevent different diseases.

In Nigeria, there is still a dearth of information on nitrate levels and its health effects. The few studies available (Mbonu et al., 1991; Onianwa et al., 1999; Arowolo, 2005) are sporadic generalized water quality assessments. Global information systems (GIS) technology allows location-specific data from different sources to be joined and mapped to display important geographic relationships (Reissman et al., 2001). It uses a location reference system such as longitude, latitude and elevation. In recent years, GIS has been used widely in different studies: land use planning (Tong and Chen, 2002; Stout and Lee, 2004), watershed management (Lively and Czapar, 2002), risk of lead exposure (Reissman et al., 2001), and ecological implications of Fulbe pastoralism (Omotayo, 2003). However, spatial data to allow targeting water sources and quality to households are presently inadequate. The study combined GIS data with nitrate concentration of wells used as household drinking water sources during the wet and dry seasons to produce nitrate and nitrate exceedence maps for the study area. This has helped to highlight areas of health risks to the inhabitants in the study area, particularly, infants and pregnant women who are more vulnerable.

MATERIALS AND METHODS

This study was carried out in Ibadan South East Local Government Area (IBSELGA), one of the five urban local government areas of Ibadan metropolis, the capital of Oyo State. Ibadan, the largest indigenous urban center in Africa south of the Sahara lies on longitude 03° 53' 47" East of the Greenwich Meridian and latitude 07° 23' 16" North of the equator (Maplandia, 2005). It has a population of 1.34 million (Federal Republic of Nigeria, 2007) and covers an area of over 100 km^2. Its elevation is 210 m above sea level with isolated ridges and peaks rising to an elevation of 274 m (Udo, 1994). The study area (Figure 1) falls in the indigenous area of the city, with high population density (11,298.8/km^2), low level of sanitation and lack of safe water supply, with most people depending on shallow wells. The study area is made up of 30 localities and a population of 266,046. From this area, 48 household wells were selected randomly for monitoring during the wet and dry seasons. Two sets of water samples were collected from each well in polyethylene bottles (after they had been scrupulously cleaned) and stored in the refrigerator. Nitrate as NO_3^- was determined using the phenoldisulfonic acid method described by APHA/AWWA/WPCF (1975). The absorbance of the yellow color developed on reacting the nitrate in the sample with phenoldisulfonic acid was measured with UV spectrophotometer at a wavelength of 410 nm and related to the concentration of prepared standards. An observation checklist was used to assess the sanitary conditions (such as nearness to pollution sources such as septic tanks, refuse heaps, animal dung, drainage) around the wells.

A hand-held global positioning system (GPS) receiver (Garmin GPS 12 Personal Navigator) was used to locate the position of the 48 wells by measuring the coordinates (longitude, latitude and altitude), which were recorded in a notebook and later transferred into Microsoft Excel™ spreadsheet. Due to the skewness of nitrate concentrations in water samples during the two seasons, data were log transformed using SPSS 15.0 for Windows, and expressed as geometric mean. Nitrate concentrations in samples were compared with WHO permissible limit. Statistical analyses were performed using descriptive statistics and spearman's correlation coefficient at 5.0% level of significant. Spatial data saved in a Microsoft Excel™ spreadsheet were imported into ArcGIS™ 9.0 (ESRI, 2004) to produce nitrate and nitrate exceedence maps for the study area.

RESULTS

The distance in meters between the selected households and drinking water sources (wells) are shown in Table 1. Majority of the households 36 (75%) were at a distance between 1 and 10 m from the drinking water sources, while nine (19%) were between 11 to 20 m. Only 3 (6%) wells were at a distance 21 m and above from the households. Nitrate was detected in 98% of the wells tested and the concentration during the wet season ranged between 0.00 and 42.80 mgL^{-1}, with a geometric mean of 5.43 mgL^{-1} and a median of 4.59 mgL^{-1}. During the dry season, the range was 0.00 to 93.30 mgL^{-1} with a geometric mean of 4.97 mgL^{-1} and a median of 5.02 mgL^{-1}. The means of the transformed values were 0.735 and 0.696 for the wet and dry seasons, respectively (Table 2). During the wet season, all the samples had nitrate concentrations that were lower than 45 mgL^{-1} NO_3, whereas, the nitrate values for some of the samples increased during the dry season. About 13% of the wells dried up completely for three months while the nitrate values for a large number of the remaining samples (52.4%) increased by 2 to 26 fold. Also, 12.5% of the wells had values which exceeded the WHO limit. The nitrate map of the study area is shown in Figure 2. The

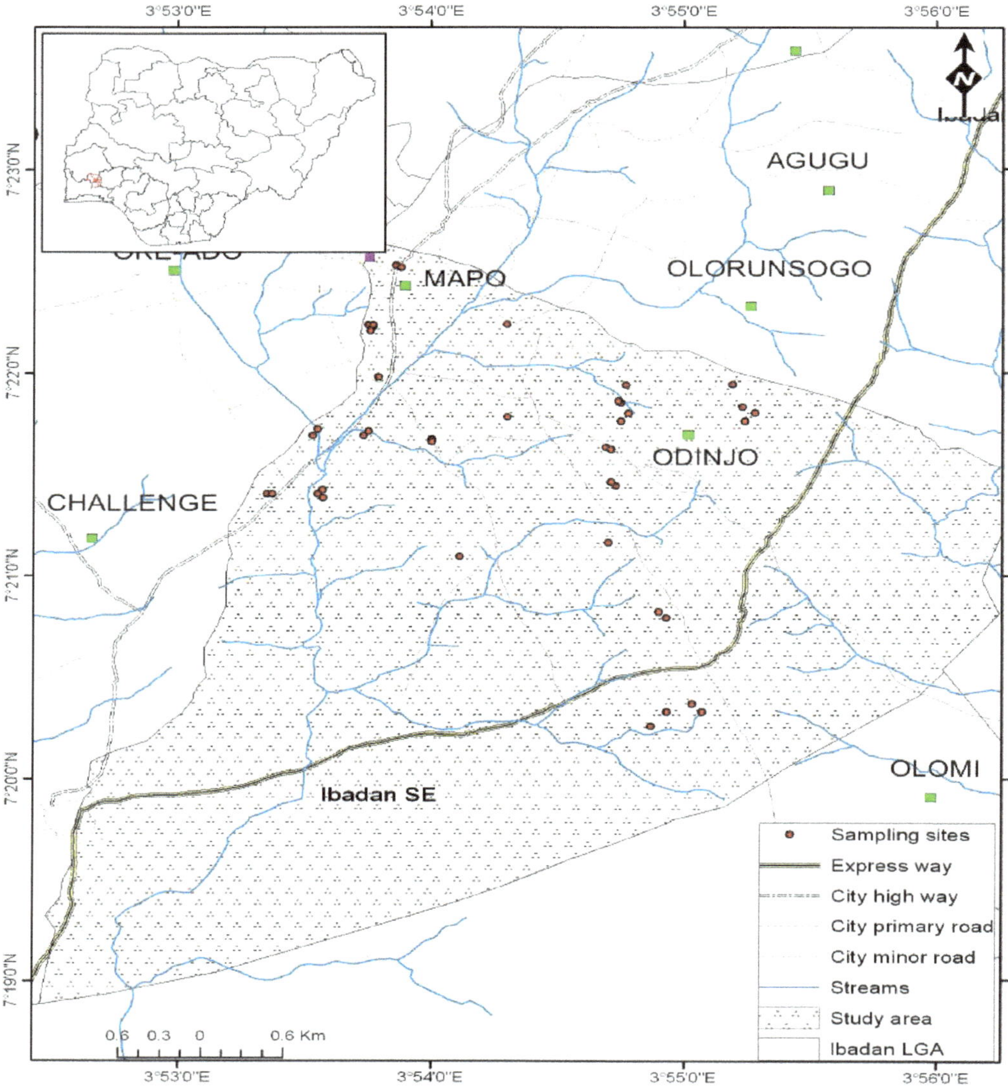

Figure 1. Map of study area showing the sampling sites.

Table 1. Distance between selected households and drinking water sources (wells).

Distance of source to household (m)	No. of wells
1 - 10	36 (75%)
11 - 20	9 (19%)
21 - 30	1 (2%)
31 - 40	1 (2%)
>40	1 (2%)

Table 2. Mean nitrate concentration and Log_{10} transformed values.

Season	Arithmetic mean of NO_3 mg L^{-1}	Range of NO_3 mgL^{-1}	Log_{10} of NO_3mgl^{-1}	Geometric mean of NO_3 mg L^{-1}
Wet	8.2675±9.047	0.00 - 42.80	0.735	5.43 mgL^{-1}
Dry	15.6638±23.3494	0.00 - 93.30	0.696	4.97 mgL^{-1}

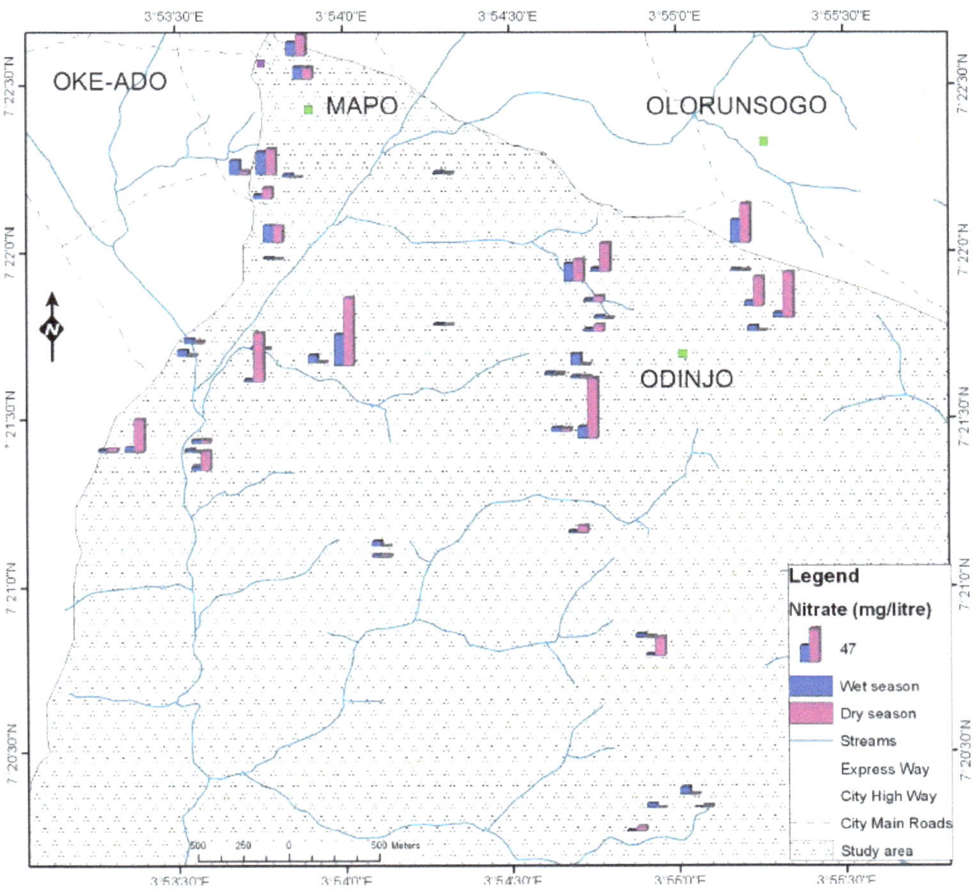

Figure 2. Nitrate map for the study area.

purple and blue bar graphs in Figure 2 represent nitrate concentrations in the sampled wells during the dry (purple) and wet (blue) seasons. The bar graph is measuring at the unit of nitrate concentration. Nitrate concentrations at different locations (during the two seasons) are indicated by the height of the bar. The number 47 is the scale which refers to the height of the purple one on the bar graph. It means that wherever the bar height is half, the size of the bar labeled 47 (purple), nitrate concentration is 47/2 = 23.5 mgL^{-1}; where it is twice the height, the nitrate concentration is also twice (47 × 2 = 94 mgL^{-1}). Also, Figure 3 shows the map of nitrate exceedence for wells in the study area during the dry season. These were wells with nitrate concentration higher than the WHO permissible limit (WHO, 1995, 1996).

The scatter plot was constructed to examine the type of pattern of relationship between the nitrate concentration (NC) during the dry and wet seasons. In Figure 4, the data revealed that slight linear pattern was displayed by the NC of the wells in the dry and wet seasons. Figures 5 and 6 represent the histogram of the distribution of the NC during the wet and dry seasons. In these figures, there was an indication of skewness in the NCs for dry and wet seasons. The data is also evidenced that NCs of the wells during the two seasons were positively correlated ($r = 0.390$; $p = 0.011$).

Variation in distance of wells to sources of pollution

Table 3 shows the location of wells used by selected households to sources of pollution. Majority (50%) of the wells used as drinking water sources were at a > 10 m from refuse dumps. Also, about 21% of the wells were at a distance less than 1 m and only 10.4% were at a distance between 6 and 10 m from refuse dumps. Among households who use pit latrine/septic tank, 4.2% of the wells were located at less than 1 m away, 56.3% were between 1 to 10 m, while 39.5% of the wells were at > 10 m away from pit latrines/septic tanks. Table 4 shows the six wells that had nitrate concentration exceeding the WHO limit in relation to their location to sources of pollution, while Plate 1 show some of the wells in the study area.

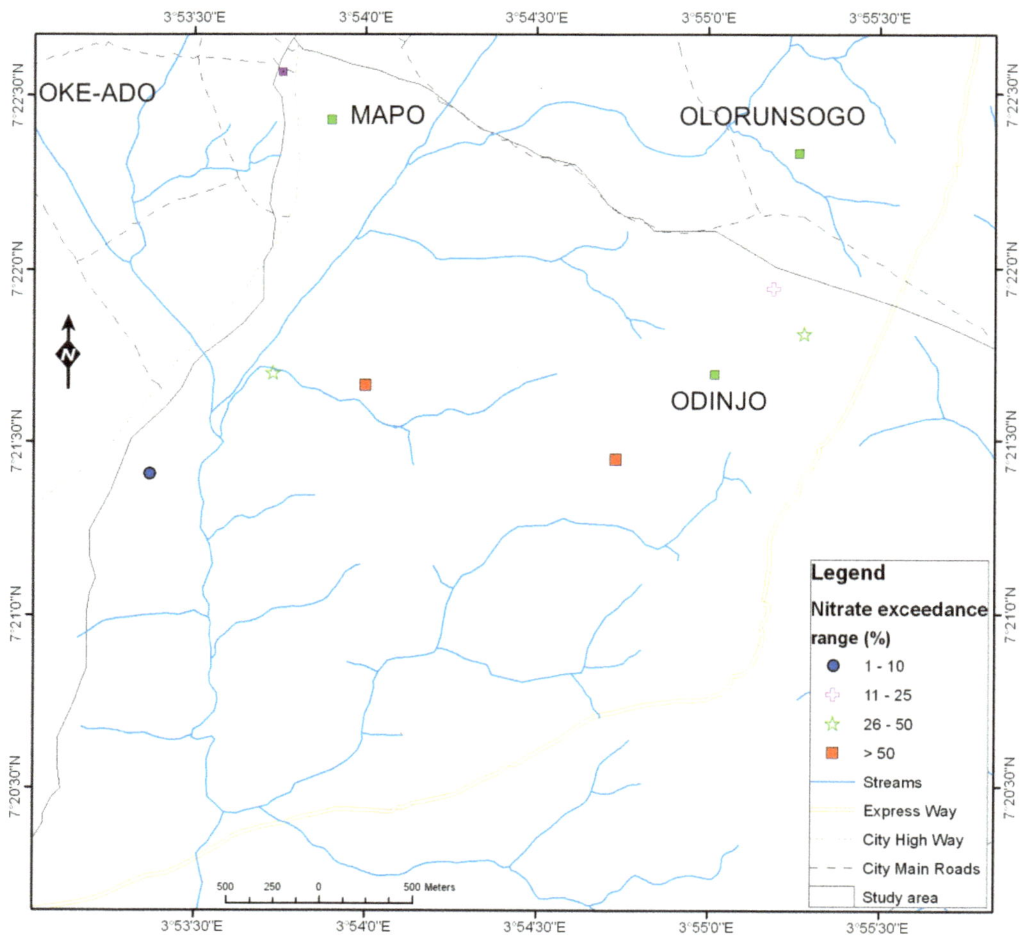

Figure 3. Nitrate exceedence map.

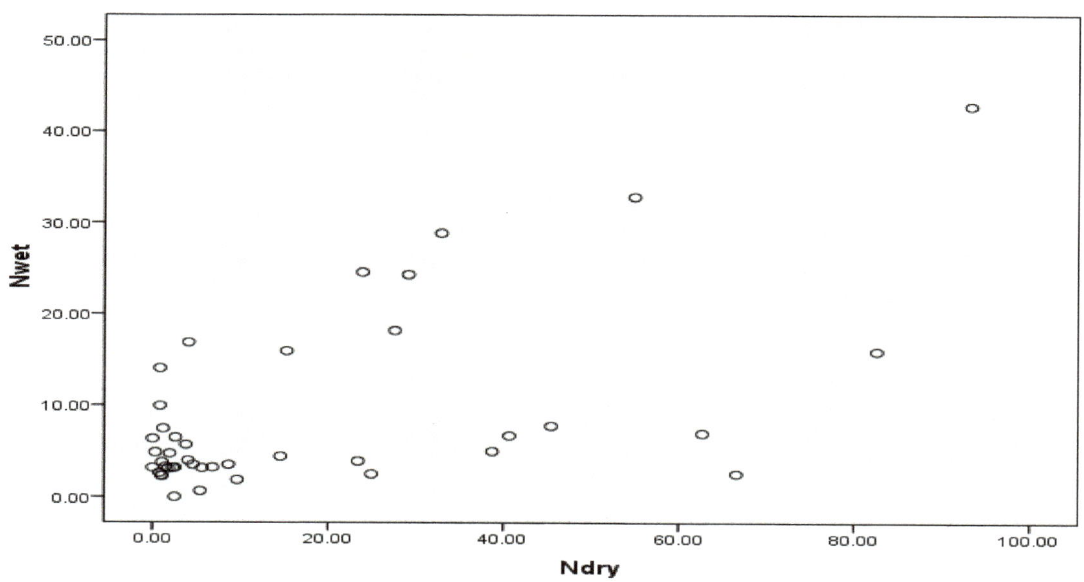

Figure 4. Scatter plot of the nitrate concentration in the analysed wells between the dry and wet seasons.

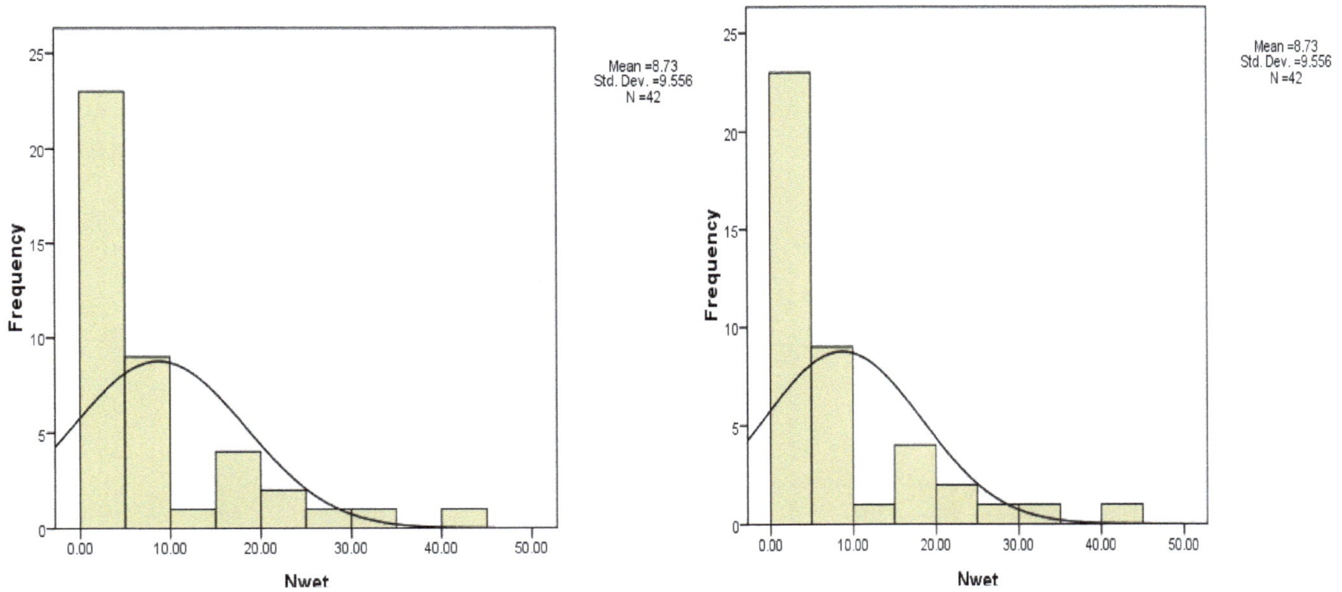

Figure 5. Histograms and superimposed normality curve of the nitrate concentration in the analysed wells between the dry and wet seasons.

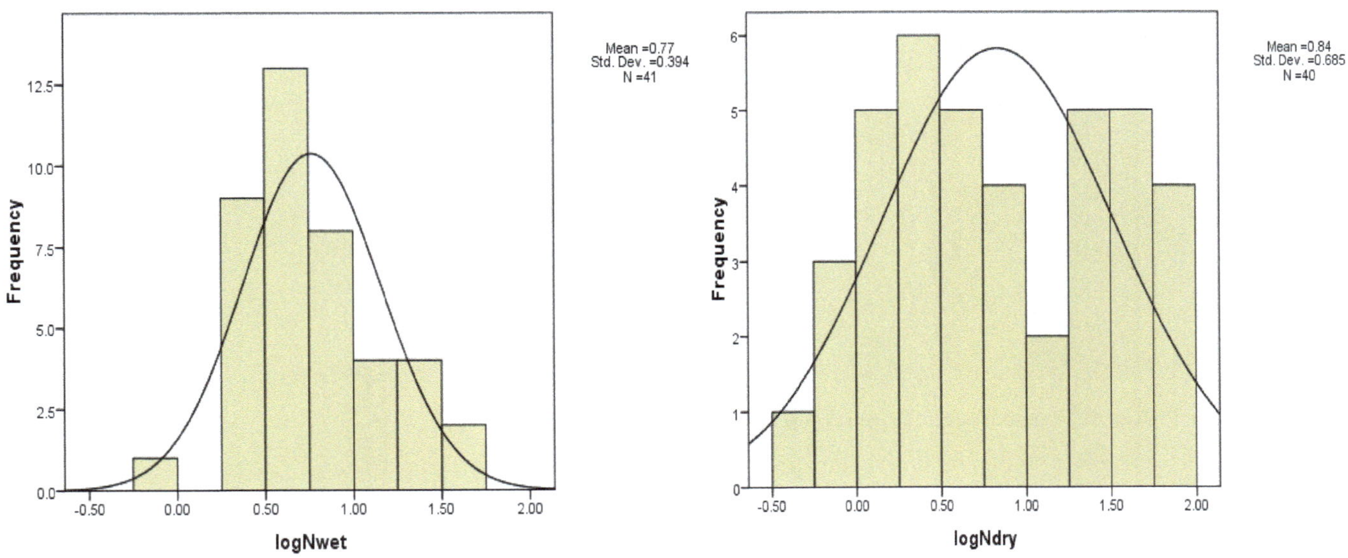

Figure 6. Histograms and superimposed normality curve of the log_{10} transformation of the nitrate concentration in the analysed wells between the dry and wet seasons.

Table 3. Distance between wells and sources of pollution.

Pollution source	Number of wells located at specified distances from pollution sources			
	<1 m	1-5 m	6-10 m	>10 m
Refuse dump	10 (20.8%)	9 (18.8%)	5 (10.4%)	24 (50.0%)
Pit latrine/septic tank	2 (4.2%)	15 (31.3%)	12 (25.0%)	19 (39.5%)

*Figures in parentheses are percentages.

Table 4. Nitrate concentration in wells relative to distance from pollution sources.

Pollution source	Nitrate concentration (as NO_3^- in mg/L) in wells located at specified distances from pollution source					
	1-5 m		6-10 m		>10 m	
Refuse dump	*W3 (NL)	62.69			IBSE W42 (NL)	66.60
	W25 (NL)	54.92	IBSE W47 (NL)	82.60		
	W39 (NL)	45.48			IBSE W31 (NL)	93.30
Pit latrine/ septic tank	W25	54.92	IBSE W3	62.69	IBSE W39	45.48
	W42	66.60	IBSE W31	93.30	IBSE W47	82.60

*W3 is a well location; NL, well not lined.

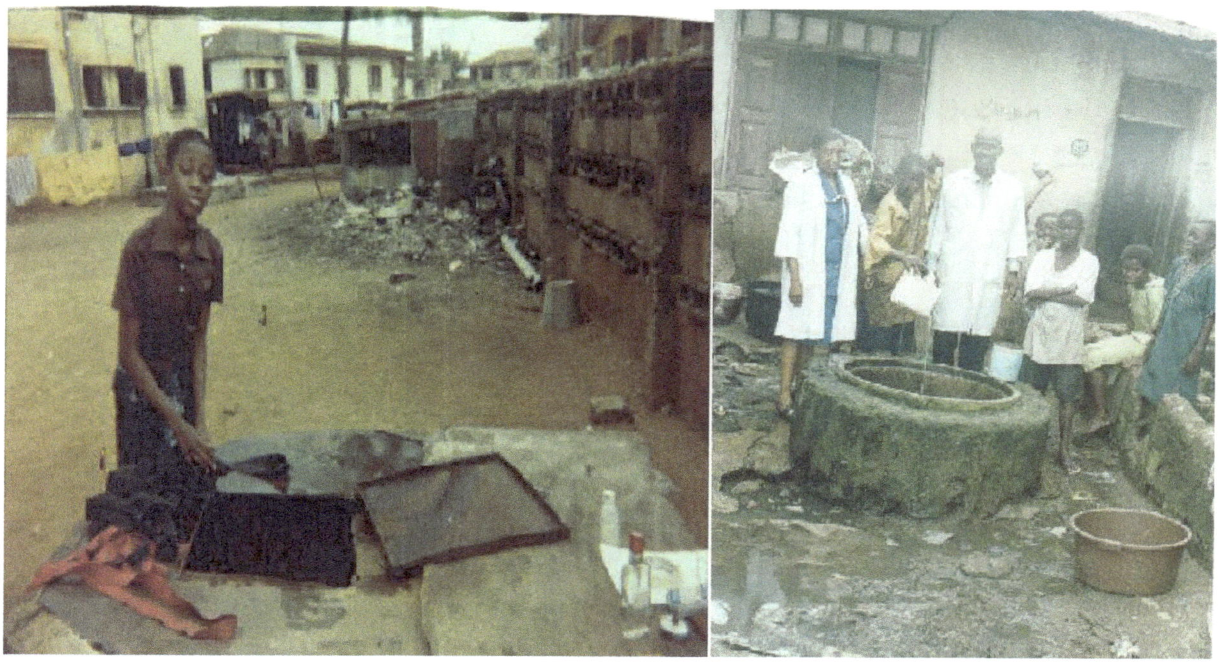

Plate 1. Picture of sampled well in the study area.

DISCUSSION

The results obtained show that only few: six (14.3%) of the wells had nitrate levels exceeding the WHO permissible limit during the dry season. Even though the range of the concentrations was wide, the findings are comparable to those in the study in the basement complex region of North-central, Nigeria (Mbonu et al., 1991) which reported that four (4) of the groundwater points sampled had nitrate levels > 100 mgL^{-1} NO_3. Similarly, Nieeko et al. (2001) and Gelberg et al. (1999) discovered nitrate levels higher than the WHO limit in wells from the parish of Szydlowiec, Swietokrzyskie province and rural New York State in USA. Mbanugo et al. (1990), however, reported nitrate levels in tap waters, boreholes and rivers ranging from 1.10 to 48.4 mgL^{-1} with a mean of 10.20 mgL^{-1} in south-eastern part of Nigeria. The study revealed that high concentration of nitrate in the wells may be due to their closeness to sources of pollution such as septic systems, refuse dumps and the fact that some of the wells lacked sanitary features such as cover, apron and lining. Nolan et al. (1998) and Alhajjar et al. (1990) reported that the water table in shallow wells is closer to the land surface and to potential sources of contamination such as septic systems and their leachates. The lower nitrate concentration in some of the wells during the rainy season may be due to the dilution or flushing action of increased water throughput in the wells. The majority of the wells are located between 1 and 10 m from sources of pollution, which is below the 30 m distance recommended as a minimum (Morgan, 1990).

Conclusion

This study presents the concentration of nitrate in 88 samples from 48 wells used as drinking water sources in both dry and wet seasons in an urban-slum setting in Ibadan Southeast local government area of Oyo state. Contaminating point sources located not farther than 10 m from wells may influence the quality of water, thus putting the lives of consumers at risk. The exceedence map identified those areas at risk; which therefore require urgent intervention. The results of this study will provide background information that will assist the local government officials to put in place appropriate regulatory actions to improve drinking water quality in their domain. The implementation of actions to prevent degradation, and to maintain and improve water quality is imperative. This study, therefore, recommends routine investigation and monitoring of nitrate levels in wells used as drinking water sources.

ACKNOWLEDGEMENTS

The authors acknowledge the contribution of the research assistants who participated in data collection, and the Geospatial Laboratory of the International Institute for Tropical Agriculture, Ibadan Nigeria for producing the nitrate maps for this study.

REFERENCES

Alhajjar BT, Chesters G, Harkin JM (1990). Indicators of chemical pollution from septic systems. Groundwater 28:559-568.

APHA /AWWA/ WPCF (1975). Standard methods for the examination of water and wastewater, 14th edition Washington DC, Maryland, USA.

Arowolo TA (2005). Seasonal variation of nitrate-N in surface and ground-waters of South Western Nigeria: A preliminary survey. Chem. Ecol. 21(2):83-89.

Bernard TN, Barbara CR, Kerrie JH, Dennis RH (1998). A national look at nitrate contamination of ground water. Water Cond. Purif. 39:76-79.

Gelberg KH, Church L, Casey G, London M, Roerig DS, Boyd J, Hill M (1999). Nitrate levels in drinking water in rural New York State. Environ. Res. 80(1):34-40.

Gupta SK, Gupta RC, Gupta AB, Seth AK, Bassin JK, Gupta A, Sharma MI (2001). Recurrent diarrhoea in children living in areas with high levels of nitrate in drinking water. Archives of Environmental Health July issue. Retrieved February 18, 2007 from http://www.findarticles.com/p/articles/mi_m0907/is_4_56/ai_79124886

Itama E, Olaseha IO, Sridhar MKC (2006). Springs as supplementary potable water supplies for inner city populations: A Study from Ibadan. Urban Water J. 3(4):215-223.

Jennings GD, Sneed RE (1996). Nitrate in drinking water, North Carolina Cooperative Extension Service, Publication No. AG 473-4.

Kelter PB, Grundman J, Hage DS, Carr JD (1997). A discussion of water pollution in the US and Mexico with high school laboratory activities for analysis of lead, anthrazine and nitrate. J. Chem. Educ. 74(12):1413-1418.

Kempster PL, Van Vliet HR, Kuhn A (1997). The Need for guidelines to bridge the gap between ideal drinking-water quality and that quality which is practicably available and acceptable. Water SA 23(2):163-167.

Lively TJ, Czapar GF (2002). Evaluating conservation practices using GIS and virtual reality modeling language. J. Nat. Resour. Life Sci. Educ. 31:38-43.

Mbanugo JI, Sridhar MKC, Anunuso CI (1990). Methaemoglobin levels in infants receiving low dietary nitrates in Eastern Nigeria. Toxicol. Environ. Chem. 25:195-199.

Mbonu M, Travi Y, Guuidaud R (1991). Groundwater quality in the basement complex region of North-Central, Nigeria: A precursory survey of nitrate concentration. In: E.O.A. Aina and N.O. Ogundipe (Eds), Water Quality Monitoring and Environmental Status in Nigeria. FEPA Monograph 6:176-190.

Morgan P (1990). Sanitary surveys of wells and tubewells. In Rural Water Supplies and Sanitation: A text from Zimbabwe's Blair Research laboratory: 58-60 Macmillan Education Ltd., London.

Nieeko J, Sobotka M, Kukulka-Nieeko U, Nieeko M (2001). The levels of nitrate in drinking water of selected water wells situated in the Parish of Szydlowiec. Przeglad Lekarski 58(7): 25-29.

Nolan BT, Ruddy BC, Hitt KJ, Helsel DR (1998). A national look at nitrate contamination of groundwater. Water Cond. Purif. 39(12):76-79.

NRDC (2003). Common tap water contaminants: Health effects, treatment and recommendations. In: What's on tap? Grading drinking water in U.S.

Nugent M, Kamrin MA, Wolfson L, D'Itri. (1993). Nitrate-a drinking water concern. Michigan State University extension, Extension Bulletin WQ-19.

Omotayo AM (2003). Ecological implications of fulbe pastoralism in Southwestern Nigeria. Land Degradation Dev. 14: 445-457

Onianwa PC, Ogunniyi OJ, Ogunnowo AA (1999). General physico-chemical analysis of drinking water in Ibadan, Nigeria. Int. Environ. Educ. Inf. 18(2):143-154.

Reissman DB, Staley F, Curtis GB, Kaufmann RB (2001). Use of geographic information system to aid health department decision making about childhood lead poisoning prevention activities. Environ. Health Perspect. 109(1):89-94.

Sandor J, Kiss I, Farkas O, Ember I (2001). Association between gastric cancer mortality and nitrate content of drinking water: Ecological study on small area inequalities. Eur. J. Epidemiol. 17(5): 443-447.

Stout HM, Lee BD (2004). Land use planning exercise using geographic information systems and digital soil survey. J. Nat. Resour. Life Sci. Educ. 33:11-15.

Tong STY, Chen W (2002). Modeling the relationship between land use and surface water quality. J. Environ. Manage. 66(4) 377-393.

Udo RK (1994). Ibadan in its regional setting. In: MO Filani, FO Akintola, C.O. Ikporukpo (Eds.), Ibadan region: 6-17. Rex Charles Publication and Department of Geography, University of Ibadan.

Weyer PJ, Smith BJ, Feng Z, Kantamnenl JR, Riley DG (2006). Comparison of nitrate levels in raw water and finished water from historical monitoring data on Iowa Municipal drinking water supplies. J. Environ. Monit. Assess. 116(1-3):81-90.

WHO (1995). Guidelines for drinking water quality, Vol. 1, Second Edition: Recommendations. Geneva, WHO: 174-181.

WHO (1996). Guidelines for drinking water quality, Vol. 2, Second Edition: Health Criteria and Supporting Information. Geneva WHO (2):1-94.

4

Hydrogeochemical analysis and quality evaluation of groundwater from Onicha-Uburu, Southeastern Nigeria for irrigation purposes

Nweke, O. M., Aghamelu, O. P. and Obasi, I. A.

Department of Geology and Exploration Geophysics, Ebonyi State University, P.M.B. 053, Abakaliki, Ebonyi State, Nigeria.

A total of 20 groundwater samples from boreholes and hand dug wells within Onicha-Uburu and its environs were assessed to determine their suitability for irrigation purposes. Physical and hydrogeochemical parameters were evaluated, while techniques such as sodium absorption ratio (SAR), sodium percent, residual sodium carbonate (RSC) and magnesium adsorption ratio (MAR) were used to assess the suitability of these groundwater samples for the intended purpose. Results indicated that temperature ranges between 20 and 26°C for hand dug wells and 19 and 23°C for borehole samples. The electrical conductivity (EC) varies from 8.4 to 36.4 µs/cm for the hand dug wells and 10.0 to 76.4 µs/cm for the borehole water samples, while pH values range between 7.0 and 9.0 for the entire samples and indicate alkaline water. Statistical analysis shows appreciable concentrations of basic ions (average value of 28.8 mg/l for Na^+, 9.1 mg/l for K^+, 16.0 mg/l for Ca^{2+}, 14.6 mg/l for Mg^{2+}, Nil for $C0_3^{2-}$, as well as 45.0 mg/l for $HC0_3^-$, 20.5 for mg/l Cl^- and 8.7 mg/l for $S0_4^{2-}$). While Na^+ and Ca^{2+} are the predominant cations, $HC0_3^-$ and Cl^{-1} are the anions that predominate. The chemical analysis revealed that the total hardness (measured in terms of $CaCO_3$) varied from 16 to 140 mg/l for both sets of samples indicating soft to moderately hard water. Based on the values of EC, percent sodium, SAR (which vary from 0.3 to 1.7 meq/l), RSC (from 0.1 to 0.99 meq/l less than the standard value of 1.25 meq/l) and the magnesium hazard (MH) values which range between 0.99 and 89.8%, revealed that water samples from the study area are generally suitable for irrigation purposes except for few locations.

Key words: Groundwater, hydrochemical facies, hydro-geochemical analysis, irrigation purposes, Onicha-Uburu area.

INTRODUCTION

It is generally recognized that the quality of groundwater is just as important as its quantity. Most groundwater contains dissolved salts in concentrations as to make them unusable for ordinary water supply purposes. Soluble salts in groundwater originate primarily from natural dissolution of common rock minerals such as silicates, clays (for example, kaolinites and montmorillonite), feldspars as well as carbonates (Todd, 1980). Apart from natural processes, other controlling factors on the groundwater quality include pollution, resulting from some uncontrolled affluent discharge from industries, harmful agricultural practices (e.g., excessive application of nitrate-based fertilizers) and acid rain. The quality required of a groundwater supply depends on its purpose, examples: drinking (that is, human consumption), industrial use, irrigation, recreation and fish culture purposes (Todd, 1980). Knowledge of irrigation water quality is critical to understanding what management changes are necessary for long-term productivity (Brady, 2002).

The suitability of groundwater for irrigation is contingent on the effects of the mineral constituents of the water on the plants and the soils as well as the salt. According to Todd (1980), poor quality irrigation (with high salt concentration) water may affect plant growth physically by limiting the uptake of water through modification of osmotic processes. According to Todd (1980), effects of salts on soils could result in alterations in soil structure, permeability and aeration, which indirectly affect plant growth. Problems of groundwater quality degradation are much difficult to overcome as it often requires long period of time to detect the true extent of degradation. Groundwater chemistry also has a potential use for tracing the origins and the history of water as well as its compositional changes through reactions with the environment. The chemical parameters such as Ca^{2+}, Mg^{2+}, Na^{+}, K^{+}, HCO_3^{-}, Cl^{-} and SO_4^{2-} play a significant role in classifying and assessing water quality for irrigation.

Considering the individual and paired ionic concentration, certain indices are proposed to find out the alkali hazards. For example, residual sodium carbonate (RSC) can be used as a criterion for ascertaining the suitability of irrigation water (Sadashivaiah et al., 2008). Other methods of representing the chemistry of water, such as Collins bar diagram (Hem, 1970), radiating vectors (Maucha, 1940), and parallel and horizontal histograms (Stiff, 1940), have been successfully used in the past in some parts of the world to study the proportion of ionic concentrations in water samples. Sadashivaiah et al. (2008) followed a series of methods to interpret and classify the chemistry of groundwater in hard rock, including coastal zones in the Southern India. Presentation of chemical analysis in graphical form makes understanding of groundwater system simpler and quicker.

In this work, the major ionic chemistry of groundwater samples in Onicha-Uburu and its environs in Southeastern Nigeria are assessed using the methods proposed by Piper (1953) and United States Salinity Laboratory (Wilcox, 1995). Based on the results, their suitability for use for irrigation purposes is highlighted. Salinity, sodicity and toxicity generally need to be considered in the evaluation of the suitability of groundwater for irrigation (Todd, 1980). Parameters such as electrical conductivity (EC), percent sodium, residual sodium carbo-nate (RSC), magnesium hazard (MH) and sodium ad-sorption ratio (SAR) were used to assess the suitability of water for irrigation. Continued use of water with a high SAR value leads to a breakdown in the physical structure of the soil caused by excessive amounts of colloidally absorbed sodium. This breakdown results in the dispersion of soil clay that causes the soil to become hard and compact when dry and increasingly impervious to water penetration due to dispersion and swelling when wet. Fine-textured soils, those high in clay, are especially subject to this action. It is hoped that this study would assist in determining the degree of salinity, sodicity and toxicity thereby removing some of the speculation con-cerning the quality of water used for agricultural purposes in the Onicha-Uburu area, and its impact on crop yield.

Geology and hydrogeology

Onicha-Uburu lies within the southeastern Nigeria in the southern Benue Trough (Figure 1). The origin of the trough has been associated with the breakup of the Gondwanaland in the early Cretaceous (Wright, 1979). The area is underlain by the Asu River Group. The Asu River Group is believed to be deposited by the first marine transgression in the trough, which started in the mid-Albian period. It is predominately shales and localized development of sandstone, siltstones and limestone (Reyment, 1965). The average thickness is about 2000 m.

Occurrence, movement and storage of groundwater within this area are influence by lithology, thickness and structure of rock formation (predominantly shales of the Asu River Group). Shale is an aquiclude and does not permit reasonable transmission of water, especially when fresh and unweathered. There are, however, intercalations of clays and sand clays which has led to artesian conditions in the study area. Nonetheless, the groundwater in the areas generally exists in zones; fractured zones within the Asu River Group, sandstone and limestone layers or members, weathered zones, and bedrock interfaces with shale group.

The area is largely drained by the Asu River and its tributaries such as River Odii, River Idee and River Ugenoma which is a continuation of Omuiyi River and the tributaries of the Cross River. Asu River takes its origin from Enugu Escarpment. The tributaries are perennial and usually overflow their banks at the peak of the rainy season. They are heavily contaminated as a result of mining activities in Okposi as well as indiscriminate waste disposal in the course of its flow.

MATERIALS AND METHODS

A total of 20 samples, 10 from hands dug wells and 10 from boreholes, were collected from different locations within the study area for hydrogeochemical analyses whereas; physical tests (pH, temperature and electrical conductivity) were carried out *in situ*. The samples were designated WH A-J for the hand dug well samples and BH A-J for the borehole water samples. All the hydrochemical analyses were carried out in the laboratories of the department of Civil Engineering, University of Nigeria, Nsukka.

Clean 1 L volume plastic containers were used for the sampling. For the pH values of the samples (determined in the field), the HACH portable pH meter was employed. Analyses of ions were

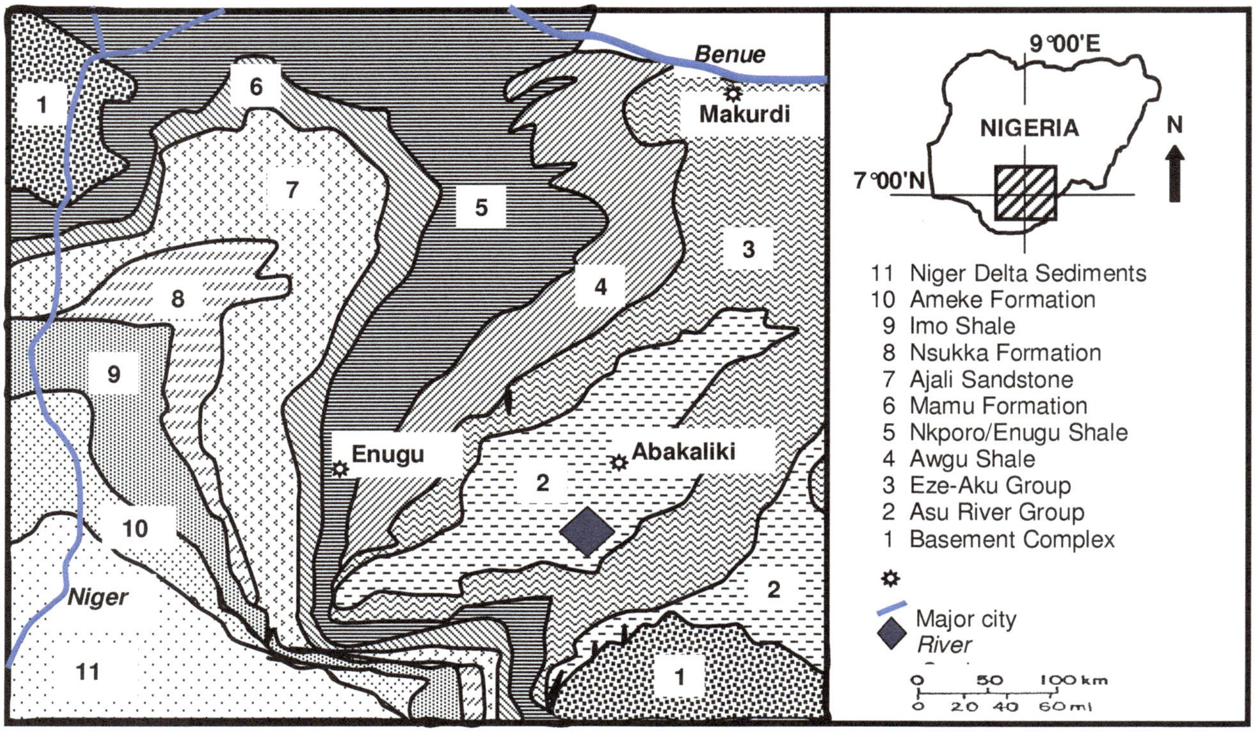

Figure 1. Geological map of Southeastern Nigeria showing the study area (Reyment, 1967).

limited to the major elements' ions such as Na^+, K^+, Ca^{2+}, Mg^{2+}, CO_3^{2-}, HCO_3, Cl^- and SO_4^{2+}. These were carried out (in the laboratory) employing standard procedures recommended by American Public Health Association (APHA 1994).

Statistical analysis

Statistical average and ranges for each of the parameters represent the mean value and the lowest and highest values, respectively, for all the locations considered. The sodium in irrigation water is usually denoted as percent sodium and is determined using Equation 1:

$$Na\ (\%) = \frac{Na^+ \times 100}{Ca^{2+} + Mg^{2+} + Na^+ + k^+} \quad (1)$$

Sodium adsorption ratio (SAR) according to Wilcox (1995) is a better measure of the sodium hazard which is used to express reactions with the soil. It was computed by using the relationship given in Equation 2:

$$SAR = Na^+/\{(Ca^{2+} + Mg^{2+})^{1/2}/2\} \quad (2)$$

The relative proportion of sodium in the water in the form of sodium carbonate was also determined using Equation 3:

$$RSC = (HCO_3^- + CO_3^{2-}) - (Ca^{2+} + Mg^{2+}) \quad (3)$$

All the computational formulas given above are possible when ionic concentrations are expressed in meq/l.

Magnesium Hazard (MH) was also calculated in percentage using Equation 4:

$$MH = \frac{Mg^{2+} \times 100}{Ca^{2+} + Mg^{2+}} \quad (4)$$

This indicator (MH) can be used to specify the magnesium hazard proposed by Szabolcs and Darab (1964) for assessing irrigation water. These analytical data not only can be used for the classification of water for utilitarian purposes but also for ascertaining various factors on which the chemical characteristics of water depend (Sadashivaiah et al., 2008).

RESULTS AND DISCUSSION

The results of the hydrochemical analyses of the water samples are presented in Table 1. Temperature ranges between 20 and 26°C for the hand dug well and 19 and 23°C for borehole samples. The groundwater temperature tends to remain mildly low and is considered an advantage for industrial uses. The electrical conductivity (EC), as presented in Table 1, varies from 8.4 to 36.4 µs/cm for hand dug well and 10.0 to 76.4 µs/cm for the borehole water samples. The EC values are relatively low except the borehole at Odii market square which has value above 76.4 µs/cm. The pH values range between 7.0 and 9.0 for the entire samples and indicate alkaline water. The pH values generally encountered may be largely due to the influence of the soils and organic

Table 1. Summarized results of the hydrogeochemical analysis of the water samples.

Location	Source	Temperature (°C)	pH mg/l	EC (us/cm)	Na^+ (mg/l)	K^+ (mg/l)	Ca^{2+} (mg/l)	Mg^{2+} (mg/l)	$C0_3^{2-}$ (mg/l)	HCO_3^{2-} (mg/l)	Cl^- (mg/l)	$S0_4^{2-}$ (mg/l)	Hardness (mg/l)
Amankwo	WH A	21	7.5	15.1	27.5	10.1	18.0	24.3	0	24.0	38.0	17.2	40
Afor Enugu	WH B	20	7.4	20.0	18.9	0.4	4.0	28.9	0	20.0	18.0	0.2	120
Amanator	WH C	20	7.5	16.0	9.7	0.6	28.0	9.7	0	22.0	25.0	1.6	40
Isu	WH D	26	7.7	10.4	24.3	0.3	16.0	14.3	0	50.0	30.0	0.9	20
Nkwoagu	WH E	20	7.4	8.4	24.6	0.6	18.0	14.3	0	20.0	15.0	2.6	120
Umunko	WH F	21	8.5	36.4	14.6	0.7	14.0	14.6	0	10.0	13.0	13.6	110
Umueziuku	WH G	22	7.5	28.0	9.0	0.3	30.0	0.3	0	50.0	18.0	8.6	36
Isuachara	WH H	20	9.0	36.0	45	4.6	20.0	14.3	0	36.8	3.0	6.9	60
Umubo	WH I	20	7.3	48	6.7	0.2	26.5	28.6	0	110	13.0	6.9	40
Onicha Hqt	WH J	22	9.2	36	44.1	0.6	8.0	9.7	0	12.0	15.0	1.2	16
Amefi	BH A	21	8.8	15.2	26.2	18.0	14.8	20.3	0	20.0	22.0	12.4	120
Isu Hgt	BH B	20	7.4	10.0	26.2	0.4	20.0	12.4	0	16.0	8.0	5.6	90
Afor Mkt	BH C	19	8.0	47.0	26.4	4.6	18.6	16.0	0	30.0	16.0	8.2	80
Odii Square	BH D	23	7.5	76.4	16.8	0.6	12.0	19.7	0	24.0	10.2	10.0	90
Ugenoma	BH E	20	7.0	8.0.1	44.8	0.3	20.0	13.6	0	80.0	12.2	16.1	40
Umunko	BH F	21	7.8	36	36.4	10.0	16.0	18.0	0	36.0	3.0	16.8	80
Nkwoagu Mkt	BH G	19	7.0	46	24.6	0.8	12.1	23.6	0	22.0	3.6	5.8	20
Ata-Nti	BH H	22	8.0	20.4	16.2	8.0	4.6	10.6	0	10.0	18.0	8.0	70
Umubo	BH I	20	8.4	18.4	18.20	4.6	12.6	12.8	0	16.0	10.0	10.0	40
Isuachara	BH J	21	9.0	16.8	4.6	4.2	18.0	24.6	0	28.0	12.0	6.4	140

matter present (Orajaka, 1972).

Water hardness is primarily caused by the presence of cations such calcium and magnesium as well as anions such as carbonate and bicarbonate. According to Sawyer and McCarty (1967) classification for hardness (Table 3), ten samples fall under soft while the remaining ten samples fall under moderately hard. The chemical analysis revealed that the total hardness as in $CaCO_3$ varied from 16 to 140 mg/l for both samples indicating soft to moderately hard water. The moderate hard water noticed in most of the samples especially the Isuachara borehole (140 mg/l) could be attributed to calcareous soil and limestone unit found in the Asu River Group that contain calcareous soil and limestone unit. The Ca^{2+} salt in most of the groundwater samples suggests calcareous aquifer. The normal ranges of concentration of the ion analyzed are presented in Table 2. The predominant cations are Na^+ and Ca^{2+}, whereas the predominant anions are $HC0_3^-$ and Cl^-. The CO_3^{2-} values for all the samples are low.

Criteria for irrigation purposes

The suitability of groundwater for irrigation purposes depend on its mineral constituent (Sadashiviah et al., 2008). Total salt concentration as measured by electrical conductivity (EC), relative proportion of sodium to other principal cation as expressed by SAR, sodium percent and residual sodium carbonate (RSC) are some of the techniques adopted by the US Salinity Laboratory of the Department of Agriculture (Wilcox, 1995) in evaluating the suitability of water for irrigation purposes. The total concentration of soluble salts in irrigation water in all the samples expressed in terms of specific conductance according to salinity hazard classes (Table 4) were found to be excellent and hence suitable for irrigation purposes.

The classification of groundwater samples with respect to percent sodium are shown in Table 5. Four samples have percent sodium values 6 to 20% which indicates values largely less than or

Table 2. Normal ranges of concentrations.

Ion (mg/l)	Range	Average	Well sample mean	Borehole mean
Na^+	4.6- 45.0	24.8	22.40	24.00
K^+	0.2- 18.0	9.1	1.80	5.2
Ca^{2+}	4.0 - 28.0	16.0	6.0	10.0
Mg^{2+}	0.3 - 28.9	14.6	15.9	17.2
CO_3^{2-}	Nil	Nil	Nil	Nil
HCO_3^-	100 - 80.0	45.0	32.5	28.2
Cl^-	3.0 - 38.0	20.5	18.8	11.5
SO_4^{2+}	0.2 - 17.2	8.7	6.0	9.9

Table 3. Classification of water based on hardness after Sawyer and McCarthy (1967).

Hardness as $CaCO_3$	Water class	Hand dug well	Borehole
0-75	Soft	7 samples	4 samples
75-150	Moderate Hard	3 samples	6 samples
150 - 300	Hard	-	-
>300	Very hard	-	-

-Not available.

Table 4. Salinity hazard class after Wilcox (1995).

Class	EC (μmoh/cm)	Remarks on quality
C_1	100-250	Excellent
C_2	250 -750	Good
C_3	750 -2250	Doubtful
C_4 & C_5	>2250	Unsuitable
Study samples	8.4 - 76.4	Excellent*

*All values fall below 100 μmoh/cm.

equal to 20%, whereas 12 samples have values ranging between 23 and 38%, with most values generally less than 40%, indicating that they are all safe for irrigation purpose. Three samples are within permissible water class of (46 and 47%). Water sample from Uburu hand-dug has value as high as 61%, suggesting that its suitability is doubtful if used for irrigation purpose. Orajaka (1972) have noted occurrences of salt ponds (with salinity of 4.5% and pH 8 at 25°C) within Uburu and its environs. Total dissolved solids are measured at 4.55% and NaCl at 4.32%. These confirmed the high concentrations of sodium in most of the studied samples. The Salinity Laboratory of the U.S Department of Agriculture established the sodium adsorption ratio as a direct method of assessing the suitability of water for irrigation. The classification of the groundwater samples from the study, with respect to SAR as presented in Table 6 indicate that the SAR values of all the samples were found to vary from 0.3 to 1.7 meq/l, revealing excellent irrigation water. The classifications of the groundwater samples based on residual sodium carbonate are presented in Table 7. The values which vary from 0.1 to 0.99 meq/l are generally less than 1.25 meq/l also revealing good irrigation water. According to USSL Department of Agriculture, water having more than 2.5 meq/l of RSC is not suitable for irrigation purposes. Based on RSC values obtained, all the 20 samples are generally less than 1.25 meq/l and therefore safe for irrigation purposes. Another indicator can be used to specify the magnesium hazard (MH) as proposed by Szabolcs and Darab (1964) for irrigation water. They proposed that if this percentage hazard was greater than 50, then the water will not be safe and will be unsuitable for irrigation purposes. From the calculated value, the magnesium hazard values range between 0.99 and 89.8% (Table 8), and can be classified as suitable for irrigation use for those that ranged from 0.99 to 50% while those that ranged from 51.9 to 89.8% are unsuitable. These were noted mostly in the water samples listed as follows: 66.1 (BH-G), 69.7 (BH-H) and 89.8% (WH-B).

The chemical analyses revealed that samples (from boreholes and hand dug wells) of groundwater from Onicha-Uburu are classified as soft water. The classification of the groundwater samples from the study, with respect to SAR indicate that the SAR values of all the samples were found to vary from 0.3 to 1.7 meq/l and classified as excellent for irrigation purposes as no sodium toxicity will arise by using the groundwater for surface irrigation. The classifications of the groundwater samples based on residual sodium carbonate vary from 0.1 to 0.99 meq/l indicating that the samples are safe as values generally less than 1.25 meq/l are good for irrigation purposes. The magnesium hazard (MH) values

Table 5. Sodium percent water class after Wilcox (1995).

Sodium (%)	Water class	Study samples (%)	Remarks
< 20	Excellent	6, 7, 16, 20,	4 samples
20-40	Good	23-38	12 samples
40- 60	Permissible	46- 47	3 samples
60-80	Doubtful	61	1 sample
>80	Unsuitable	-	-

-Not available.

Table 6. Sodium hazard classes based on sodium adsorption ratio after Wilcox (1995).

Sodium hazard class	SAR	Remark on quality	Study samples	Remarks
S_1	10	Excellent	0.3-1.7	All the 20 samples
S_2	10-18	Good	-	
S_3	18-26	Doubtful	-	
S_4 and S_5	>26	Unsuitable	-	

-Not available.

Table 7. Water quality based on residual sodium carbonate after Wicox (1995).

RSC (meq/l)	Remark in quality	Study sample	Remark
< 1. 25	Good	< 0.99	All the 20 samples
1. 25 -2.5	Doubtful	-	
> 2.5	Unsuitable	-	

-Not available.

Table 8. Water quality based on magnesium hazard after Szabolcs and Darab (1964).

MH (%)	Remark in quality	Study sample
< 50	Suitable	0.99% (WH-G), 25.7%, WH-C), 38.4% (BH-B), 40.5% (BH-E), 42% (WH-H), 44.2% (WH-E), 47.1% (BH-C), 47.19% (WH-D), 48% (BH-I), 50% (WH-F)
>50	Unsuitable	51.9% (WH-I), 52.9% (BH-F), 54.8% (WH-), 57.1% (BH- J) 57.4% (WH-A),57.8% (BH- A), 62.1% (BH-D.14), 66.1% (BH-G), 69.7% (BH-H), 89.8% (WH-B)

range between 0.99 and 89.8% (Table 8). However, water samples from the following locations: Nkwoagu Mkt borehole, BH-G (66.1%), Ata Nti borehole, BH-H (69.7%) and Afor Enugwu well, WH-B (89.8%) may not be unsuitable for irrigation purpose using the proposed parameter (MH).

Generally, the analyzed water samples are suitable for irrigation purposes except for few samples from Nkwoagu Mkt borehole, Ata Nti borehole and Afor Enugwu well. To maximize the benefits for irrigation purpose, it is suggested that groundwater utilization and management policies should be formulated and implemented to overcome future conflict in utilization of the groundwater in the area and also to maintain the sustainability of the irrigation scheme.

REFERENCES

American Public Health Association (1994). Standard method for examination of water and wastewater. NW, DC 20036.

Brady NC (2002). The nature and properties of soil. Prentice Hal, 10th Edition, New Delhi.

Hem JD (1970). Study and interpretation of the chemical characteristics of natural water. 2nd ed., U.S. Geological Survey water supply. p. 1473.

Maucha R (1940). The graphic symbolization of the chemical composition of natural waters. Hiderol Kozlony. p. 29.

Orajaka S (1972). Salt water resource of East central state of Nigeria. Journal Min ∝ Geol. In Geology of Nigeria, C.A. Kogbe (ed). Eliz. Pub. Co. pp. 455-477.

Piper AM (1953). A graphic procedure in the geochemical interpretation of water analysis. USGS, Groundwater Note p.12.

Reyment RA (1965). Aspect of Geology of Nigeria. Ibadan University press. p. 145.

Sawyer GN, McCarthy DL (1967). Chemistry of sanitary Engineer, 2nd Ed, McGraw Hill, New York. p. 518.

Sadashivaiah C, Ramakrishnaiah CR, Ranganna G (2008). Hydrochemical analysis and evaluation of groundwater quality in Tumkur Taluk, Karnataka State, India. Int. J. Environ. Res. Public Health 5(3):158-164.

Stiff Jr HA (1940). Interpretation of chemical water analysis by means of patterns. J. Petrol. Technol. 3:15-16.

Szabolcs I, Darab C (1964). "The influence of irrigation water of high sodium carbonate content of soils". In: Proceedings of 8th International Congress of Isss, Trans, 2:803-812.

Todd DK (1980). Groundwater Hydrology. Wiley, New York, 2nd edn., p.315.

Wright JB (1976). Origin of the Benue Trough. A critical review In: C. A. Kogbe (ed.) Geology of Nigeria. Elizabethan Publishing Company, Lagos- Nigeria, p. 187-295.

Wilcox LV (1995). Classification and use of irrigation waters. US Department of Agriculture, Washington DC. p. 19.

5

Climate change and decline in water resources in Kikuletwa Catchment, Pangani, Northern Tanzania

Linus K. Munishi[1] and Pudensiana C. Sawere[2]

[1]School of Life Sciences and Bio-Engineering, Nelson Mandela African Institute of Science and Technology, P O Box 447, Arusha, Tanzania.
[2]Pangani Basin Water Office, P O Box 3020, Arusha, Tanzania.

The sensitivity of hydrology and water resources to climate variation in Kikuletwa Catchment, Pangani Basin, northern Tanzania was assessed using 30 years of river level and climate data as well as questionnaire, focused group discussion (FGD) and participant observation. The results show a significant association between mean annual river level for Kikuletwa River and mean annual rainfall in the past thirty years. The results further indicate an inverse relationship between river level and temperature in the catchment suggesting the effects of climate change on water resources at Kikuletwa Catchment area. Although the results of our study indicate an upward trend in precipitation (from the two rain stations) over the past three decades, there was a consistent decline in river level in the main rivers. The eight villages covered by this study face a variety of water shortage and environmental challenges that are intertwined with the causes and consequences of a changing climate. Access to water is the primary natural resource concern in Pangani Basin and Kikuletwa catchment villages and a lack of infrastructure for storing and directing water during rainy periods limits opportunities for harvesting water for irrigation and other household uses. More than seventy percent of heads of household are farmers whose crop production depends mainly on rainfall. While water deficit remains a major concern, its severity is not immune to the challenges of shifting climate and environmental destruction resulting from livelihood activities in the study area.

Key words: Climate change, water resources, agriculture.

INTRODUCTION

Observational records and climate projections provide abundant evidence that freshwater resources are vulnerable and have the potential to be strongly impacted by climate change, with wide-ranging consequences for human societies and ecosystems (Bates et al., 2008). The first decade of the 21st century was the warmest decade recorded since modern measurements began around 1850, which was also marked by dramatic climate and weather extremes such as long-term droughts in various regions of the world which includes Australia and East Africa (Oxfam, 2012; WMO, 2013). Climate change has become a source of uncertainty for planners and decision-makers in climate-sensitive economic sectors. For ample quantities of water available in all seasons has been declining (in most places) due to reduced stream flow especially in dry season because of evapotranspiration and increase in water demand for various uses (Neff et al., 2000). Additionally, observed warming over several decades has been linked to changes in the large-scale hydrological cycle such as: increasing atmospheric water vapour content; changing precipitation patterns, intensity and extremes; reduced snow cover and widespread melting of ice; and changes in soil moisture and runoff (Neff et al., 2000; Bates et al., 2008).

The perennial water of Kikuletwa catchment is a source of life to millions of people living in the area. The catchment is renowned for its agricultural values, centre of biodiversity, as well as an important source of water supply for various uses (Munishi et al., 2009). Water resource management (including northeast Tanzania in which the Kikuletwa catchment is found) clearly impacts on many policy areas, for example, energy (towards generation of hydroelectric power at the Nyumba ya Mungu dam, Hale, New Pangani Falls and Old Pangani). It also affects agriculture sector (especially on part of the country's major staples such as maize, rice and wheat) which employs 80% of the workforce and accounts for 45% of the country's GDP and 55% of foreign exchange earnings (Noel, undated). Other policy areas that are affected by water resource management include health and nature (biodiversity) conservation.

The montane forests (including East African mountains e.g., Meru and Kilimanjaro) are heavily threatened by global climate change impacts (Neff et al., 2000; Hemp, 2009). Many studies from around the world have considered the impacts of future climate changes on water resources to exhibit several significant trends. While decline in sream flow and catchment areas (which consequently affect water resources) in other regions of the world have been associated with observed changes in temperature and rainfall (Neff et al., 2000; de Wit and Stankiewicz, 2006; Bates et al., 2008), the effects of these past climate changes on river level (the depth of water at a monitoring station) in Kikuletwa Catchment remain unclear.

Also, the gradual increase in drought and human population has made water become a scarce resource not only in arid and drought-prone areas but also in humid or sub-humid zones, leaving its sustainability to be threatened by various human activities (Noel, undated; WWF, 2006; de Wit and Stankiewicz, 2006). While creating awareness among different stakeholders dealing with water use and management in these areas becomes important, further detailed study is necessary to provide quantitative information that would support and provide a framework upon which water could be used and managed. Considering the inadequate of this information and understanding of the effect of climate change on water resources, this study therefore aimed to assess the relationship between climate change and water resources at Kikuletwa catchment, Pangani Basin in Tanzania.

Assessing the consequences of climate change on water resources is therefore an important step in generating the information that will be used to provide solution to the water resource management challenges. Specific objectives were; (i) to assess rainfall and river level patterns of Kikuletwa catchment at Pangani over the past thirty years. (ii) To determine the relationship between river level and rainfall patterns in the selected rivers over the past thirty years, and (iii) to assess the extent of water scarcity in the villages around Kikuletwa Catchment in Pangani Basin.

METHODS

The study area

The research study was conducted in Kikuletwa catchment at Pangani basin northern Tanzania. The Pangani basin has two main river catchments with different tributaries, originating from the slopes of Mount Kilimanjaro and Meru. Both rivers rise in the basin's northernmost portions (Figure 1). The Pangani River Basin covers an area of 43,650 km^2 of which 3,914 km^2 lies in Kenya. In Tanzania, the basin is distributed through Kilimanjaro, Arusha, Manyara and Tanga administrative regions (PBWB, 2009).

Climate

Mt Kilimanjaro and Meru areas are characterized by a typical equatorial climate. The distribution of precipitation over the year follows the Intertropical Convergence Zone and is affected by elevations. Due to its equatorial location, two distinct rainy seasons occur in the study area: the long rains from March to May, and the short rains around November. The driest period is from August to October, while April and May are the wettest months. The average rainfall ranges from 1000 to 1700 mm varying with elevation and aspect. And precipitation decreases from the lower forest boundary down to the plains, where it is less than 700 mm annually (Munishi et al., 2009). The annual precipitation reaches its maximum in the mid montane zone, and at higher elevation, precipitation declines, mainly starting near the upper forest border at higher altitudes (Hemp, 2009). The mean annual temperature decreases linearly upslope with a lapse rate of 0.56°C per 100 m starting at the foothills (Hemp, 2009) and the maximum and minimum temperature on the lower slopes (settlement areas) ranges are 15 - 30°C and 12 - l7°C, respectively (Munishi et al., 2009).

Data collection and analysis

This study used questionnaire, focused group discussion (FGD) and participant observation to assess changes in water resources as well as identifying challenges that communities are facing as a result of climate change. We conducted separate focus groups of men, women and village leaders to provide context to the information gathered in household surveys. Participants were asked to identify the three most acute problems faced by villagers. Communities were selected based on their type of livelihood and degree of experiences on the changes in the climate that have been observed in the area for the past years. Assessments took place in three villages within Kikuletwa Catchment and five villages in Pangani main basin area. For villages around Kikuletwa Catchment were: Manyata, Mbuguni, and Olbil and for Pangani Mainstream Catchment area, the selected villages were; Sange, Ruvu Mferejini, Longoni, Ruvu Jiungeni and Makanya. Participant observation took place in the protected (catchment) areas, destroyed area and also some places along the rivers were used as study area for the purpose of confirming the real situation and type of environmental destruction in the area.

The climate data (temperature and rainfall) as well as river level for the past 30 years were obtained from the Department of Hydrology in Pangani Office and Arusha AirPort Meteorology Centre. River levels and rainfall data were collected from records on

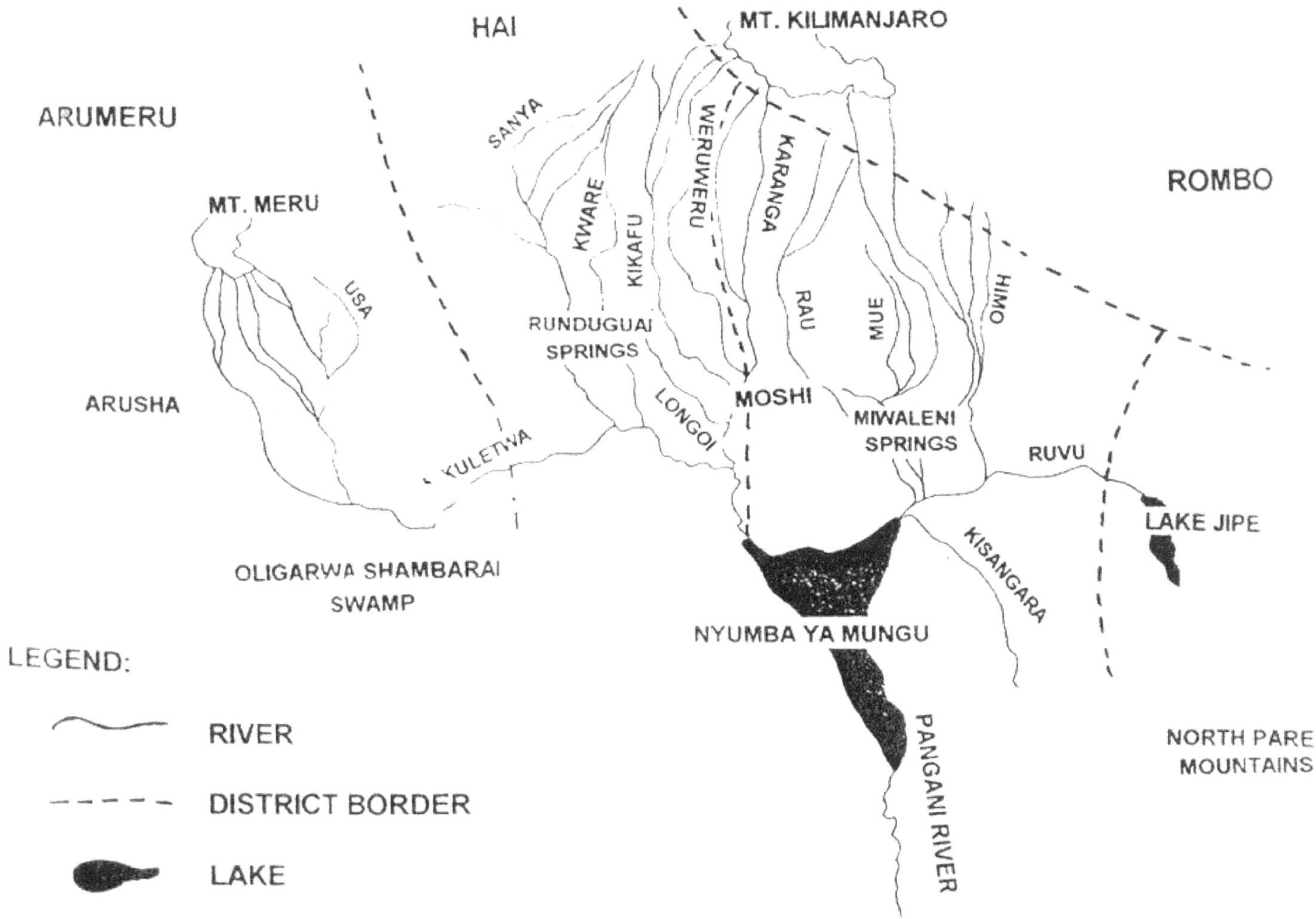

Figure 1. Pangani Basin and the Kikuletwa Catchment area.

Table 1. Mean annual rainfall data (mm) for ten-year interval from the two rain stations in Kikuletwa Catchment area.

Rain gauge station (with rain amount in mm)/year	1980-1990	1991-2000	2001-2010
Tengeru	905	887	954
Themi	855	876	1012

water levels from five major rivers (Malala, Maji ya Chai, Magadrisho, Karangai and Kikuletwa) and two (Tengeru and Themi) rainfall stations located within the study area. Thus, Pangani Water Office and Arusha Airport Meteorology Centres facilitate the gathering of data from these field stations for monitoring and research purposes.

The long term river level and rainfall data collected were subjected to statistical analysis for significance levels using correlation analyses in R 2.13.0 (R Development Core Team, 2011). The relationships between climate change (rainfall and temperature) and water resources (river level) for the Kikuletwa River were compared in correlation analysis. Qualitative data from FGD were summarized in key statements and presented in a table for interpretation and discussion.

RESULTS

Evaluation of long-term rainfall and river level data for the past thirty years

The mean annual rain-fall from the two (Themi and Tengeru) rain stations located on the study area was summarized on a ten year interval and is presented in the Table 1. The rainfall pattern seemed to have variation with no directional pattern on Tengeru rain gauge station. The mean annual rainfall reached a maximum of over 950 and 1000 mm for Tengeru and Themi rainfall stations,

Table 2. Mean river level data (m) for the past 30 years in the major rivers at Kikuletwa Catchment area.

River	Year		
	1980-1990	1991-2000	2001-2010
Malala	0.14	0.12	0.15
Maji ya Chai	0.12	0.13	0.16
Magadrisho	0.12	0.13	0.21
Kikuletwa	0.86	0.84	0.75
Karangai	0.61	0.60	0.56

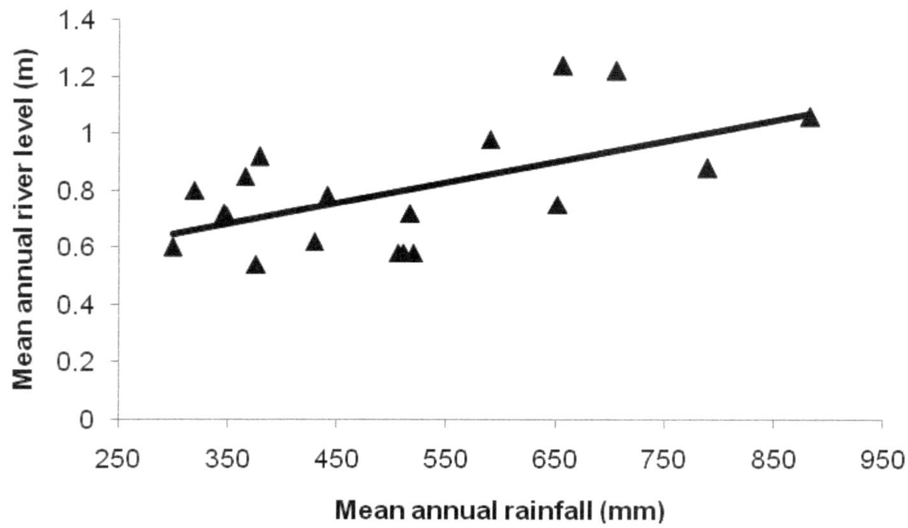

Figure 2. Correlation between water level and mean rainfall for Kikuletwa River.

respectively in the last decade (2001-2010). The rainfall pattern for Themi station showed an increase in the amount of rainfall for the past 30 years (Table 1).

The data on average river level for the major selected rivers were summarized on the ten year interval and are presented in Table 2. The river level data at the catchment fluctuated over the study period (Table 2). In particular, there was a consistent decline in river level in the two major rivers, Kikuletwa and Karangai (Table 2). There was a significant positive correlation between mean annual river level for Kikuletwa river and mean annual rainfall in the past three decades (Pearson's correlation: = 0.58, df =17, p = 0.009; Figure 2).

Although, the annual mean river level showed a decrease with increased mean annual temperature for Kikuletwa river, this was not significant (Pearson's correlation: = -0.31, df =15, p = 0.22; Figure 3).

Key findings from questionnaire and FGD

The eight villages covered by this study face a variety of water shortage and environmental challenges. Overall, water shortage was the most frequently cited village problem (Table 3). The reason for low water supply was reportedly in part due to the large fluctuations in rainfall patterns, including severe drought in some years.

Environmental destruction is another major source of environmental stress. Trees are commonly cut to make charcoals which are then used as cooking fuel. Wood harvesting is also done in the area and used for construction materials. Cleared land is used for crop cultivation and for pasture, providing short-term economic opportunity for households suffering from poverty and food insecurity. More than 70% of the heads of household in the study area are farmers (Figure 4). The next largest occupational category in the Kikuletwa Catchment area is livestock keeping, followed by fishing.

DISCUSSION

Kikuletwa and Ruvu catchments have a series of large springs in a semi-circular band to the south of Mt Meru and Mt Kilimanjaro. These springs are recharged by groundwater from volcanic aquifers that act like artesian wells (PBWB, 2002). The major rivers in the Kikuletwa catchment are Kikuletwa and Karangai (Figure 1). The

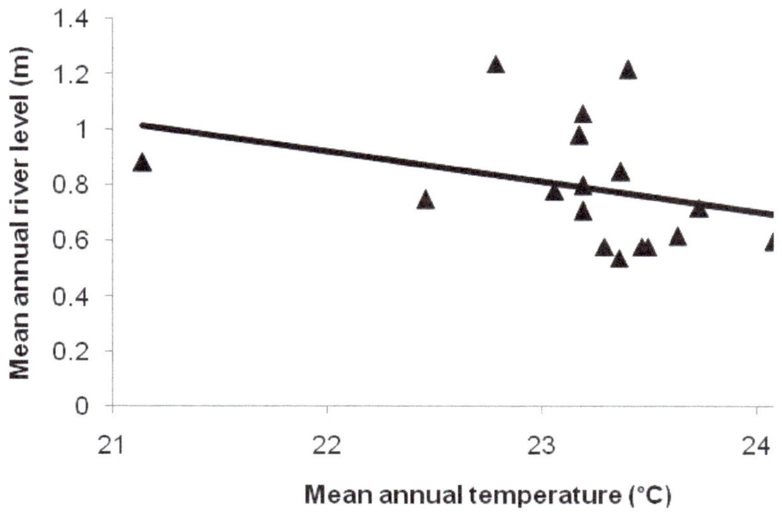

Figure 3. Correlation between river level and mean temperature for Kikuletwa River.

Table 3. Ranking of problem in top three by villagers from Kikuletwa Catchment area.

Problem	FGD responses (%)		
	Village leaders	Men	Women
Water shortage	19	25	28
Deforestation	3	5	7
Soil erosion	8	2	5
Food shortage	7	1	6

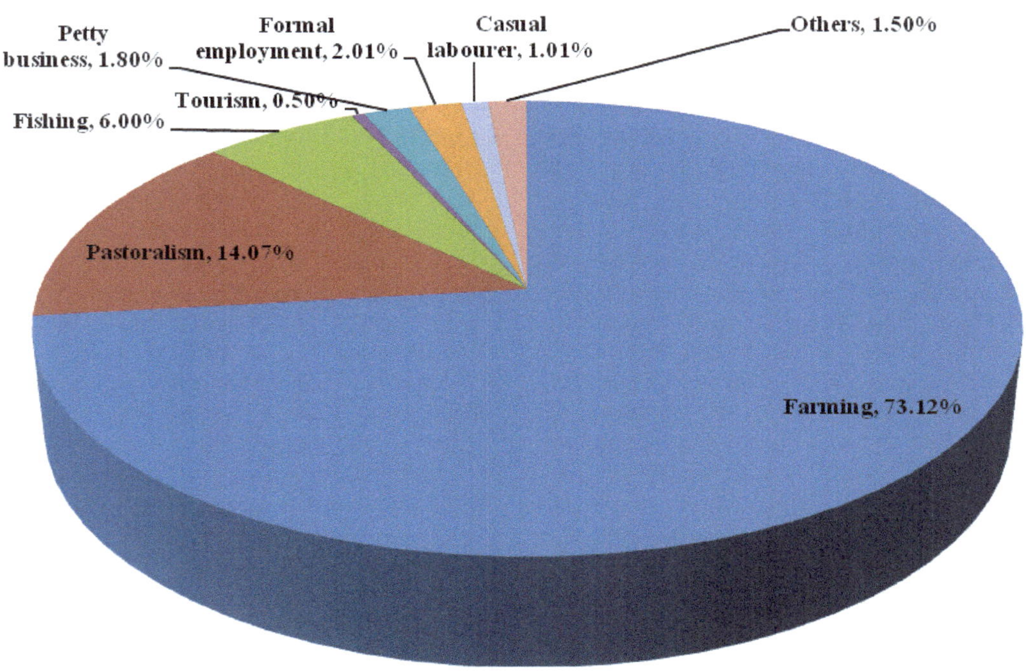

Figure 4. Occupation of heads of the household in Kikuletwa Catchment area.

two rivers have their major tributaries that include Maji ya chai, Magadrisho, Karangai, Malala, Usa, Kikafu, Kware and Makumira.

As noted, climate variation has a large effect on hydrology and, consequently, on the water available for human and ecosystem use in Kikuletwa Catchment area. The close association between river level and temperature in the catchment suggest that there are significant effects of climate change on water resources at Kikuletwa Catchment area. The results are consistent with other previous studies (IPCC, 1998, 2001a,b, 2007) that showed the projections of decreased mean annual rainfall and decreased stream flow in Africa and water flow in the two other key basins (Wami-Ruvu and Pangani) in the northern parts of Tanzania (de Wit and Stankiewicz, 2006). The de Wit and Stankiewicz's (2006) study further indicated that other drivers of climate and water resources, such as land use change and environmental destruction could often be considered as additional factors that would potentially reduce water resources in Pangani Basin.

Also, according to other climate change studies done in Tanzania, it is predicted that climate change will provoke a general shift in forest ecosystems, in terms of changing forest types and species and distribution of forests (Mwandosya, 2006). Indirect impacts are also expected as the carbon dioxide concentration in the atmosphere doubles: subtropical thorn woodlands will be completely replaced and subtropical dry forests and subtropical moist forests will decline by 61.4 and 64.3%, respectively (URT, 2003).

In their study on the slopes of Mount Kilimanjaro, Munishi et al. (2009) indicated that the combined cover of closed and open forests decreased by 68% while closed and open forests decreased by 56 and 64%, respectively, due to forest clearing and climate change.

Another study by Hemp (2009) indicated that during the last 70 years, Mount Kilimanjaro has lost nearly one-third of its forest cover, in the upper areas caused by climate-driven fire and clearing. The loss of 150 km^2 of cloud forest- the most effective source in the upper montane and subalpine fog interception zone- caused by fire during the last three decades means a considerable reduction in cloud forests and water yield (Hemp, 2009). The cloud forests are of great importance for watersheds in East Africa (Hemp, 2006, 2009). In addition to the function of filtering and storing water, the upper montane and subalpine cloud forests have a high potential of collecting cloud water. Fog interception increases with altitude, and so does its contribution to water yield on higher altitudes of Kilimanjaro (Hemp, 2005). Thus, the loss of cloud forests due to climate induced fires as well as the loss of montane forest due to clearing, causes a considerable reduction and enhanced variability of water yields of the Kilimanjaro catchments.

As a consequence, Hemp (2005) study showed that annual precipitation on Mount Kilimanjaro has decreased by 600-1200 mm over the last 120 years while since 1976, temperatures have increased drastically. Both temperature and rainfall changes do not only expose rivers to vulnerability of water yields but also water demand to become more aggressive and thus, increasing irrigation demand along the slopes of the mountain, resulting in a potentially much decrease in the river level on the main rivers in the catchment area.

Although, the results of our study indicated an upward trend in precipitation (from the two rain stations) during the past three decades, there was a consistent decline in river level in the main rivers of Kikuletwa Catchment area (Table 2). The reasons for decrease in river level in Kikuletwa and Karangai rivers in the study area could be increase in streamflow water demand for both domestic and agricultural/livestock uses.

Agricultural activities in the Kikuletwa catchment depend on rainfall and irrigation and since precipitation decreases from the lower forest boundary of the mountains down to the plains, then most of the areas just few kilometers from the lower elevations are semi-arid in nature requiring irrigation for agriculture production. Irrigation activities are also practiced in highlands, through traditional furrows (mifongo), and in the lowlands where most of the large scale plantation occurs (Munishi et al., 2009).

The agricultural sector is the leading sector of the economy of Tanzania and accounts for over half of the gross domestic product (GDP) and export earnings (World Bank, 2002; Agrawal et al., 2003). The livelihood of more than 80% of the population that lives in rural areas depends on agriculture (URT, 2001; World Bank, 2002). The performance of agriculture is therefore a major factor in determining livelihood fortunes. Given the importance of agriculture for gross domestic product (GDP), employment, and livelihoods in many developing countries, the impacts of climate change on water resources (and ultimately on agriculture) agriculture are likely to reverberate throughout the economies of these countries including Tanzania.

In combination with an increase in the frequency and severity of drought in lowland areas, limited water supply may force human and livestock population to move upstream towards higher altitudes, with consequences more harmful extending for long term due to the loss of the catchments and water sources in the upper elevation.

The increase in human and livestock numbers and related augmented water demand, particularly in the catchment area and large fluctuations in rainfall patterns, including severe droughts, are expected to be an ongoing problem in rural northern Tanzania, particularly if forest and grassland resources continue to be used intensively by the human population.

Drought is expected to increase in frequency and severity in the future as a result of climate change, mainly as a consequence of not only decreases in regional precipitation but also because of increasing evaporation driven by forest clearing that exacerbates global warming (Sheffield et al., 2012).

Global warming, climate change and rising sea level are expected to intensify the resource sustainability issue in many water stressed regions of the world by reducing the annual supply of renewable fresh water (Neff et al., 2000; WMO, 2013). Deforestation on mountain foothills raises mean cloud condensation level that results in a gradual shrinking of the cloud zone. A similar effect is caused by global warming and drying of the air (Hemp, 2009; Seneviratne, 2012). In addition to changes in the water balance of the mountain, loss of cloud cover may have added to the observed general decreasing trend in precipitation during the last century.

The eight villages covered by this study face a variety of water shortage and environmental challenges that are intertwined with the causes and consequences of a changing climate. Severe droughts and periodic flooding are expected to affect the region in the following years (Oxfam, 2012; WMO, 2013). Access to water is the primary natural resource concern in Pangani Basin and Kikuletwa catchment villages and a lack of infrastructure for storing and directing water during rainy periods limits opportunities for harvesting water for irrigation and other household uses.

As in much of sub-Saharan Africa, water shortages, loss of wooded area, and land degradation present daily obstacles to these households' pursuit of economic well-being which in turn often contributes to environmental deterioration (Oxfam, 2012; Sheffield et al., 2012). With majority of the households in the area being farmers (more than two-thirds) and livestock keepers in the study area (Figure 4), environmental degradation combined with climate change are likely to fundamentally alter the water (primary) resources used by those who engage in agriculture. Shifting weather patterns, deteriorating soil quality and loss of pasture may often result in lower yields and increase the uncertainty faced by farming households.

If environmental destruction coupled with loss of cloud cover continued over a long time, these trends are likely to affect millions of people and livestock living along the Pangani River and its basin, by far accelerating the problem of water scarcity. Research on the trends in migration and increase in the population along the river stream could be a major focus to provide more evidence on this aspect.

In general, agricultural activities were cited as the main contributors of the water shortage problem. As reported in Table 3, nearly every village that participated in focus groups ranked poor water supply in the top three village problems. Also, the results from focused group discussion revealed that access to water for drinking and other household uses is often limited, can require traveling significant distance to collect and can be of poor quality. According to interviews from the eight villages in the study area, the households from one village can have access to two nearest sources of water and most villagers (from five villages) reported only one source of water (shallow wells); and three reported two sources.

While some villages have a water source within the village, others must travel long distances to get water from underground pipes. During the FGD, it was also observed that the quality and quantity of water in Pangani basin was diminishing. For example in the FGD two village leaders expressed the challenges their families face in getting water for domestic use. The village leaders said that the shallow well serves almost half the population of villagers, so members of the household almost wake up as early as 2:00 am, to collect water and return home in the afternoon. Sometimes it does not matter if one is as early as that because there would be a long queue of women waiting to fetch water and by the time their turn comes, there would be no more water in the well. And whoever comes later than that usually does not get water and will have to wait until when enough would have collected in the well again. The quality of water also has also been dropping overtime in the area. 85% of household respondents reported the water from the wells is muddy, salty, brownish and/or polluted by debris. In areas near the coast where Pangani River discharges its water in the Indian Ocean, piped water is reported to be salty, due to rise in sea levels as a result of climate change.

Several measures for mitigating the effects of climate change in the area were identified from the discussion held with the stakeholders at Kikuletwa Catchment. The most measures that were identified as common strategies in the area included: migration, cultivation of drought resistant and early maturing crops, diversification of income generating activities, conservation and protection of river catchments. The results from FGD clearly shows that stakeholders and communities in the study area are very much willing to respond to long term mitigation measures that would ensure the sustainability in their livelihoods.

Conclusion

The results from this study suggest an association between rainfall and river level in the Kikuletwa River. Considering the aspect of rainfall and river water level, the study showed significant effects of climate change on water resources in the study area. Also, data from household surveys and focus groups provide a snapshot of population living in water scarcity areas in spite of the continued decrease in the available water resources. The environmental destruction continues to increase in the catchment, while agriculture and fuel wood consumption is the major cause. This has been linked to a loss of clean water resources in some parts of the Pangani Basin. Based on the questionnaire survey, FGD and participant observation, this study has also provided baseline information on household use of water resources under the climate change and how this affects economic activities of local people. To mitigate climate change and provide effective adaptation measures, it is imperative to periodically repeat community surveys in the climate-change affected areas.

ACKNOWLEDGEMENTS

This work was funded and supported by the Pangani Basin Water board. We thank the staff of Department of Hydrology in Pangani Basin and Arusha Airport Meteorology Centre, Selian Agricultural Institute who kindly provided us with hydrological and meteorological data for this research. We are also very thankful to three anonymous reviewers for their useful comments that led to the improvement of this manuscript.

REFERENCES

Agrawal S, Moehner, Hemp A, Valst M, Hitz S, Smith J, Meena H, Mwakifwamba SM, Mwaipopo TOU (2003). Development and Climate Change in Tanzania: Focus on Mount Kilimanjaro. Paris: OECD.

Bates BC, Kundzewicz ZW, Wu S, Palutikof JP Eds. (2008). Climate Change and Water. Technical Paper of the Intergovernmental Panel on Climate Change, IPCC Secretariat, Geneva. p. 210.

Burke MB, Miguel E, Satyanath S, Dykema JA, Lobell DB (2009). Warming increases the risk of civil war in Africa. Proc. Natl. Acad. Sc. 106: 20670 - 20674.

de Wit M, Stankiewicz J (2006). Changes in Surface Water Supply across Africa with Predicted Climate Change. Science 311: 1917-192.

Hemp A (2005). Climate change driven forest fires marginalizes the ice cap wasting on Mt. Kilimanjaro. Glob. Change Biol. 11: 1013 - 1023.

Hemp A (2006). Continuum or zonation? Altitudinal gradients in the forest vegetation of Mt. Kilimanjaro. Plant Ecol. 184: 27 - 42.

Hemp A (2009). Climate change and its impact on the forests of Kilimanjaro. Afr. J. Ecol. 47: (Suppl. 1): 3 - 10.

Intergovernmental Panel for Climate Change (IPCC) (1998). Global Warming and Hydrologic changes that affect fresh water resources, Coroland. Bulder Press.

Intergovernmental Panel for Climate Change (IPCC) (2001a). Millennium Ecosystem Assessment, United Nation, Geneva Publications.

Intergovernmental Panel for Climate Change (IPCC) (2001b). Millennium Ecosystem Assessment, United Nation, Geneva Publications.

Intergovernmental Panel for Climate Change (IPCC) (2007). World Meteorological Organization (WMO) Guide to climate Change Practise. World Meteorology. Organization general Publisher, New York.

Munishi, PKT, Hermegast, AM, Mbilinyi BP (2009). The impacts of changes in vegetation cover on dry season flow in the Kikuletwa River, northern Tanzania. Afr. J. Ecol. 47 (Suppl. 1): 84-92.

Mwandosya MJ (2006). Mainstreaming Environment and Climate Change Concerns in National Planning in Tanzania. Ministry of the Environment. Dar-es-Salaam, Tanzania.

Neff, R, Chang, H, Knight, CG, Najjar, RG, Yarnal, B, Walker HA (2000). Impact of climate variation and change on Mid-Atlantic Region hydrology and water resources. Clim. Res. 14(3): 207-218.

Noel S (Undated). The Economics of Climate Change Tanzania. Water Resources. Stockholm Environment Institute SEI-Africa Centre. Institute of Resource Assessment, University of Dar es Salaam, Tanzania. p.32.

Oxfam (2012). Food crisis in the Horn of Africa Progress report July 2011-July 2012. p.30.

PBWB (2002). Pangani Basin annual report, Moshi, Tanzania.

PBWB (2009). The role of non- farm activities in household economies: a case study of Pangani Basin, Tanzania: Dar Es Salaam University press.

R Development Core Team (2011). R: A language and environment for statistical computing. R Foundation for Statistical Computing, Vienna, Austria. ISBN 3-900051-07-0, URL http://www.R-project.org.

Seneviratne SI (2012). Climate science- Historical drought trends revisited. Nature 491: 338 -339

Sheffield J, Wood EF, Roderick ML (2012). Little change in global drought over the past 60 years. Nature 491: 435 - 438. United Nations Development Programme. (2008). Climate change and ecosystem Management. World Bank, Washington.

United Republic of Tanzania (URT) (2001). Agricultural Sector Development Strategy (ASDS). Ministry of Agriculture, Dar-es Salaam, Tanzania.

United Republic of Tanzania (URT) (2003). Initial National Communication under the United Nations Framework Convention on Climate Change. Dar-es-Salaam, Tanzania.

Vice President Office (2009). African Resilience to climate change: Protecting Biodiversity and enhance traditional knowledge. Dar-Es-Salaam University press, Dar es Salaam.

World Bank (2002). Climate Change Series. Published Jointly with the Agriculture and Rural Development Department. The World Bank Environment Department.

World Meteorological Organization (WMO) (2013). The global climate 2001- 2010. A decade of climate extremes summary report. Geneva 2, Switzerland.

WWF-World Wide Fund for Nature (2006). Climate Change Impacts on East Africa. A Review of the Scientific Literature.

Simulation of oil spill infiltration and redistribution in a shallow aquifer

Luqman K. Abidoye and Maina R. Wairagu

Department of Chemical Engineering, Loughborough University, LE11 3TU, Loughborough, Leicestershire, England.

Simulations of oil spill infiltration and redistribution in a shallow aquifer were conducted in this work to quantify their effects on groundwater. Results showed that flow of oil spills can be adequately described by infiltration and redistribution patterns within the domain. Without the influence of rainfall, the contaminant plume pervaded the water table within one year of the spill through the influence of dissolution, diffusive and convective mechanisms. The plume advanced in the domain first by domain wall trailing in a manner suggestive of oil-wetting characteristics followed by bulk plume movement. Simultaneous infiltration with rainfall increased the rate of contaminant penetration and aided the plume front further into the domain. Without influence of rain, uncontaminated water can be found at a depth beyond 30 m from the spill surface at 30 years after the spill, while the possibility of clean water with influence of rainfall, at the same period of time, became remote even up to the bottom of the aquifer as the rainfall drives the plume into the furthest depth though, this is accompanied by cleaner water body at the upper portion of the aquifer. It was found that the impacts of contaminants on aquifer water are influenced by contaminant components, wettability of the aquifer, annual rates of rainfall, boundary conditions of the source and possibly size of the domain. Findings in this report point to the needs for quicker clean up exercises and identification of multi-approach procedures for different spill scenarios.

Key words: Simulation, oil spills, infiltration, redistribution, Niger Delta.

INTRODUCTION

Worldwide, oil spills are recurrent problems with serious consequences both on the living and non-living in the ecosystem. Soil or geological contaminations, owing to spills, are global phenomenon which are eyesores at many sites worldwide. The atmosphere is similarly not spared as volatile and less dense components easily evaporate. Many organic liquids existing as a separate phase are in fact often slightly miscible with water and their solubility can exceed the drinking water standard by orders of magnitude (Van Genuchten, 1980).

The dire consequences of these oil spills are, majorly, borne by the host communities where exploration takes place. Also, remote from the exploration sites, the pipe networks for transporting oil to different terminals have further drawn the danger closer to several communities as the cases of ruptured pipes and vandalisation have resulted in casualties from these remote places.

In the Niger Delta area of Nigeria, recurrent cases of spills have drawn international attention. This area is the flood plain area of the country which accounts for over 7.5% of total Nigeria land mass. Kadafa (2012) reported estimates of 9 to 13 million barrels of spill since the beginning of exploration in 1958. Many of these cases have been attributed largely to, pipeline corrosion and faulty tankers. The area consists of highly networked pipelines connecting several small oil fields.

These pipelines are highly prone to rupture as a consequence of poor material structure and maintenance. Vandals and saboteurs are highly involved in this menace as they engage in vandalisation for financial gains or to prove political points by holding government to ransom. Many households, in the exploration areas, whose means of livelihood are based on subsistent farming, have been reduced to paupers due to contamination of their land by persistent spills of oil which destroy the soil fauna.

A further and major consequence of the spills is the health impairment due to the presence of the contaminants in the soil, air and drinking water. Court cases have been initiated against the oil-producing companies, at both local and international courts, with some still remaining inconclusive till date.

Nwankwo and Emujakporue (2012), made it clear that even years after the clean-up of oil-contaminated sites, the contaminants proved difficult to eliminate. Problems of water contamination are particularly more serious in Niger Delta areas owing to shallow aquifers with the topmost groundwater levels occurring anywhere between close to the surface and a depth of 10 m[3]. Nwankwo and Emujakporue (2012) carried out electrical resistivity surveys on a spill site in the Niger Delta Area of Nigeria to evaluate subsoil conditions and impacts of contamination on groundwater quality.

This was done as a follow up to the post-spill clean-up exercise. The results showed that, 3 years after clean-up was done, the plume was still strongly present up to a depth of 49 m although they concluded that clean aquifer could still be accessed at an adjacent depth of about 30 m.

This work presents a numerical simulation to investigate the infiltration and redistribution of oil, a non-aqueous phase liquid (NAPL), into the shallow aquifer as found in the Niger Delta area of Nigeria. The Ahoada region of the Niger Delta area of Nigeria has an annual rainfall of about 2400 mm and mean temperature of 27°C (Ehirim et al., 2009). The resistivity values of the sedimentary rock are controlled by parameters such as water contents, salinity, texture, matrix conductivity and the presence of clay materials (Nwankwo and Emujakporue, 2012).

Niger Delta basin is characterised by over 90% sandstone with shale intercalations and thickness of about 2100 m[4]. The domain is composed of coarse gravel and locally fine grained, poorly sorted, sub angular to well-rounded sand samples which give the formation good porosity and effective permeability (Nwankwo and Emujakporue, 2012). Edwards and Santogrossi (1990) described the primary Niger Delta reservoirs as Miocene paralic sandstones with 40% porosity and permeability of 2 Darcy's having 100 mm thickness. However, Molua et al. (2011) found a porosity of 0.22 for the Niger-Delta basin using Wyllie Time-Average equation. Our work made use of intermediate porosity between these reported values.

Niger Delta aquifers are highly vulnerable to oil spills in the geological environment and in area where there is heavy dependence on groundwater for its water supply, effects of spill can be catastrophic. Flores et al. (2007) assessed the suitability of image analysis method to measure LNAPL saturation distribution in the fluctuating groundwater. They evaluated the effects of fluctuating groundwater on the oil migration. They used a 1-dimensional column in a small domain (3.5 × 3.5 × 50 cm). In this work, a 1000 × 1 × 50 m aquifer dimension was utilised to simulate the site at Ahoada where the spill history was known.

GOVERNING EQUATIONS

In multiphase flow models, a key element that signifies them all is the functional relationship between relative permeability, k, fluid saturation, S and capillary pressure, P_c. Leverett (1941) was among the first to discuss the theory of capillary pressure proposing a Leverett J function attempting to convert all capillary pressure data to a universal curve. In practice, capillary pressure and its relations with fluid-fluid and fluid- solid properties is shown in Equation 1. This is commonly known as the Young Laplace equation.

$$P_c = P_n - P_w = \frac{2\sigma \cos\theta}{r} \tag{1}$$

where, P_n is the pressure in non-wetting phase, P_w is the pressure in wetting phase, σ is the interfacial tension, θ is the contact angle and r is the effective pore radius. Generally, wettability is related to the interface between the solids and fluid or fluid and fluid as shown in Figure 1. The location, flow and distribution of a fluid in porous media is governed by wettability (Anderson, 1987).

The wettability of a NAPL-water system is rather complex in nature due to particle size, the disposition of the NAPL as well as the presence of surfactants (Mercer, 1990).

Models describing capillary pressure/saturation relationships have been in existence but the most established ones were those proposed by Brooks and Corey (Brooks, 1964) and Van Genuchten (Van Genuchten, 1980) as seen in Equations 2 and 3, respectively.

$$S_e = \frac{S_w - S_{rw}}{1 - S_{rw}} = \left(\frac{h_d}{h_c}\right)^{\lambda} \tag{2}$$

where, S_e is the effective saturation, S_w is the water saturation, S_{rw} is the residual water saturation, h_d is the displacement pressure head, h_c is the pressure head

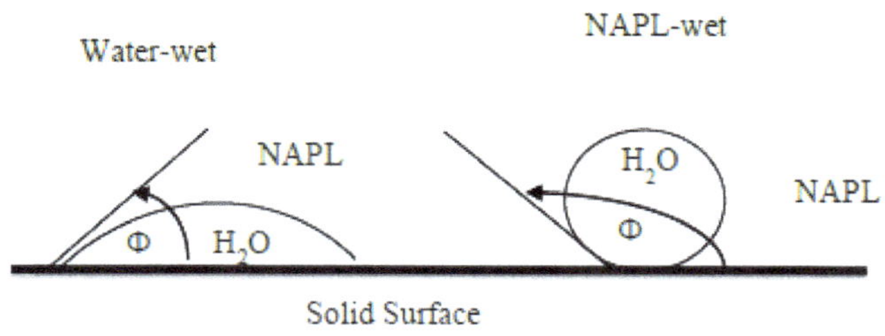

Figure 1. Contact angle, wetting and non-wetting interface (Mercer, 1990).

and λ the fitting parameter for the pore size index.

$$S_e = \frac{S_w - S_{rw}}{1 - S_{rw}} = \left(\frac{1}{1 + (\alpha h_c)^n} \right)^m \tag{3}$$

Where, $\alpha = \frac{1}{h_b}\left(2^{1/m} - 1\right)^{1-m}$ and $n = \frac{1}{1-m}$ h_b is the bubbling pressure and α, m, n are soil parameters.

Relative permeability-saturation relationship

It is inherently difficult to experimentally determine this relationship. This has led to numerous propositions with the popular ones being those proposed by Brooks and Corey (1964) and Van Genuchten (1980). This work utilised the Van Genuchten model as shown in Equations 4 and 5, respectively.

$$k_{rw} = \bar{S}_w^{1/2}\left[1 - \left(1 - \bar{S}_w^{1/m}\right)^m\right]^2 \tag{4}$$

$$k_{rn} = \left(1 - \bar{S}_n\right)^{1/2}\left[1 - \left(1 - \bar{S}_n^{1/m}\right)^m\right]^{2m} \tag{5}$$

Where, k_{rw} is the aqueous relative permeability, k_{rn} the NAPL relative permeability, $\bar{S}_w$ the effective water saturation and $\bar{S}_n$ the NAPL saturation.

Saturation

Saturation is the fraction of pore space that is engrossed by the fluid phase as represented in Equation 6.

$$S_i = \frac{\textit{volume of the fluid}}{\textit{volume of the void space}} \tag{6}$$

Mass and momentum balances

The main equation for consideration was the continuity equation that is shown in Equation 7.

$$\frac{\partial}{\partial t}(\eta \rho_i S_i) = -\nabla.(\rho_i q_i) \text{ for } i = w, o \tag{7}$$

Where,

$$q_i = -\frac{k_{ri} k}{\mu_i}.(\nabla P_i + \rho_i g e_g) \text{ for } i = w, o$$

Where, η is the porosity, ρ the density (ML^{-3}), S saturation, q the fluid flow velocity (LT^{-1}), k_r the relative permeability, k the intrinsic permeability (L^2), μ the viscosity (ML^{-1} T^{-1}), P the pressure (ML^{-1} T^{-2}), g the magnitude of gravitational acceleration (LT^{-2}) and e_g is a unit vector in vertical direction.

The mass balance equation and the Van Genuchten model were solved using a multivariable residual based Newton Raphson iteration technique with maximum number of Newton iterations set at 16 and convergence at 10^{-6}.

MODEL DESCRIPTION

The subsurface transport over multiple phases (White and Oostrom, 1996) was used for simulation in this paper. It was chosen due to its versatility in multiphase flows and has been successfully validated and applied in the simulation of a wide range of multi-phase flow problems (White and Oostrom, 1996; Schroth et al., 1998; Ataie-Ashtiani et al., 2001; Das et al., 2004). It is composed of FORTRAN subroutines that solve the governing equations. A series of output files were post-processed with Perl scripts and Tecplot was finally used to generate the plots.

Simulation domain

The non-hysteretic two phase version of water-oil mode of STOMP was used for simulation purposes where the governing equations are discretised into algebraic form using the integrated finite difference method (Patankar et al., 1980). A 2-D Cartesian coordinate system was used to identify the model domain of the Niger Delta region being studied. The 1000 × 1 × 50 m domain was divided into a grid containing 25 and 20 grid blocks in the x and z direction, respectively. Figure 2 shows the 2-D domain of the modelled aquifer. An automatic time step that was dependent on relative component property changes was used to discretise the total time step.

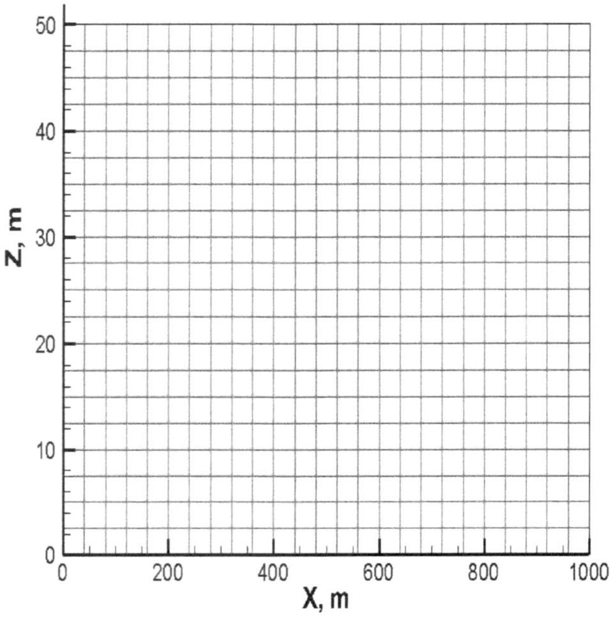

Figure 2. 2-D domain model of aquifer.

Table 1. Soil properties.

Property	Value
Bulk density, ρ (kg m^{-3})	2740
Porosity, η	0.37
Van Genuchten, α	0.1 (White and Oostrom, 1996)
Van Genuchten, *n*	2.0 (White and Oostrom, 1996)
Displacement head, h_c (m)	10 (White and Oostrom, 1996)
Residual water, (S_{rw})	0.1 (White and Oostrom, 1996)

Properties for homogenous soil system

Generally, density, permeability, porosity and saturation are input soil parameters in the STOMP application. Table 1 lists the properties adapted for the current numerical simulation. It is worth noting that a constant porosity of 0.37 was used for the entire simulation which was within the average values reported for the Niger Delta region (Nwankwo and Emujakporue, 2012; Molua et al., 2011).

Fluid properties

Data on the properties of the various fluids involved in the contaminant transport is required for the modelling of LNAPL flow. The fluid properties are listed in Table 2.

Initial and boundary conditions

In the Niger Delta, majority of the water table is found at around 10 m from the surface (UNEP, 2011), thus the simulation was conducted by setting the water table at the value. Subsequently, aquifer thickness was set at 40 m and was used to calculate the initial boundary NAPL and aqueous pressures with the addition of atmospheric pressure to a value of 493725 Pa.

The boundary conditions were set up as follows. At the top of the domain, a constant atmospheric pressure was applied. The spill zone was along the entire x direction at the top boundary. To conserve the mass of the oil and make its movement finite within the domain, the west and east boundaries were designated as zero flux for NAPL. A table of boundary conditions used for the simulation is shown in Table 3.

RESULTS AND DISCUSSION

Infiltration without influence of rain

At the beginning of the simulation with no oil saturation present, oil infiltration began at all the specified surface nodes as specified in input simulation card. After about one year of infiltration it was observed that the oil front was on the surface of the water table. A similar trend was observed by Nwankwo and Emujakporue (2012). The saturation at this instance is below 0.4 and the highest point of concentration lies behind the front plume as seen in Figure 3. It can also be seen that the oil plume travelled in phases ranging from a very light zone at the front phase and the thickest zone close to the spill source. These layers were thought to be of varying oil properties such as density and viscosity as well as relative permeability due to the different components of oil. As a result, each component travels at a different rate.

It is widely acknowledged that infiltration and redistribution at spill sites occurs over a long period of time. In the current simulation, it was seen that at one year of infiltration, the oil was present in the water table. This can be understood from the nature of aquifers in the Niger Delta area of Nigeria. They are typically shallow whereby boreholes and wells are able to obtain fresh water within a few meters of digging (UNEP, 2011).

In the present simulation, it was shown that within a short period of infiltration, the oil was able to invade the water table. This would impact negatively on the suitability of the water for human consumption. It can be seen that the concentration of oil close to the water table was about 0.1. The more concentrated front was observed close to the spill surface as infiltration proceeded.

The diffusive nature of the plume is discernible as the concentration gradient progresses outwardly from the infiltration zone. While lighter zone components can be found below the water table, the water table retards the movement of the heavier insoluble constituents. However, it was later shown that the diffusive behaviour of the plume broke resistance to penetration as lighter or less viscous components separated out. Five years after

Table 2. Fluid properties.

Property	Value
Molecular weight of oil (g mol^{-1})	165.83 (White and Oostrom, 1996)
Oil density, ρ (kg m^{-3})	830 (Isehunwa and Olubukola, 2012; 1996)
Oil viscosity, μ (Pa s)	0.002 (Isehunwa and Olubukola, 2012; 1996)
Critical compressibility of oil, bar	0.28 (White and Oostrom, 1996)
Critical pressure of oil, bar	47.6 (White and Oostrom, 1996)
Boiling temperature of oil, *K*	394 (White and Oostrom, 1996)
Freezing temperature of oil, *K*	251 (White and Oostrom, 1996)

Table 3. Initial and boundary conditions for different simulation conditions.

Simulation condition	Time (days)	Top boundary		Time (days)	West boundary		Time (days)	East boundary	
		NAPL	Aqueous		NAPL	Aqueous		NAPL	Aqueous
1. Oil infiltration	365	Neumann	Neumann (0.0 mm/year)	21900	Hydraulic gradient	Zero flux	21900	Hydraulic gradient	Zero flux
2. Rainfall with oil redistribution	365	Neumann	Neumann (2400 mm/year)	21900	Hydraulic gradient	Zero flux	21900	Hydraulic gradient	Zero flux

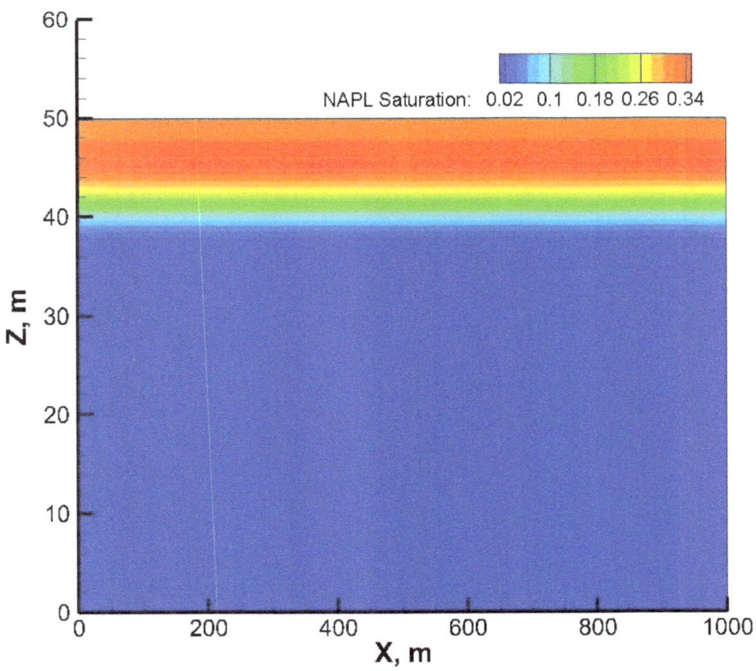

Figure 3. Oil infiltration without influence of rain at 1 year.

the spill, the water table level was observed to be fully contaminated and further infiltration into the water table level was visible as seen in Figure 4. This would suggest typical behaviour in shallow aquifer regions such as the Niger Delta of Nigeria. However, this behaviour might be different in deep lying aquifers. Clear evidence of the influence of time on the percolation mechanism is evident as the 1 year of infiltration showed the oil plume below the 10 m depth from surface as seen in Figure 3.

In contrast, Figure 4 showed the plume to be a few me-

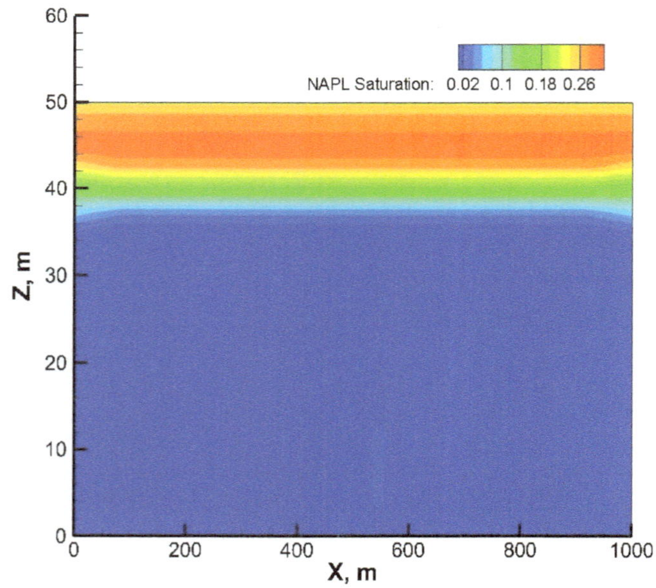

Figure 4. Oil redistribution after 5 years of spill.

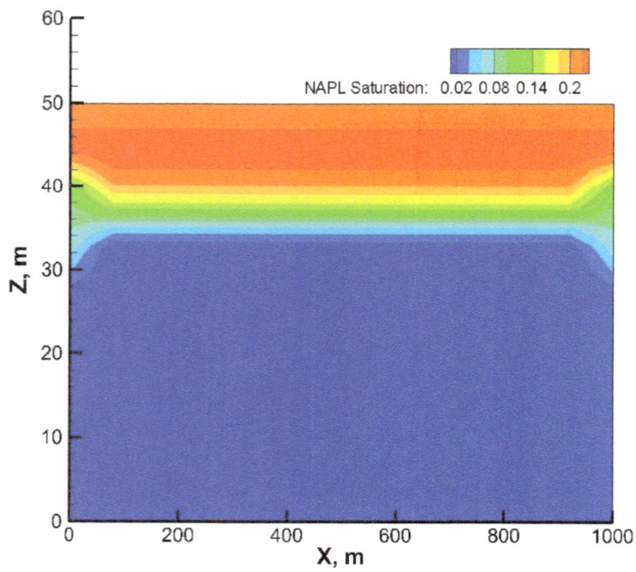

Figure 5. Oil redistribution after 10 years of spill.

ters below the water level table mark.

As redistribution proceeds, it is observed that the penetration force of the advancing plume starts reducing as is seen in Figure 5 10 years into the spill. This is due to oil infiltration at the surface of the spill stopping after one year. However, the fringe redistribution in the domain showed the tendency of the plume to trail the domain wall at the east and west ends which precedes the bulk plume movement into the water body. This could be as a result of the changing wettability of the medium which tends to be more oil-wetting or weak-wetting as oil saturation

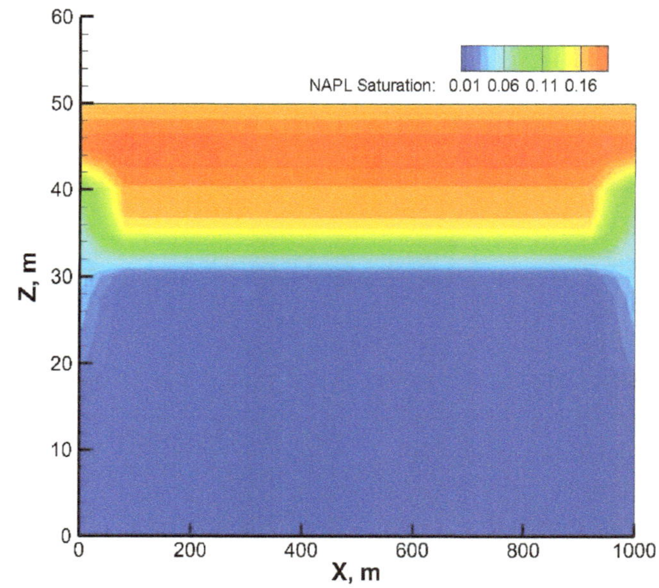

Figure 6. Oil redistribution after 20 years of spill.

increases.

It was seen that there was a marked drastic movement of the plume in the domain after 20 years of simulation as shown in Figure 6. The movement of the plume at the wall boundaries was seen to be moving at a faster rate into the domain. This resulted in more bulk oil penetrating into the water body. It is noteworthy that the concentration of the oil in these areas had become weaker owing to the discontinuity of infiltration of oil at the surface. This resulted in limited clean water 20 m below the surface of the oil after 30 years of the spill. It was seen that oil infiltration advanced slowly and the concentration reduced as seen in Figure 7. The wall boundaries of the plume still continued advancing thus showing the possibility of oil wetting conditions or a weak wetted medium.

Nwankwo and Emujakporue (2012) reported three major resistivity zones in their domain; low, medium and high as seen in Figure 8. They considered the low resistive zones as uncontaminated. Contamination was found as far as thirty meters below the surface though clean water was still available ten to fifteen metres towards the west of the aquifer. This was three years after the actual spill as reported in their paper. In the same line, the simulations carried out in this paper highlight the danger of delayed remediation which would result in contaminant presence within the entire water body over a period of time.

Influence of rainfall on redistribution

In investigating the behaviour and effects of pollutants,

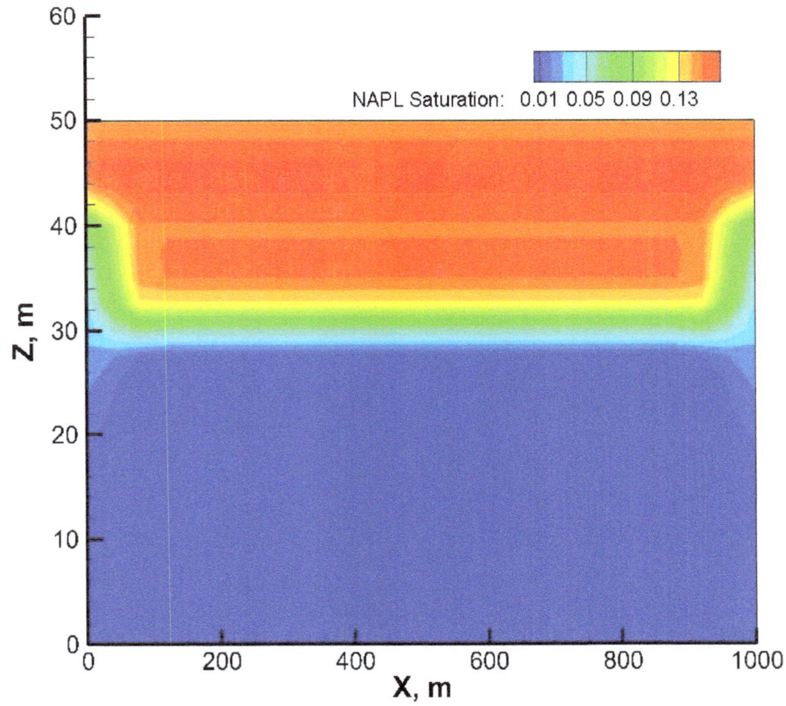

Figure 7. Oil redistribution after 30 years of spill.

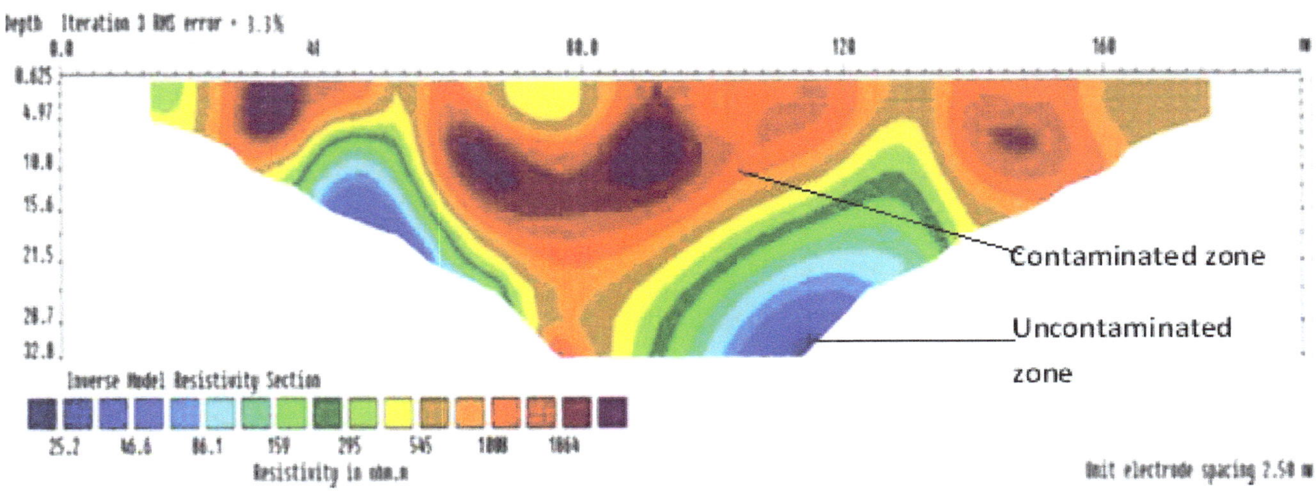

Figure 8. Aquifer resistivity zones at oil spill site in Niger Delta (Nwankwo and Emujakporue, 2012).

one should also investigate the effects of certain environmental conditions particular on the area of study which have direct and indirect influence on the behaviour and effects. The Niger Delta is reported to have a mean rainfall of 2400 mm per annum (Ehirim et al., 2009). Influence of rainfall was investigated on the infiltration and redistribution of the oil plume. This condition was implemented in our simulation.

It was observed that five years after the spill and with inclusion of rainfall, the movement of the plume within the domain was more pronounced as compared to dry conditions. For the same length of time, that is, 5 years, under dry conditions (Figure 4) the plume was at about 10 m from the spill surface while under the influence of rainfall (Figure 9), the plume was already at about 30 m from the same surface. The concentration of the oil had thinned out from 0.1 in Figure 4, to 0.001 in Figure 9. This was as a result of advanced dispersion of the oil phase by the rainfall.

Concentration distributions were observed within the domain as shown in Figure 9. The region closest to the spill surface remained light and uniform in concentration

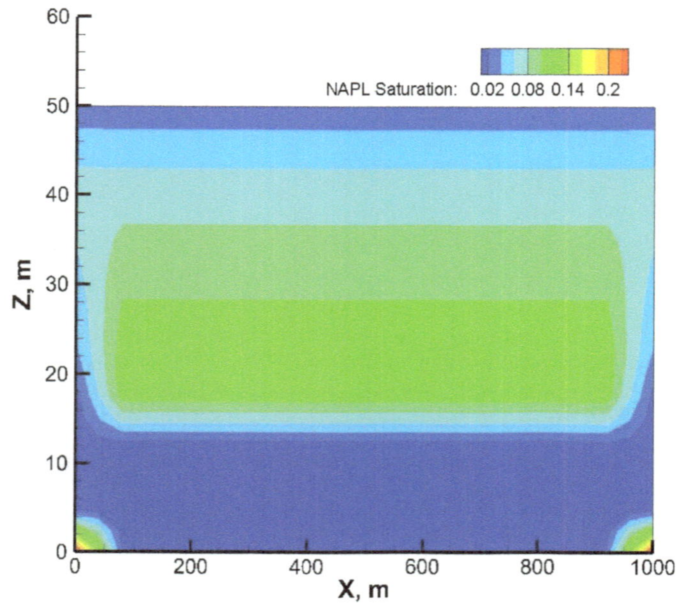

Figure 10. Oil infiltration with influence of rain at 10 years.

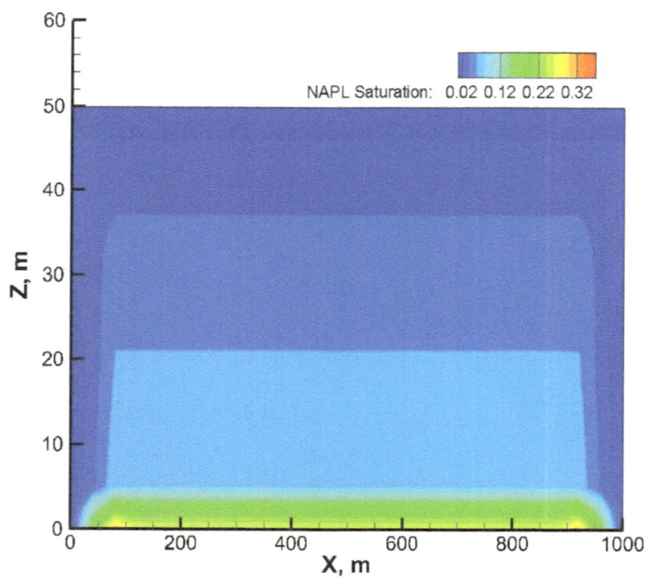

Figure 11. Oil infiltration with influence of rain at 30 years.

at about 0.09 while the region immediately following this, shown with yellow band has concentration in the region of 0.1. The thickest region of the plume had a concentration of about 0.13. Thus, it was deduced that the concentration increased away from the spill surface. The region closest to the spill surface remained lighter owing to the discontinuity of oil infiltration after a year. Meanwhile, the least oil-saturated region of the plume occurs at the front where the oil body interfaces with the fresh water body. This is thought to occur due to the rapid dissolution of the oil components at this interface or differential speed of oil components as it enters different medium.

This behaviour can also be explained to arise from the effect of convective transport in which the incoming rainfall transports the oil components further into the domain. The viscous forces could be higher for the light components of the oil thus they are transported further into the domain leaving behind the more viscous samples which tend to be evenly distributed throughout the top portion of the aquifer.

The significant influence of rainfall was more evident 10 years after the spill. Redistribution of the oil was under the influence of rainfall as seen in Figure 10. The layer closest to the spill surface had further thinned out owing to the washing off by rainfall into the domain. This led to aggregation which increased the concentration of the oil plume further into the domain. It was also observed that aggregates of oil were found at the bottom of the domain close to the bottom boundary surfaces with further advancement of the main plume in comparison with Figure 5. Rainfall was shown to speed up the contamination within the domain. At later simulations times, it was observed that rainfall led to further infiltration within the domain as seen in Figure 11. The oil plume had advanced to the bottom of the domain as compared to Figure 7 at the same thirty years period. Thus, it is overwhelmingly evident that rainfall speeds up aquifer contamination making the rain-prone areas more susceptible.

CONCLUSION

Simulations of oil spill infiltration and redistribution in a shallow aquifer were conducted in this work to quantify their effects on groundwater. A case study of the Niger Delta in Nigeria was used to obtain the input conditions for the simulation. Results showed that flow of oil spills can be adequately described by infiltration and redistribution patterns within the domain.

Without influence of rainfall, the infiltration profile showed that the penetrating plume of oil had moved up to the water table within one year of active penetration with average NAPL saturation of about 0.3 in the unsaturated portion of the domain. The redistribution of oil contaminants, at the end of infiltration, was mainly characterised by domain wall-plume movement which later drew the bulk fluid of the plume further down into the domain. Bulk oil was found at around 20 m from the spill surface 30 years after the spill.

The plume was observed to be thinning out after the stoppage of infiltration. This showed that redistribution continues long after the infiltration stops. This served to carry the contaminants further into the water body. It was also observed that the medium showed oil-wetting or weak-wetting condition as evidenced with the plume trailing along the domain walls which continued to look like norm before the bulk plume advanced. Experimental work is needed to investigate this further and could point direction on the control of wettability to check contaminant progress in an aquifer.

Rainfall was shown to drive the contaminant plume further into the water zone. The interpretation of this is that contaminant transport was faster with rainfall when compared with the infiltration without rain at the same length of time.

Thus, the impacts of contaminants on aquifer water are influenced by contaminant components, wettability of the aquifer, annual rates of rainfall, boundary conditions of the source and possibly size of the domain.

REFERENCES

Anderson WG (1987). Wettability Literature Survey- Part 5: The effects of wettability on Relative Permeability. J. Petrol. Eng. 1453-1468.

Ataie-Ashtiani B, Hassanizadeh SM, Oostrom M, Celia MA, White MD (2001). Effective parameters for two-phase flow in a porous medium with periodic heterogeneities." J. Contaminant Hydrol. 49(1-2): 87-109.

Brooks RH, Corey AT (1964). Hydraulic Properties of Porous Media. Hydrology Papers, Colorado State University (March).

Das D, Hassanizadeh S, Rotte, B, Ataie-Ashtiani B (2004). A Numerical Study of Micro-Heterogeneity Effects on Upscaled Properties of Two-Phase Flow in Porous Media. Transport in Porous Media. 56(3): 329-350.

Edwards JD, Santogrossi PA (1990). Summary and conclusions, in, Edwards JD, Santogrossi PA, eds., Divergent/passive Margin Basins, AAPG Memoir 48: Tulsa. Am. Assoc. Petrol. Geol. 239-248.

Ehirim CN, Ebeniro JO, Ogwu DA (2009). A geophysical and hydro-physiochemical impact of a Solid Waste Landfill (SWL) in port harcourt municipality, Nig. Pacific J. Sci. Technol. 10 (2) 596 – 603.

Flores G, Katsumi T, Kamon M (2007). Evaluation of LNAPL migration under fluctuating ground water by image analysis. Ann. Disas, Prev. Res. Kyoto Univ. 50 B.399-405.

Isehunwa SO, Olubukola O (2012). Interfacial Tension of Crude oil brine systems in the Niger Delta. IJRRAS. 460-465.

Kadafa AA (2012). Oil exploration and spillage in the Niger Delta of Nigeria. Civil Environ Res. 2:38-51.

Leverett MC (1941). Capillary behaviour in porous Media. Trans. AIME. 142:341-358.

Mercer JW, Cohen RM 1990. A review of immiscible fluids in the subsurface: Properties, models, characterization and remediation. J. Contaminant Hydrol. 6(2):107-163.

Molua OC, Igherighe E, Ighrakpata FC (2011). “Porosity Estimation in the Niger-Delta Basin of Nigeria using Sonic Log”. Pacific J. Sci. Technol.12(1):493-501

Nichols WE, Aimo NJ, Oostrom M, White MD (1997). STOMP: Subsurface Transport Over Multiple Phases, Application Guide. Pacific Northwest National Laboratory, PNNL-11216, Richland, WA.

Nwankwo CN, Emujakporue GO (2012). Geophysical Method of Investigating Groundwater and Sub-Soil Contamination: A Case Study. Environ. Res. J. 6:124-129.

Patankar SV (1980). Numerical heat transfer and fluid flow. Hemisphere Publishing Corporation, Washington, D. C.

Schroth MH, Istok JD, Selker JS (1998). Three-phase immiscible fluid movement in the vicinity of textural interfaces. J. Contaminant Hydrol. 32(1-2):1-23.

UNEP (2009). Environmental Assessment of Ogoniland, United Nations Environmental Programme, Nairobi, Kenya, 2011.pp 262.

Van Genuchten MTH (1980). A closed form equation for predicting the hydraulic conductivity of unsaturated soils. Soil Sci. Society Am. J. 44:892-898.

White MD, Oostrom M (1996). STOMP: Subsurface Transport over Multiple Phases. User’s Guide.PNNL-11218. Pacific Northwest National Laboratory, Richland, WA.

Evaluation of physico-chemical and microbial parameters on water quality of Narmada River, India

Virendra Kumar Soni[1], Manish Visavadia[1], Chirag Gosai[1], Mohammad Dilshad Hussain[2], Mukesh Singh Mewada[3], Suresh Gor[4] and Khwaja Salahuddin[5]

[1]Department of Zoology, Bahauddin Science College, Junagadh-362001, Gujarat, India
[2]School of Studies in Zoology & Biotechnology, Vikram University, Ujjain-456010, M.P, India
[3]Department of Zoology, Indira Gandhi National Tribal University, Amarkantak-484886, M.P,India
[4]Department of Botany, Government Science College, Gandhinagar-382010, Gujarat, India
[5]Department of Botany, Bahauddin Science College, Junagadh-362001, Gujarat, India

The present study was carried out to ascertain the quality of water at western zone of Narmada River of Madhya Pradesh (India). Due to heavy discharge of harmful and deleterious substances in river, the biological, chemical and physical characteristics of water have changed to a considerable extent. The objectives of this study were to find out the changes in physico-chemical nature as well as biological health of Narmada River. A thorough study was done on the basis of prevailing seasons. The deterioration of water quality was noticed at lower site of water body in a particular season as low temperature, dissolved oxygen and higher concentration of content of chlorine, etc., all pointing towards the nutrient enrichment. This evaluation will be immensely helpful to estimate the effect of impoundments on the quality of water, fisheries development and reservoir management policy. The microbial analysis was also conducted in terms of most probable number (MPN) of total coliforms in water sample and its highest value (1239) was reported from downstream of Omkareshwar dam site of Narmada River. Water quality index of all sites of Narmada River was calculated for three seasons, viz. summer, monsoon and winter and it was found to be highest (123) at Koteshwar dam site of river in monsoon season and the lowest (70) was observed in water of upstream of Omkareshwar dam during winter. The suitable correlation coefficients were calculated for 7 pairs of variables and correlation matrices were then established seasonally.

Key words: Impoundments, water quality index, dissolved oxygen, most probable number

INTRODUCTION

Water is the most indispensable natural resource in the world for the sake of existence of life. Narmada is the fifth largest river of India. It is commonly known as the Life line of Madhya Pradesh. Water quality is critical public health concern in India. Water pollution in Narmada River has come to alarming proportion in recent past. Thus, the estimation of water quality is very important for proper assessment of associated hazards (Warhate et al., 2006). Recently, several indigenous and useful species are dwindling day by day. To cope with these problems, there is need for application of oriented limnological research so that we can better utilize the most important resource of ours. In this context, change in several physico-chemical features of water of Narmada River was

Table 1. Physico-chemical and microbial parameters and their methodology.

pH	-----	Digital pH meter
Alkalinity	mgl^{-1}	Titrimetry with H_2SO_4
Total hardness	mgl^{-1}	Titrimetry with EDTA
Chlorine content	mgl^{-l}	Argentometric method
Total Dissolved Solids	mgl^{-l}	TDS method
Dissolved Oxygen	$mgl-^{1}$	Winkler method
Biochemical Oxygen Demand	mgl^{-l}	BOD_5 method
Total coliform	Cfu/100 ml	Membrane filtration method

ascertained. These physico-chemical characteristics in many ways have significant influence and impact on aquatic life (Agarwal et al., 1976). Any alteration in these parameters may disturb the quality of water. Dissolved oxygen is of great importance to all the living organisms and is considered to be the sole parameter which to a large extent can reveal the nature of whole water body. Eutrophic water bodies have a wide range of dissolved oxygen and as such oligotrophic water bodies have narrow range of dissolved oxygen (Rucinski et al., 2010). Biological oxygen demand (BOD) also indicates the amount of organic compounds in water as measured by the volume of oxygen required by bacteria to metabolise it under aerobic condition. Biochemical oxygen demand is represented by the amount of organic matter present in water. For more organic matter, more oxygen is required by bacteria for its decomposition. This results in release of organic nutrients in water bodies resulting in death of organisms thriving on water (Rustum et al., 2008). The magnitude of BOD is related to the amount of organic material in waste water. The strength of waste water is expressed in terms of BOD level. Water pollution is generally indicated by the presence of harmful and harmless microbes. Microbial examination of water is a direct measurement of deleterious effects of pollution. Most probable number is the most common microbial parameter for the sanitary analysis of water. The test is used to detect coliforms, a group comprising of all the aerobic and facultative anaerobic, Gram negative, non spore forming and rod-shaped bacteria. These inhabit the intestines of all warm-blooded animals. The discharge of waste water from municipal sewers containing human faecal matter is hazardous to human health. Faecal contamination was routinely detected by microbiological analysis (Nogueira et al., 2003). The aim of this study was to ascertain the impact of several physico-chemical parameters on water quality of Narmada River and to assess further its nature in terms of microbial growth. Assessment of water quality is done to analyse the physical, chemical and biological characteristics of water (Kazi et al., 2009). Water quality index aims at giving a single value to the quality of a source reducing great amount of parameters into a simpler expression and enabling easy interpretation of monitoring data. In this study, water quality index (WQI) is used for assessing surface water quality. Water quality index is a tool used to determine conditions of water quality and, like any other tool, require knowledge on principles and basic concepts of water and related issues (Khan et al., 2003).

MATERIALS AND METHODS

Narmada River is the largest western flowing river in the Indian peninsula (Table 1). In the present investigation, water samples were collected from five different sites viz. Omkareshwar upstream, Omkareshwar downstream, Maheshwar dam site, Mandleshwar and Koteshwar sites (Figure 1). These study sites were situated in a stretch of 220 km of the western zone of Narmada River. The study was based on spatial analysis (Yusuf and Shah, 1986). Water samples were taken in plastic bottles in triplicates for all seasons such as winter, summer and monsoon and analysed onsite and offsite for selected physico-chemical parameters such as pH, alkalinity, total hardness, chlorine content, total dissolved solids, dissolved oxygen and biochemical oxygen demand (Pejman et al., 2009). Temperature and pH were noted onsite, while other parameters were analysed in the laboratory after the samples were properly transported using a pre-cleaned 1 L plastic container in an icebox (NEERI, 1986; APHA 2000)

Microbial analysis was performed in terms of most probable number (MPN)

MPN is performed sequentially in three stages: presumptive, confirmed and completed test.

Presumptive coliform test

It is used to detect coliforms in a water sample. In this lactose fermentation, tubes are inoculated with different water volumes and production of acid and gas from the fermentation of lactose in any of the tubes is a presumptive evidence of coliforms in water sample. The lactose broth (lactose- 5.0 g, peptone- 5.0 g, beef extract- 3.0 g and distilled water- 1 L) used in this test is selective for the isolation of coliforms. A pH indicator Bromocresol purple is also added to lactose broth for the detection of acid. The colour of indicator changed to yellow with the production of acid from lactose.

Confirmed coliform test

In this, water samples from all the positive presumptive lactose broth were inoculated into tubes of brilliant green lactose broth and incubated at 35°C for 48 h. This test is used to determine MPN. MPN = Number of coliforms per colony count x 100 / Volume of sample used

Water quality index

The calculation of water quality index was done using weighted

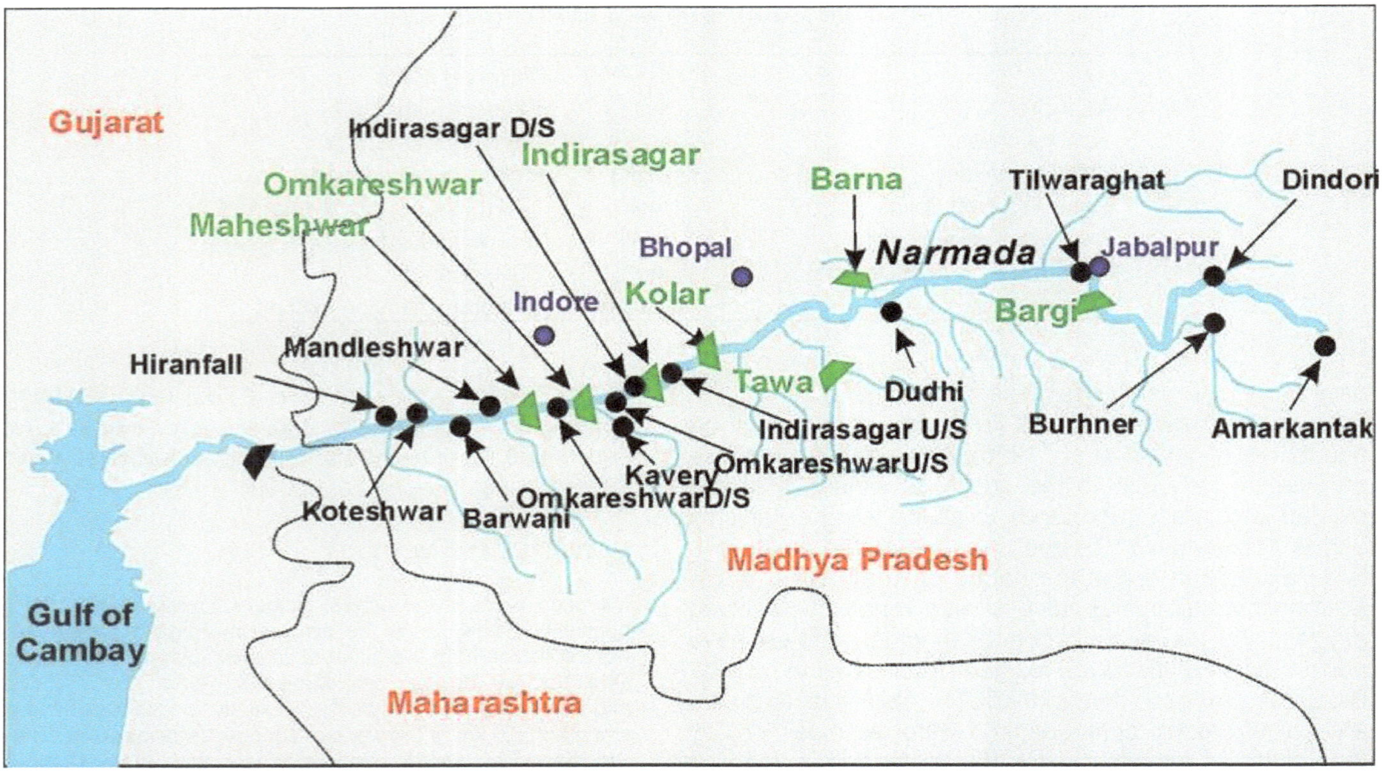

Figure 1. Map showing the exact location of different sites of Narmada river.

arithmetic index method (Brown et al., 1972). They proposed multiplicative form of the index where weights to individual parameters were assigned based on a subjective opinion based on the judgement and critical analysis of the author. The weight assigned reflected a parameter's significance for use and had considerable impact on the index. This index is based on the comparison of the water quality parameters to regulatory standards and gives a single value to the water quality of a source (Abbasi, 2002).

$$\mathrm{WQI}\ \sum_{n=1}^{n} q_n W_n \,/\, \sum_{n=1}^{n} W_n$$

Where, qn = 100 [V_n - V_{io}]/(S_n - V_{10})]

Where q_n = Quality rating for the nth water quality parameter; V_n = estimated value of the nth parameter at a given sampling station; S_n = standard permissible value of nth parameter; V_{io} = ideal value of nth parameter in pure water (pH = 7, DO = 14 mg/l and for other parameter = 0); W_n = unit weight for n^{th} parameter; S_n = standard value for nth parameter; K = constant for proportionality.

RESULTS AND DISCUSSION

The present result depicted the true picture of water quality on the basis of seasonal variations. The pH of different water sample shows little variation from slightly acidic to slightly alkaline. The greater pH may be caused by increase in dissolved oxygen produced as a result of photosynthesis (Wetzel, 1975). The acidic pH is a characteristic of oligotrophic water bodies whereas the neutral and alkaline pH are shown by eutrophic and mesotrophic nature of water bodies, respectively (Kumar et al., 2008). The pH is directly dependent on the amount of CO_2 present and inversely proportional to the activity of photosynthesis (Pandit et al., 2001). At all sites of Narmada River, alkaline pH was reported and its value was found to be maximum (8.4) at Koteshwar dam during winter season. (Figure 2)

Alkalinity is total measure of the substances in water that have acid neutralizing ability (Hoko, 2008). Its level showed greater variation at all sites. The highest alkalinity (138 mgl^{-1}) was reported from the site of Koteshwar study site during monsoon season, whereas the lowest (115 mgl^{-1}) was found to be at Maheshwar dam site during the same season. The amount of alkalinity is dependent on the nature of materials discharged in water bodies (Figure 3).

The total hardness was found to be high in all water bodies. This is highest (172 mgl^{-1}) in Mandleshwa site during monsoon season and lowest (136 mgl^{-1}) at Omkareshwar downstream study in winter season. The hardness is due to dominance of salts of calcium and magnesium which indicated surge in eutrophication of river resulting to greater pollution level. The presence of lower pH and higher hardness may affect their continued use (Chatterjee and Raziuddin, 2002) (Figure 4).

Chloride that dissolves easily in water is toxic to most aquatic organisms because it reacts quickly with other substances in water (Padmanabha and Belagali, 2001).

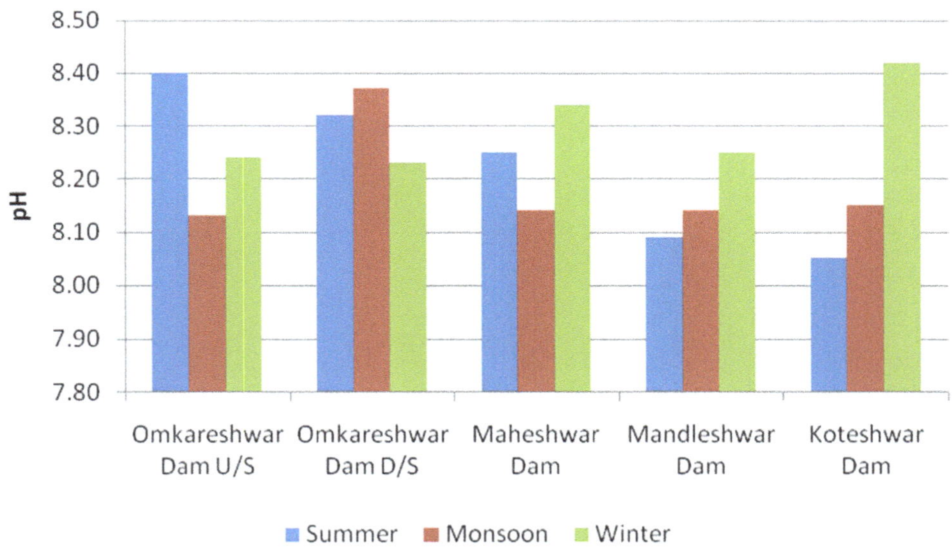

Figure 2. Fluctuation in the value of pH at different sites of Narmada River in different seasons.

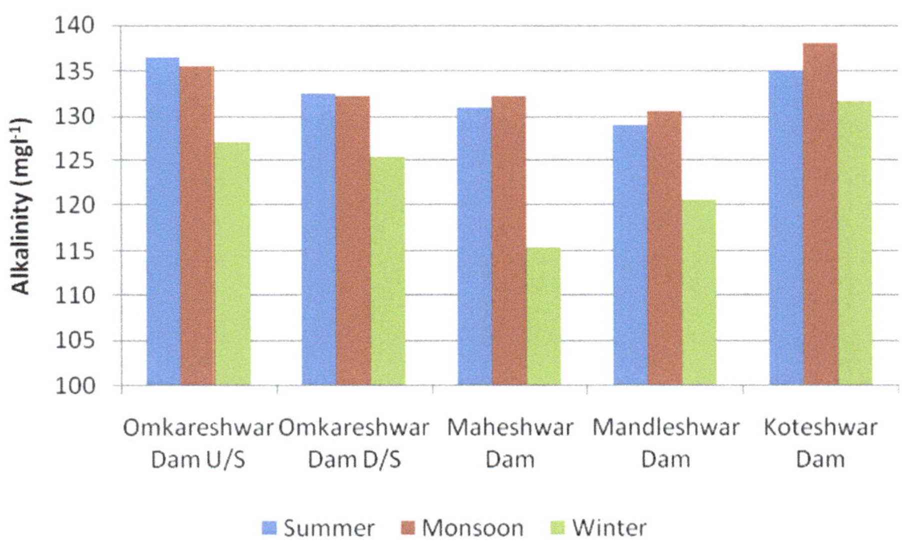

Figure 3. Fluctuation in the value of alkalinity at different sites of Narmada River in different seasons.

The amount of chloride in all water bodies was found to be higher in all sites. At Maheshwar dam site, the maximum content of chloride (26 mgl^{-1}) was noticed during monsoon season, while the minimum (18 mgl^{-l}) was reported from upstream of Omkareshwar dam in winter. The high amount of chloride is an indication of burgeoning anthropogenic pressure on water bodies. The increased chlorine is generally due to the salts of sodium, potassium and calcium (Figure 5). Chloride content showed negative correlation with pH (r = - 0.32) and positive correlation with alkalinity (r = + 0.35) and hardness (r = + 0.49) (Table 2).

The total dissolved solid (TDS) is the sum of the cations cations and anions concentration. A high content of solids elevates the density of water, reduces solubility of gases like oxygen and mitigates the utility of water for drinking, irrigation and other purposes. TDS was found to be highest (357 mgl^{-1}) at Mandleshwar dam site during monsoon season and the lowest (228 mgl^{-1}) was reported from downstream of Omkareshwar dam during winter season (Figure 6). TDS was found to be positively correlated with alkalinity (r = + 0.17), total hardness (r = + 0.73) and chloride contents (r = + 0.45).

The depletion of oxygen can result from a number of natural factors but is most often a concern as a result of pollution and eutrophication in which plant nutrients enter

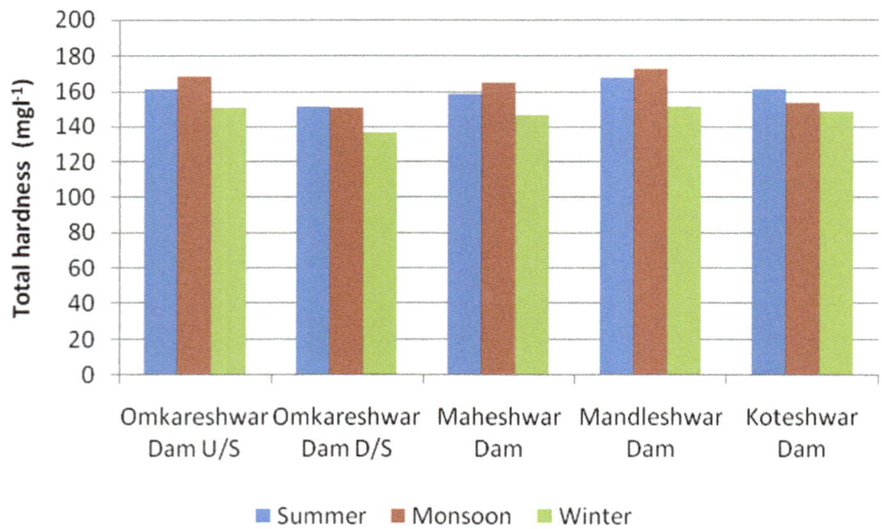

Figure 4. Fluctuation in the value of total hardness at different sites of Narmada River in different seasons.

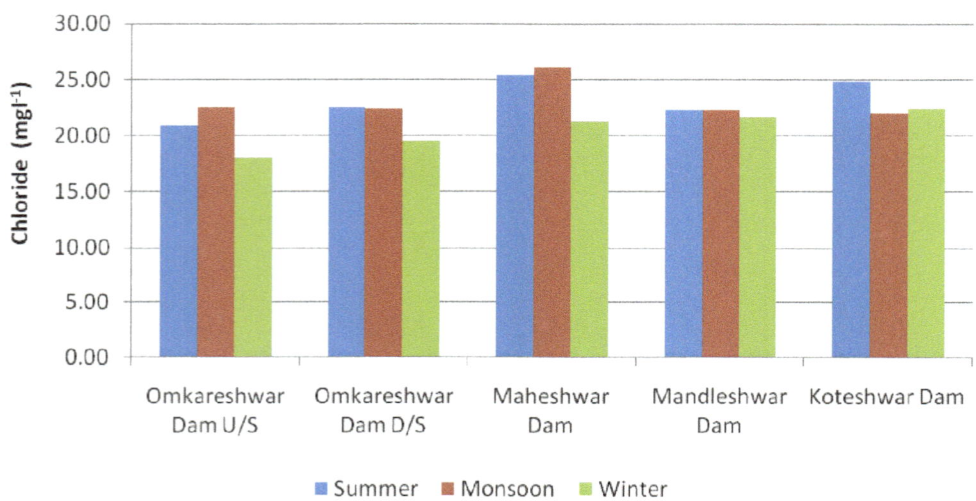

Figure 5. Fluctuation in the value of chloride contents at different sites of Narmada River in different seasons.

Table 2. Correlation matrices between different physicochemical and biological parameters.

	pH	Alkalinity	Hardness	Chloride	TDS	DO	BOD	MPN
pH	0.00							
Alkalinity	-0.20	0.00						
Hardness	-0.54	0.46	0.00					
Chloride	-0.32	0.35	0.49	0.00				
TDS	-0.36	0.17	0.73	0.45	0.00			
DO	0.28	-0.67	-0.56	-0.66	-0.32	0.00		
BOD	-0.51	0.59	0.43	0.48	0.47	-0.67	0.00	
MPN	0.13	0.40	-0.04	0.41	0.30	-0.42	0.57	0.00

a water source (Heinz et al., 1995).of same site showing the inverse relationship with temperature. This happens due to heavy amount of organic matter discharged which needs oxygen for decomposition. As such, low dissolved

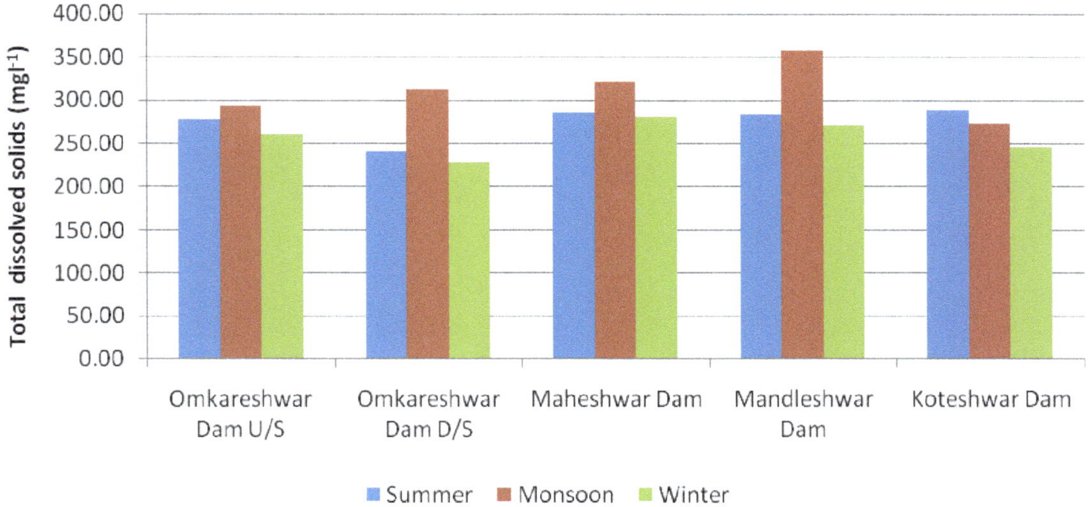

Figure 6. Fluctuation in the value of total dissolved solids at different sites of Narmada river in different seasons.

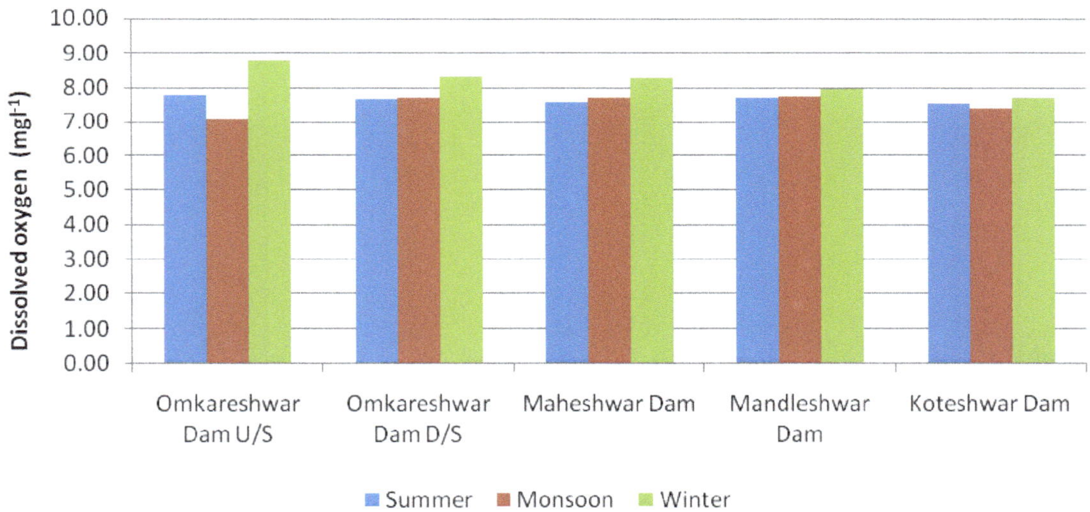

Figure 7. Fluctuation in the value of dissolved oxygen at different sites of Narmada River in different seasons.

oxygen indicates the biodegradation of organic matter (Abowei et al., 2010) (Figure 7).

Likewise, the biochemical oxygen demand also indicates the amount of organic compounds in water as measured by the volume of oxygen required by the bacteria to metabolise it under aerobic condition. For more organic matter, more oxygen is required by bacteria for its decomposition. This results in release of organic nutrients in water bodies resulting in death of organisms thriving on water (Jonasson et al., 2012). The highest degree of biochemical oxygen demand (10 mgl^{-1}) was reported from Koteshwar study site during monsoon season, while lowest level (4.0 mgl^{-1}) was observed from upstream of Omkareshwar dam during summer season (Figure 8). Dissolved oxygen was found to be positively correlated with pH (r = + 0.28) and showed negative correlation with chloride (r = - 0.66), alkalinity (r = - 0.67), hardness (r = - 0.56) and biochemical oxygen demand (r = - 0.67).

Water is considered safe for drinking if it contains less than four coliforms per 100 ml of water. The occurrence of coliforms detected in water is a direct measurement of deleterious effects of pollution of human health. In laboratory, to confirm the water quality to drinking water standards, the actual number of coliforms was not reported but they are mentioned as an approximate count-MPN. These dam sites are present in close vicinity of human settlements so the abundance of coliforms is immi-

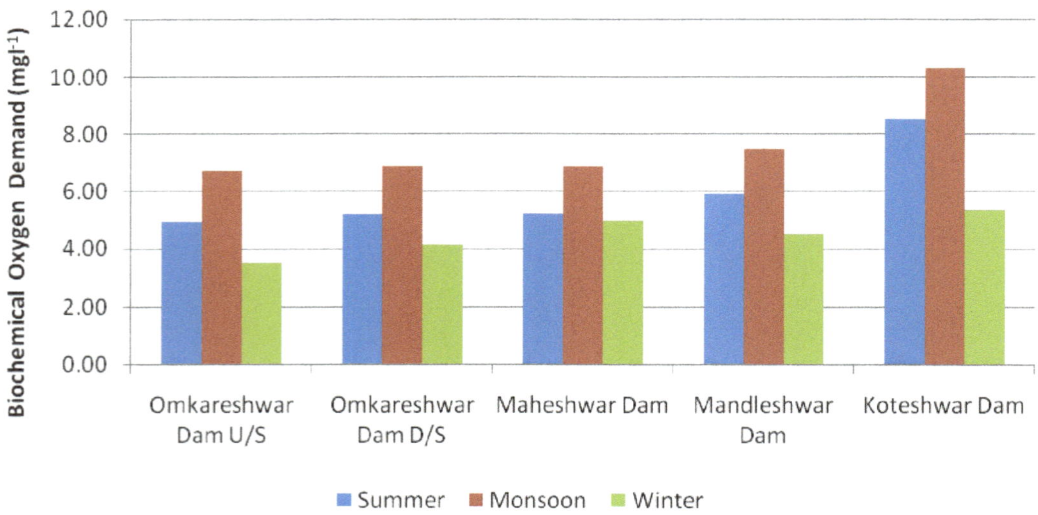

Figure 8. Fluctuation in the value of BOD at different sites of Narmada River in different seasons.

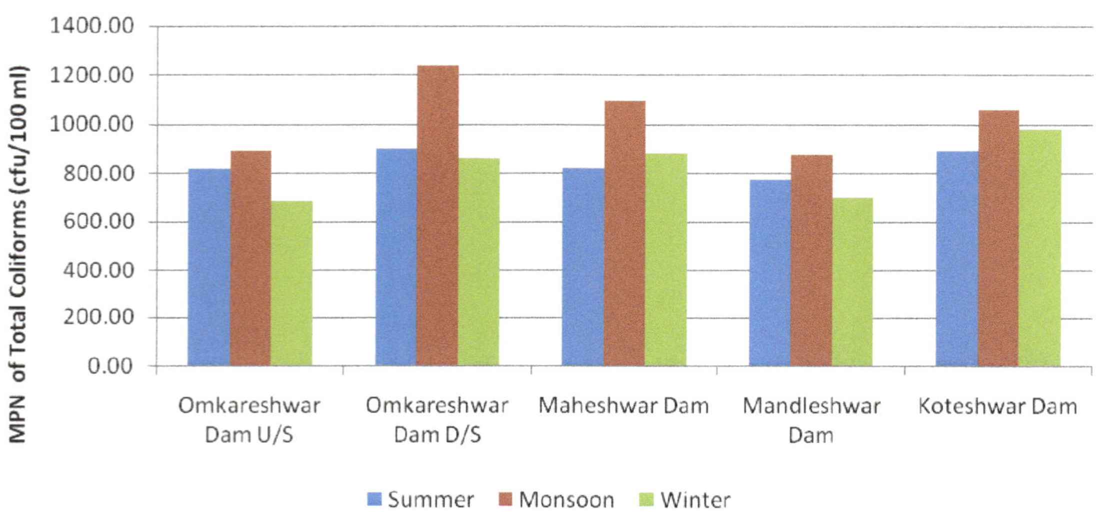

Figure 9. Fluctuation in the value of total coliforms (cfu/100 ml) at different sites of Narmada river in different seasons.

nent. In this study, the MPN value was found to be highest (1239 cfu/100 ml) at downstream of Omkareshwar dam site during monsoon season and lowest (682 cfu/100 ml) was reported at Mandleshwar dam site of Narmada River in winter (Karafistan and Arik-Colakoglu, 2005) (Figure 9). MPN showed negative correlation with DO (r = - 0.42) and positive correlation with BOD (r = + 0.57).

The expression of water quality that offers a stable and reproducible unit of measure which responds to changes in the principal characteristics of water was responded (Faisal et al., 2003; Kannel et al., 2007). Water quality index (WQI) is a mechanism for presenting a cumulatively derived numerical expression defining a certain level of water quality (Bordalo et al., 2006). In other words, WQI summarizes large amounts of water quality data into simple terms e.g., excellent, good, bad, etc. for reporting to management and the public in a consistent manner. In the present study, water quality index was found to be the highest (123 mgl^{-l}) at Koteshwar study site of Narmada River during monsoon season, whereas the lowest (70 mgl^{-l}) was observed at upstream of Omkareshwar dam of Narmada River during winter (Figure 10).

ACKNOWLEDGEMENTS

Financial assistance from Narmada Valley Development Authority (NVDA), Bhopal (Madhya Pradesh) is gratefully acknowledged. The corresponding author is also thankful to Ms. Shabana Nabi and Ms. Teena Soni for statistical analysis of data.

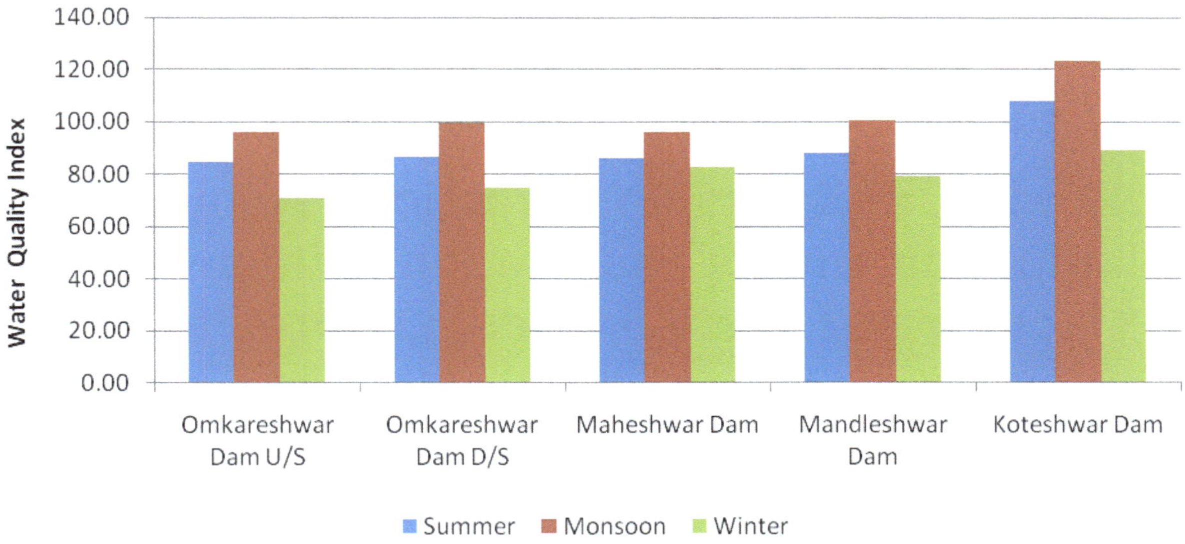

Figure 10. Fluctuation in the value of water quality index at different sites of Narmada River in different seasons.

REFERENCES

Abbasi SA (2002). Water quality indices, state of the art report, National Institute of Hydrology, scientific contribution no. INCOH/SAR-25/2002, Roorkee: INCOH. pp: 73.

Abowei JFN (2010). Salinity dissolved oxygen, pH and surface water temperature conditions in Nkoro River, Niger Delta, Nigeria. Adv. J. Food Sci. Technol. 2(1): 36-40.

Agarwal DK, Gour SD, Tiwari IC, Narayan Swami N, Narwash SM (1976). Physico-chemical characteristic of Ganga water in Varanasi. Ind. J. Environ. Health 18: 201-20

APHA (2005). Standard methods for examination of water and wastewater (21st ed). American Public Health Association, Washington D.C., USA

Bordalo AA, Teixeira R, Wiebe WJ (2006). A water quality index applied to an international shared river basin: The case of the Douro River. Environ. Manage. 38:910–920.

Brown RM, Mc Cleiland NJ, Deininger RA, O'Connor MF (1972). A Water Quality Index-crossing the psychological barrier (Ed Jenkis, S.H). Proc. Intl. Conf on water Poll, Jerusalem. 6:787-797.

Chatterjee C, Raziuddin M (2002). Determination of water quality index of a degraded river in Asansol industrial area, West Bengal. J. Environ. Pollut.1(2):181-189.

Faisal K, Tahir H, Ashok L (2003). Water quality evaluation and trend analysis in selected watersheds of the Atlantic region of Canada. Environ. Monit. Assess. 88:221-248.

Heinz G, Stefan, Xing Fang (1995). Stimulation of dissolved oxygen profiles in a transparent, dimictic lake. Limnol.Oceanogr.40 (1):105 – 118.

Hoko Z (2008). An assessment of quality of water from boreholes in Bindura District, Zimbabwe. Phys. Chem. Earth 33: 824–828.

Jonasson L, Hansen JLS, Wan Z, She J (2012). The impacts of physical processes on oxygen variations in the North Sea-Baltic Sea transition zone. Ocean Sci. 8: 37–48.

Kannel PR, Lee S, Lee YS, Kanel SR, Khan SP (2007). Application of water indices and dissolved oxygen as indicators for river water classification and urban impact assessment. Environ. Monit. Assess. 132:93–110.

Karafistan A, Arik-Colakoglu F (2005).Physical, chemical and microbiological water quality of the Manyas lake, Turkey. Migration and Adaptation Strategies for Global Change.10:127-143.

Kazi TG, Arain MB, Jamali MK, Jalbani N, Afridi HI, Sarfraz RA, Baig JA, Shah AQ (2009). Assessment of water quality of polluted lake using multivariate statistical techniques: A case study, Ecotox. Environ. Safe. 72(20): 301-309.

Khan F, Husain T, Lumb A (2003). Water quality evaluation and trend analysis in selected watersheds of the Atlantic region of Canada. Environ. Monit. Assess. 88:221–242.

Kumar J, Singh S, Pal A (2008). Water quality of Turamdih and Jaduguda uranium mines and adjacent areas, East Singhbhum, Jharkhand, J. Ecophysiol. Occup. Health 8 : 7-14

NEERI (1986). Manual on water and waste water analysis.National Environment Engineering Research Institute, Nagpur

Nogueira G, Nakamura VC, Tognim CBM, Filho AAB, Filho PDB (2003). Microbiological quality of drinking water of urban and rural communities, Brazil. Rev. Saude Publica. 37(2):232-236.

Padmanabha B, Belagali SL (2007). Water quality index of Kabini river in the kallahally village of Nanjangud Taluka of Mysore District, Karnataka. J. Environ. Sci. Eng. 49(1):48-50.

Pandit AK, Rathore SA, Bhat SA (2001). Limnological features of freshwater of Uri, Kashmir. J. Res. Dev. 1: 22-29.

Pejman AH, Bidhendi NGR, Karbassi AR, Mehrdadi N, Bidhendi EM (2009). Evaluation of spatial and seasonal variations in surface water quality using multivariate statistical techniques. J. Environ. Sci. Technol. 6(3):467-476.

Rucinski DK, Beletsky D, DePinto JV, Schwab DJ, Scavia D (2010). A simple1-dimensional, climate based dissolved oxygen model for the central basin of lake Erie. J. Great Lakes Res. 36: 465-476.

Rustum R, Adeloye AJ, Scholz M (2008). Applying Kohonen Self organising map as a software sensor to predict the BOD. Water Environ. Res. 80(1): 32-40

Warhate SR, Yenkie MKN, Chaudhari MD, Pokale WK (2006). Impacts of Mining Activities on Water and Soil. J. Environ. Sci. Eng. 48(2):81 - 88.

Yousuf, A.R and Shah, G.M (1988). Comparative Limnology of some fresh habitats of Kashmir. Geobios, 7: 58-61.

Knowledge, attitudes and practices study of dengue viral infection and its association with environmental factors and health issues, Lahore Pakistan

Mariam Zameer[1] , Maryum shuja[1], Arooj Ashraf[1], Nadia Mukhtar[1] and Bilal Munir Ahmad[2]

[1]College of Earth and Environmental Sciences, University of Punjab, Lahore.
[2]Microbiology Diagnostic and Research Laboratory, Mayo Hospital/Kemu, Lahore, Pakistan.

This study was designed to assess and compare the awareness, approach, and practices of dengue positive and dengue negative patients regarding dengue fever in a tertiary care hospital of Lahore district, Pakistan from April to May 2012. A total of 250 samples (125 dengue positive and 125 dengue negative) were collected and tested for Dengue IgM and IgG antibodies test by ELISA technique. Pre-tested questionnaire was used for data collection after the informed consent of the patients. Out of 250 study participants, 78% stated fever as major symptom of dengue virus, 70% had knowledge about the vector, 76% had idea about mosquitoes breeding sites and 80% considered dengue a treatable disease. For 50%, the main source of information about dengue was their TV/Radio. Regarding preventive practices, out of 80% who stored water at home, 52% used to cover their water containers, 64% used to wear protective clothing and 22% used mosquito repellents. Overall, 32% were satisfied from Governmental efforts against the spread of dengue fever. The study indicated that the community was quite familiar with Dengue, but there were some confusion about vaccination and water storage for domestic use.

Key words: Dengue fever, knowledge, attitudes, environment, IgM, IgG.

INTRODUCTION

Dengue is a viral mosquito-borne infection. About 24,000 deaths are reported annually because of dengue and it is prevalent in more than 100 countries. Flu-like illness is caused by the infection, which if persists can lead to the development of potentially lethal complication known as severe dengue or dengue hemorrhagic fever. Transmission has increased predominantly, in recent years, in the urban and semi-urban areas which become a major international public health concern (World Health Organization 2012). Dengue is caused by single-stranded RNA viruses which are about 40 to 50 mm in diameter and belong to the flavivirus group. Four species have been described known as serotypes: DEN-1, DEN-2, DEN-3 and DEN-4. Three structural proteins (capsid C, membrane protein M, and envelope glycoprotein E) and seven non-structural proteins (NS1, NS2a, NS2b, NS3, NS4a, NS4b and NS5) are encoded by viral genome. In the E protein, the main biological properties of the viruses are located. Some of the nonstructural proteins play part in viral replication (Singhi et al., 2007).

The suitable temperature for development of the *Aedes* mosquito ranging from 15 to 30°C at this temperature face lower mortality rate (Yang et al., 2009). The symptoms of dengue infection are high fever, severe headache, painful joints and muscles, vomiting, nausea, pain behind the eyes and skin rashes. These symptoms almost last for about one week, but weakness and tiredness may last for several weeks. In some patients dengue fever leads to development of Dengue Hemorrhagic Fever (dhf), and at the time of subsidence of the fever, the patient may reflect problems with blood circulation. These can include blood in the urine or stool, bleeding gums or bloody nose. These symptoms may lead to death if untreated (Ohio Department of Health 2011). The primary vector of dengue is mosquito, *Aedes aegypti* and transmission may also cause by *A. albopictus*. Through the bites of infected female mosquitoes, the virus is transmitted to humans. Female mosquito feeds on blood because it needs protein for laying eggs. After incubation of virus for about 4 to 10 days, an infected mosquito is capable of transmitting the virus to the host for the rest of its life. The main carriers and multipliers of the virus are infected humans. The mosquito *A. aegypti* found in urban habitats and breeds commonly in man-made containers like flower vases, water storage jars, unused toilets bowls and chocked roof gutters. *A. aegypti* is a daytime feeder; its peak periods of biting are early in the morning and evening before dusk (World Health Organization 2012). DHF is endemic in Asian countries; the epidemics went on increase in the last 15 years. At now, dengue becomes an important viral mosquito borne disease affecting humans.

In 1994, first outbreak of dengue fever was reported in Pakistan. In Pakistan, dengue virus is endemic, now kept on circulating throughout the year with a peak incidence in the period of post monsoon. Flood of 2010 in Pakistan made the situation worse (Jawad et al., 2001). The increase of global warming which is associated with increased precipitation actually creates hospitable habitat for immature mosquitoes. Survival of eggs and adult mosquitoes is based on higher humidity rates. Because of droughts, people store drinking water in containers that can also act as breeding sites for mosquitoes (Gage et al., 2008). Higher temperatures shorten the "extrinsic incubation period" time required by virus to replicate within the mosquito. So, virus in the mosquito within short time can reach its salivary glands and can be transmitted to humans (Johansson et al., 2009). Because of over crowded cities, inadequate sanitation, unsafe drinking water and such other factors, Pakistan is at high risk of dengue fever. Such conditions are suitable for the infectious diseases spreading and also a cause of dengue outbreak in different parts of the country which is associated with increased morbidity and mortality, each year. (Jahan, 2011).

This study is questionnaire based performed in a tertiary care hospital in order to assess the knowledge, attitude and practices of dengue positive and dengue negative patients related to dengue viral infection and it was found that overall knowledge of patients (both dengue positive and dengue negative) was fair.

METHODOLOGY

The plan of work was based on the selection of a total of 125 dengue positive patients and 125 dengue negative patients. Blood samples were collected and tested for the detection of antibodies against dengue virus with the help of 'third generation enzyme linked immunosorbent assay'. All dengue positive and dengue negative patients were interviewed and questionnaires were filled after taking informed consent from each dengue positive and dengue negative patient. Data was entered in SPSS on daily basis and analyzed for comparison of knowledge, attitude and practices regarding prevention and control of dengue fever practices in concerned areas.

Study area

The present KAP study was conducted to evaluate the knowledge, attitude and practices of Dengue positive and Dengue negative patients visiting tertiary care hospital in Lahore. The hospital is situated in Lahore and had more than 2399 beds. It is a multidisciplinary and a referral hospital. This hospital is also a teaching hospital of a medical university. The hospital has various departments.

Proposal of work

The study was done for evaluation of potential risk factors association and identification of possible routes of transmission through Dengue antibodies positive patients to Dengue antibodies negative patients. A total of 25 Dengue positive patients were randomly selected from the patients visiting for Dengue immunoglobulin M and immunoglobulin G test by ELISA method and 25 Dengue negative patients were selected from OPD of tertiary care hospital, Lahore (Figure 1). Blood samples were collected and questionnaire was filled and analyzed by the help of Statistical Package for Social Sciences (version 16.0; SPSS Inc., Chicago, IL, USA). The study was conducted during the month of April to May 2012; during the entire study period, various visits were made to collect the blood samples of patients for the detection of dengue antibodies by third generation ELISA method (Lange and Sarrazin, 2009).

A total of 25 dengue positive patients and 25 dengue negative patients were selected. Blood samples were collected and tested for the detection of antibodies against dengue virus with the help of third generation enzyme linked immunosorbent assay.

i) All dengue positive and dengue negative patients were interviewed and questionnaire was filled after taking informed consent from each dengue positive and dengue negative patient.
ii) Data was entered in SPSS on daily basis and analyzed for comparison of knowledge, attitude and practices regarding prevention and control of dengue.

Materials

The following equipments were used for both IgM and IgG testing for Dengue:

i) ELISA microwell plate reader, equipped for the measurement of absorbance at 450/620 nm.

No. of Dengue IgM/ IgG positive patients: 25
No. of Dengue IgM/ IgG negative patients: 25
Study settings: College of Earth and Environmental sciences, Punjab university & Main Clinical lab of Mayo Hospital, Lahore

Blood sample collection

KAP Study

Diagnosis By: ELISA technique

Data collection

(Through questionnaire by face to face interview or telephone (whichever is suitable)
Parameters: Demographic, occupation, socioeconomic, family size, marital status, immune system, knowledge, attitude and practices

Data analysis

Figure 1. Random Test Sample

ii) Incubator at 37°C.
iii) Manual or automatic equipment for rinsing wells.
iv) Micropipettes to deliver volumes between 10 and 1000 µl.
v) Vortex tube mixer.
vi) Deionized or (freshly) distilled water.
vii) Disposable tubes.
viii) Timer.
ix) Dengue IgM ELISA Kit (Nova Lisa TM).
x) Dengue IgG ELISA Kit (Nova Lisa TM).
xi) 5 mm BD syringes.
xii) Centrifuge machine.

Material supplied with kit

i) 1 Strip holder.
ii) 1 Cover foil.
iii) 1 Test protocol.
iv) 1 distribution and identification plan.

Principle of the assay for Dengue IgG/IgM test

The qualitative immunoenzymatic determination of IgG-class/IgM-class antibodies against Dengue virus is based on the ELISA (enzyme-linked immunosorbent assay) technique. Microtiter strip wells are precoated with Dengue virus antigens type 2 to bind corresponding antibodies of the specimen. After washing the wells to remove all unbound sample material, horseradish peroxidase (HRP) labeled anti-human IgG/IgM conjugate is added. This conjugate binds to the captured Dengue virus-specific antibodies. The immune complex formed by the bound conjugate is visualized by adding tetramethylbenzidine (TMB) substrate which gives a blue reaction product. The intensity of this product is proportional to the amount of Dengue virus-specific IgG/IgM antibodies in the specimen. Sulphuric acid is added to stop the reaction. This produces a yellow end point color. Absorbance at 450 nm is read using an ELISA microplate well reader.

Dengue enzyme immunoassay was used for qualitative determination of antibodies (IgM/IgG) in human serum or plasma samples. The serum obtained through centrifugation was further processed for the detection of antibodies against Dengue by using ELISA Kit. Tests for the Dengue were carried out in 'main clinical pathology lab' of a tertiary care hospital in Lahore.

RESULTS

Practices to avoid dengue fever

With regards to personal preventive practices, the storage of water in covered containers for domestic usage was widespread among dengue negative patients. Among dengue positive patients, 105(84%) used to store water at home, but only 45(36%) practiced to cover their containers; and in case of dengue negative patients, 95(36%) used to store water at home and 85(68%) practiced to cover their containers. However, 75(60%) dengue positive and 85(68%) dengue negative patients stated that they used to wear protective clothing that is, full sleeves shirts and long trousers. The use of mosquito repellents and bed nets was not found common among

Table 1. Preventive practices against dengue fever.

Variable	Number (%)		
	Dengue positive	Dengue negative	Total patients
A) Personal preventive practices			
Storage of water at home			
Yes	105(84%)	95(76%)	200(80%)
No	20(16%)	30(24%)	50(20%)
Coverage of water containers			
Yes	45(36%)	85(68%)	130(52%)
No	80(64%)	40(32%)	120(48%)
Wearing of long sleeved clothing			
Yes	75(60%)	85(68%)	160(64%)
No	50(40%)	40(32%)	90(36%)
Application of mosquito's repellents			
Yes	5(4%)	50(40%)	55(22%)
No	120(96%)	75(60%)	195(78%)
B) Practices taken by Government			
Role in leveling defective ground surfaces			
Yes	20(16%)	20(16%)	40(16%)
No	105(84%)	105(84%)	210(84%)
Role in providing insecticides, repellents and bed nets			
Yes	0(0%)	5(4%)	5(2%)
No	125(100%)	120(96%)	245(98%)
Role in providing spraying facility			
Yes	110(88%)	90(72%)	200(80%)
No	15(12%)	35(28%)	50(20%)

patients. Regarding practices taken by Government, 110(88%) dengue positive and 90(72%) dengue negative patients stated that Government provided spraying facility in their areas, 105(84%) dengue positive and dengue negative patients stated that Government did not play part in leveling defective ground surfaces whereas 125(100%) dengue positive and 120(96%) dengue negative patients stated that pesticides, repellents and bed nets were not provided by Government. The detailed data of practices to avoid dengue has been shown in Table 1.

Socio demographic information

Out of 250 patients, 125 were dengue positive and 125 were dengue negative. The sample comprised of 70(56%) dengue positive, 55(44%) dengue negative males and 55(44%) dengue positive, and 70 (56%) dengue negative females. The mean age of dengue positive patients was 32 years and of dengue negative patients was 30 years. Illiteracy rate was observed high among respondents, about 85(68%) of dengue positive and same number of dengue negative patients had education up to secondary level and only 10(8%) of dengue negative patients were completed tertiary education that is, up to post college. In terms of income 20(16%), dengue positive patients and 50(40%) dengue negative patients had their income between 1,000 to 10,000 rupees and only 15(12%) dengue positive and 10(8%) dengue negative patients had income above 30,000 rupees, the detailed information regarding socio demographics is given in Table 2.

Knowledge of dengue fever

High percentages of surveyed respondents such as 110(88%) dengue positive, 85(68%) dengue negative patients correctly identified main symptom of dengue virus that is fever; whereas low percentages of respondents identified skin rashes and bleeding from

Table 2. Socio-demographic characteristics of study participants.

Characteristic	Number (%)		
	Dengue positive	Dengue negative	Total patients
Gender			
Male	70(56%)	55(44%)	125(50%)
Female	55(44%)	70(56%)	125(50%)
Educational level			
Secondary school	85(68%)	85(68%)	170(68%)
Under graduate	40(32%)	30(24%)	70(28%)
Post graduate	0(0%)	10(8%)	10(4%)
Household income			
5000-10000	20(16%)	50(40%)	70(28%)
11000-20000	60(48%)	30(24%)	90(36%)
21000-30000	30(24%)	35(28%)	65(26%)
>30000	15(12%)	10(8%)	25(10%)

nose as symptoms of dengue virus given in Table 2. Most participants, 90(72%) dengue positive and 85(68%) dengue negative patients were aware that mosquito transmits dengue virus to host; whereas, only 10(8%) dengue positive patients stated that flies are the main vector of transmission. When asked about breeding sites, 45(36%) dengue positive, 25(20%) dengue negative patients stated it is garbage and 45(36%) dengue positive, 75(60%) dengue negative patients had idea that water is required for breeding of mosquitoes, the remaining ones did not have idea about mosquitoes breeding sites. 90(36%) of all respondents (45 dengue positive and 45 dengue negative) correctly identified biting times of mosquito that is sunset/sunrise. Regarding sources of information, majority of 50(40%) dengue positive, 75 (60%) dengue negative patients reported they came to know about dengue from TV/Radio; whereas only 15(12%) dengue positive and 20(16%) dengue negative patients obtained information from newspapers/magazines. Other sources of information found among patients were health personnels and relatives/friends, detailed percentages are mentioned in Table 3.

Attitude towards dengue fever control

Regarding treatability of majority of respondents, 105(84%) dengue positive, 95(76%) dengue negative patients showed positive attitude that dengue is treatable and when asked about responsibility of maintaining hygienic environment positive attitude observed among respondents 75(60%) dengue positive and 90(72%) dengue negative patients started to cooperate with neighborhood in leveling defective ground surfaces. Attitude of self medication or self-treatment was also observed among respondents (Table 4).

DISCUSSIONS

In this questionnaire based survey, the ratio of male to female respondents was same; this may be due to random selection of patients. Among all 250 of respondents, 170(68%) patients had education up to secondary level which may be due to poverty, under developed education system and may be due to the fact that mostly visited persons belongs to rural or under developed areas, with low income and less number of educational facilities in their areas. The survey further revealed that only 25(10%) of respondents belongs to upper middle class because this class commonly prefer private hospitals.

Majority of patients 195(78%) identified fever as a major symptom of dengue fever and about half of respondents 120(48%) had idea about exact mosquito breeding site that is stagnant water, only a few 10(4%) patients identified flies as actual transmission route of dengue virus, these are actually the dengue awareness campaigns after the epidemic outbreak of virus in 2011 which make the people aware about disease. 35(14%) of study participants came to know about dengue from newspaper and magazine, the reason behind this is lack of interest of participants to read newspaper or reliance on radio/TV as a source of information. Most of respondents, 200(80%) said that dengue is a treatable disease. This is actually due to provision of proper medical facilities by Government in well-developed health centers after the emergence of disease. This survey revealed that 80(64%) respondents took paracetamol in

Table 3. Knowledge variables regarding dengue fever among study participants.

Variable	Number (%)		
	Dengue positive	Dengue negative	Total patients
Symptoms			
Fever	110(88%)	85(68%)	195(78%)
Skin rash	10(8%)	0(0%)	10(4%)
Bleeding from nose	5(4%)	5(4%)	10(4%)
All of the above	0(0%)	25(20%)	25(10%)
Do not know	0(0%)	10(8%)	10(4%)
Transmission route			
Mosquito bite	90(72%)	85(68%)	175(70%)
Flies	10(8%)	0(0%)	10(4%)
Sharing food	0(0%)	10(8%)	10(4%)
Do not know	25(20%)	30(24%)	55(22%)
Breeding site			
Garbage	45(36%)	25(20%)	70(28%)
Stagnant water	45(36%)	75(60%)	120(48%)
Do not know	35(28%)	25(20%)	60(24%)
Biting time			
Sunset/Sunrise	45(36%)	45(36%)	90(36%)
Day Time	10(8%)	15(12%)	25(10%)
Night	70(56%)	65(52%)	135(54%)
Sources of information on dengue			
Radio/TV	50(40%)	75(60%)	125(50%)
Newspapers/magazines	15(12%)	20(16%)	35(14%)
Health personnel	15(12%)	10(8%)	25(10%)
Relatives and friends	45(36%)	20(16%)	65(26%)

Table 4. Attitude towards dengue fever.

Variable	Number (%)		
	Dengue positive	Dengue negative	Total patients
Treatability of dengue			
Yes	105(84%)	95(76%)	200(80%)
No	5(4%)	10(8%)	15(6%)
Do not know	15(12%)	20(16%)	35(14%)
Cooperation with neighborhood in maintaining hygienic environment			
Yes	75(60%)	90(72%)	165(66%)
No	50(40%)	35(28%)	85(34%)
Self medication or self treatment			
Take Panadol	90(72%)	70(56%)	160(64%)
Drink a lot of water	15(12%)	10(8%)	25(10%)
Take medicine and drink Water	5(4%)	5(4%)	10(4%)
Do not know	15(12%)	40(32%)	55(22%)

first stage of fever. This trend of self-medication is due to the hesitation among patients to consult the doctor because of their personal believes or misconceptions and also due to economic instability of the patients or may be due to availability of medicines on medical stores without prescription. The practice of storing water was found among 200(80%) of respondents and out of these, 130(52%) used to cover water containers tightly. This may be due to dengue awareness campaigns in reported areas of this country, in which information is provided to people that open water is the breeding site for the mosquitoes; because of that useful information, poor personal practices against the spread of the dengue fever were found among few of respondents. Among all respondents, only 90(36%) refused to wear the long sleeved clothing in order to protect themselves from dengue because of the hot weather.

In this study, 245(98%) of respondents stated that the control of dengue fever is the issue to be solved by the government. This is due to insufficient involvement of the Government against the spread of the dengue fever because in our country only spraying facility is provided properly and awareness campaigns were arranged by Government; whereas, Government did not play an important role in leveling defective ground surfaces to avoid accumulation of stagnant water and also did not provide bed nets, insecticides and repellents to reported communities, it is the matter of Government to participate actively in making the situation better. Compared to our results, more than half of the parents (54%) had good knowledge about signs, symptoms and modes of transmission of dengue. Approximately, 47% considered dengue to be a serious but preventable disease to which they are vulnerable. Nevertheless, a majority (77%) did not use effective dengue preventive methods such as screening of homes and 51% did not use bed nets.

Educational attainment (OR, 2.98; CI, 1.23 to 7.23) was positively associated with knowledge of dengue. There was no correlation between knowledge about dengue and preventive practices (p = 0.34). Radio and TV were the predominant sources of information about dengue fever (Faisal et al., 2010).

Conclusions

Among all respondents, illiteracy rate was observed high. Majority of respondents belonged to joint families. Most of the respondents identified fever as a major symptom of disease, mosquito bite as causal agent of disease and got knowledge from TV/radio about dengue fever. Majority of respondents had knowledge that dengue fever is severe and curable. Regarding preventive practices, maximum respondents used to store water at home and used to cover water container. Wearing of long sleeved clothing is common personal preventive practice observed among respondents. It is stated by majority of respondents that Government did not play a part in leveling defective ground surfaces to avoid accumulation of stagnant water and did not provide insecticides, repellents and bed nets to their communities, but spraying facility provision by Government found common among respondents. Overall, the knowledge and preventive practices taken by dengue negative patients found good as compare to dengue positive patients. Climate change including increase in temperature, precipitation and humidity is considered as main factor to increase the epidemics of dengue virus in different parts of the world.

REFERENCES

Gage KL, Burkot TR, Eisen RJ, Hayes EB (2008). Climate and vector borne diseases. Am. J. Prev. Med. 35:436-450.

Jahan (2011) Dengue Fever (DF) in Pakistan. Asia Pacific Family Medicine10:1-4.

Jawad KA, Masood S, Tassawar H, Inam B, Waheeduz ZT (2001). Outbreak of Dengue Hemorrhagic Fever in Karachi. Pak Armed Forces Med. J. 51:94-98.

Johansson MA, Cummings DAT, Glass GE (2009). Multi-year climate variability and Dengue-El Niño Southern oscillation, weather, and dengue transmission in Puerto Rico: A longitudinal data analysis. PLoS Medicine 6:1000168.

Johansson MA, Dominici F, Glass GE (2009). Local and global effects of climate on dengue transmission in Puerto Rico. PLoS Neglected Tropical Diseases 3:382.

Ohio Department of Health (ODH) (2011). Dengue fever fact sheet: 1-6.

Singhi, S., Kissonn, N, Bansal, A. (2007) Dengue and dengue hemorrhagic fever . J. Pediatria 83:22-35.

World Health Organization (2012). Dengue and Severe Dengue fact sheet.

World Health Organization (2012). Dengue and Severe Dengue fact sheet.

Yang HM, Macoris MLG, Galvani KC, Andrighetti MTM, Wanderley DMV (2009). Assessing the effects of temperature on the population of *Aedes aegypti*, the vector of dengue. Epidemiol. Infect. 137:1188-1202.

Reginol interpretation of river Indus water quality data using regression model

Aziz-ur-Rahman and Muhammad Owais Chughtai

Department of Energy and Environment, Hamdard University, Karachi, Pakistan.

Water samples were collected from river Indus over 591 km^2 from Kashmor to Keti Bandar/Shah Bandar in the province of Sindh, Pakistan, during 2008 and 2009 on seasonal bases. These samples were analyzed for 12 water quality variables including physical and chemical parameters. Then correlation study was carried out and correlation co-efficient "r" was determined using correlation matrix to identify the highly correlated and interrelated parameters. Regression model were developed to test the significant 'r' and P-value test was carried out to test the significance of the pair of parameters. F-test was also used to examine the joint-effects of several independent variables without taking into account the separate effects of each variable. The comparison of the observed and predicted values of different parameters using regression equation suggested that the regression model provide useful means of rapid and easier monitoring of water quality of a river in a region.

Key words: Regression model, correlation coefficient, river indus, monitoring, physico-chemical parameters.

INTRODUCTION

Water is undoubtedly the most precious natural resource that exists on our planet. Although we recognize this fact, we disregarded it by polluting our rivers, lakes and oceans. It is the basic duty of every individual to conserve water resources (Jothivenkatachalam et al., 2010; Vega et al., 1998). The riverine system is most important resources of water supply in different countries of the world. At the source of a river, the water is relatively pure as it flows towards downstream. In Pakistan riverine system are getting polluted day by day. A primary reason for this is that all three major sources of pollution (industry, agriculture and domestic) are concentrated along the river belt and work together to reduce water quality which is a cause of alarm.

According to world wide fund for nature (WWF), five rivers in Asia serving 870 million people are among the most threatened in the world, as dams, water extraction and climate change all take their toll. The Indus River is among the top five most threatened river basins. Various studies were carried out to investigate the quality of Indus River water (Sadia et al., 2013; Ali et al., 2004; Tassaduqe et al., 2003; Khuhawar et al., 2000).

The environment, economic growth and development of Pakistan especially in province of Sindh are highly influence by water, it regional and seasonal availability, and quality of surface water and microbiological parameters. A few number of research articles are available regarding the analysis of water quality data using regression techniques (Mulla et al., 2007; Patil and Patil, 2010, 2011; Kumar and Sinha, 2010; Obiefuna and Orazulike, 2010; Mehta, 2010; Sami et al., 2011; Jain and Sharma, 2000; Joarder et al., 2008). These parameters are closely interlinked. All the research so far reported on river water quality of Pakistan based on physico-chemical analysis, no attempt has been made to predict the river water quality by using regression analysis.

In the present study, river water quality was determined by measuring the concentration of some physico-

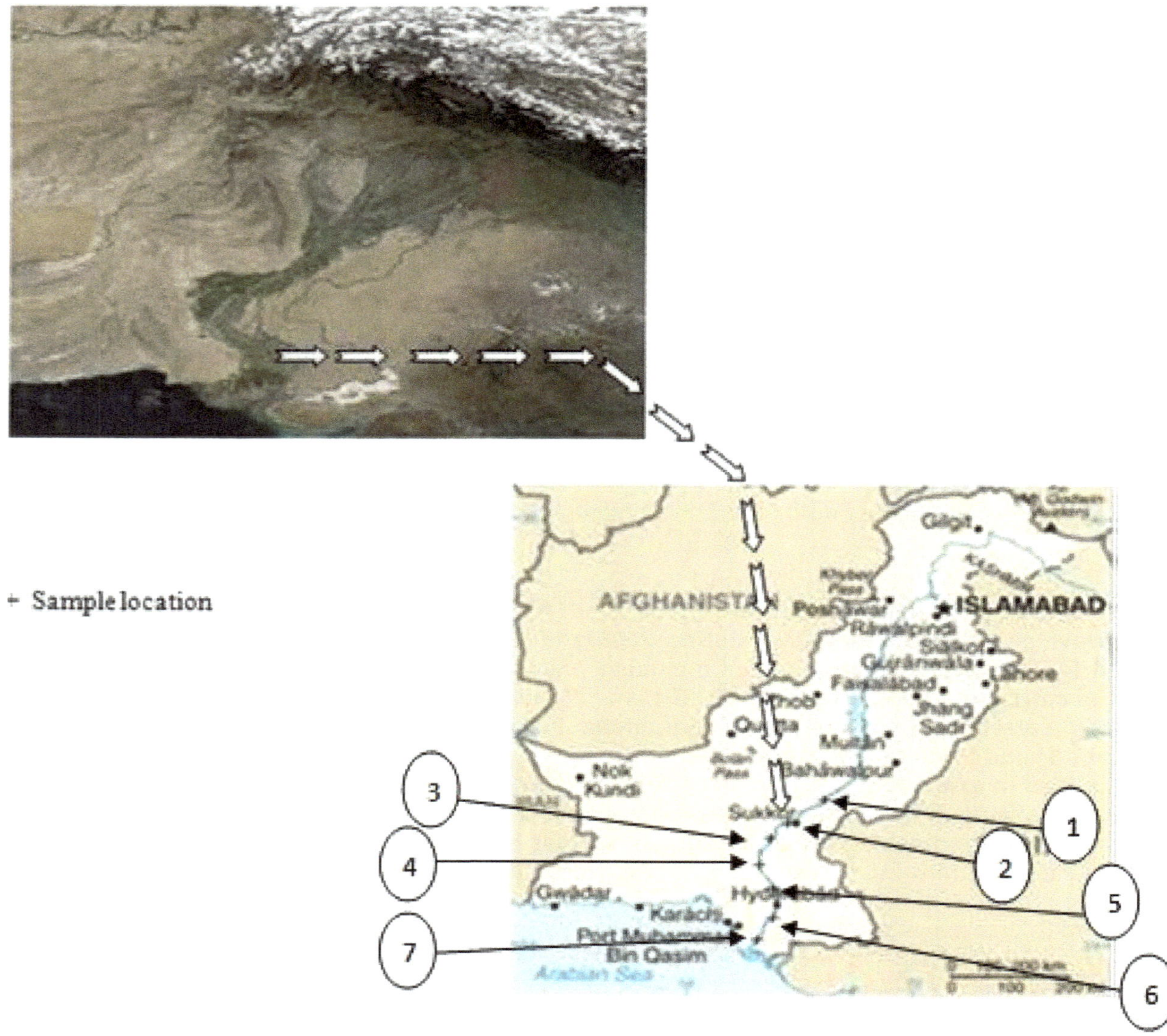

Figure 1. Location map. 1. Guddu Barrage; 2. Sukkur barrage; 3. Dadu-Moro Bridge; 4. Indus link 1 km downstream from Latifabad; 5. River Indus after Indus link outfall 6. Kotri Barrage at main bridge; 7. Sajawal bridge.

chemical parameters and comparing them with drinking water standards (WHO). Regression equations for the parameters having significant correlation were used to estimate the concentration of other constituents. Regression model were developed to test the significant correlation coefficient to overall water quality relative concentration of various pollutants in water to provide necessary clue for implementation of rapid water quality management.

MATERIALS AND METHODS

Water quality parameters of Indus River were monitored at seven different sampling points: Guddu Barrage; Sukkur barrage; Dadu-Moro Bridge; Indus link 1 km downstream from Latifabad; River Indus after Indus link outfall ; Kotri Barrage at main bridge and Sajawal bridge, considering sampling locations (receiving bulk quantity of effluent either by municipal/industrial/agriculture return flow) (Figure 1).

Chemical analysis

The samples were collected using boat from the middle of the flow within 3-9 inches from the surface of water. Two to four sub samples of equal volume were also collected from vertical section. The samples were mixed well and saved in a 2.5 L glass bottle. The samples were analyzed for physicochemical parameters such as alkalinity, bi-carbonate, calcium, pH, conductivity, total dissolved solid, and total suspended solid, chloride, chemical oxygen demand, biological oxygen demand, sulfate, sodium, potassium and magnesium using standard methods (APHA, 2005). Results obtained were subjected to regression statistical analysis using Statistical Package for social Scientist (SPSS) 14th version.

Table 1a. Estimated values of water quality parameters and WHO prescribed limits (pre-monsoon).

Parameter	ST-1	ST-2	ST-3	ST-4	ST-5	ST-6	ST-7	St.Dev	WHO limit
Conductivity	0.381	0.323	0.323	0.417	0.380	0.396	0.494	0.059	0.3
pH	7.7	7.5	7.6	7.8	7.6	7.7	7.7	0.098	7 to 8.5
Alkalinity	71.5	61.1	77	89	91.2	45	76	16.021	100
BOD	28	24	22	33	57	21.3	39	12.711	06
COD	67	56	53.3	41.7	131	50	93	31.449	10
TSS	324	321	187	63.3	167	60.3	24	123.342	500
TDS	220.3	200.5	224.7	233.7	237.1	207.1	287	28.372	500
Cl^{-1}	40	38.7	38.4	39.3	39.3	37	50.3	4.454	200
HCO_3^{-1}	87.3	74.7	93.6	93.3	92.7	73.0	97.7	9.798	-
SO_4^{-2}	30.3	28	29.3	33.3	35	37.3	53	8.520	-
Ca^{+2}	35.3	29.3	35.6	37	37.7	37.5	40	3.358	100
Mg^{+2}	5.2	6	5.9	6.1	6	2.9	8.5	1.645	-
Na^{+1}	21.8	22	21.6	22.1	23	24.6	31.3	3.476	-
K^{+1}	1.9	1.8	1.8	2.2	2.2	3	6	1.513	-

Statistical screening

The regression analysis was used to estimate two water quality parameters to describe realistic ground water situations. In addition, this analysis attempts to establish the nature of the relationship between the variables and thereby provides a mechanism for prediction or forecasting (Kumar and Sinha, 2010; Agarwal and Sexena, 2011; Mulla et al., 2007). To find the relationship between two physicochemical parameters X and Y, the Karl Pearson's correlation coefficient, r is used and it is determined as follows:

$$r = \frac{n\Sigma xy-(\Sigma x.\ \Sigma y)}{\sqrt{[n\Sigma x^2-(\Sigma x)^2][n\Sigma y^2-(\Sigma y)^2]}} \quad (1)$$

where, n = number of data points; x = values of X-variable; y = values of Y-variable.

To evaluate the straight-line by linear regression, the following equation of straight line can be used:

$$Y = a X + b \quad (2)$$

where, Y = dependent variable ; X = independent variable ; a = slope of line; b = intercept on y-axis

$$a = \frac{n\Sigma xy-(\Sigma x.\ \Sigma y)}{n\Sigma x^2-(\Sigma x)^2} \quad (3)$$

and

$$b = \overline{y} - a\,\overline{x} \quad (4)$$

Where, x = arithmetic mean of all values of x; y = arithmetic mean of all values of y.

For good correlation, value of r should be between $-1< r < 1$.

RESULTS AND DISCUSSION

The values of physicochemical parameters, standard deviation and WHO prescribed limits are presented in Table 1a and b during pre and post-monsoon, respectively. The relationship between alkalinity with HCO_3^{-1} is highly correlated and with BOD and Mg^{+2} are moderately correlated where as alkalinity is weakly correlated with COD and TDS during pre-monsoon (Table 2a).

The relationship between alkalinity with Na^{+1}, Ca^{+2} and TDS during post-monsoon which give correlation co-efficient are equal to 0.448, 0.546 and 0.531, respectively (Table 2b) showing less moderate to moderate as values of 'r' lies between -1 and +1.

The terms alkalinity and hardness are often used interchangeably when discussing water quality. These aggregate properties of water share some similarities but are distinctly different. Alkalinity is a measure of the acid-neutralizing capacity of water.

Alkalinity in most natural waters is due to the presence of carbonate (CO_3^{-2}), bicarbonate (HCO_3^{-1}), and hydroxyl (OH^{-1}) anions. However, borates, phosphates, silicates and other bases also contribute to alkalinity if present. This property is important when determining the suitability of water for irrigation and/or mixing some pesticides and when interpreting and controlling wastewater treatment processes.

In the present study, the alkalinity values were maximum during pre-monsoon. This may be attributed to increase in the rate of organic decomposition during which CO_2 is liberated, which reacts with water to form HCO_3^{-1}, thereby increasing total alkalinity in pre-monsoon. The increased alkalinity during pre-monsoon was due to the concentration of nutrients in water. The decrease was due to dilution caused by the rainwater during monsoon. Similar observation was reported by

Table 1b. Estimated values of water quality parameters and WHO prescribed limits (post-monsoon).

Parameter	ST-1	ST-2	ST-3	ST-4	ST-5	ST-6	ST-7	St.Dev	WHO limit
Conductivity	0.377	0.408	0.342	0.378	0.415	0.330	0.119	0.102	0.3
pH	7.8	7.8	7.8	7.4	7.6	7.3	7.4	0.219	7 to 8.5
Alkalinity	63.5	67.5	70.8	62.4	49.3	53	75.4	9.336	100
BOD	14.3	25.2	18	35.6	36	15.9	9.7	10.447	06
COD	38	67.3	40.3	98.3	67.7	46	21	25.572	10
TSS	239.7	260.3	160.3	153.3	321.3	210.3	3.3	101.735	500
TDS	229.2	228.6	234.5	399.2	276.7	227.8	805.3	213.055	500
Cl^{-1}	42.3	40.7	42.3	35	81.4	35.7	35.7	16.484	200
HCO_3^{-1}	75.7	77	80.3	82.5	97.1	80	101.7	10.239	-
SO_4^{-2}	40	43	39	38.7	89.4	34	87.3	24.262	-
Ca^{+2}	38	37.1	38.7	36.8	37.5	42	80	15.838	100
Mg^{+2}	8.3	7.6	8.4	6.3	71.6	5.2	22.7	24.11	-
Na^{+1}	20.6	20.3	22	19.7	45.3	18.7	129.3	40.731	-
K^{+1}	4	2.7	3.7	18.3	15.1	3.3	6.3	6.365	-

Table 2a. Statistical regression analysis data (pre-monsoon).

R -Equation	I-Var.	D-Var.	n	r	F value	S.E
Alkalinity =30.650 x pH - 161.720	pH	Alkalinity	7	0.187	0.181	17.241
Alkalinity = 0.234x COD+56.499	COD	DO	7	0.460	1.342	15.583
Alkalinity = 0.880x BOD + 44.785	BOD	DO	7	0.698	4.748	12.569
Alkalinity= - 0.012 x TSS + 74.926	TSS	DO	7	-0.092	0.043	17.476
Alkalinity = 0.278 x TDS +8.905	TDS	DO	7	0.493	1.607	15.267
Alkalinity = 45.23 x Conductivity+55.44	Conductivity	DO	7	0.166	0.141	17.307
Alkalinity = 0.841x Cl^{-1} + 38.97	Cl^{-1}	DO	7	0.234	0.289	17.063
Alkalinity = 1.391 x HCO_3^{-1} - 48.66	HCO_3^{-1}	DO	7	0.850	13.070	9.232
Alkalinity = 0.069 x SO_4^{-2} + 70.55	SO_4^{-2}	DO	7	0.037	0.007	17.538
Alkalinity = 1.421 x Ca^{+2} + 21.74	Ca^{+2}	DO	7	0.298	0.487	16.754
Alkalinity = 5.952 x Mg^{+2} + 38.450	Mg^{+2}	DO	7	0.611	2.982	13.890
Alkalinity = -0.359 x Na^{+1} + 81.512	Na^{+1}	DO	7	-0.078	0.031	17.497
Alkalinity = -0.531x K^{+1} + 74.404	K^{+1}	DO	7	-0.050	0.013	17.528

Table 2b. Statistical regression analysis data (Post-monsoon).

R-Equation	I-Var.	D-Var.	n	r	F value	S.E
Alkalinity = 11.134 xpH-21.328	pH	Alkalinity	7	0.262	.367	9.871
Alkalinity = -0.153 x COD +71.379	COD	DO	7	-0.418	1.057	9.292
Alkalinity=-0.477 x BOD +73.664	BOD	DO	7	-0.533	1.989	8.650
Alkalinity = -0.067 x TSS + 76.119	TSS	DO	7	-0.735	5.868	6.937
Alkalinity = 0.023 x TDS +55.142	TDS	DO	7	0.531	1.966	8.664
Alkalinity = -54.03xConductivity+81.4	Conductivity	DO	7	-0.588	2.645	8.271
Alkalinity = -0.341 x Cl^{-1} + 78.395	Cl^{-1}	DO	7	-0.603	2.851	8.161
Alkalinity = 0.25 x HCO_3^{-1} + 61.024	HCO_3^{-1}	DO	7	0.027	0.004	10.224
Alkalinity = -0.010 x SO_4^{-2} + 63.662	SO_4^{-2}	DO	7	-0.026	0.003	10.224
Alkalinity = 0.322 x Ca^{+2} + 48.880	Ca^{+2}	DO	7	0.546	2.119	8.571
Alkalinity= -0.192 x Mg^{+2} + 66.689	Mg^{+2}	DO	7	-0.495	1.620	8.888
Alkalinity = 0.103 x Na^{+1} + 59.083	Na^{+1}	DO	7	0.448	1.254	9.145
Alkalinity = -0.562 x K^{+1} + 67.415	K^{+1}	DO	7	-0.383	0.860	9.447

(Brion, 1973; Wetzel, 1983).

Alkalinity is important because it buffers the pH of water within the system. Without this buffering capacity, small additions of acids or bases would result in significant changes of pH, which could be deleterious for aquatic life. Alkalinity also influences the distribution of some organisms within aquatic systems. The pH range was 7.5 to 7.8 during pre-monsoon and 7.3 to 7.8 during post-monsoon. The pH of most natural waters falls in the 6 to 9 range because of the bicarbonate buffering.

In the present study, the TDS and TSS values were maximum during post-monsoon and minimum during pre-monsoon. High values of TSS during post-monsson may be due to siltation, deterioration, heavy precipitation and mixing runoff rain water which carried mud, sand, etc. Similar observations were made by Jawale and Patil (2009) and Salve and Hiware (2006).

COD is the amount of chemical oxidant required for the oxidation of the organic matter present in the waste. River Indus receives high amount of organic matter which generally originate from domestic and industrial effluents on the bank of Indus River. In the present study, the COD value vary from 41.7 to 131 during pre-monsoon and 21 to 98.3 during post-monsoon. For biodegradation, this organic waste requires oxygen, causing significant depletion of dissolved oxygen in river Indus water. The oxygen depletion not only effect biotic community of the river but also affects the purification capacity. High value of COD indicates that river received high amount of organic matter during pre monsoon. Hence the low value of alkalinity indicates that the compound responsible for decrease alkalinity is working as chemical oxidant for COD and hence significant increase in the value of COD. As regression equation, $Y = 0.477\ X + 73.664$ and $Y = -0.153\ X + 71.379$ was used to estimate the values of BOD and COD, respectively. It also helps to find the value of BOD/COD ratio to analyze the extent of pollution and biodegradability of water. The high value of BOD suggest that oxygen present in water is consumed by aerobic bacteria which makes fish, blind dolphins and other aquatic species to find it difficult to survive during pre monsoon.

REFERENCES

Agarwal A, Sexena M (2011). Assessment of pollution by physico-chemical parameters using regression analysis: A case study of Gagan River at Moradabad-India. Adv. Appl. Sci. Res. 2 (2):185-189.

Ali M, Salam A, Ahmed N, Bakhtyawar AK, Khokhar MY (2004). Monthly Variation in Physico-Chemical Characteristics and Metal Contents of Indus River at Ghazi Ghat Muzaffargarh Pakistan. Pak. J. Zool. 36(4):295-300.

APHA (2005). Standard methods for the examination of water and waste water. American Public Health Association, Washington D.C. (21st edn).

Brion M (1973). The influence of environmental factors of the distribution of fresh water algae on experimental Study. The role of pH, Carbon dioxide and biocarbonate system. J. Ecol. 6(1):157.

Jain CK, Sharma MK (2000).Regression analysis of groundwater quality data of Sagar District Madhya Pradesh. Indian J. Environ. Health 42(4):159-168.

Jawale AK, Patil SA (2009). Physico-chemical characteristics and Phytoplanktons abundance of Mangrul Dam Dist-Jalgaon Maharashtra. J. Aqua. Biol. 24(1):7-12.

Joarder M A, Raihan F, Alam J B, Hasanuzzaman S (2008). Regression analysis of Groundwater quality data of Sunamjang District, Bangladesh. Int. J. Environ. Res. 2(3):291-296.

Jothivenkatachalam K, Nithya A, Chandra Mohan S (2010). Correlation analysis of drinking water quality in and around Perur block of Coimbatore District, Tamil Nadu, India. Rasayan J. Chem. 3(4):649-654.

Khuhawar MY, Leghari SM, Bozdar RB, Mastoi GM, Jahanger TM (2000). Variation in water quality Assessment of River Indus from Kotri Barrage to Arabian Sea. Proc. National Seminar on Drainage in Pakistan, Mehran University of Eng. Tech. Jamshoro August 16-18.

Kumar N, Sinha DK (2010). Drinking water quality management through correlation studies among various physicochemical parameters: a case study. Int. J. Environ. Sci. 1(2):253-259.

Mehta KV (2010). Physicochemical characteristics and statistical study of groundwater of some places of Vadgam taluka in Banaskantha district of Gujarat state (India). J. Chem. Pharm. Res. 2(4):663-670.

Mulla JG, Farooqui M, Zaheer A (2007). A correlation and regression equations among water quality parameters. Int. J. Chem. Sci. 5(2):943-952.

Obiefuna GI, Orazulike DM (2010).Physicochemical characteristics of groundwater quality from Yola Area, Northeastern Nigeria. J. Appl. Sci. Environ. Manage. 14(1):5 - 11.

Patil VT, Patil PR (2010).Physicochemical analysis of selected groundwater samples of Amalner town in Jalgaon district, Maharashtra, India. Electron. J. Chem. 7(1):111-116.

Patil VT, Patil PR (2011).Groundwater quality of open wells and tube wells around Amalner town of Jalgaon, district, Maharashtra, India. Electron. J. Chem. 8(1):53-78.

Sadia A, Feroza HW, Imran Q , Muhammad Hamid SW, Tirmizi SA, Muhammad AQ (2013). Monitoring of anthropogenic influences on underground and surface water quality of Indus River at district Mianwali-Pakistan.Turk. J. Biochem. 38(1):25-30.

Salve BS, Haware CJ (2006). Studies on water quality of Wanparakalpa Reservoir Nagapur near Parli Vaijath Dist Beed Marathwada region. J. Aquat. Biol. 21(2):113-117.

Sami GD, Ahmed SW, Ahmed SB, Manal HO (2011). Linear correlation analysis study of drinking water quality data for AlMukalla City, Hadhramout, Yemen. Int. J. Environ. Sci. 1(7):692-1701.

Tassaduqe K, Ali M, Salam A, Latif M, Zahra T (2003). Study of the Seasonal Variations in the Physico Chemical and Biological aspects of Indus River Pakistan. Pak. J. Biol. Sci. 6(21):1795-1801.

Vega M, Prado R, Barrado E, Deban L (1998). Assessment of seasonal and polluting Effects on the quality of river water by exploratory data analysis. Water Res. 32 (2):3581- 3592.

Wetzel RG (1983). Limnology. Second Edition. Edited by Wetzel LG. Michigan State University. CRS College Publishing Philadelphia New York Chicago. p. 789.

10

Assessment of groundwater pollution by nitrates using intrinsic vulnerability methods: A case study of the Nil valley groundwater (Jijel, North-East Algeria)

BOUFEKANE Abdelmadjid[1] and SAIGHI Omar[2]

[1]Water and Environment Engineering Laboratory, Higher National School of Hydraulics, BP 31, 09000, Blida, Algeria.
[2]Laboratory of Geo-Environment, Faculty of Earth Sciences and Country Planning, University of Sciences and Technology, Houari Boumediene (U.S.T.H.B), El Alia, BP 32, Bab Ezzouar, 16111 Algiers, Algeria.

Scientists are deeply concerned with the state of vulnerability of groundwater reservoirs. It is a complex task because of the difficulties in determining the degree of pollution of the ground water. Many methods have been adopted like DRASTIC, GOD, SI, SINTACS, etc. The present article targets the determination of the vulnerability of groundwater reservoirs of a climatic Mediterranean region (Nil valley region, Jijel). The excessive use of fertilizers in agriculture has been the main reason behind the increase of the Nil valley groundwater pollution with nitrates (Jijel, North-East Algeria). In fact, the use of fertilizers in high quantities relatively to the needs of the plants lead to the leaching and infiltration of the excess fertilizers into groundwater which increases the nitrates percentage; as a result, the allowed norms of water consumption are exceeded. Relevant to this, the aim of this study was to assess the aquifer vulnerability caused by pollution with nitrates using DRASTIC, GOD and SI methods. The spatial distribution of the found nitrates in groundwater shows that the DRASTIC method is the most appropriate method in this case with a percentage of 71% vs. 54 and 63% for GOD and SI methods, respectively. Moreover, it was found that the studied water is characterized by a medium to high degree of vulnerability. Pertinent to this, it is highly recommended to find solutions to better protect and preserve the Nil valley groundwater.

Key words: Aquifer vulnerability, pollution, nitrate, agriculture, groundwater protection.

INTRODUCTION

Groundwater reservoirs are easily affected by pollution. The process is slow but its effects are very dreadful (Baghvand et al., 2010) (from the ground to the unsaturated zone). Thus, whatever the nature of the physical pollution (radioactivity), be it chemical (Mineral pollutants) (Lain et al., 2007 in Attoui et al.2012, Lake et al., 2003), organic (pesticides) (Worrali et al., 2004) or bacteriological (bacteria, viruses) (Schijven et al., 2010), the aquifers are affected.

In agricultural areas, in particular, an excessive use of fertilizers has directly or indirectly affected the groundwater quality (Huang et al., 2012). But beyond the quantitative aspect, it is also advisable to remain vigilant on the level of the water quality consumed by the populations (Diodato et al., 2013).

Nitrate is a common compound, naturally generated from the nitrogen cycle. However, anthropogenic sources have greatly increased the nitrate concentration, particularly in groundwater (Chand et al., 2011). Nitrate losses from non point agricultural sources, mainly originated by fertilizers application, have been recognized as one of the most serious threats for pollution of groundwater (Salemi et al., 2012). An improvement of knowledge is however essential to make the water services more powerful and to reinforce the policy for the access to safe water in the country (Diodato et al., 2013).

However, the prevention against groundwater pollution constitutes an important phase to which scientists are doing their best notably in studying the vulnerability of the groundwater. They therefore, created classical scientific methods (Etienne et al., 2009) and numerical (Boufekane and Saighi, 2010), to facilitate the identification of the state of these groundwaters and to control the pollutants in the reservoirs such as DRASTIC, GOD, SINTACS, etc. These different methods are presented under the form of numerical quotation systems based on the consideration of the different factors influencing the hydrogeological system (Rouabhia, 2004, in Attoui et al., 2012).

Prevention of aquifers pollution is considered as an important factor in the management of groundwater resources; also, the assessment of aquifer vulnerability by scientists is an essential factor which gives us solutions to protect groundwater resources. The Nil valley groundwater (Jijel, Algeria) which is characterized by a favorite vegetables and greenhouse gardening activity undergo a contamination by nitrates due to the use of fertilizers in agriculture. In order to prevent groundwater resources pollution, the present work studies the aquifer by using three parametric methods named DRASTIC, GOD and SI, by examining the surface activities which characterize the aquifer. The obtained vulnerability maps via the three methods are validated by comparison with the observed nitrates in the groundwater.

Description of the studied area

The studied area is located in the north-east of Algeria. The alluvial aquifer of this area forms part of the coastal plains region of Jijel (Figure 1); it covers an area of 83 km^2 and opens to the north of the Mediterranean Sea. It corresponds to the lower Nil valley and its two tributaries, Boukaraa valley (left bank) and Saayoud valley (right bank). This region is characterized by a few natural lakes and swamps, the most notable are downstream of the plain: Ghedir Beni Hamza in the north-east and Ghedir el Merdj in the north-west.

The maritime location of this region gives to it a mild and damp climate. The average air temperature is 17°C/year, while rainfall, relatively high, reaches 900 mm/year. In addition, the geological substratum of the area consists of gneiss and the schist. However, the parts of swallow, sedimentary formations mainly marly Oligocene age, Miocene and Pliocene cover these metamorphic facies. Finally, the depressions and valleys are filled with quaternary alluvial deposits which are interstation terraces aquifers.

The groundwater recharge is mainly directed by infiltration of rainfall and the low water situation by the various rivers which cross the plain. The aquifer forms part of the socio-economic development of the region by the exploitation of the domestic wells and boreholes (36 million m^3/year).

MATERIALS AND METHODS

Working environment

The study is based on the obtained measurements from the field surveys that were conducted during the hydrological year, 2010-2011 and supplemented by the compilation of the information collected from various technical services of the willaya of Jijel (National Agency Resources Water, Direction of the Water Resources, Direction of the Agricultural Services). The data was collected from:

1. Piezometric campaign conducted in September 2010,
2. Results from pump tests in 36 wells; in fact, these results allow to deduce the transmissivity values (T ranging between 10^{-3} m^2/s and 10^{-2} m^2/s) and permeability values (K = 10^{-4} to 10^{-3} m/s).
3. Data sheets and logs of the geological drilling.
4. Results of geophysical interpretations (maps of apparent resistivity, resistance transversal, cuts geoelectrical).
5. Cartographic documents on the scale 1/50 000 (geological map and soil map, topographic maps of Jijel, Sidi Merouane) and with the 1/25 000 (topographic maps of Jijel and Texanna).
6. Digital maps of land at 1/10 000 (Chekfa, Taher and EL Kennar),
7. Slope map and Digital Elevation Model (DEM).
8. The meteorological data (hydrological year 2010/2011) in order to assess the water balance and estimate the blade infiltrated water.

Data preparation

The documents of the different parameters allowing the map development of vulnerability by groundwater pollution have been created accordingly to values grid by using regular grid of 25 x 25 m by cutting the 83 km^2 of the study area (elementary units of this size).

Presentation of the used methods for vulnerability assessment

DRASTIC method

DRASTIC method was developed by the staff of the U.S. Agency for Environmental Protection USEPA (Aller et al., 1987). It allows the assessment of vulnerability of the vertical aquifer pollution caused by parametric systems. It is based on the estimation of 7 parameters in particularly, the percentage of the effective recharge, the soil type and the characteristics of the saturated and unsaturated zones of the aquifer. Each parameter, divided to an interval of significant values, is assigned by a numerical rating based on its growing importance in the vulnerability. These seven

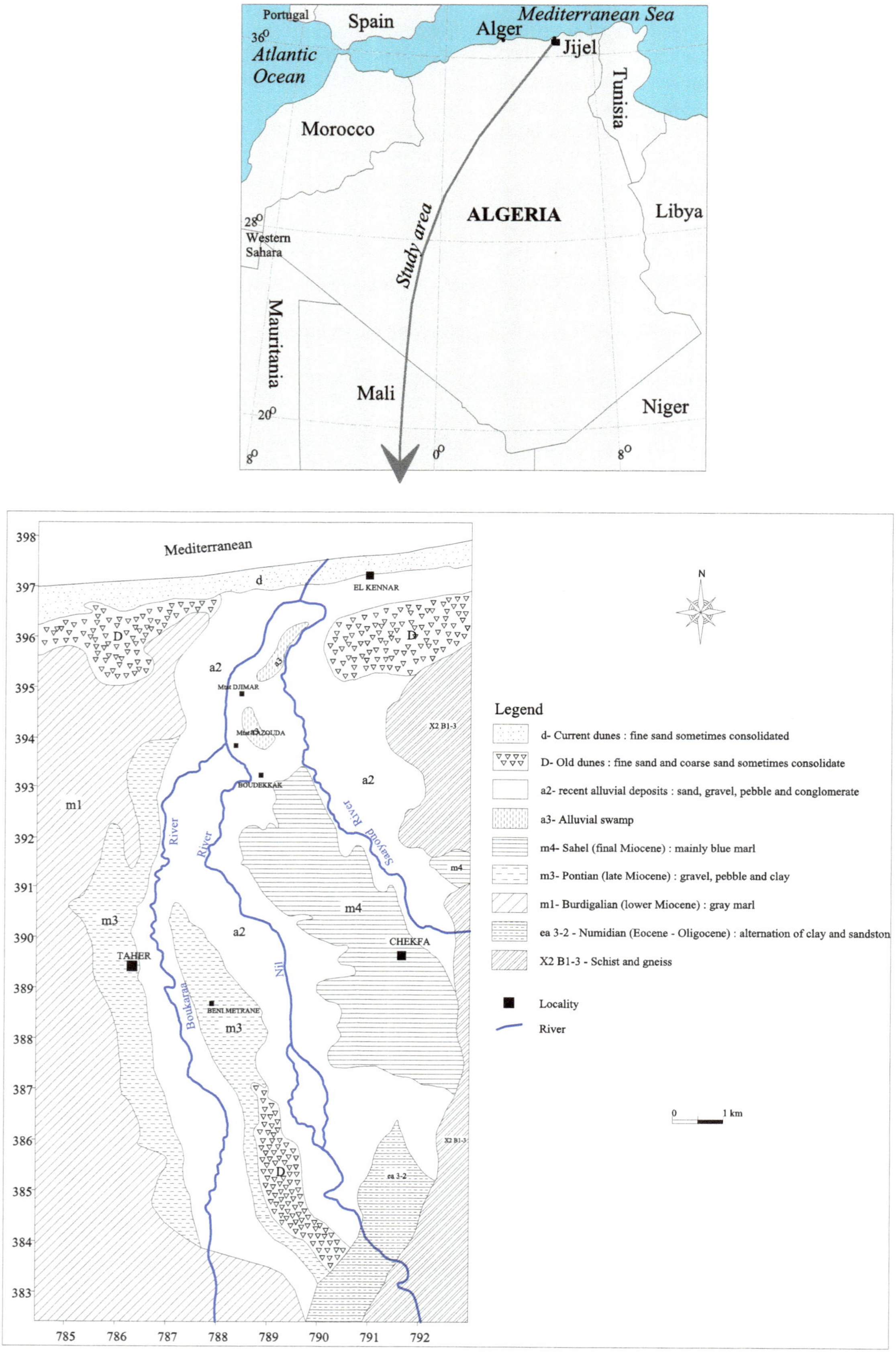

Figure 1. Map showing geology of the studied area.

Table 1. Weight settings of DRASTIC (Aller et al., 1987).

Symbol	Parameter	Weight
D	Water depth	5
R	Effective recharge	4
A	Middle aquifer	3
S	Soil type	2
T	Topography	1
I	Impact of the vadose zone	5
C	Hydraulic Conductivity	3

Source: Aller et al. (1987).

Table 2. Criteria of the vulnerability assessment by using DRASTIC method (Engel et al., 1996).

Class vulnerability	Low	Average	High	Very high
Index	<101	101-140	141-200	>200

Source: Engel et al. (1996).

parameters are used to define the different hydrogeological units, variously influenced by transport processes and attenuation of contaminants in the soil. Numerical value called a weight parameter, between 1 and 5 is assigned to each parameter and reflects its influence degree. Each parameter is listed based on the associated scores ranging from 1 to 10, the lowest score represents the conditions of the lower vulnerability contamination; also, to describe the vulnerability degree of each hydrogeological unit, the numerical value called DRASTIC vulnerability index should be determined.
The DRASTIC vulnerability index was calculated by addition of the different products (score × weight of the corresponding parameter):

$$\text{DRASTIC Index} = D_w D_r + R_w R_r + A_w A_r + S_w S_r + T_w T_r + I_w I_r + C_w C_r$$

Where: D, R, A, S, T, I and C are the seven parameters and the subscripts r and w are the corresponding rating and weights, respectively.

The partial index of each parameter is then calculated using the equation:

$$\text{Partial index} = \text{weight} \times \text{rating}$$

So, the DRASTIC index is defined by the scores of the all vulnerability parameters multiplied by their respective weights.

The DRASTIC weight parameters, defined according to Aller et al (1987), are shown in Table 1.

The extent of vulnerability of the aquifer hydrogeological is defined by the indices DRASTIC. These indices are divided into four classes and vary between the extreme values ranging from 23 to 226 (Engel et al., 1996), (Table 2).

GOD method

This method is characterized by a rapid assessment of the aquifer vulnerability; it was developed by Foster in 1987 and 1998 (Ferreira, 2004) for studying the vulnerability of the aquifer against the vertical percolation of pollutants through the unsaturated zone, without considering their lateral migration in the saturated zone.

The approach used in this model takes in consideration three parameters:

1. Groundwater occurrence
2. Overall aquifer class
3. Depth table of the groundwater.

The GOD index which is used to evaluate and map the aquifer vulnerability caused by the pollution, was calculated by multiplication of the influence of the three parameters using the followed equation:

$$\text{GOD Index} = Cl \times Ca \times Cd$$

Where: Ca is the type of aquifer, Cl is the lithology of the unsaturated zone and Cd is the depth on the water surface.

The GOD indexes are divided into five classes and vary between the extreme values ranging from 0 to 1 (Table 3).

SI method

The SI method (Susceptibility Index) was developed in 2000 in Portugal by Ribeiro; it is used to assess the vulnerability of vertical agricultural pollution generated mainly by nitrates and secondarily by pesticides. This method is based on five parameters:

The first four (D: water depth, R: effective recharge, A: middle aquifer, T: topographic slope of the land) are identic with those used in the DRASTIC method. The fifth parameter (LU) represents Land. The dimensions of five parameters and their corresponding individual classes are given in Table 4.

The lands are classified accordingly to the Corine Land Cover (European Community, 1993). Each land class is assigned by a factor land use (LU) ranging from 0 to 100 (Table 5). The assigned values to the parameters of the different classes have been multiplied by 10 to facilitate the reading results, they range from less

Table 3. Intervals values of GOD index and corresponding classes.

Class vulnerability	Very low	Low	Average	High	Very high
Index	0 - 0.1	0.1 - 0.3	0.3 - 0.5	0.5 - 0.7	0.7 - 1

Source: Murat et al. (2003).

Table 4. SI method parameters and their corresponding weight.

Symbol	Parameter	Weight
D	Water depth	0.186
R	Effective Recharge	0.212
A	Middle aquifer	0.259
T	Topography	0.121
LU	Land Use	0.222

Source: Ribeiro (2003).

Table 5. Main soil occupation classes and their correspondant land use (LU) values.

Land classification according to the Corine Land Cover	LU (land use factor)
Industrial waste discharges, landfill	100
Irrigated rice fields	90
Caries, shipyards, open-air mines	80
Areas covered artificial, green spaces	75
Permanent crops (vine yards, orchards, olive, etc.)	70
Discontinuous urban areas	70
Pasture and agro-forestry	50
Aquatic environments (salt marshes, salinas, etc.)	50
Forests and semi-natural zones	0

Source: Ribeiro (2000) .

Table 6. Criteria for the evaluation of vulnerability in the SI method.

Class vulnerability	Low	Average	High	Very high
Index	< 45	45 - 64	65 - 84	85 - 100

Source: Ribeiro (2000)

vulnerable to most vulnerable. The assigned weights to the SI parameters vary from 0 to 1 depending on the size of the parameter in vulnerability.

According to their index values, the SI method has four degrees of vulnerability (Table 6).

Validation of vulnerability maps

The developed vulnerability maps by each method are then compared with the results of the hydrochemical analyzes of pollutants. In this study, we are only interested in studying the nitrates percentage and their distribution in the aquifer. For this purpose, 35 samples are analyzed and compared with respect to the nitrates map in the water with the distribution of the vulnerability classes obtained by the three methods.

RESULTS

DRASTIC method

The obtained values of the DRASTIC index vulnerability

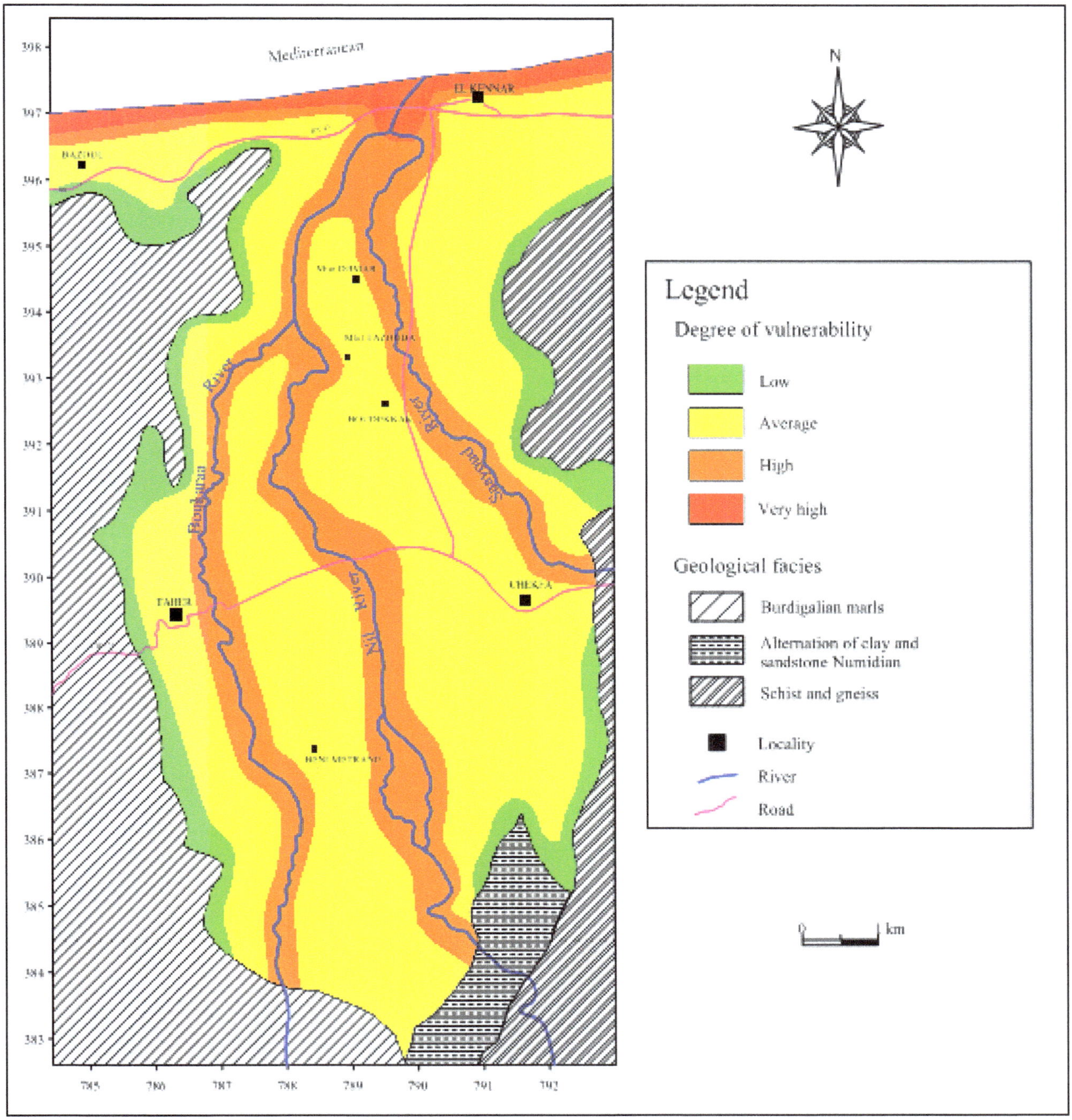

Figure 2. Aquifer vulnerability map (DRASTC method).

by this method vary from 85 to 210. Their spatial distribution allows distinguishing, within the studied area, four zones with different degrees of vulnerability (Figure 2), closely related to the hydrographic network.

a. The zone with very high vulnerability at the coast affects more particularly the mouth of Nil valley. It occupies around 3% of the studied plain. This high degree of vulnerability can be explained by the shallow aquifer as well as the existence of a permeable massif of dunes, allowing the infiltration of pollutants present at the surface.

b. The zone with high vulnerability encompasses the previous one and penetrates inside the land along the topographic drains of the Nil valley and its two tributaries, Saayoud and Boukaraa. This second zone represents 26%

26% of the plain surface area, and juxtaposes the extended areas of the major waterways, where the major sediments composing the alluvial terraces are dominant, and the piezometric level is shallow (1.5 to 5 m). Such characteristics favor the penetration of pollutants to the aquifer, particularly in the depressed parts.
c. The zone with an average vulnerability covers about two-thirds (2/3) upstream the plain, where, despite the importance of gravelly and sandy fraction of the aquifer, it is less protected from pollution because of its shallow depth from the surface (generally more than 5 m).
d. The zone which is a bit vulnerable corresponds to the edges of the plain and covers 10% of the area. Its low vulnerability is due first to the depth of the aquifer (about 10 m) and secondly, to the low permeable covering soil and the unsaturated zone.

GOD method

GOD indices show values ranging from 0.16 to 0.54. Their low dispersion allow differentiating just three zones of vulnerability (Figure 3), whose spatial distribution, somewhat different from the DRASTIC case, is as follows:

a. The highest degrees of vulnerability are located, as before, in coastal areas, especially at the Nil river. However, they remarkably invade inland up along Boukraa and Saayoud valley, around 4 km. This area covers 18% of the plain area and, as already mentioned, corresponds to shallow-aquifer areas (1.5 to 5 m), associated with a highly permeable sediments at surface (dune, sandy and gravelly aquifer, etc.).
b. Moderate vulnerability zones contain the previous one and go up further along the valley, until covering 47% of the plain surface area. The aquifer is situated between 5 and 9 m deep and consists of gravel and sand that promote the spread of pollutants, despite the sporadic presence, in the airy area, clay and marly intercalations.
c. Regarding DRASTIC case, though somewhat wider, low-vulnerability band surrounds the plain. On the other hand, the valley intervals also indicate low vulnerability levels because the underground water there is more than 12 m deep, and saturated and unsaturated zones are a bit permeable. The cumulative surface area of low vulnerability areas reaches 35%.

SI method

As a result of its analogy with DRASTIC method (DRST parameters identity), it is sufficient in this case to define the land use parameter (LU) in order to immediately develop the vulnerability map, SI. The results reveal index values ranging from 33 to 91, defining four (4) areas on the vulnerability map (Figure 4). The configuration of this map exhibits similarities with both GOD map in highly vulnerable areas (coast and Nil valley mouth) and the Drastic map, concerning the rest of the plain. This third map highlights the particular behavior of the major valley beds which act as preferential paths of contamination. According to a descending vulnerability degree, the four individualized cover areas of 21, 25, 44 and 10% respectively.

Vulnerability maps validation to nitrate pollution

Nitrate spatial distribution into groundwater

Nitrate concentrations spatial variation into groundwater depends on agricultural activities, covering soil lithology and airy area thickness. Recorded contents in 35 samples (19 drillings and 16 wells) collected in September 2010 (Figure 5) show the following:

1. The highest values are observed in the northeast corner of the plain. These values frequently exceed the threshold of 50 mg/l tolerated by the WHO (World Health Organization) for drinking water. Nitrate high-concentration in this corner is attributed to dense market and greenhouse gardening activity. Furthermore, this northeast area is characterized by shallow groundwater and permeable aquifer.
2. Outside this corner area, nitrate contents progressively decrease upstream, while still remaining relatively high at the Nil valley mouth (40 mg/l) and along two of its tributaries, Saayoud and Boukeraa, revealing leaching bands of highly fertilized agricultural fields.
3. In the rest of the plain, nitrate concentrations are close to 20 mg/L. Traditional soil fertilization practices and clay covering soil, allow protection of groundwater against pollution. It is in this area, and more precisely, between Djimar and Tazouda, where the main field of water gathering area is found. Further upstream, nitrate levels are below 20 mg/l.

Nitrate distribution and vulnerability maps comparison

Comparison of nitrate-concentration distribution map and that vulnerability levels obtained by each approach, shows the following:

a. DRASTIC map case

i. The three samples whose concentrations exceed 45 mg/l correspond to average-vulnerability area;
ii. From nine samples whose contents vary between

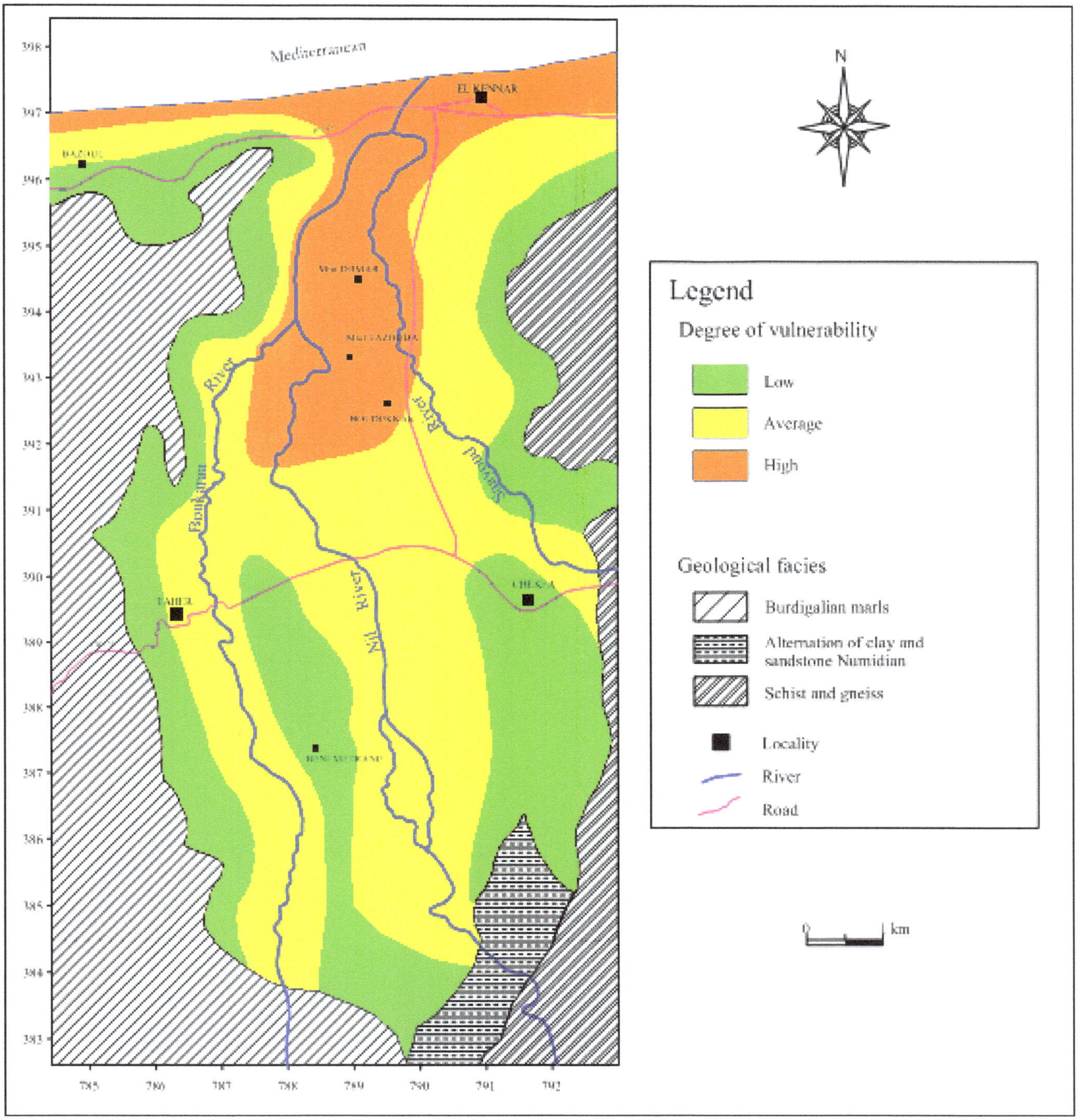

Figure 3. Aquifer vulnerability map (GOD method).

15 and 30 mg/l, two (2) samples coincide with average-vulnerability area, and seven (7) left, with high-vulnerability zone;

iii. Concerning the twenty-two (22) samples with contents below 15 mg/l, water point lies in the low-vulnerability zone, 15 in the average and six (6) in the high one.

b. GOD map case

i. The three samples with contents exceeding 45 mg/l correspond to the average-vulnerability zone;

ii. The nine samples exhibit contents ranging between 15 and 30 mg/l, equally spread among the three vulnerability

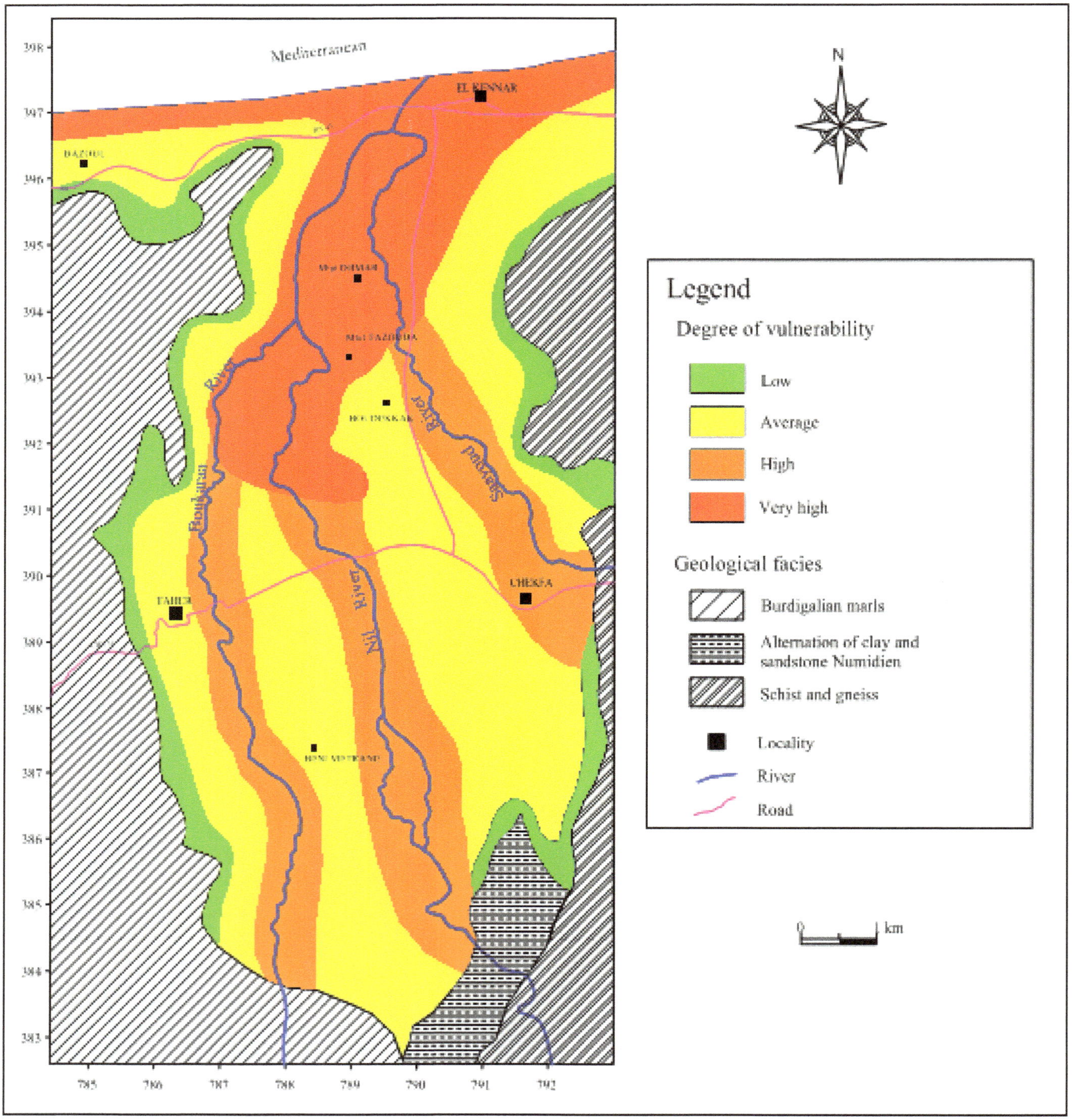

Figure 4. Aquifer vulnerability map (SI method).

zones being identified by this method (low, average and high) ;

iii. As for the 22 samples with low concentrations, five samples belong to low-vulnerability area, eight to the average, and nine to the high one.

c. SI map case

i. Of the three samples with contents above 45 mg/l, only one sampling point belongs to very high vulnerability zone. However, for the others, they fall within medium-

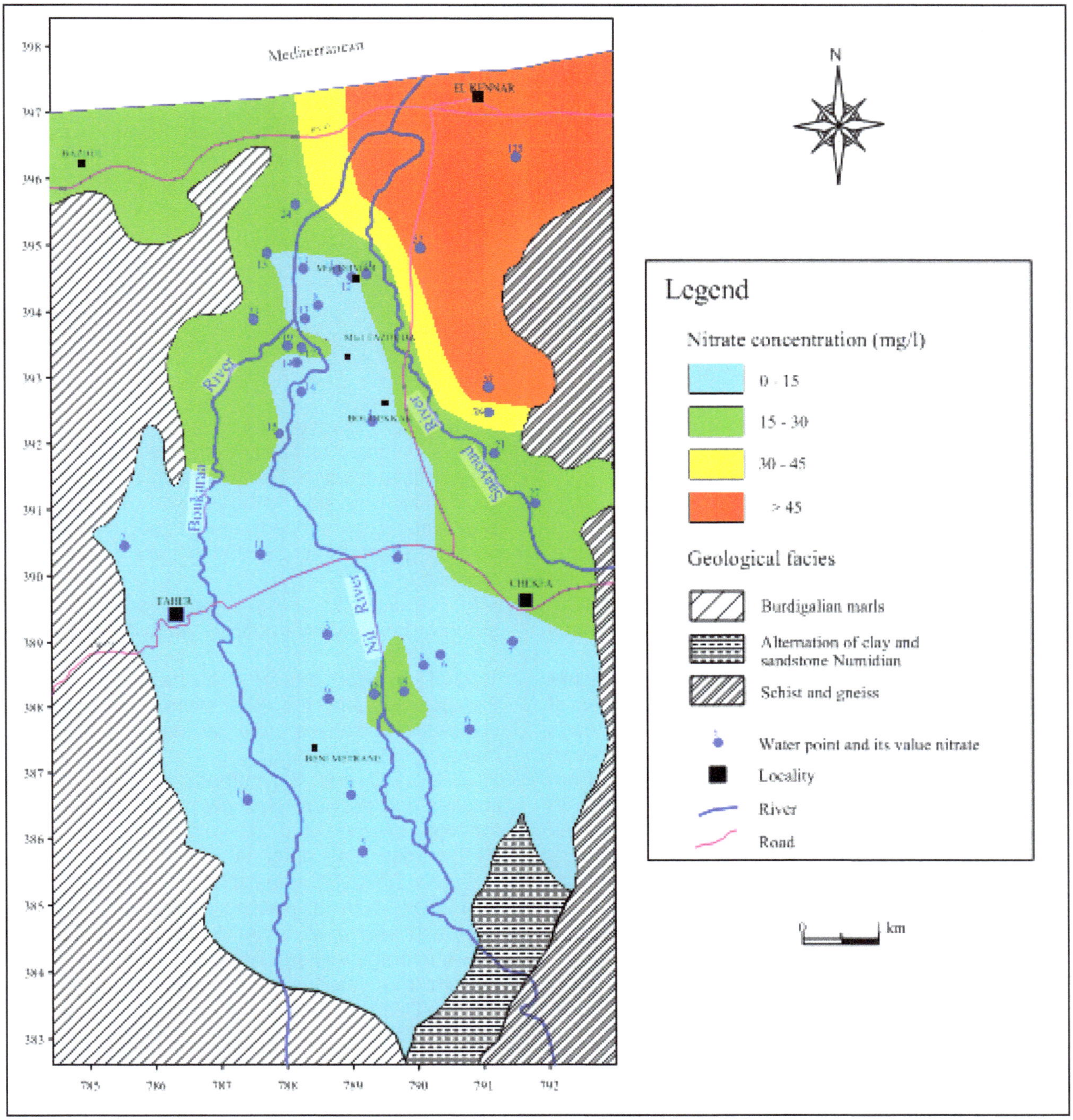

Figure 5. Nitrate distribution map.

vulnerability area;

ii. Of the nine samples displaying contents varying between 15 and 30 mg/l, the third belong to very high vulnerability area, four fall within the high, and the others in the average one;

iii. Finally, concerning 22 samples with low contents, eight samples correspond to high-vulnerability area and 13 to the average one.

Table 7 shows, firstly, that average-vulnerability class is predominant whatever the approach used. Secondly, the table results reveal that DRASTIC method seems to be the most reliable in reflecting reality on the ground, with coincidence rate of 71% (25 of 35 values), against 63% for SI method (22 values of 35) and 54% for GOD method

Table 7. Nitrate spatial distribution into groundwater, with respect to vulnerability classes being defined by the three methods.

Vulnerability	DRASTIC Method				GOD Method				SI Method			
	Samples number per nitrate content class (mg/l)											
	0-15	15-30	30-45	> 45	0-15	15-30	30-45	> 45	0-15	15-30	30-45	> 45
Low	1	0	0	0	5	3	0	0	1	0	0	0
Average	15	2	1	3	8	3	1	3	13	2	0	2
High	6	7	0	0	9	3	0	0	0	4	1	0
Very high	0	0	0	0	0	0	0	0	8	3	0	1
Total	22	9	1	3	22	9	1	3	22	9	1	3

Vulnerability: Low, average, high, very high. Nitrate concentration (mg/l): 0-15, 15-30, 30-45, > 45. Samples number: 0, 1, 2, 6, 15.

method (19 values of 35).

DISCUSSION

This study was done in an area clearly defined by natural conditions and hydrogeological limitations. However, the studied field is characterized by intensive agricultural activity, using chemical fertilizers which are potential sources of groundwater pollution.

The combined use of three approaches allows better understanding of mechanism and representativeness of pollution vulnerability in the examined area. This leads also to the distinction, depending on the case, between sometimes three, sometimes four classes, ranging from low to very high vulnerability degree. The difference in classes' number is linked to the fact that class boundaries and dimensions that are assigned to different parameters are not absolute, but instead relative. This implies that standard classes' limits may not reflect reality on the ground where a class can encompass different hydrogeological units within. Moreover, different works in literature showed that the limits may vary from one method to another, from one study to another or from one region to another. For example, GOD method does not give enough importance to recharge parameters and aquifer permeability.

Despite the deficiencies noticed in development of pollution-vulnerability maps using these intrinsic methods, reliability of results is slightly altered. Obtained results help to acquire a good idea regarding the most sensitive areas, and to subsequently prescribe necessary protection measures. Vulnerability maps validation would have been more representative if the number of nitrate measurements was greater and well distributed over the whole plain. Finally, analysis of other pollutants would have provided extra values for validation.

In the case of Nil valley groundwater, DRASTIC method proves to be the one that best reflects reality on the ground. However, it takes into account seven parameters including reliability that depends on data used for their implementation. Many of them as recharge, hydraulic conductivity and vadose zone impact, are either approximate or produced via scaling-up. This would consequently generate errors during parameters assessment process.

After validation using nitrate contents observed in groundwater, we have noticed the following:

1. GOD method provides results largely different from that generated by the other ones, and can define three vulnerability classes: low, average and high, with a matching rate of 54%.
2. DRASTIC and SI methods raise four vulnerability classes and provide relatively similar results, particularly for high, average and low-vulnerability areas (Table 7). By contrast, they differ in very high-vulnerability areas (3.1% for DRASTIC against 21.4% with SI).
3. Finally, with a matching rate of 71%, DRASTIC approach seems to better reflect reality on the ground where average vulnerability class (61%) predominates.

From the results of this work, the authors concluded that the DRASTIC method could be adequate for vulnerability mapping in this region. It confirms the results of the work already completed by several authors in the countries of the basin of the Mediterranean Sea. For example, Salemi et al. (2012) in Italy, Hamza et al. (2010) in Tunisia, Stigter et al. (2006) in Portugal.

Conclusion

Vulnerability-study results of Nil valley groundwater show that vulnerability degree increases upstream and downstream due to rising of the aquifer and the important sandy fraction in the airy zone. In addition, this relates to intensive anthropogenic soil-activity. High index areas are on the coast where groundwater is shallow and agricultural activity is dense. DRASTIC approach is more reliable in representing nitrate distribution into water, with a matching rate of 71%, against 63% in SI approach case and 54% in that of GOD. Nevertheless, whatever the method used, results reveal that Nil valley groundwater, mainly downstream, is highly vulnerable. Indeed, very high-vulnerability class covers 20% of the plain, while that having a high level reaches 30%. GOD method is less accurate and, less alarmist, however. Comparison of results provided by the three methods applied to the aquifer, clearly defined by its geometry and natural conditions, evidences discrepancies in the phenomenon assessment. Variations results from differences between the scoring systems of parameters.

Since it has been proved that this vital groundwater reservoir is vulnerable and present a risk of pollution, the following protection measures must be taken with the results of the vulnerability:

1. Protection perimeters of groundwater must be installed.
2. Protection perimeters supported by a general evacuation plan of used water must be installed.
3. Continuous monitoring of fertilizer use in agriculture.

REFERENCES

Aller L, Bennet T, Lehr JH, Petty RJ, Hachet G (1987). DRASTIC: A standardised system for evaluating groundwater pollution potential using hydrogeologic settings (EPA 600/2-87). Environmental Research Laboratory, Office of Research and Development, U.S. Environmental Protection Agency Report. pp. 622.

Attoui B, Kherci N, Bousnoubra H (2012). State of vulnerability to pollution of the big reservoirs of groundwater in the region of Annaba-Bouteldja (NE Algeria). Geographia Technica. 2:1-13.

Babiker IS, Mohamed AA, Tetsuya Hiyama, Kikuo Kato (2005). A GIS-based DRASTIC model for assessing aquifer vulnerability in Kakamigahara Heights, Gifu Prefecture, central Japan. Sci. Total Environ. 345:127-40.

Baghvand A, Nasrabadi T, Nabibidhendi G, Vosoogh A, Karbassi A, Mehradadi N (2010). Groundwater quality degradation of an aquifer in Iran central desert. Desalination 260(3):264-275.

Boufekane A, Saighi O (2010). Use of Gis to Characterize the Vulnerability to Pollution - Application to the Alluvial Aquifer of the wadi Nil (Jijel, Northeast Algeria). International Network Environmental Management Conflicts 1:267-274.

Chand S, Ashif M, Zargar MY, Ayub BM (2011). Nitrate Pollution :A Menace to Human, Soil, Water and Plant. Universal Journal of Environmental Research and Technology. 1:22-32.

Diodato N, Esposito L, Bellocchi G, Vernacchia L, Fiorillo F, Guadagno FM (2013). Assessment of the Spatial Uncertainty of Nitrates in the Aquifers of the Campania Plain (Italy). Am. J. Clim. Chang. 2:128-137.

Engel BA, Navulur KCS, Cooper BS, Hahn L (1996). Estimating groundwater vulnerability to non point source pollution from nitrates and pesticides on a regional scale. Int. Assoc. Hydrol. Sci. Publi. 235:521-526.

Etienne DK, Boyssoro H, Dibi B, Mahaman B, Biemi J (2009). Contribution Methods Intrinsic Vulnerability DRASTIC and GOD in The Study of Pollution by Nitrates in Bonoua Region (South-east of Ivory Coast). Eur. J. Sci. Res. 31(1):157-171.

Ferreira JPL, Oliveira MM (2004). Groundwater vulnerability assessment in Portugal. Geofísica Internacional. 43(4):541-550.

Foster SSD (1998). Groundwater recharge and pollution vulnerability of British aquifers:a critical review. In:Robins, N.S. (Ed.), Groundwater Pollution, Aquifer Recharge and Vulnerability. Geological Society, London, Special Publications. 130:7-22.

Hamza MH, Maâlej A, Ajmi M, Added A (2010). Validity of the vulnerability methods DRASTIC and SI applied by GIS technique to the study of diffuse agricultural pollution in two phreatic aquifers of a semi-arid region (Northeast of Tunisia). AQUAmundi - Am01009:57-64.

Huang T, Pang Z, Edmunds WM (2012). Soil profile evolution following land-use change: Implications for groundwater quantity and quality. Hydrol. Process 27(8):1238-1252.

Lake LR, Lovett A, Hiscock KM, Beston M, Foley A, Sunnenberg G, Evers S, Fletcher S (2003). Evaluating factors influencing groundwater vulnerability to nitrate pollution: developing the potential of Gis. J. Environ. Manage. 68(3):315-28.

Salemi E, Mastrocicco M, Colombani N, Aschonitis VG (2012). Comparison of Three Different Methods for Groundwater Intrinsic vulnerability Mapping in the Ferrara Province, Italy. BALWOIS 2012-Ohrid, Republic of Macedonia, 28 May-2 June 2012.

Schijven JF, Hassnizadeh SM, Husman AM (2010). Vulnerability of unconfined aquifers to virus contamination. Water Res. 44(4):1170-1181.

Stigter TY, Ribeiro L, Carvalho Dill AMM (2006). Evaluation of an intrinsic and a specific vulnerability assessment method in comparison with groundwater salinisation and nitrate contamination levels in two agricultural regions in the south of Portugal, Hydrogeol. J.14:79-99.

Thirumalaivasan D, Karmegam M, Venugopal K (2003). AHP-DRASTIC: software for specific aquifer vulnerability assessment using DRASTIC model and GIS. Environ. Model. Softw. 18:645-656.

Worrali F, Kolpin DW (2004). Aquifer vulnerability to pesticide pollution-combining soil, land-use and aquifer properties with molecular descriptors. J. Hydrol. 293 (1-4):191-204.

Rate of absorption and interfacial area of chlorine into aqueous sodium hydroxide system

K. S. Agrawal

Department of Chemical Engineering, Faculty of Technology and Engineering,
The M. S. University of Baroda, Vadodara, Gujarat, India.

Due to excellent mass transfer characteristics with energy efficiency jet ejectors can be used in place of conventional countercurrent systems, namely, packed bed contactors as well as venturi scrubbers, cyclones and airlift pumps. The removal of chlorine from certain gases by absorption in aqueous solutions of sodium hydroxide is industrially important in several chemical processes particularly in pollution control. Although, a number of papers have been published in the past, none of them provided a theoretical basis for the prediction of rate of absorption of chlorine from certain gases by absorption in aqueous solutions of sodium hydroxide in jet ejector. In this work, the rates of absorption of chlorine from different concentration of gas into aqueous sodium hydroxide solutions of various concentrations were measured at 30°C using a liquid jet ejector. The experimental results were analyzed on the basis of the penetration theory for gas absorption. The theoretical model to calculate rate of absorption is developed. The rate of absorption predicted from developed model is compared with experimental results. They were in good agreement. In this work, an attempt also has been made to develop mathematical model to estimate enhancement factor for jet ejector applying Higbie penetration theory.

Key words: Higbie penetration theory, jet ejector, chlorine, aqueous solutions, gases.

INTRODUCTION

There has been different reaction systems used to remove chlorine from gas. The commonly used systems are chlorine-water, chlorine-aqueous solution of alkaline carbonate, chlorine-aqueous alkaline hydroxide solution, chlorine- aqueous sulfite/bisulfite etc. The most commonly used system for chlorine removal is chlorine-aqueous sodium hydroxide solution.

Table 1 shows the summary of the rate constant developed by different researchers for absorption of chlorine in NaOH. Ashour et al. (1996) studied the absorption of Cl_2 into aqueous bicarbonate and aqueous hydroxide solutions both experimentally and theoretically. They estimated the reaction rate coefficient of reaction between Cl_2 and OH^- over the temperature range of 293 to 312 K and fitted by the Arrhenius equation:

$$k_2 = 3.56 \times 10^{11} \exp\left(\frac{-1{,}617}{T}\right) \quad (1)$$

Table 1 indicates that there is variation in the value of the forward reaction rate coefficient of absorption between Cl_2 into aqueous solutions of NaOH from 10^7 to 10^{14}. Attempt has been made to develop the co-relation to predict k_2 on the basis of our experimental results.

Table 1. The values of rate constant for Cl_2 - NaOH reaction system.

Author	Rate constant	Temperature (K)
Morris (1946)	$5 \times 10^{14} m^3/(kmol.s)$	298
Lifshitz and Perlmutter–Hayman (1961)	$(9 \pm 2) \times 10^{10} m^3/(kmol.s)$	283
Lifshitz and Perlmutter–Hayman (1962)	Of the order of $10^{10} m^3/(kmol.s)$	283
Spalding (1962)	Of the order of $10^{10} m^3/(kmol.s)$	298
Sandall et al.(1981)	$2.7 \times 10^7 m^3/(kmol.s)$	273
Ashour et al. (1996)	$1.2 \times 10^9 m^3/(kmol.s)$	298

REACTION MECHANISM AND RATE OF ABSORPTION OF CHLORINE INTO AQUEOUS NaOH SOLUTIONS

When Cl_2 is absorbed in aqueous NaOH solutions, the following reactions may take place:

$$Cl_2 + H_2O \xleftrightarrow{K_1,k_1} HOCl + H^+ + Cl^- \quad (2)$$

$$Cl_2 + OH^- \xleftrightarrow{K_2,k_2} HOCl + Cl^- \quad (3)$$

$$HOCl + OH^- \xleftrightarrow{K_3,k_3} OCl^- + H_2O \quad (4)$$

$$H_2O \xleftrightarrow{K_4,k_4} OH^- + H^+ \quad (5)$$

In this model, all reactions are assumed to be reversible. However reactions (2) and (3) have finite reaction rates, whereas reactions (4) and (5) are assumed to be an instantaneous.

Here three equilibrium constants K_1, K_3 and K_4 are independent and remaining K_2 can be obtained by following equation:

$$K_2 = \frac{K_1}{K_4}$$

The chemical species that are present in aqueous NaOH solutions are renamed as follows:

$$w_1 = C_{Cl_2}$$
$$w_2 = C_{HOCl}$$
$$w_3 = C_{Cl^-}$$
$$w_4 = C_{H^+}$$
$$w_5 = C_{OCl^-}$$
$$w_6 = C_{OH^-}$$
$$w_7 = C_{H_2O}$$
$$w_8 = C_{Na^+} = C_{NaOH,initial}$$

Bulk concentration of an individual chemical species in liquid

Assuming all the reactions are at equilibrium, the following equations can be derived by overall mass balance. Defining, L as molar ratio of chlorine to NaOH at initial condition.

Chlorine balance:

$$2w_1^0 + w_2^0 + w_3^0 + w_5^0 = C_{Cl_2,initial} = LC_{NaOH,initial} \quad (6)$$

Hydrogen balance:

$$w_2^0 + w_4^0 + w_6^0 + 2w_7^0 = LC_{NaOH,initial} + C_{H_2O,initial} \quad (7)$$

Oxygen balance:

$$w_2^0 + w_5^0 + w_6^0 + w_7^0 = C_{H_2O,initial} + C_{NaOH,initial} \quad (8)$$

Electro neutrality balance:

$$w_4^0 + w_8^0 - w_3^0 - w_5^0 - w_6^0 = 0 \quad (9)$$

$$w_8^0 = C_{Na^+} = C_{NaOH,initial} \quad (10)$$

As the reactions are at equilibrium the independent equilibrium constants are:

$$K_1 = \frac{w_2^0 w_4^0 w_3^0}{w_1^0} \quad (11)$$

$$K_2 = \frac{w_2^0 w_3^0}{w_1^0 w_6^0} \quad (12)$$

$$K_3 = \frac{w_5^0}{w_2^0 w_6^0} \quad (13)$$

$$K_4 = w_6^0 w_4^0 \quad (14)$$

We have $w_1^0\ to\ w_8^0$ unknowns and equation 6 to 11, 13 and 14 algebraic independent equations. Thus we have 8 unknowns and 8 algebraic independent equations.

These equations are 'linear system' of equations and can be solved by minimal residual technique using MATLAB to get $w_1^0\ to\ w_8^0$ (which converge to the solution). It may be noted that in case of aqueous NaOH solution do not contain any chlorine initially (means L = 0) then:

$$w_1^0 = w_2^0 = w_3^0 = w_5^0 = 0 \tag{15}$$

$$w_6^0 = C_{NaOH,initial} \tag{16}$$

Partial differential equations and nonlinear algebraic equations describing diffusion/ reaction process

By assuming all reactions reversible, the following reaction rate expression can be given:

$$R_1 = -k_1 w_1 + \frac{k_1}{K_1} w_2 w_3 w_4 \tag{17}$$

$$R_2 = -k_2 w_1 w_6 + \frac{k_2}{K_2} w_2 w_3 \tag{18}$$

The rate of reaction R_3 and R_4 are instantaneous having large values of rate of reaction and are eliminated. We also assume that:

i) Reactions are at equilibrium.
ii) The diffusivity of ionic spices are equal.
iii) The fluxes of the nonvolatile species at interface are equal to zero.

By considering the mass balance following differential equation are derived:

Cl_2 balance

$$\frac{\partial w_1}{\partial t} = D_1 \frac{\partial^2 w_1}{\partial x^2} + R_1 + R_2 \tag{19}$$

Cl_2/ Cl^- balance

$$\frac{\partial w_1}{\partial t} + \frac{\partial w_3}{\partial t} = D_1 \frac{\partial^2 w_1}{\partial x^2} + D_3 \frac{\partial^2 w_3}{\partial x^2} \tag{20}$$

Total chlorine balance

$$2\frac{\partial w_1}{\partial t} + \frac{\partial w_2}{\partial t} + \frac{\partial w_3}{\partial t} + \frac{\partial w_5}{\partial t} = 2D_1 \frac{\partial^2 w_1}{\partial x^2} + D_2 \frac{\partial^2 w_2}{\partial x^2} + D_3 \frac{\partial^2 w_3}{\partial x^2} + D_5 \frac{\partial^2 w_5}{\partial x^2} \tag{21}$$

Electro neutrality balance:

$$w_4 + w_8 - w_3 - w_5 - w_6 = 0 \tag{22}$$

Where $w_8 = C_{Na^+} = C_{NaOH,initial}$

As it is assumed all reactions are reversible; hence, the instantaneous reactions are also at equilibrium and their equilibrium constant may be given as follows:

$$K_3 = \frac{w_5}{w_2 w_6} \tag{23}$$

$$K_4 = w_4 w_6 \tag{24}$$

There are 6 unknowns and 6 partial differential equations/algebraic equations which can be solved for the concentrations of all chemical species.

Initial condition and boundary condition

At t = 0 (for all $x \geq 0$) and at $x = \infty$ (for all $t \geq 0$), the concentration of chemical species are equal to bulk concentrations in liquid.

$$w_i = w_i^0, i = 1\ to\ 6 \tag{25}$$

$$i.e. (w_1 = w_1^0,\ w_2 = w_2^0,\ w_3 = w_3^0,\ w_4 = w_4^0,\ w_5 = w_5^0,\ w_6 = w_6^0)$$

Boundary conditions at interface

At the interface of gas-liquid $x = 0$
For non volatile species

$$\frac{\partial w_j}{\partial x} = 0, at\ x = 0, t > 0 \tag{26}$$

for all *j* except j =1 (Cl_2)

For volatile species (Cl_2)

$$-D_j \frac{\partial w_j}{\partial x} = k_{g,j}[P_j - H_{ei}\, w_j(0,t)] \tag{27}$$

Here in our system there is only one volatile species that is, chlorine and hence j = 1. The equation may be re-written as:

$$-D_1 \frac{\partial w_1}{\partial x} = k_{g,1}[P_1 - H_{e1}\, w_1(0,t)] \tag{28}$$

Where H_1 is the physical equilibrium constant (Henry's law constant) of Cl_2. In case of pure Cl_2 in the gas phase, mass transfer resistance in gas phase can be neglected.

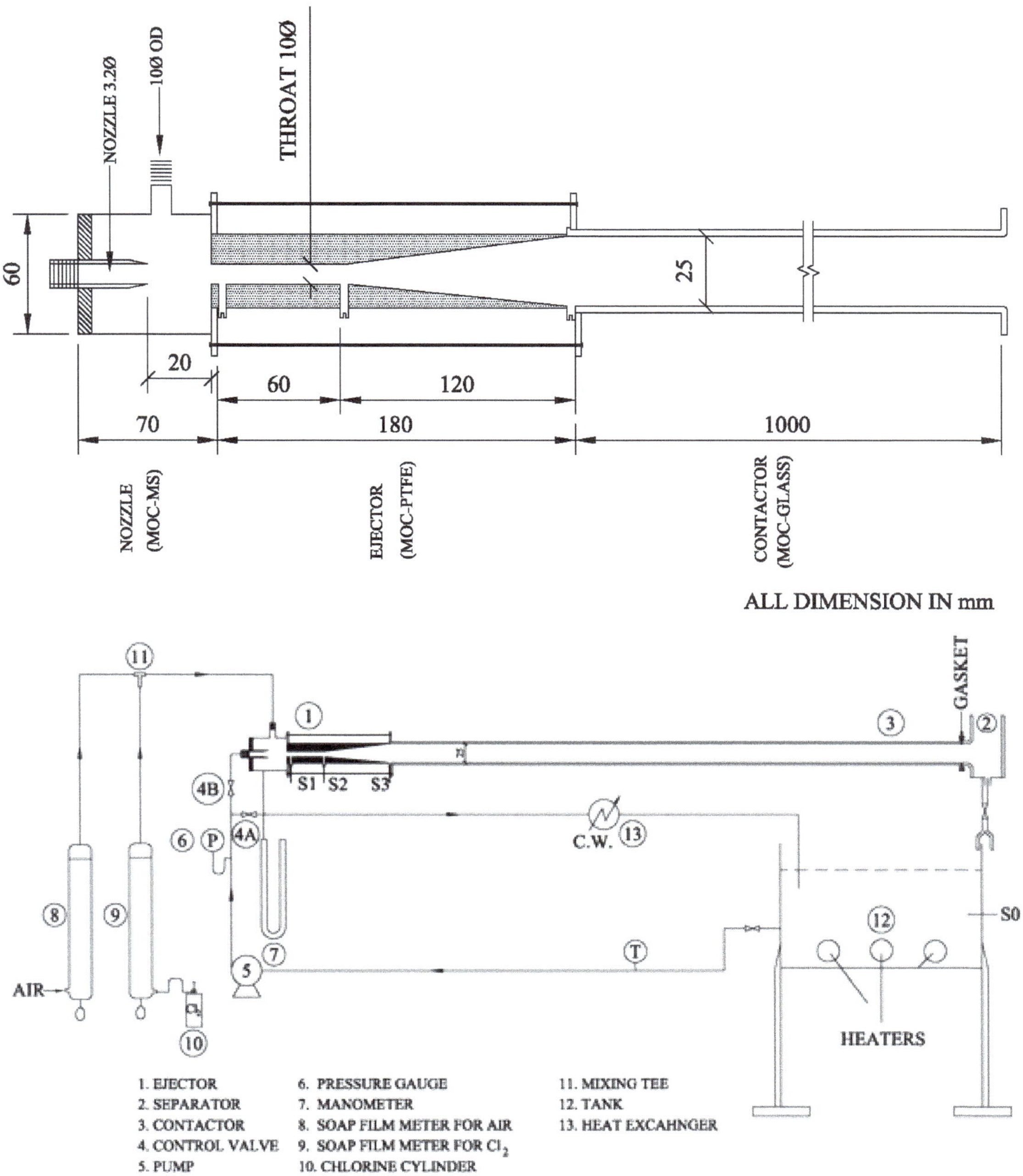

Figure 1. Detail of jet ejector and schematic diagram of experimental setup.

Hence $P_1^* = P_1$ and $(k_{g,1} \rightarrow \infty)$, the boundary condition for pure Cl_2 at the gas liquid inference reduces to:

$$w_1(0,t) = w_1^* = \frac{P_1}{H_1} \quad at\ x = 0\ t > 0 \qquad (29)$$

EXPERIMENTALS

A detailed experimental set up for the ejector is shown in Figure 1. It consists of an ejector having detail as in Figure 1, an extended contactor, a gas liquid separator, aqueous caustic soda tank and other accessories. The ejector-head and extended contactor are made of transparent glass for visual observation of the flow and mixing patterns. The outlet of ejector and inside diameter of the contactor is 25 mm and length of the contactor 1 m. A high velocity liquid jet is produced through the nozzle (orifice) of the ejector which entrains air along with it. Known flow rate of aqueous solution of caustic soda whose concentration is predetermined is passed through the orifice of ejector. Then free suction of atmospheric air is allowed to enter. After measuring air flow rate, the chlorine flow rate is adjusted to achieve required gas concentration. Once steady state is achieved, samples are withdrawn from sample point S1, S2, S3 and S4. The available chlorine was analyzed by the iodometric method. The detailed experimental procedure has been discussed in an earlier paper (Agrawal, 2012). The rate of absorption of

Table 2. Densities and viscosities of aqueous NaOH solutions, Henery's law constants of Cl_2, and diffusion coefficients of Cl_2 (D_1), HOCl (D_2) and OH^- (D_6) (Ashour et al., 1996).

T K	$[NaOH]_{initial}$ $kmol/m^3$	$10^{-3}\rho$ kg/m^3	$10^{+3}\mu$ $kg/m.s$	H_3 $atm.m^3/kmol$	$10^9 D_1$ m^2/s	$10^9 D_2$ m^2/s	$10^9 D_6$ m^2/s
293	0.09985	0.9985	1.010	13.44	1.29	1.34	2.99
298	0.10000	0.9973	0.903	16.36	1.47	1.54	3.43
303	0.09960	0.9960	0.802	19.55	1.68	1.75	3.89
312	0.09970	0.9940	0.676	26.98	2.05	2.14	4.76

chlorine is computed using experimental result as follows:

$$R_{Cl_2} = N_{Cl_2} a = \left[C_{Cl_2,gas,in} - C_{Cl_2,gas,out}\right]/t_e \; kmol/m^3 s \quad (30)$$

Numerical implementation

The gas liquid exposure time for the jet ejector may be stated as:

$$t_e = \frac{Total\ volume\ of\ Jet\ ejector}{Sum\ of\ volumetric\ flow\ rate\ of\ aqueous\ solution\ and\ gas} \quad (31)$$

$$t_e = \frac{V_J}{F_L + F_G}$$

The average rate of absorption of Cl_2 per unit interfacial area may be written as:

$$N_{Cl_2} = \frac{D_A}{t_e}\int_0^{t_e}\left(\frac{\partial w_1}{\partial x}\right)(0,t)\,dt \quad (32)$$

and the enhancement factor of Cl_2 is determined from the following equation:

$$\beta = \frac{N_{Cl_2}}{k_L(w_1^* - w_1^0)} \quad (33)$$

Where, w_1^* *and* w_1^0 are interfacial and bulk concentrations of Cl_2 in the liquid, respectively. k_L is the liquid phase mass transfer coefficient for physical absorption of Cl_2 is given by:

$$k_L = 2\sqrt{\frac{D_1}{\pi t_e}} \quad (34)$$

The Finite Difference Method (FDM) is used to transform each partial differential equation into the system of ordinary differential equation in *t*. We choose following finite difference expressions to approximate the partial derivatives:

$$\frac{\partial w_{i,j}}{\partial x} = \frac{1}{2h_j}\left[w_{i,j+1} - w_{i,j-1}\right] + higher\ order\ terms \quad (35)$$

$$\frac{\partial^2 w_{i,j}}{\partial x^2} = \frac{1}{h_j^2}\left[-2w_{i,j} + w_{i,j+1} + w_{i,j-1}\right] + higher\ order\ terms \quad (36)$$

Where, i refer to the ith chemical species, j, refers to the spatial node number and $h_j = x_{j+1} - x_j$. Typical values for the initial nodal spacing h_0 at the gas-liquid interface are about 10^{-4} mm. The system of partial differential equations /algebraic differential equations were transformed into large system of differential and algebraic equations that can be solved by MATLAB software by using ODE solver, BVP4c and preconditioning technique with special Jacobi pre-conditioner.

Physicochemical properties

The mathematical models is used to predict the rates of absorption and reaction (enhancement) factors of Cl_2 .We need the properties like densities and viscosities of the aqueous sodium hydroxide solutions, Henry's law constants of Cl_2 and the diffusion coefficients of all chemical species in these solutions which are tabulated in Table 2. We also need the equilibrium constants of all chemical reactions and the forward rate coefficients of reactions (2) to (5) which are tabulated in Table 3.

RESULTS AND DISCUSSION

The penetration model has been used to develop mathematical model for absorption of Cl_2 into aqueous solution of sodium hydroxide. The mathematical model to predict absorption rate is presented by equation (19) to (29). To solve this model the value of k_1, k_2, k_3 and k_4 were required. The value of k_1 was determined by the correlation given by Brian et al. (1966):

$$k_1 = 1.4527 \times 10^{10}\, exp.\left(\frac{-6138.6}{T}\right)$$

As the reactions are instantaneous, k_3 and k_4 are eliminated. There is large variation in the value of k_2 in the literature as clear from Table 1. Hence, attempt has been made to estimate value of k_2 by using data obtained for absorption of Cl_2 in aqueous sodium hydroxide solution in the jet ejector. The value of k_2 was adjusted until the theoretically predicted rate of absorption was

Table 3. Values for equilibrium constants of reactions 1 to 4 at various temperatures(Ashour et al., 1996).

T K	$10^4 K_1$ $(kmol/m^3)^2$	$10^{-10} K_2$	$10^{-6} K_3$ $m^3/kmol$	$10^{14} K_4$ $(kmol/m^3)^2$
293	3.890	5.722	3.726	0.6798
298	4.500	4.491	2.790	1.002
303	5.181	3.580	2.109	1.447
308	5.938	2.895	1.609	2.051
313	6.776	2.371	1.238	2.858

Table 4. The values of k_2 at atmospheric pressure for Cl_2- aqueous NaOH system.

T K	$(NaOH)_{initial}$ $kmol/m^3$	Estimated k_2 $\frac{m^3}{kmol.s}$ [Ashour et al. (1996) Equation (1)]	Experimental value k_2 $\frac{m^3}{kmol.s}$	Estimated k_2 $\frac{m^3}{kmol.s}$ present model Equation (37)
293	0.09985	1.42×10^9	-	1.399×10^9
298	0.10000	1.566×10^9	-	1.669×10^9
303	0.09960	1.71×10^9	-	1.669×10^9
312	0.09970	1.99×10^9	-	1.945×10^9
303	0.03	1.399×10^9	1.662×10^9	-

within 1% of the experimentally measured rate of absorption of Cl_2. Thus, the following correlation is developed to predict the value of k_2:

$$k_2 = 3.39 \times 10^{11}\ exp.\left(-\frac{1610}{T}\right) \quad (37)$$

The predicted values of k_2 from equation (1) reported by Ashour et al. (1996) and predicted value of k_2 from equation (37) by proposed model, along with the value obtained from the experimental result of present work are presented in Table 4 and plotted in Figures 2 and 3. The experimental value and value predicted by present model (equation 37) are comparable. The values estimated by Ashour et al. (1996) are a little higher by 2.7% at maximum and the present experimental k_2 value differs only negligibly from Equation 37. As the error is very less, it may be concluded that the proposed model is good. Figure 4 presents N_{Cl2} vs. $C_{Ag,in,}$ (C_{BO} as parameter). The values obtained by experiment result and proposed model are in good agreement. Thus, the chemical absorption mechanism proposed in the present work may be considered to be correct.

Conclusion

The value N_{Cl2} obtained by experiment and predicted from the proposed model are in good agreement. Hence, the proposed mathematical model may be used to predict the value of reaction rate (flux), N_{Cl2}. These values may be further utilized to predict enhancement factor β and interfacial area using following co-relations:

$$\beta = \frac{N_A}{k_L^0(C_A^* - C_A^0)}$$

$$a = \frac{N_A . a}{N_A}$$

The correlation obtained to estimate rate constant for forward part of absorption of chlorine in aqueous NaOH is:

$$k_2 = 3.39 \times 10^{11}\ exp.\left(-\frac{1610}{T}\right)$$

ACKNOWLEDGEMENT

The author would like to thank Professor Vasdev Singh for their constant help and encouragement during this work.

Nomenclature

Latin letters: a, Interfacial area (m^2/m^3); **C,** concentration

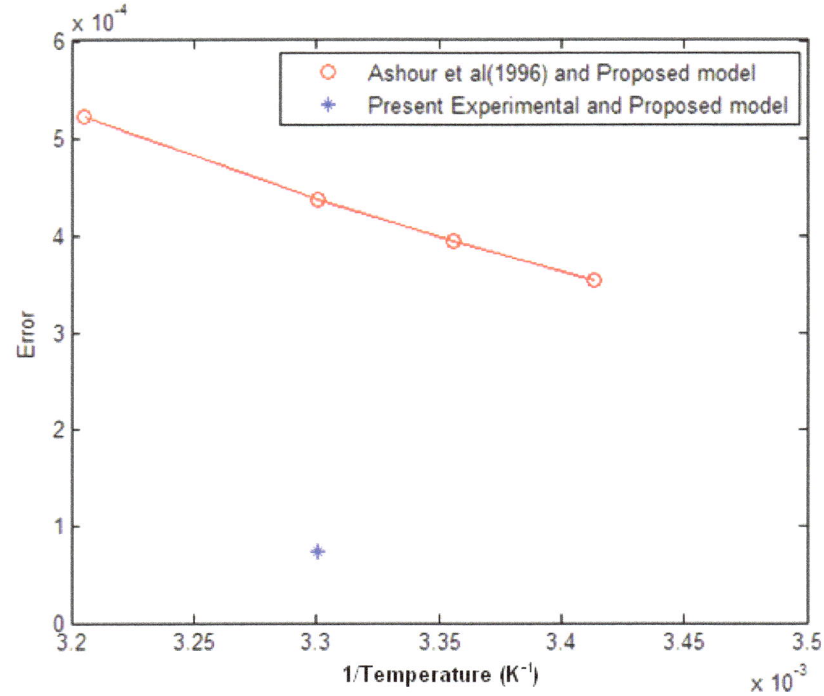

Figure 2. Comparison of k_2 obtained by proposed mathematical model with present experimental result.

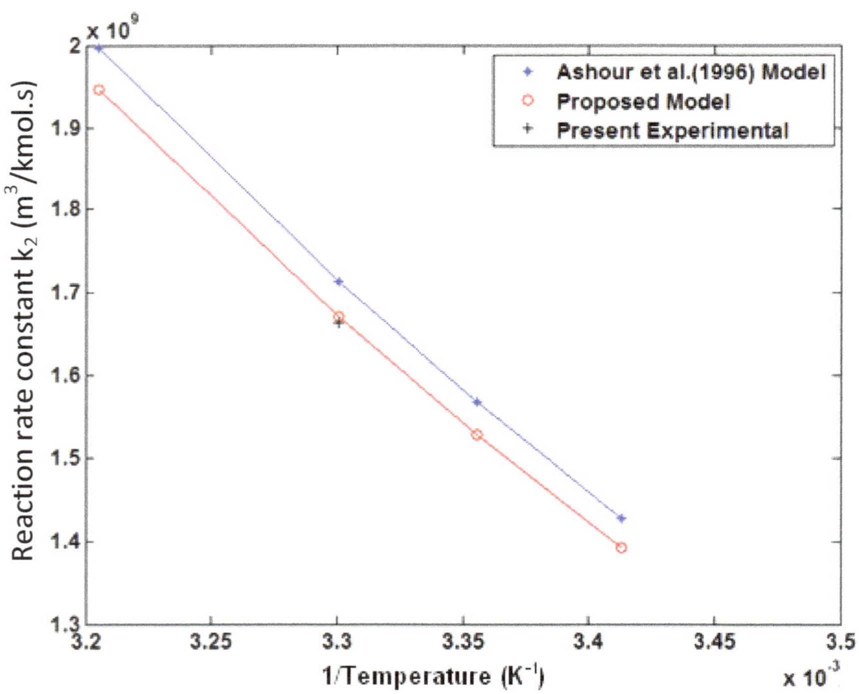

Figure 3. Error estimates for k_2 and proposed mathematical model.

of species indicated as subscript (kmol/m^3); **D,** diffusion coefficient of species indicated as (m^2 s^{-1}) subscript in the liquid phase; **F,** volumetric flow rate (m^3 s^{-1}); **H,** Henry's law coefficient (Pa m^3 mol^{-1}); **K,** equilibrium constants (-) k chemical reaction rate constant (as per order of reaction); **N_A*,** average rate of physical absorption (flux)

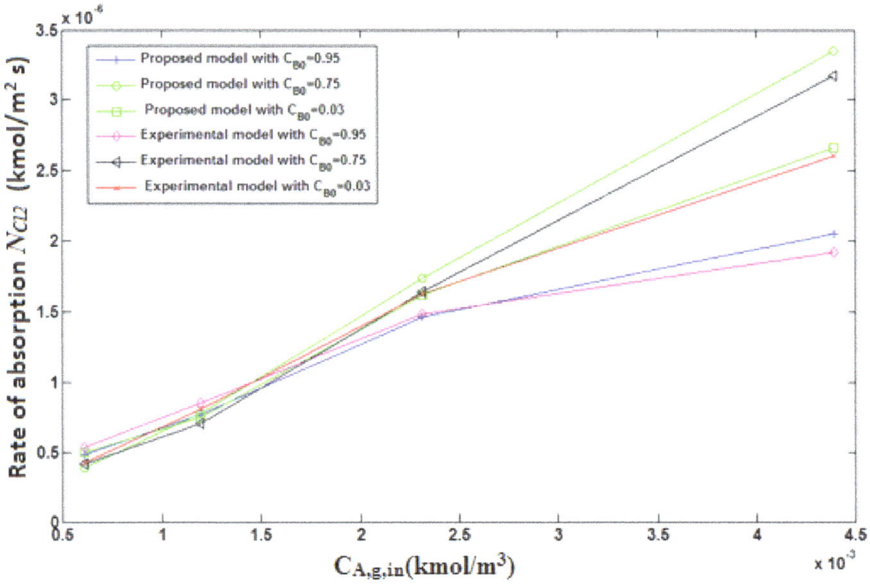

Figure 4. Comparison between the values for rate of absorption obtained experimentally and predicted by proposed model with respect to C_{A0} at different C_{B0}.

of A (kmol m^{-2} s^{-1}); **N,** rate of molar absorption with chemical reaction (kmol m^{-2} s^{-1}) (flux) of the species indicated by subscript; **k_G,** gas sided mass transfer coefficient (kmol m^{-3} Pa^{-1} s^{-1}); **k_L,** liquid sided mass transfer coefficient (m s^{-1}); **P,** system pressure (Pa); **R,** rate of reaction (kmol m^{-3} s^{-1}); **t,** time (s); **t_e,** exposure time (s); **T,** temperature (K); **V,** volume (m^3); **V_J,** volume of reactor/ejector (Figure 1) (m^3); **w_i,** concentration of species *i* in the liquid phase (kmol/m^3); **w_{ij},** concentration of species *i* in the liquid phase at x_j (kmol/m^3); **x_j,** spatial variable at node j (mm).

Greek letters: β, Enhancement factor (-).

Subscripts: A, B…., Components A, B…..

Superscripts: ***,** Equilibrium, physical solubility; **i,** species number; **j,** node number; **g,** gas; **in,** at inlet of ejector; **out,** at outlet of ejector.
6

REFERENCES

Agrawal KS (2012). Rate of absorption in laboratory scale jet ejector. Nat. J. Appl. Sci. Engr. 1(2):60-64.

Ashour SS, Edward RB, Orville C (1996). Absorption of chlorine into aqueous bicarbonate solutions and aqueous hydroxide solutions. AIChE J. 42(A3): 671-682.

Brian PLT, Vivian JE, Piazza C (1966). The effect of temperature on the ratio of absoraption of chlorine in to water. Chem. Eng. Sci. (21): 551-558.

Hikita H, Asai S, Himukashi Y, Takatsuka T (1973). Absorption of chlorine into aqueous sodium hydroxide solutions. Chem. Eng. J. (5):77-84.

Lahiri N, Yadav GD, Sharma MM (1983). Absorption of chlorine in aqueous solutions of sodium hydroxide desorption of hypochlorous acid followed by its dissociation to chlorine monoxide. Chem. Eng. Sci. 38(7):1119-1133.

Lifshitz A, Perlmutter HB (1962). The kinetics of the hydrolysis of chlorine III. The reaction in the presence of various bases and discussion of the mechanism. J. Phys. Chem. 66:701-705.

Lifshitz A, Perlmutter HB, (1961). The kinetics of the hydrolysis of chlorine II. The hydrolysis in the presence of acetate. J. Phys. Chem. 65:753-757.

Morris JC (1946). The mechanism of the hydrolysis of chlorine. J. Am. Chem. Soc. 68(9):1692-1694.

Roy S, Rochelle GT (2004). Chlorine Absorption in Sulfite Solutions. Sep. Sci. Technol. 39(13):3057-3077.

Sandall OC, Goldberg IB, Hurlock SC, Laeger HO, Wagner RI (1981). Solubility and rate of hydrolysis of chlorine in aqueous sodium hydroxide at 273K. AIChE J. 27(5): 856–859.

Spalding CW (1962). Reaction kinetics in the absorption of chlorine in to aqueous. AIChE J. (8):685-689.

Takahashi T, Hatanaka M, Konaka R (1967). Absorption of chlorine into still liquid in a simple stop-cock type absorber. Can. J. Chem. Eng. (45):145-149.

Biomonitoring of heavy metals pollution in Lake Burullus, Northern Delta, Egypt

Elsayed M. Ali. Nafea[1] and Mohamed A. Zyada[2]

[1]Aquatic Environment Department Faculty of Fish Resources, Suez University, Egypt.
[2]Environmental Sciences Department Faculty of Science Dameitta University, Egypt.

Aquatic macrophytes and benthos are unchangeable biological filters and they carry out purification of the water bodies by accumulating dissolved metals and toxins in their tissues. In view of their potential to entrap several toxic heavy metals, 3 groups of benthos and 6 macrophytes (submerged species: Potamogeton pectinatus, Ceratophyllum demersum and Najas armata); (floating species: Lemna gibba and Eichhornia crassipes root and shoot) and (emergent species: Phragmites australis shoot) were collected from 15 different locations on Lake Burullus and analyzed for determination of 6 heavy metals (Fe, Zn, Ni, Cu, Pb and Cd) contenets. The study was aimed at understanding the importance of these benthos and macrophytes in accumulation of toxic metals and suggesting preservation and restoration of Lake Burullus ecosystem. The distribution of the investigated metals in water, sediments, benthos and aquatic plants of the lake showed that, the eastern and eastern southern parts of the lake have generally higher concentrations of heavy metals than the western and middle one. Potamogeton pectinatus showed high contents of Pb, Cd and Zn respectively. On the other hands Eichhornia crassipes showed high level of copper while in Ceratophyllum demersum high concentration of Iron was detected. The present study reveals that the aquatic macrophytes and benthos play a very significant role in removing of the different metals from the aquatic environments and they probably reduced the effect of high concentrations of these metals on the lake ecosystem. Bioaccumulation factor values showed that the trend of accumulation of most metals in the benthos was as follows: Mullusca > Arthropoda > Annelida > and in aquatic plants as: Lemna gibba> Potamogeton pectinatus > Ceratophyllum demersum > Eichhornia crassipes Root > Najas armata>phragmites australis shoot> Eichhornia crassipes shoot, which make them suitable candidates to be used in biomonitoring surveys as a good tools for heavy metal pollution markers, in the biological treatment of the polluted water and in sustainable development of Lake Burullus.

Key words: Biofilter, bioaccumulation, sustainable developments, biological treatment, macro benthos, aquatic plants, sustainable management.

INTRODUCTION

The term heavy metal refers to any metallic chemical element that has a relatively high density and is highly toxic or poisonous at low concentrations (Harris and Santos, 2000). Macrophytes are aquatic plants, growing in/or near water that are emergent, submerged or floating. Macrophytes are considered as important component of the aquatic ecosystem not only as food source for aquatic invertebrates, but also act as an efficient accumulator of heavy metals (Devlin, 1967; Chung and Jeng, 1974). They are unchangeable biological filters that play an important role in the maintenance of aquatic ecosystem.

Some sources of heavy metals are industry, municipal wastewater, atmospheric pollution, urban runoff, river dumping, and shore erosion. High levels of Cd, Cu, Pb, and Fe can act as ecological toxins in aquatic and terrestrial ecosystems. Some of these metals (Cu, Ni, Cr and Zn) are essential trace metals to living organisms, but become toxic at higher concentrations. Others, such as Pb and Cd have no known biological function but are toxic elements (Guilizzoni, 1991; Balsberg-Påhlsson, 1989). Many of the aquatic macrophytes and benthos are found to be the potential scavengers of heavy metals from water and wetlands sediments (Gulati et al., 1979). The present investigation was planned and executed considering the potentials of benthos and macrophytes as biological filters for metals that become bound to living materials.

Biomonitoring of pollutants using accumulator species is based on the capacity which has some plants and animal taxa have to accumulate relatively large amounts of certain pollutants by concentration many times higher than those of the surrounding waters, (Nafea, 2005). In addition, the pollutants concentration in sediments and the organisms are the result of the past as well as the recent pollution level of the environment in which the organism lives, while the pollutant concentrations in the water only indicate the situation at the time or seasons of sampling (Ravera et al., 2003).

Although, Lake Burullus attracts attention of many authors because of its economic and scientifically importance to study its unique ecosystem but, the studies dealing with the accumulation of heavy metals in different ecosystem components are still scarce except few studies (Elsaraf, 1995a; Radwan and Lotfy, 2002; Nafea, 2005). The present study deals with the aquatic plants and macro benthos as biomarkers and bio accumulators for heavy metals, in order to use these aquatic plants and benthos in sustainable development and management of Lake Burullus.

MATERIALS AND METHODS

The study area

Al Burullus Lake is situated along the Mediterranean coast. It occupies more or less, a central position between the two branches of the Nile and extends between 31° 22' - 31° 26' N and 30° 33' - 31° 07' E. It's a shallow brackish lake connected with the sea by a small outlet (Boughaz), about 44 m width and 150 m length. The length of the lake was about 65 km, and its width varies between 6 and 16 km, with an average of about 11 km. The depth of the lake anges between 0.42 and 2.07 m. The eastern sector of the lake is the shallowest, showing an average depth of 0.8 m. The present area of the lake is about 410 km^2 (100,000 Feddan), of which 370 km^2 is open water. The capacity of the lake is about 330 million cubic meters. The eastern and southern parts of the lake receives agricultural sewage drainage water through 8 drains and one brackish water canal, while saline water enters the lake from the sea through El-Boughaz (Figure 1).

Methods

The study focused primarily on metal investigation in water, sediments, benthos and aquatic plants. The sampling program was carried out in the summer and winter of 2013. Aquatic plants and benthos were collected from the 15 sites and five samples were prepared for each species at every sampling site. At the same time water and sediment samples were collected at the corresponding sampling locations. The collected water was filtered through a Whatman glass-fiber filter (0.45 μm). The filtered water was stored in a 0.5-L polypropylene bottle to avoid any adsorption of metals on the wall of the sample bottles; the filtered water was preserved by acidification with 1.0 ml concentrated nitric acid. Water analysis for heavy metals was according to Solvent extraction method (APHA, 1992). The sediment samples were air dried at room temperature (25°C) for 10-15 days, then ground in a mortar and sieved in 0.5 nm sieve, the samples were then finally stored for analysis according to Moore and Chapman (1986).

Five plant samples were mixed with each other's and analyzed for heavy metals where the dry samples of Macrophytes were wet-digested in a mixture of concentrated nitric acid and perchloric acid (4:1 v/v) (Sawicka-Kapusta, 1978). Samples were analyzed with a Perkin Elmer model 2380 atomic adsorption spectrophotometer (A.A.S.). Bottom fauna were classified to three main groups (Mullusca, arthropoda and Annelida), and digested after drying according to Metcalfe-Smith (1994). Method metals concentrations were determined using atomic absorption (Perkin Elemer Model 3700) with flameless graphite furnace (GA-2). The bioaccumulation factor was calculated according to Klavinš et al. (1998) as follow: BAF = M_tissue / M_water or sediments where M_tissue is metal concentration in plant tissue and M_water is the metal concentration in water or sediments.

RESULTS

Water and sediments

Copper content in water of Lake Burullus ranged between19.7 μg/l in the western parts and 35.8 μg/l in the eastern part. Its values in sediment ranged from 19.4 (μg/g) dry wt and 47.9 μg/g dry wt. in the western and eastern sites, respectively. Iron content in water ranged between 25.3 μg/l in station 8 and 60.4 μg/l in station 4. Its Iron values in sediments ranged between 42.4 μg/g dry wt.in site 15 and 97.5 μg/g dry wt in site 4. Cadmium contents ranged between 2.9 μg/l in station10 and 8.5 μg/l in station 4 while in sediments ranged between 3.2 μg/g in station 10 and 8.5 μg/g in station 6. Zinc contents in water ranged between 20.6 μg/l in station15 and 55.3

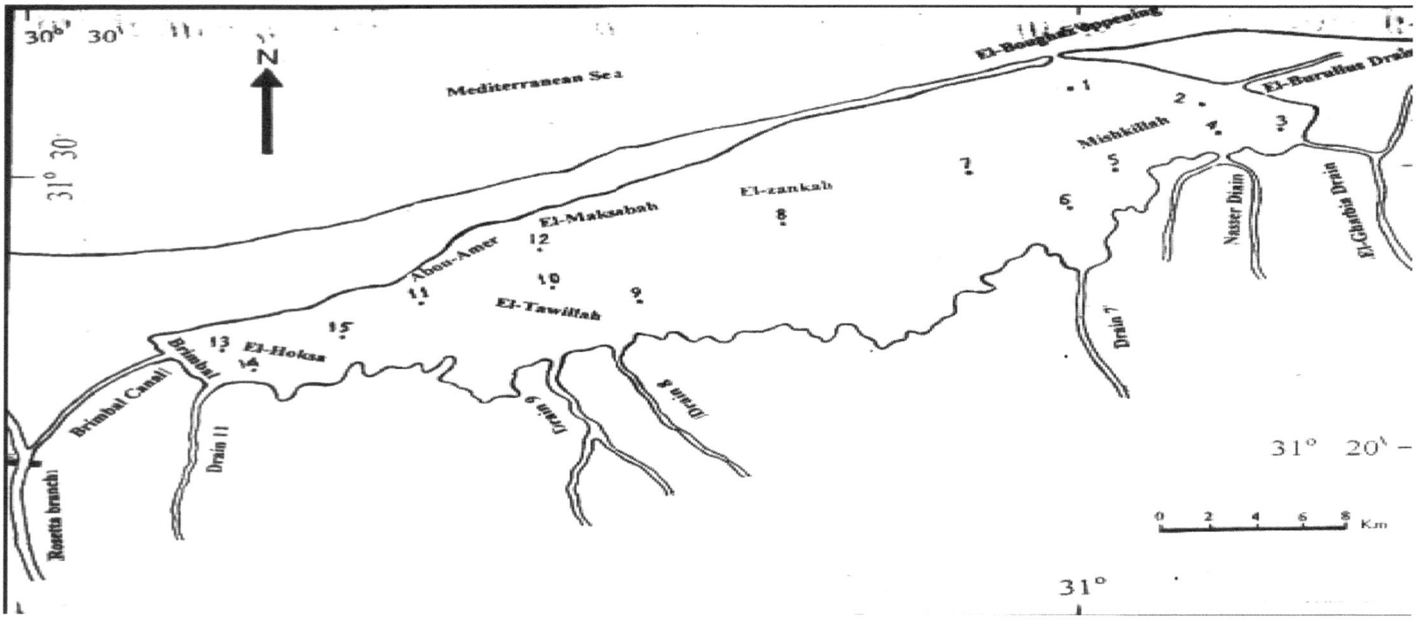

Figure 1. Location map for the study area.

Table 1. The concentration of heavy metals in water (µg/l) at 15 stations at Lake Burullus.

Stations	Metal concentration					
	Cu (µg/l)	Fe (µg/l)	Cd (µg/l)	Zn (µg/l)	Pb (µg/l)	Ni (µg/l)
1	24.5	32.3	4.1	28.2	7.4	6.2
2	33.6	45.2	5.2	35.7	7.2	10.3
3	18.2	52.3	6.1	33.1	8.5	5.3
4	30.1	60.1	8.5	55.3	10.1	5.4
5	20.2	33.2	6.1	50.6	7.1	5.2
Mean eastern	25.32	44.6	6.0	40.58	8.06	5.64
6	22.3	30.2	7.3	42.3	8.6	5.2
7	23.5	34.5	6.5	36.2	4.3	6.5
8	24.6	25.3	5.3	42.3	8.5	4.3
9	18.6	35.2	4.2	26.4	5.1	5.3
10	11.3	55.3	2.9	25.3	6.6	5.9
Mean middle	20.6	44.45	5.24	18.8	6.62	5.38
11	21.6	40.2	5.5	23.5	6.9	5.2
12	26.4	42.3	4.6	22.6	4.5	4.6
13	18.3	59.2	5.1	27.5	5.3	5.3
14	12.9	41.3	4.2	25.7	7.2	6.1
15	16.2	39.2	4.0	20.6	6.3	4.9
Mean western	19.1	44.4	4.6	24	6.1	5.22

µg/l in station 4 while in sediments it ranged between 24.2 µg/g in station15 and 97.2 µg/g in station 4.

Lead content showed its high values in water at site 4 (10.1 µg/l) and low value (4.5 µg/l) in station 15 while in sediments high value was 27.5 µg/g in station 1 and 6.5 µg/g in station15. Nickel showed its high value in water (10.3 µg/l) in station 2 and the low one (4.3 µg/l) in station 6 while in sediments high value recorded 19.7 µg/g in station 2 and 7.1 µg/g in station 15. The metals concentrations are given as shown in Tables 1 and 2.

Table 2. The concentration of heavy metals in sediment (µg/g) dry weight at 15 stations at Lake Burullus.

Stations	Metal concentration					
	Cu (µg/l)	Fe (µg/l)	Cd (µg/l)	Zn (µg/l)	Pb (µg/l)	Ni (µg/l)
1	29.5	71.3	5.2	58.6	27.5	18.4
2	47.6	92.4	6.1	72.4	18.2	19.7
3	29.6	94.5	5.9	65.3	27.1	10.2
4	39.5	97.5	7.8	97.2	19.7	14.2
5	32.2	58.9	7.2	94.7	20.4	11.6
Mean eastern	35.7	83	6.5	77.64	22	15
6	39.1	81.1	8.2	81.2	17.4	12.9
7	35.4	44.2	5.9	57.9	11.4	10.2
8	29.5	61.3	6.1	64.3	18.6	8.7
9	22.7	57.5	5.2	36.1	15.7	9.7
10	20.8	75.7	3.1	25.3	17.4	12.8
Mean middle	29.5	64	5.7	53	16.1	11
11	31.2	52.1	5.3	37.2	12.3	9.2
12	34.1	51.3	5.2	34.4	9.7	8.7
13	27.3	60.2	5.4	31.2	11.2	9.1
14	19.7	52.3	4.1	27.1	10.9	8.2
15	19.4	42.1	5.0	24.1	9.6	7.1
Mean western	26.4	51.6	5	30.1	10.8	8.5

Aquatic plants

Copper contents in the aquatic plants showed high range of variation which ranged between 13.9 µg/g in *Eichhornia* root in eastern site and 5.1 µg/g in *Ceratophyllum* in western site. Iron contents in aquatic plants ranged between 107 µg/g in potamogeton and 50 µg/g in *Najas*, while cadmium content ranged between 1.0 µg/g in Ceratophyllum and 4.7 µg/g in potamogeton. On the other hand nickel ranged between 15.3 µg/g in Lemna and 5.8 µg/g in Najas. High value of Lead observed at *potamogeton* 15.4 µg/g and low value 5.5 µg/g at *Eichhornia* shoot. Zinc contents ranged between 98 µg/g in Lemna and 43 µg/g in *Eichhornia* shoot as shown in Table 3.

Benthos mullusca

Benthos mullusca showed high ranges of metal content in their bodies and tissues more than Arthropoda and Annelida as shown in Table 4.

Bioaccumulation factor

The bioaccumulation values of heavy metal by aquatic plants in relation to water ranged between (260-548, 1256-5370, 200-913,1111-2392,1171-2722 and 850-2220 for Cu, Fe, Cd, Zn, Pb and Ni, respectively (Table 3), while in benthos in relation to sediments ranged between 0.59-0.69, 0.96-1.3, 0.32-0.62, 1.1-1.44, 0.63-0.75 and 0.93-1.2 for Cu, Fe, Cd, Zn, Pb and Ni respectively (Table 4).

DISCUSSION

Generally the high concentrations of heavy metals were measured in the sediments, benthos and macrophytes in the eastern sites of the lake, while the low concentrations were detected in the middle and western sites of the lake. The concentration of lead varied from site to site where the high concentrations were detected in water, sediments, Mullusca and *Potamogeton pectinatus* in the eastern site (8.06 µg/l, 15, 15 and 13 µg/g dry wt) respectively while the low concentrations were measured in the western parts of the lake (6.1 µg/l, 8.5,10 and 5.5 µg/g dry wt) respectively (Tables 1, 2, 3 and 4). The variation of lead content in benthos sediments and macrophytes depends on the inflow of many sources of pollution from sewage, agricultural and industrial wastes into the lake. El-Sarraf (1995a) mentioned that in Lake Manzala *P. pectinatus* had high lead concentration (26 .6 µg/g dry wt) and thus it is a good lead contamination indicator. This agrees with the conclusion of Abo-Rady (1977) that *P. pectinatus* may be considered as an indicator for lead in Lake Manzala. Our results in burullus confirm this conclusion (Radwan and shokier, 2005).

High levels of cadmium contents were found in *L. gibba*

Table 3. The mean concentrations of heavy metals (µg/g) dry weight and the bioaccumulation factor values for the aquatic plants in the (east, middle and west) of Lake Burullus.

Plants	Metal concentration (µ g/g)					
	Cu	Fe	Cd	Ni	Zn	Pb
Ceratophyllum demerssum E	6.6	94	1.2	10.5	98	8.5
M	5.4	84	1.1	9.5	92	7.8
W	5.1	81	1.0	8.2	84	6.4
Bioaccumulation factor	263	3268	368	1853	2234	1102
Potamogeton pectinatus E	10.2	107	4.7	13.4	64	15.4
M	9.1	78	2.7	12.5	61	11.4
W	8.3	83	2.3	10.2	52	9.7
Bioaccumulation factor	426	3469	679	2116	1459	1780
Lemna gibba E	11.8	104	4.2	14.2	92	14.2
M	10.4	101	3.6	15.3	98	14.7
W	9.2	87	4.2	11	87	14.2
Bioaccumulation factor	484	3852	766	1822	2247	1807
Phragmites aus. Shoot E	9.1	72	2.0	8.5	75	10.2
M	8.2	71	1.7	5.8	71	10.1
W	7.2	66	2.0	8.5	64	8.9
Bioaccumulation factor	392	2323	401	1418	1698	1480
Eichhornia crass. Shoot E	8.6	62	1.5	7.6	53	7.5
M	7.3	63	1.6	6.4	52	6.9
W	6.5	54	1.5	7.6	43	5.5
Bioaccumulation factor	436	2879	300	1344	1223	1266
Eichhornia crass. Root E	13.9	95	3.5	8.7	86	9.1
M	8.3	91	2.7	7.4	75	9.3
W	8.7	82	3.5	8.7	56	5.1
Bioaccumulation factor	417	3081	575	1558	1724	980
Najas armata E	11.5	51	1.4	6.2	76	11.6
M	9.5	46	1.5	5.8	78	10.2
W	9.3	50	1.4	6.2	66	10.0
Bioaccumulation factor	469	2371	274	1132	1868	1615

E, M, W means east, middle and west, respectively.

Table 4. The mean concentrations and bioaccumulation of heavy metal in the benthos (µg/g) dry weight in the (east, middle and west) of lake burullus in relation to sediment concentration.

Benthos	Metal concentration					
	Cu (µg/g)	Fe (µg/g)	Cd (µg/g)	Zn (µg/g)	Pb (µg/g)	Ni (µg/g)
Mullusca : east	27	72	4.5	83	13	15
Middle	22	69	3.2	76	12	13.5
West	13	63	2.9	73	11.6	12.7
Bioaccumulation	0.69	1.03	0.62	1.44	0.75	1.2
Arthropoda: east	23	72	2.3	75	11	12
Middle	19	65	1.4	69	10.5	11.7
west	11	57	1.9	71	10.1	11.3
Bioaccumulation	0.59	0.98	0.32	1.34	0.65	1.02
Annelida :east	21	68	3.2	67	11	11
Middle	19	62	1.2	58	10	10.9
west	17	60	1.1	49	9.7	10
bioaccumulation	0.63	0.96	0.32	1.1	0.63	0.93

and Eichhornia root, Mullusca and sediments of the eastern parts (4.5, 6.5 and 4.2 µg/g dry wt.) while the lowest concentrations of cadmium were observed in *C. demersum* (1 µg/g dry wt). El-Sarraf (1995b) mentioned that there was high significant correlation between lead and cadmium concentration in aquatic plants which is probably attributed to their association in the same phase during assimilation (Tables 2, 3 and 4).

The macrophytes and benthos have different levels of zinc concentrations in their organs where the high levels were recorded in *L. gibba*, *P. pectinatus* and Mullusca (83, 77.6, 107 and 104 µg/g dry wt.) respectively, Alloway and Davis, (1971); while the low concentrations were found in *N. armata*, *E. crassipes* shoot and Annelida (50, 46 and 49 µ g/g dry wt.) respectively. Heydt (1977) found that *P. pectinatus* has high zinc content which ranged between 16.5 and 517µ g/g dry. For weight in Elsenz River, Bauda et al. (1981) recorded that in Lake Manzala the mean concentration of Zinc level in the same species is 168 µg/g dry wt. El-Sarraf (1995a) found that the concentration of Zinc in *P. pectinatus* was 117 µg/g dry weights whereas Abo-Rady (1977) found that the zinc content of *P. pectinatus* ranged between 137 and 213 µg/g dry weight in Leine River.

The copper concentrations fluctuated between 5.4 and 35 µg/g dry wt. in plants, benthos and sediments. The copper content in Lake Manzala varied from 5.0 to 37.6 µg/g wt. in potamogeton *pectinatus* (Boudo et al. 1988). On the other hand El-Sarraf (1995a) found that copper content showed a small range of fluctuation with irregular concentration in aquatic plants. In *Ceratophyllum demersum* the highest level was 18.5 µg/g dry wt. The positive correlation between Cu and Zn was attributed to the same biological behaviors during the assimilation in macrophytes (Alloway and Davis, 1971; El-Sarraf, 1995b).

The concentrations of trace metals (Cu, Zn, Pb, Fe, Ni and Cd) in the aquatic macrophytes, benthos and sediments varied according to their locations at the different parts of the lake; this depends on the source of pollution invading the lake from several directions. Seidal (1996) and Ozimek (1978) recorded high contents of trace metals in the macrophytes growing in habitats affected by industrials effluents and effect of sewage and industrial wastes on the chemical composition of aquatic macrophytes is very obvious. The magnitude of aquatic plants and benthos to assimilate heavy metals would be largely dependent upon the levels of these metals in the water and sediment. The removal of certain mineral from water reservoirs by submerged macrophytes and benthos is observed as practical methods for water purification (Hillman and Cully, 1978). The high variations found in the element content of the aquatic macrophytes both between species and within species were related to different location. Crowder and Painter (1991) inferred that the variation of metals content in macrophytes is not necessarily bioaccumulations or biomagnified these metals from the sediment and it may be attributed to site-specific and species specific differences in metals uptake. From this hypothesis it is important to mention that the non-essential trace metals such as lead and cadmium were highly concentrated in Lake Burullus eastern side (Nafea, 2005) and the industrial wastes may also be responsible for the elevation of the Pb and Cd in Lake Burullus. The order of abundance of the trace metals in the macrophytes and benthos of Lake Burullus were:

(1) Lead: *Lemna > Potamogeton > Eichh. Root > Ceratophyllum > phragmites sh. >Najas > Eichh. Sh., Mullusca>Arthropoda> Annelida.*

(2) Cadmium: *Lemna > potamogeton > Najas>phragmites > Eichhornia Root > Eichh. Shoot > Ceratophyllum.*

And in benthos: Mullusca>arthropoda> Annelida

(3) Zinc: *Lemna > Ceratophyllum > Potamogeton > Eichh. Root >phragmites shoot > Eichh. Shoot > Najas*

And in benthos: Mullusca>arthropoda> Annelida

(4) Copper: *Lemna > Eichh. Root > Najas >phrag. shoot >Eichh.Shoot >Ceratophyllum.> Potamogeton*

And in benthos: Mullusca > Annelida> arthropoda.

(5)Nickel: *Lemna > Potamogeton > Ceratophyllum > Eichh. Root > Najas > phrag. Shoot > Eichh. Shoot and in benthos: Mullusca > arthropoda > Annelida.*

(6)Iron: *Lemna > Ceratophyllum >Eichh. Root > Najas > phrag.shoot> Potamogeton > Eichh. Shoot and in benthos: Mullusca>arthropoda> Annelida.*

Lemna gibba, *potamogeton pectinatus* and Mullusca showed the higher capacity of heavy metal accumulation than the other aquatic plants and benthos groups. Aquatic macrophytes and benthos can be used as bio-indicator and biomarkers for water and sediment pollution as they can trap micro- and macro-elements (inorganic pollutions) as investigated by Fayed and Abdel-Shafy (1985). El-Khatib and Sawaf (1998) reported that the concentrations of heavy metals in macrophytes were positively related to the concentration in the environment and the macrophytes have high potential power for pollution monitoring (Yamanowska et al., 1999).

Depending on the heavy metals concentration in the aquatic macrophytes and benthos it can be concluded that the aquatic macrophytes and benthos can accumulate heavy metals and have a restricted role in the treatment and control of pollution of the aquatic ecosystems. Accordingly, the macrophytes and benthos can be considerable as reliable way for Biomonitoring the heavy metals contamination in Lake Burullus. Trace metals concentration in macrophytes and benthos species widely differ. This can confirmed if a species is used for heavy metals monitoring within one or different areas. Ghobrial (2000b) reported that *Ceratophyllum demersum* can accumulate zinc more than Cu, Pb and Cd and acts as a potential biological filter for trace metals removal from domestic effluents and has a capacity to retain heavy metals in its tissues.

High range of the heavy metals concentration in the studied aquatic plants and benthos indicates different extent of pollution; this high variability is associated with the different absorption rate for the heavy metals by the aquatic plants and benthos (Pajević et al., 2003; Klink, 2004; Maria et al., 2006).

Recently, there has been growing interest in the use of metal-accumulating plants or benthos for the removal of heavy metals from contaminated aqueous streams, in the biological purification of waste water and in Biomonitoring of heavy metals pollution in the Egyptian lakes (Nafea, 2005).

We conclude that there is a uniform pattern of heavy metal variation in the macrophytes, sediments and benthos of Lake Burullus. In general, values of some metals like iron, zinc and copper are higher in almost all the specimens. This shows the universal importance of these macrophytes, sediments and benthos in cleaning up of the aquatic environment. The results presented here could be very useful for environmental monitoring and checking the health of the water body. The aquatic macrophytes and benthos were found to be the potential source for accumulation of heavy metals from water and sediments and act as biofilters for metals. Accordingly they could be used in sustainable development, management and pollution assessment program in the northern deltaic lakes of Egypt especially lake burullus.

Conflict of Interests

The author(s) have not declared any conflict of interests.

REFERENCES

Abo-Rady MD (1977). Die Belastungder oberch Leine mits chwermetallen. Durc Kommunate and. Industrielle Abwasser, ermittelt and hand von wasser, sediment, F.shared prlanzenunt, enchungen. Dess. Univ. Gyttingen (FRG) 120.

Alloway, B.J., Davis, B.E (1971). Heavy metal content of plants growing on soils contaminated by lead mining. J. Agric. Sci. Camb. 76:321-323.

APHA (1992). Standard methods for the examination of water and waste water, Washington, 18 Ed.

Bauda R, Galanti P, Guilizzoni P, Varini P (1981). Relationship between heavy metals and aquatic organisms in Lake Manzala hydrograph systems (Northern Italy): Mem. IST. Ital. Hdrobiol. 39:203-225.

Chung IH, Jeng SS (1974). Heavy metal pollution of Ta-Tu River. Bull. Inst. Zool. Acad. Sci.13:69-73.

Crowder A, Painter P (1991). Submerged macrophytes in Lake Ontario: current know edge, importance, threats to stability and needed studies. Can. J. fish. Aquat. Sci. 48:1539-1545.

Devlin RM (1967). Plant Physiology. Reinhold, New York. p. 564.

El-Khatib A, El-Sawaf N (1998). Differential trapping of heavy metals by macrophytes in different water bodies near Sohag, Upper Egypt. Acta Hydrobiologica 40:67-73.

El-Sarraf WM (1995a). Trace metals concentration of aquatic Macrophytes in Lake Manzalah, Egypt. Bull. Nat. Inst. Ocean. and Fish. A.R.E. 21(1):171-181

El-Sarraf, W.M. (1995b): Chemical analysis of Some macrophytes in Maruit and Idku Lakes, Egypt. Alex. J. Agr. Res.40 (1): 255-271.

EPA (1983). Methods of chemical analysis of water and waste water USA. Environmental Protection Agency (EPA) 690:4-79.

Fayed S, Abd-El-Shafy H (1985). Accumulation of Cu, Zn, Cd and Pb by aquatic Macrophytes. Environ. Int. 11(1):77-88.

Gbaruko, B. C.and. Friday, O.U. (2007): Bioaccumulation of heavy metals in some fauna and flora Int. J. Environ. Sci. Tech., 4 (2): 197-202,

Ghobrial MG (2000b). Treatment of Cadmium, Copper, Zinc and Iron in waste water by the Horn wort Ceratophyllum demerssum. Egypt. J. Aquat. Biol. Fish. 4(1):35-46.

Guilizzoni P (1991). The role of heavy metals and toxic materials in thePhysiological ecology of submersed macrophytes. Aquat. Bot. 41:87-109.

Gulati KL, Nagpaul KK, Bukhari SS (1979). Uranium, boron, nitrogen, phosphorus and potassium in leaves of mangroves, Mahasagar Bull. Nat. Inst. Oceangra. 12:183-186.

Harris RR, Santos MCF (2000) Heavy metal contamination and physiological variability in the Brazilian mangrove crabs Ucides cordatus and Callinectes danae (Crustacea: Decapoda). Mar. Biol. 137:691-703.

Heydt G (1977). Schwerrnetallg chalteven Wasser, wasser.p franzen, chironomikae and Mullusca der. Elsenz. Dipl. Arbeituciv, Heidelberg. 143.

Hillman WS, Culley DD (1978). The used of Duck weed. Am. Sci. 66:442-451.

Klavinš M, Briede A, Parele E, Rodinov V, Klavina I (1998). Metal accumulation in sediments and benthic invertebrates in Lakes of Latvia. Chemosphere 36 (15): 3043-3053.

Metcalfe-Smith JL (1994). Influence of species and sex on metal residues in fresh water mussels (family Unionidae) from the St. Lawrence River, with implications for biomonitoring programs. Environ. Toxicol. Chem. 13: 1433-1443.

Moore P, Chapman S (1986): Methods in plant ecology. Second edition, Blackwell Scientific Publication.

Nafea EMA (2005). On the ecology and sustainable development of the northern delta lakes, Egypt. PhD Thesis, Mansoura University Faculty of Science.

Radwan AR, Lotfy IH (2002). On the pollution of Burullus Lake water and sediments by heavy metals. Egypt. J. Aquat. Biol. Fish 6(4):147-164.

Radwan, A., shokier, L. (2005): study on the heavy metals in some fishes and aquatic plants of Burullus Lake Bull. Nat. Inst. Ocean. Fish. AR.E, 32: 215-231.

Sawicka- Kapusta K (1978). Estimation of the content of heavy metals in atlas of rae-deer from silesian Woods. Arch. Ochr. Sord. 1:107-121.

Seidal K (1966). Biologischer seenschutzpfan zer wasser filter-foederation Europaischer Gewasser schutz. Symp. 76: 357 - 195.

Basement and alluvial aquifers of Malawi: An overview of groundwater quality and policies

Harold W. T. Mapoma[1,2] and Xianjun Xie[1]

[1]School of Environmental Studies, China University of Geosciences (Wuhan), Lumo Road 388, Wuhan 430074, Hubei, China.

[2]Department of Physics and Biochemical Sciences, University of Malawi, The Polytechnic, P/ Bag 303, Blantyre 3, Malawi.

This paper highlights the quality of groundwater in basement and alluvial aquifers of Malawi through literature assessment. Groundwater in these aquifers serves about 60% of Malawian population. Alluvial aquifers yield high groundwater in excess of 10 L/s and more mineralized than basement aquifers. The values from literature are presented as ranges. The geochemical quality of both aquifer types are classified as good. However, in some cases values higher than maximum permissible limits (MPL) are worrisomely apparent. Significant levels for some elements have been demonstrated. Although groundwater policies and instruments are available, more groundwater research, monitoring, data archiving is needed.

Key words: Alluvial aquifer, basement complex, groundwater quality, borehole water, shallow wells, Malawi.

INTRODUCTION

Potable water is a core to sustainable development. Most of the surface water in use for domestic and industrial purposes is highly contaminated, thus requiring a great deal of treatment before reaching the end user. In some regions, even this surface water is scarce owing to the great distances covered just to get to the nearest water point. Besides, a great deal of treatment required to make surface water fit for use makes it a difficult task for communities, especially in developing countries. The alternative is groundwater, which is probably the cheapest and safest potable water source (Foster, 1986). Groundwater is perceived to be cleaner in terms of contaminants. However, depending on various factors, groundwater may be contaminated as well. Its physic-chemical characteristics may exceed recommended quality standards. This may arise from either geogenic control or anthropogenic interference. Various quality parameters in groundwater have been assessed (Adekunle et al., 2007; Rowland et al., 2008; Muneer et al., 2010). In some instances, dangerous high quantities of health threatening elements have been documented (Smedley and Kinniburgh, 2002; Xie et al., 2013; Wen et al., 2013) Correlations and impacts of groundwater contaminants on human health have been highlighted (Lyman et al., 1985; Hall et al., 2002). To safeguard human health, policies and standards regarding the use of groundwater for domestic and commercial purposes have been formulated in various countries. Most countries use the WHO standards (WHO, 2008) or have modified them to suit local conditions.

Most Malawians in rural areas depend on groundwater for domestic purposes (Government of Malawi (GOM), 2011). Open shallow wells have been in use since time immemorial. Recently, a lot of boreholes have been sunk through government initiatives, non-governmental organizations and international agencies (Stoupy and Sugden, 2003; Grimason et al., 2013). Borehole wells offer contamination freer water and guarantee safety to the drawer. Open wells on the other hand are prone to deliberate contamination and present physical dangers of falling in. Urban areas in Malawi depend much on surface water for potable water use, however this water flow originate from base flow in hills and mountains (Ngongondo, 2006). As such, variations in aquifer water quantity and quality has a significant contribution to variations in river water volume and quality.

Various policy documents have been formulated in Malawi as regards water resources use and degradation (GOM, 2011). The scientific community has been keen on quantity and quality of groundwater in Malawi as well as the SADC region. However, research work is still fragmented and data is not easy to come-by. This paper seeks to identify literature on Malawi's groundwater and use the same to summarize physico-chemical and biological values and related policies. At the same time, it will highlight various strides undertaken in Malawi as regards groundwater research and present areas where most study results have been published. The goal is to bring together values in one paper that will act as a starting point when identifying references about groundwater literature in Malawi.

SOURCES OF INFORMATION

A desk review approach was employed in this study. Google scholar, PubMed and Elsevier Scorpus were used to search for journal articles. General information about Malawi and policy documents were sourced from various government websites and published materials related to groundwater. Besides, some information came from theses available on university websites. The use rights were considered.

GEOLOGY OF MALAWI

Geography, general relief and climate

Malawi is a sub-Saharan African country south of the equator situated between latitudes 9° 22′ and 17° 7′ south and between longitudes 32° 40′ and 35° 55′ east. It is bordered by Mozambique to the south east and south west, Zambia to the west and north-west and Tanzania to the north east and north (Figure 1A). It has a geographical area of 118480 km^2. According to 2012 world bank records, Malawi has a population of 15.91 million (World Bank, http://data.worldbank.org/indicator/SP.POP.TOTL).

Mkandawire (2004) explained the geologic and climate characteristics of Malawi as follows. Malawi has an array of relief that strongly influences climate, hydrology and groundwater occurrence. The topography consists of (1) Plateaus that are gently undulating surfaces with broad valleys and large level areas on the interfluves generally located at 900-1,300 m above mean sea level (msl). They are extensively covered by a thick weathered material; (2) Mountainous areas that rises abruptly from the plateau, located 2,000–3,000 m above msl covered by a more erosion resistant underlying strata; (3) Escarpment that fall steeply from the plateau areas underlain by poor and fractured aquifer where erosion significantly strips away the weathering products; and, (4) The Rift valley floor (hosting the alluvial plain) located below 600 m above msl, is gently sloping and of very low relief (Figure 1B). There is considerable potential for groundwater in these areas. Based on topography, the country is divided into Water Resource units (Figure 2).

In terms of climate, Malawi's seasons are influenced by the migration of the Inter-Tropical Convergence Zone (ITCZ). However, climatic conditions are complex due to the wide topographic variation and the influence of Lake Malawi. Much higher rainfall is experienced in uplands as compared to low lands. The mean annual temperature is 22–23°C with less latitudinal effect.

Rainfall variability has a significant impact on variations in both water availability and total potential discharge in a particular catchment (McCarthy et al., 2001; Bloomfield, 2002; Ngongondo, 2006). However, processes other than rainfall play a significant role in the hydrologic processes of any catchment (Ngongondo, 2006). The numerous rivers, lakes, dams, lagoons and marshes existing in Malawi (GOM, 2011) connect with groundwater in either recharge zones or mainly in discharge zones.

Aquifer geology and characteristics

The geological setting of Malawi is mainly characterized by crystalline metamorphic and igneous rocks of Precambrian to lower Palaeozoic age (Carter and Bennett, 1973; Chilton and Smith-Carington, 1984). This is referred to as the basement complex that covers approximately 70% of Malawi's landscape and supplying water to about 60% of the population (National Statistics Office, 2002). The major lithological units of the basement aquifer complex are syenitic granites, charconoctic and ultra-basic

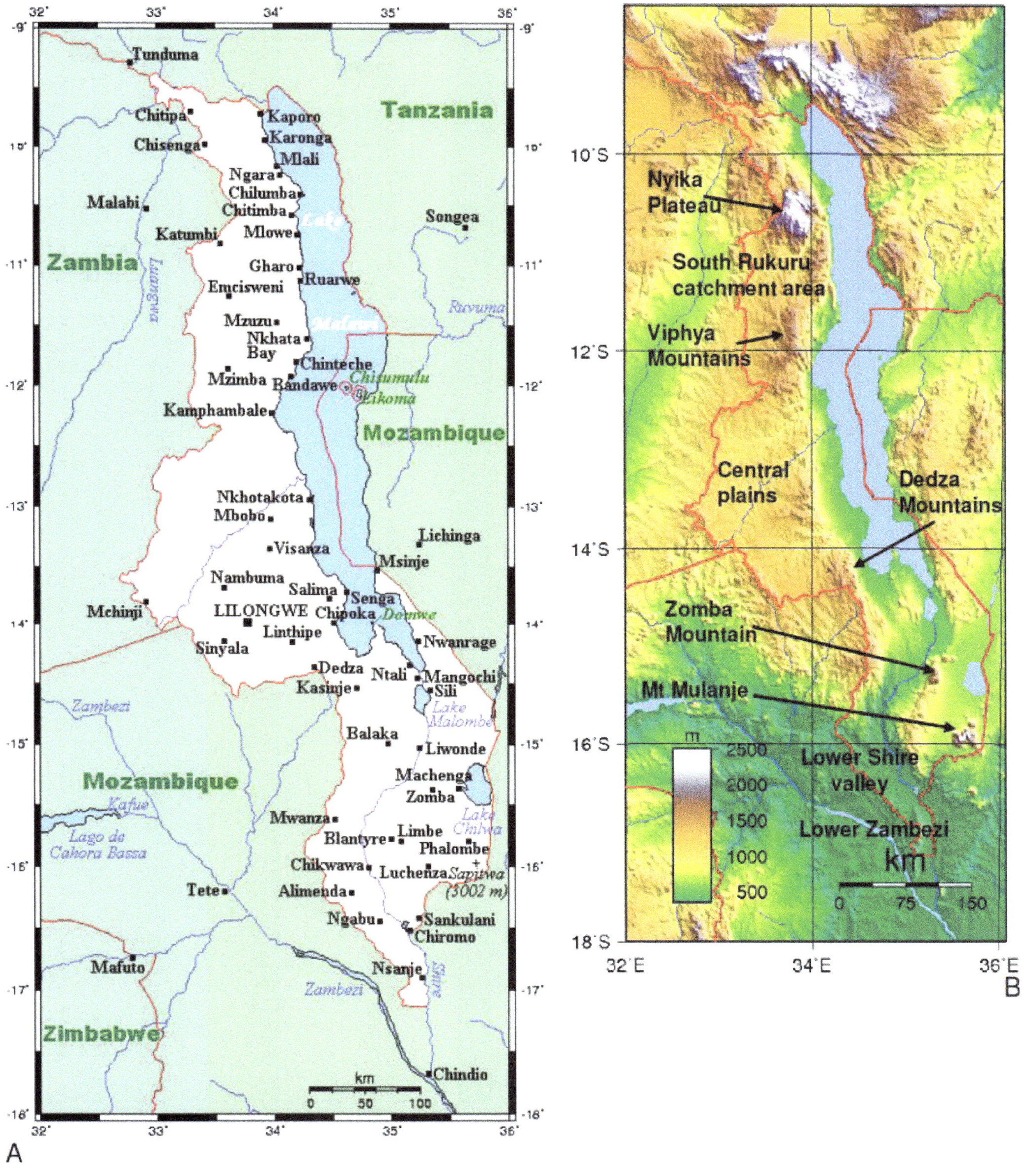

Figure 1. Map of Malawi illustrating (A) the political and administrative regions and (B) topographic zones.

gneisses, schistis, granulalite and quartzites (Mkandawire, 2004; Monjerezi and Ngongondo, 2012). Large inselberg of these rocks rises above the plateau as a result of epeirogenic events (Chilton and Smith-Carington, 1984). The lift valley has modified and interrupted this landscape. Groundwater is of variable qualities and quanti-

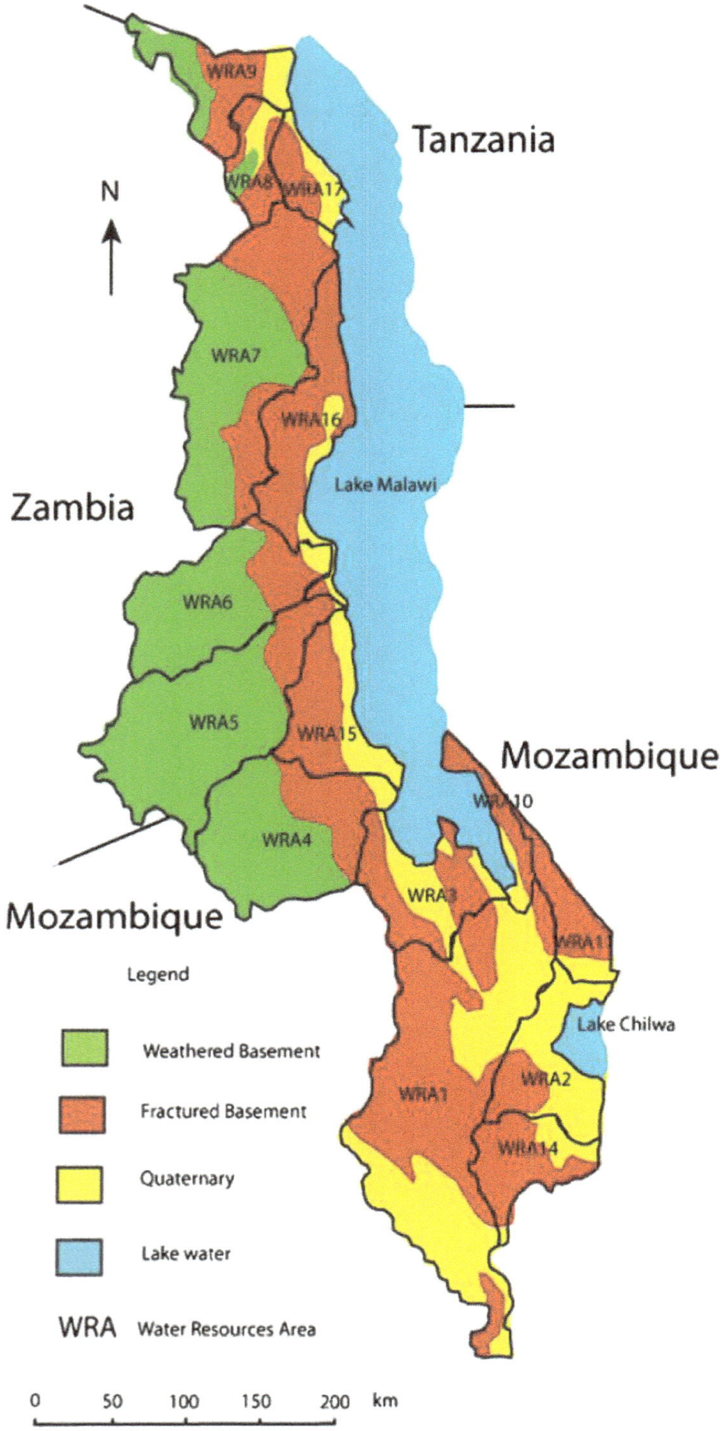

Figure 2. Outline of the geology of Malawi and Water Resource Areas (WRA) (Robins et al., 2013).

ties, unevenly distributed in time and space and is subjected to poor conservation and management (GOM, 2007). Aquifer recharges from broad interfluves and groundwater discharge in surface depressions. In terms of chemistry, the shallow component is less mineralized than the deeper component (McFarlane and Bowden, 1992). This basement is divided into weathered and fractured basement aquifers (Robins et al., 2013) (Figure 2). Besides basement complex, the quaternary alluvial aquifer exists in Malawi (Chimphamba et al., 2009; Robins et al., 2013) located mainly along the Lake Malawi shores and Lower Shire valley within the Great Lift Valley floor (Figure 2).

According to Makandawire (2004), transmissivity of the weathered basement aquifers are generally in the range of 5 to 35 m^2/day with estimated hydraulic conductivities ranging from 0.01 to 1 m/d. The storage coefficient, on the other hand, has been assumed to range from 0.01 to 0.001 (Mkandawire 2004). It is suggested that the groundwater is under unconfined to confined conditions. The water table is closer to the surface in the valley region and deeper in higher grounds (Mkandawire, 2004).

Borehole yields are generally highest where the saturated thickness of the weathered zone is greatest and the parent bedrock coarsest (Chilton and Smith-Carington, 1984). In general, however, the weathered basement aquifer produces low borehole yields (GOM, 2007; Chilton and Smith-Carington, 1984). Despite this, domestic water use (either shallow wells or hand-pump boreholes) can be sufficiently obtained (Chilton and Smith-Carington, 1984). On the other hand, alluvial aquifers are high yielding with recorded yields in excess of 10 L/s (Chimphamba et al., 2009; GOM, 2007). McFarlane et al. (1992) observed and concluded that very flat land at least one kilometer from the berg is generally likely to be high yielding than sites close to the berg in the plains of central region of Malawi. This was done through statistical analysis using data from about 1500 boreholes in this region.

Despite the relative expanse of the aquifer system, significant physical and chemical characteristics are observed on a local scale due to differences in mineralogy and structure (Chilton and Smith-Carington, 1984). The heterogeneity in terms of hydraulic characteristics and lithology is due to fractures, weathered zones and intrusions that control the occurrence of aquifers (Chimphamba et al., 2009; Pritchard et al., 2007).

GROUNDWATER QUALITY IN MALAWI

Introductory remarks

As mentioned earlier, spatial variation in chemical and physical quality is attributed to the heterogeneity of the aquifer system, groundwater flow regime and weathering processes (Chilton and Foster, 1995). Some shallow wells dry up in dry season forcing people to use surface water (e.g. rivers) which are grossly contaminated. Contamination of shallow wells is due to sanitation facilities being close to the wells (Dzwairo et al., 2006). The Water Resources Act of Malawi stipulates that water points be upstream of sanitation facilities. Fourie and van Ryneveld (1995) wrote that pollution from on-site is influenced by (1) varying subsurface conditions especially saturated versus unsaturated zones, (2) varying contaminant characteristics mainly mobility and persistence, and (3) varying mechanism of movement through materials.

Determination of quality of groundwater is very vital. For instance, determination of nitrates is considered significant as an indicator for water quality due to its association with blue baby syndrome effect, possible carcinogen and crude fecal pollution indication (Fourie and van Ryneveld, 1995). In general, determination of physical and chemical quality serves to provide a guide on the quality of groundwater for all purposes. In the case of poor quality, the results help in identifying and prescribing the water borne health implications in an area.

Aquifer quality characteristics

Various studies on groundwater quality have been carried out in several districts in Malawi. Table 1 summarizes the range of biological, physical and chemical quality parameters observed in various districts. The values in Table 1 are quoted from various papers published in refereed journals, conference proceedings and reports. However, due to limited papers found so far on Malawian groundwater, the values presented were all found from the search engines. The ranges are presented as they are found in the cited papers. Some large values where rounded off to remove decimal points. Some significantly higher range values are explained by the various authors themselves. However, this paper will try to highlight some of the issues already explained by the various authors on the values. The summarized results are for both boreholes and shallow wells. Shallow wells can be open unprotected or closed protected wells (Pritchard et al., 2007, 2008, 2009; Mkandawire, 2008). As such, high variations in range of values are expected if protected shallow wells are sampled together with unprotected open wells. Thus, the combined range of values will definitely result in a higher range. Most of the highest values for shallow wells are expected for open unprotected wells. Makandawire (2008) and Pritchard et al. (2008, 2009) observed higher values for fecal coliforms (FC) in unprotected open shallow wells both during dry and wet seasons. Similar observations were made for other parameters such as sulfates (SO_4), turbidity, pH and electric conductivity (EC). Some parameters improve during wet season while others worsen (Prichard et al., 2007, 2008, 2009). The difference in range values for the same district is an indication of the complexity of the aquifer. This depends on sampling location within the district even though the aquifer is either basement complex or alluvial. A synopsis of the results is summarized in Table 1a and b.

Table 1a. Range of expected Physical, Chemical and Biological quality of groundwater in Balaka, Blantyre, Chikhwawa, Chiradzulu, Karonga and areas in Central plains of Malawi.

Parameter	Balaka[a]	Blantyre	Chikhwawa	Chiradzulu	Dedza[k]	Karonga[l]	Central Plains[m]
pH	6.7 – 8.1	6.38 – 7.85[b] 6.42 –10.3[c] 6.56 – 8.81[d]	6.6 - 8.00[f] 5.0 – 9.0[h] 6.9 – 7.7[j]	6.14 – 7.91[c] 6.25 – 7.47[e]	6.4 – 7.7	6.3 – 8.1	6.4 – 7.0
TDS, mg/L	152 - 686	176 – 342[d;e] 172 – 338[c]	713- 2250[a] 16 – 26539[g]	74 – 179[c] 71 – 162[e]	99 – 2120	50 – 950	N/A
Turbidity, NTU	0.0 - 5.1	1.3 – 114[b] 0.0 – 62[e] 0.0 – 86[c]	0 – 209[i] 0 - 5.6[j]	0.0 – 812[c] 1.9 – 30.3[e]	1.5 - 12.5	N/A	N/A
EC, µS/cm	254 – 1428	180 – 383[b] 287 – 577[c] 294 – 570[e]	1450 – 2800[f] 254 – 15540[g] 221 – 6574[j]	174 – 299[c] 106 – 290[e]	195 – 4156	120 - 1735	N/A
Chloride, mg/L	N/A	13.6 – 35[b] 0 - 0.22[c] 0 - 0.14[e]	3.8 –11767[i] 8.1 –426[j]	0.0 – 0.18[c] 0.0 – 0.09[e]	8.5 – 248	4.7 – 990	2 – 19
Fluoride, mg/L	0.4 - 10.0	0.37 - 0.87[b]	0.59 - 1.93[f] 0.1 – 4.8[j]		0.2 - 1.1	N/A	0.5 – 7.02[n]
NO_3, (mg/L	N/A	0 - 3.2[c] 0 - 0.61[d] 0 - 1.92[e]	0.0 –178[j]	0.0 – 4.8[c] 0.0 – 0.03[e]	0.0 - 4.7	0.0 – 12.3	N/A
SO_4 , mg/L	0.0 - 33	0 – 83[c] 0 – 79[d] 0 - 18.5[e]	1.6 – 985[h] 1.1 – 2600[i]	0.0 – 113[c] 0.0 – 11.0[e]	5.1 – 887	0.0 – 505	114 – 2490
HCO_3, mg/L	N/A	N/A	143 –1314[h] 65 – 3110[i]	N/A	96 – 728	7.9 – 666	N/A
CO_3, mg/L	N/A	N/A	0.0 – 569[g]	N/A	0.0 – 49	0.0 - 72	N/A
Sodium, mg/L	N/A	15.3 – 24.3[b]	15.0 – 1178[j] 20.2 – 2050[h] 10 – 8320[i]	N/A	11.5 – 314	2.4 – 166	28 – 163
Potassium, mg/L	N/A	N/A	0.5 – 24.3[j] 0.4 – 13.8[h]	N/A	0.3 – 21.9	1.2 – 11.6	0.8 – 1.4
Calcium, mg/L	N/A	N/A	9.9 – 309.8[h] 6 – 876[i] 20.3 – 121[j]	N/A	15.8 – 351	3.6 – 412	62 – 555
Magnesium, mg/L	N/A	N/A	5.9 – 410.4[h] 1.0 – 1302[i] 12.8 – 54.3[j]	N/A	7.7 – 50.1	1.2 – 117	16 – 343
Iron , mg/L	N/A	0.1 – 0.8[b]	0.0 – 0.6[j]	N/A	0.0 – 1.1	N/A	0.02 – 0.23
Hardness, mg $CaCO_3$/L	108 - 1080	47 – 325[c] 47 – 210[d] 56 – 220[e]	6.0 – 1370[i]	3 – 165[c] 5.0 – 120[e]	71 – 1036	N/A	N/A
Fecal coliforms, cfu/100 mL	0 - 4230	0 – 11000[b] 0 - 28450[c] 0 – 17175[d]	0 – 9250[i]	0 – 29600[c] 13 – 4550[e]	N/A	N/A	N/A

Superscript denotes the main source of data for the district except where noted within the table: [a]Pritchard et al., 2008; [b]Palamuleni, 2002; [c]Pritchard et al., 2007; [d]Mkandawire, 2008; [e]Pritchard et al., 2009; [f]Sajidu et al., 2008; [g]Monjerezi et al., 2012a; [h]Monjerezi et al., 2012b; [i]Monjerezi and Ngongondo, 2012; [j]Grimason et al., 2013; [k]Kushe, 2009; [l]Wanda et al., 2013; [m]McFarlane and Bowden, 1992; [n]Msonda, 2003.

Table 1b. Range of expected Physical, Chemical and Biological quality of groundwater in Machinga, Mchinji, Mulanje, Mzimba, Nkhatabay, Rumphi and Zomba-Phalombe plains of Malawi.

Parameter	Machinga[o]	Mchinji[p]	Mulanje[c]	Mzimba[q]	Nkhatabay[r]	Rumphi[l]	Zomba-Phalombe
pH	6.70 – 9.37 7.05 – 9.50[f]	6.4 - 7.4	5.22 – 6.93	6.0 – 8.7	5.8 – 7.1	6.5 – 7.8	5.5 – 7.0[a] 6.30-7.15[f] 6.5 – 7.2[t]
TDS, mg/L	43 – 1772	39 - 215	27 – 161	29 - 1896	501 - 108	24 - 810	29 – 185[a]
Turbidity, NTU	Trace – 4.00	1 - 11	0.32 – 502	Bdl - 740	N/A	N/A	2.0 – 147[a]
EC, µS/cm	1050 - 31000 55 – 1676[f]	76 - 416	28.4 – 268	50 - 316	71 – 168	40 - 1630	129-805[f] 126 - 1806[t]
Chloride, mg/L	0.003 – 0.59	<0.01-17	0.0 – 0.96 0.01 - 0.17[e]	4.7 - 990	3.5 - 12.3	6.0 – 399	0.004-0.59[s] 0.0 – 40.5[t]
Fluoride, mg/L	0.35 – 10.3 0.24-7.51[f]	0.01-1.21	N/A	Bdl - 1.7	0.40 - 1.33	N/A	0.26-6.51[f] 0.0 – 4.5[t]
NO_3, mg/L	Trace – 58.3	0.01-0.39	0.0 – 2.0 0.0 – 0.01[e]	Bdl - 12.3	0 - 0.90	0.0 – 3.7	0.003 – 1.36[a] 0.0 – 48[t]
SO_4, mg/L	2.9 – 110	4.2-58.5	0.0 – 7 0.0 – 5.0[e]	Bdl - 505	0 - 0.91	0.0 – 53	2.9-110[s] 0.0 – 10[t]
HCO_3, mg/L	44.4 – 497	38.6-164	N/A	7.9 - 666	8.5 - 148	25 - 551	N/A
CO_3, mg/L	0.00 – 26.7	0	N/A	Bdl - 72.0	-	0.0 - 31	N/A
Sodium, mg/L	N/A	6 - 13	N/A	2.4 – 166	1.39 – 11.2	3.1 – 40	4 – 53[t]
Potassium, mg/L	N/A	0.7 - 6.8	N/A	1.0 – 11.6	0.84 – 6.16	1.0 – 12	N/A
Calcium, mg/L	N/A	9 - 45.4	N/A	3.6 – 412	6.94 – 43.9	7.4 – 142	4 – 47[t]
Magnesium, mg/L	N/A	2.7 - 12.6	N/A	0.97 – 117	1.23 – 4.32	1.0 – 45	N/A
Iron, mg/L	N/A	0.01 - 3.5	N/A	Bdl – 2.7	0.15 – 1.98	N/A	N/A
Hardness mg $CaCO_3$/L	3.49 – 199	25 - 161	0.0 – 325	14 - 1489	N/A	N/A	3.0 – 220[t]
Fecal Coliforms, cfu/100 mL	N/A	0 - 43	0 – 4020	N/A	0 – 0.5	N/A	0 – 6550[a]

Superscript denotes the main source of data for the district except where noted within the table: [a]Pritchard et al., 2008; [c]Pritchard et al., 2007; [e]Pritchard et al., 2009; [f]Sajidu et al., 2008; [l]Wanda et al., 2013; [o]Sajidu et al., 2007; [p]Mleta, 2010; [q]Wanda et al, 2011; [r]Kanyerere et al., 2012; [s]Chimphamba et al., 2009; [t]von Hellens, 2013.

Physical characteristics

Average values of pH in most cases indicate near-neutral groundwater compositions with a range of 6 - 8. The permissible levels of pH in groundwater for Malawi is 6 – 9 (Malawi Standards Board (MSB), 2005a). According to Table 1a and b, groundwater pH values lower than 6 for some places in Chikhwawa, Mulanje, Nkhatabay and Zomba can be encountered as well as those beyond pH 9 for some places in Machinga. Msilimba and Wanda (2013) reported a range of pH for groundwater in Mzuzu as 4.9 – 6.3 an indication of acidic waters in this area. Most of the groundwater recorded so far have pH range within the acceptable levels.

Total dissolved solids (TDS) were quoted as low, 24 mg/L with most of the higher values below the MPL value of 1000 mg/L (MSB, 2005a). However values higher than 1000 mg/L are reported in some areas such as those found in Chikhwawa, Dedza, Karonga, Machinga and Nkhatabay. These are same places where higher electrical conductivities were also observed. Chilton and Smith-Carington (1984) observed values of total dissolved solids up to around 2500 mg/L in Livulezi (central) and Dowa West (south-central). On the other hand, Bath (1980) concluded that total dissolved solids up to 2900 mg/L can be expected in Nkhotakota and surrounding lake shore areas in colluviums and weathered basement.

Higher turbidity range (Table 1) were observed for studies done in Blantyre (Palamuleni, 2002; Pritchard et al., 2007; Mkandawire, 2008; Pritchard et al., 2009), Chikhwawa (Monjerezi and Ngongondo 2012), Chiradzulu (Pritchard et al., 2007), Mulanje (Pritchard et al., 2007; Pritchard et al., 2009), Mzimba (Wanda et al., 2011) and Zomba-Phalombe plain (Pritchard et al., 2008). As mentioned earlier some of these observations were made for open unprotected wells (Pritchard et al., 2007, 2008, 2009; Mkandawire, 2008). The maximum permissible level for groundwater in Malawi is 25 NTU (MSB, 2005a) as such some places in the studied areas recorded in Table 1 had values higher than the MPL value. However, the average values indicated by authors themselves indicate that groundwater has values within acceptable levels in terms of turbidity.

The electrical conductivity (EC) values are in the range of 35 – 2800 μS/cm (Table 1). However, in some instances values higher than the maximum permissible level (MPL) of 3500 μS/cm (MSB, 2005a) can be expected for some places in Chikhwawa, Dedza and Machinga. Bath (1980) reported that electric conductivity as high as 7700 μS/cm can be observed in some areas in Malawi. Chilton and Smith-Carington (1984) also found mostly low-conductivity groundwater in basement aquifers from the Livulezi (central) and Dowa West (south-central) areas. In the Nkhotakota area on the western shores of Lake Malawi, Bath (1980) reported electrical conductivity values of 180–4600 μS/cm.

Chemical quality

The chloride (Cl) MPL for borehole and shallow well drinking water in Malawi is 750 mg/L (MSB, 2005a). However, chloride concentrations up to 4000 mg/L have been recorded in Shire Valley and lower than 60 mg/L in Nkhotakota area (Bath, 1980). BGS (2004) observed concentrations of chlorides up to 2100 mg/L in groundwater from alluvial deposits close to the edge of the Karoo sediments. In South Rukuru catchment area, chloride concentrations of 4 – 2000 mg/L were reported by Bath (1980). Except for Chikhwawa (Monjerezi et al., 2012b; Monjerezi and Gongondo, 2012), Karonga (Wanda et al., 2013), Mzimba (Wanda et al., 2011), and Rumphi (Wanda et al., 2013), all the areas in Table 1 show values below the MPL of 750 mg/L. Nevertheless, the high values recorded in these areas are not rampant, an indication that the average majority of sampled groundwater had values less than 750 mg/L with very few exceptions. This is what is making the range to be so highly spread.

It is expected that areas located in Rift Valley floor will have high fluoride (F) levels (BGS, 2004). As such, Malawian groundwater in the alluvial plains, are likely to be most affected. Sporadically, some groundwater in the weathered basement may also have high concentrations greater than Malawi standards of 6 mg/L (MSB, 2005a). Groundwater fluoride levels less than 1 mg/L in areas of the basement complex in Malawi and concentrations between 2 mg/L and 10 mg/L from the alluvial regions have been reported by UN (1989). Bath (1980) reported a median concentration of fluoride <1 mg/L in basement aquifers of the Nkhotakota area. However, one of the samples recorded a value higher than 1 mg/L that resulted in an inflated range of <1 – 7 mg/L. Elsewhere, concentrations in the basement aquifer of South Rukuru are reported in the range <0.1–3.3 mg/L, nonetheless most of the results were below 0.6 mg/L (Bath, 1980). Much higher fluoride concentrations, up to 15 mg/L in some samples, have been reported by Bath (1980) in the lower Shire valley. At Ulongwe in Machinga District, fluoride levels of 8.6 mg/L have been shown to exist (Sibale et al., 1998). Carter and Bennett (1973) also reported existence of fluorides in Karonga (8.0 mg/L), Mwanza (3.4 mg/L), Mazengera in Lilongwe (7.0 mg/L), Nathenje in LiLongwe (7.0 mg/L), Nkhotakota (9.6 mg/L) and Nsanje (5.8 mg/L). Spatial and temporal analyses of fluorides levels in Nathenje in Lilongwe (part of central plains) indicate higher values for central part of the area and during dry season, respectively (Msonda et al., 2007). From Table 1, higher values of fluoride concentrations are expected in some places in Balaka, Central plains, Machinga, Nsanje and Zomba-Phalombe plains. There is evidence of fluorosis problem in these areas (Msonda et al., 2007, Sajidu et al., 2007). Values of fluorides higher than 6 mg/L are worrisome in Malawi. However the problem of fluoride is sporadic in the basement aquifer and not widely spread in the alluvial aquifer (UN, 1989).

The nitrates (NO_3) MPL for groundwater used for drinking in Malawi is 45 mg/L (MSB, 2005a). Bath (1980) summarized fluoride data showing fluoride concentrations much less than the MPL value in Nkhotakota, Bua, South Rukuru catchement area and Lower Shire valley. Sporadically high concentrations of nitrates in groundwater from the lower Shire valley (as high as 18.5 mg/L) have been observed (Bath, 1980). Nevertheless, the study concluded that nitrate concentrations were mostly less than 5 mg/L and many were below 0.7 mg/L. Also, UN (1989) reported groundwater nitrate values less than 1 mg/L from both the basement and alluvial aquifers in Malawi. Observations from Table 1 indicate that nitrates higher than 45 mg/L are expected in some places in Chikhwawa (Grimason et al., 2013), Machinga (Sajidu et al., 2007) and Zomba-Phalombe plain (von Hellens 2013).

In terms of sulfates, Chilton and Smith-Carington (1984) reported concentrations higher than 2000 mg/L in Dowa west of central plains of Malawi. Sulfate up to 2400 mg/L were recorded in groundwater from alluvial deposits close to the edge of the Karoo sediments (BGS, 2004). The MPL for sulfates in borehole and shallow wells drinking water for Malawi is 800 mg/L (MSB, 2005a). According to Table 1a, higher values were observed in some places in Chikhwawa (Monjerezi et al., 2012b; Monjerezi and Gongondo, 2012), Dedza (Kushe, 2009) and Central plains of Malawi (McFarlene and Bowden, 1992). In general, sulfate levels in weathered basement and alluvial aquifers are mainly below the MPL value.

Chimphamba et al. (2009) indicate that the expected bicarbonate (HCO_3) concentrations in weathered basement aquifer are 100 – 500 mg/L while for alluvial aquifers across the country it is expected that the values fall within 200 – 1000 mg/L. Most values for the districts recorded in this study are below 500 mg/L in the weathered basement except for some samples in Dedza (Kushe, 2009), Karonga (Wanda et al., 2011) and Mzimba (Wanda et al., 2013). In Chikhwawa (Lower Shire valley floor), the values in some places were above the expected maximum value of 1000 in alluvial aquifer. Carbonate (CO_3) concentrations were as high as 569 mg/L (Table 1) in places recorded in this review so far.

The nationwide expected value for sodium (Na) is in the range 5 - 70 mg/L in weathered basement and 20 - 1500 mg/L in alluvial aquifer. On the other hand, expected values for potassium (K) are in the range of 1 - 6 mg/L for weathered basement aquifer and alluvial aquifer (Chimphamba et al., 2009). The MPL for sodium in Malawian groundwater used for drinking is 500 mg/L (MSB, 2005a). However, that of potassium is not stipulated. Nevertheless, potassium MPL for drinking water (tap water) is stipulated as 50 mg/L (MSB, 2005b). Bath (1980) reported sodium concentrations for Nkhotakota (0.4 - 720 mg/L) and Lower Shire (14 - 3550 mg/L). From Table 1, sodium concentrations falling outside nationwide expected range are reported in various places within the recorded districts in both weathered basement and alluvial aquifers such as Chikhwawa (Alluvial aquifer mainly), Dedza, Karonga, Central plains and Mzimba. In terms of drinking water quality for sodium, some places in Chikhwawa (Lower Shire) and Nkhotakota area are expected to have sodium values higher than MPL value of 500 mg/L (Bath, 1980; Monjerezi et al., 2012b; Monjerezi and Gongondo, 2012; Grimason et al., 2013). Similarly, some places in Chikhwawa, Dedza, Karonga, Mzimba and Rumphi exhibit high concentrations of potassium (Table 1) than the nationwide expectation of 1 - 6 mg/L but below 50 mg/L MPL (MSB, 2005b) stipulated for tap water. This indicates that in most places in these areas, the values will be below the acceptable maximum normal drinking water standard for potassium.

The MPL for calcium (Ca) and magnesium (Mg) in groundwater used for drinking in Malawi is 250 and 200 mg/L, respectively (MSB, 2005a). In Nkhotakota region, the groundwater is mainly of calcium-(magnesium)-bicarbonate type. In South Rukuru catchment area, concentrations of calcium up to 500 mg/L and magnesium up to 280 mg/L have been reported (Bath, 1980). According to Chimphamba et al. (2009), the nationwide expected value for calcium is in the range of 10 – 100 mg/L in weathered basement and 50 – 150 mg/L in alluvial aquifer. On the other hand, expected value for magnesium is in the range of 5 – 50 mg/L for weathered basement aquifer and 20 – 100 mg/L in alluvial aquifer (Chimpahmba et al., 2009). Based on the range indicated in Table 1, some places both in basement and alluvial aquifers had calcium values outside the expected nationwide range and in some instances higher than the recommended maximum permissible value of 250 mg/L, e.g. in some places found in Chikhwawa (Monjerezi et al., 2012b; Monjerezi and Gongondo, 2012), Dedza (Kushe, 2009), Karonga (Wanda et al., 2013), Central plains (McFarlene and Bowden, 1992), Mzimba (Wanda et al., 2011) and Rumphi (Wanda et al., 2013). Similarly, magnesium values in some places had values outside the expected nationwide range and in some instances higher than the recommended MPL of 200 mg/L such as in Chikhwawa and Central plains (Table 1).

The MPL for total iron (Fe) in groundwater for Malawian aquifers is 3 mg/L (MSB, 2005a). The nationwide expected range is 1 – 5 mg/L in both weathered basement and alluvial aquifers (Chimphamba et al., 2009). According to UN (1989), the range for total iron in weathered basement rocks and alluvial sediments is 1 – 5 mg/L. In Lower Shire Valley, a range of <0.1 – 84 mg/L have been reported and 0.8 - 82 mg/L in the weathered basement aquifers of Nkhotakota area by Bath (1980). Also, Bath (1980) recorded a range of total iron of <0.1 – 59 mg/L in Bua area of Central Malawi and <0.2 – 65 mg/L in South Rukuru catchment area. From Table 1, the highest value is 3.53 mg/L for Mchinji which falls below the maximum permissible level and within the expected range of 1 – 5 mg/L. Chilton and Smith-Carington (1984) explains the high values observed by Bath (1980) as possibly emanating from sampling methodology employed at that time that might have included particulate iron in the sample and likelihood of contamination from the pump material.

The MPL of total hardness (TH) in groundwater used for drinking is 800 mg $CaCO_3$/l (MSB, 2005a). The results in Table 1 indicate most values less than the MPL. However in some places the values recorded by cited authors in Table 1 show values above 800 mg/L in some places like Balaka, Chikhwawa, Dedza and Mzimba. It is therefore expected that some places in these areas will produce groundwater exhibiting characteristics of high hardness and unsuitable for drinking.

Manganese (Mn) concentrations of 0.1 – 0.4 mg/L have been reported in Central plains (McFarlane and Bowden, 1992), <0.001 mg/L in Mchinji (Mleta, 2010), <0.4 mg/L in Mzimba (Wanda et al., 2011) and 0 – 0.29 mg/L in Nkhatabay (Kanyerere et al., 2012). The values are below the MPL of 1.5 mg/L (MSB, 2005a).

Trace elements such as copper, lead, strontium, cadmium, boron, barium, beryllium, chromium, mercury have been analyzed in Nkhatabay and Lower Shire (McFarlene and Bowden, 1992; Kanyerere et al., 2012; Monjerezi et al., 2012; Grimason et al., 2013). However, no significant values were identified in these areas.

Phosphates of 0.16 – 1.16 were recorded in Nkhatabay (Kanyerere et al., 2012) while Sajdu et al. (2007) indicated that phosphates are not an issue in Machinga. Ammonia (0.0 – 0.5 mg/L) and nitrite (0.0 – 0.06 mg/L) were examined in samples in Blantyre (Mkandawire,

Table 2. Standards for drinking water in arid regions adopted in Malawi (Government of Malawi-UNDP, 1986).

Parameter	Good	Fair	Moderate	Poor
EC, µS/cm	0-750	750-1500	1500-3000	3000-6000
Na, mg/L	0-115	115-230	230-460	460-920
Mg, mg/L	0-30	30-60	60-120	60-120
Hardness, mg $CaCO_3$/L	0-250	250-500	500-1000	1000-2000
Cl, mg/L	0-180	180-360	360-710	710-1150
SO_4, mg/L	0-145	145-290	290-580	580-1150

2008) and were found to be lower than the MPL set by WHO (2008) of 1.5 and 3 mg/L, respectively. Arsenic is not a problem currently based on spot checks carried out by the ministry as well as studies done by various researchers in Malawi (Pritchard et al., 2007, 2008; Mkandawire, 2008; Kanyerere, 2012). All the arsenic values reported are below the MPL of 0.5 mg/L (MSB, 2005a).

Generally, it has been observed that alluvial aquifers probably have high salinity values particularly in lower Shire valley, eastern Bwanje valley and around Lake Chilwa (Bath, 1980; Monjerezi et al., 2011). In contrast, low salinity values for groundwater from weathered basement in the Bua catchment of western Malawi are on record (Bath, 1980). Elsewhere, in colluviums and basement aquifer, a very variable salinity has been observed in South Rukuru catchment area and in some instances high salinity values were recorded around Emcisweni in Mzimba District (Bath, 1980).

Fecal coliforms

Aquifer microbial contamination is from pit latrines, septic tanks, cesspool, leaky sewers and landfill leachate (Pedley and Howard, 1997). Major variations in fecal coliforms are observed between wet and dry seasons. Wet seasons are expected to produce shallow well waters with high fecal coliforms, especially if open and unprotected. However boreholes exhibit lower and below MPL for fecal coliform. Fecal coliform values of 129 – 920 colony forming units (cfu)/100 mL were observed in shallow wells in Mzuzu (Msilimba and Wanda, 2013). The authors concluded that the wells are contaminated since the values exceed the maximum recommended value of 50 cfu/100 ml (MSB, 2005a). Similar situations were observed by authors in Balaka, Blantyre, Chikhwawa, Chiradzulu, Mulanje and Zomba-Phalombe plain (Table 1).

Overall quality of groundwater

In general, high contents of cations, anions and total dissolved solids (TDS) are limiting factors to the use of groundwater (BGS, 2004; Kundell, 2008). However, based on Table 2, groundwater quality in both basement and alluvial aquifer in Malawi can be classified on average as good in terms of major cations, total hardness and sulfates. But, there are sporadic exceptions in few cases in some places within the studied districts where the groundwater quality in some parameters ranges from moderate to poor as can be observed from Table 1 when compared with Table 2.

POLICIES ON GROUNDWATER IN MALAWI

GOM (2008) identified challenges facing the water sector as (1) unharmonized policies and laws, (2) inadequate stakeholder coordination, (3) poor catchment management, (4) lack of capacity and (5) inadequate water supply and sanitation. Policies and legislation related to water resources, on the other hand, have been formulated such as the National Environmental Policy (NEP), National Strategy for Sustainable Development (NSSD), Environmental Management Act (EMA), National Environmental Action Plan (NEAP), District Environmental Action Plans (DEAPs), National Water Policy (NWP), Water Resources Act (WRA), Water Works Act (WWA) and other sectoral policies which also focus on the water resource such as the Agriculture Policy and the Forestry Policy (GOM, 2011). Standards on groundwater are available and the Malawi Standards Board is responsible for their formulation and updates. However, there are still coordination, enforcement and monitoring challenges that reduce the impact of these instruments (GOM, 2011).

Malawi is a member of the Southern Africa Development Community (SADC) that has a regional groundwater vulnerability initiative (Robins et al., 2007). The SADC established a water sector in 1996 (Molapo and Puyo, 2004). Besides, SADC has a regional strategic action plan for integrated water resource development and management that establishes rules and procedures to implement joint management of water resources between countries. There is also the Zambezi river basin commission within SADC for which Malawi is party to (Molapo and Puyo, 2004). The regional grouping established a subcommittee for hydrogeology that serves as supervisor or steering group for implementation of projects of regional magnitude (Molapo and Puyo, 2004). SADC protocol on shared watercourse systems of 1995 (and revised in 2000) places great emphasis on governance of shared water (Ramoeli, 2002; Pietersen 2005; Molapo and Puyo, 2004). It should be noted that the aquifers in Malawi are shared among adjacent states (Mkandawire, 2004; Turton et al., 2004). However, the geographic extent of the transboundary aquifers is not known which leads to management problems (Turton et al., 2004). Above all, Government of Malawi is trying its best to achieve the targets set in the millennium development goals (MDGs) and other policies. It is conclusively clear that Government of Malawi is striving at best in groundwater resources management despite the existence of various challenges.

SUMMARY OF INADEQUACIES AND POSITIVE STRIDES ON GROUNDWATER IN MALAWI

There have been some studies on groundwater in Malawi as shown in the preceding sections. Besides quality studies, some cited examples include (1) investigation of the weathered basement aquifers (Jones, 1985; Acworth, 1987; Chilton and Foster, 1995), (2) major village water supply programme featuring the introduction of village level operation and maintenance of hand pumps (Smith-Carington and Chilton, 1983), (3) Dambo research (McFarlane and Whitlow, 1990; McFarlane and Bowden, 1992) and (4) improving community based management of boreholes (DeGabriele, 2002).

However, no time series hydrologic data of any type are being collected and archived except for data collected during the drilling of boreholes (Robins et al., 2006). There is no regular groundwater monitoring done indicating a deficiency in sustainability interventions of the resource (Mleta, 2010; GOM, 2011). On a positive note, percentage of rural population with improved water source including boreholes, protected dug wells, is 82% as at 2011 as compared to 75% in 2008 (World Bank, http://data.worldbank.org/indicator/SH.H2O.SAFE.RU.Z). The responsibility of monitoring groundwater quality is vested in the hands of Department of Water through its regional and district offices.

Some significant studies on improvement of ground water quality in Malawi have been carried out such as defluoridation using clays (Msonda, 2003), gypsum (Masamba et al., 2005) and bauxite (Sajidu et al., 2008 and 2012). Pritchard et al. (2009) studied the potential of using indigenous methods of purifying water by utilizing plant extracts. These studies have yielded remarkable results. Various groundwater modeling application studies have been done in some areas (Monjerezi et al., 2012b; Wanda et al., 2013). Besides, a comprehensive groundwater resource assessment has not been examined for Malawi yet (Mkandawire, 2004).

CONCLUSIONS

Groundwater quality in Malawi can be classified as good except for some sporadic zones where some elements exceed recommended drinking water guideline values; Chimphamba et al. (2009) quoted standards adopted by Government of Malawi for drinking water in arid regions. Alluvial aquifers have high values of fluorides in some places, a situation that leads to fluorosis problems. Traces of similar incidences have been observed in weathered basement aquifers. High content of cations, anions and total dissolved solids have led to well abandonment. Saline groundwater conditions have been observed in some places like eastern Bwanje valley, Lake Chilwa and Lower Shire. However serious incidences of groundwater contamination are not imminent as indicated by fecal coliform, nitrate and sulfate observations, except in shallow wells. The amount of chemical elements found in these aquifers can be attributed mainly to the geological setting of the aquifer complex itself. The paper has shown that most of the studies are concentrated in southern region especially the Shire Valley. Thus, the conclusions are based on district level studies. More flow modeling and reactive groundwater modeling are needed to understand the hydrochemistry of the aquifer as well as mobilization and speciation of chemical species within. Transboundary groundwater data sharing, archiving and monitoring need to be in place for a better management of the shared aquifer. Multi-country research on the basement aquifer is necessary to promote sharing of knowledge between research establishments and water management institutions.

Conflict of Interests

The author(s) have not declared any conflict of interests.

ACKNOWLEDGEMENTS

We owe our gratitude to China University of Geosciences, China Scholarship Council and University of Malawi for the full support of this work. We also acknowledge the Malawi Government for making the policy documents and relevant papers readily available for academic use. We don't take for granted the expert reviewers for their invaluable input. To the authors of reviewed manuscripts (in part or in full), we are very grateful for your work. Without your work this paper would not have materialized.

REFERENCES

Acworth RI (1987). The development of crystalline basement aquifers in a tropical environment. Q. J. Eng. Geol. 20(4): 265–272

Adekunle IM, Adetunji MT, Gbadebo AM, Banjoko OB (2007). Assessment of groundwater quality in a typical rural settlement in southwest Nigeria. Int. J. Environ. Res. Public Health 4: 307-318.

Bath AH (1980). Hydrochemistry in groundwater development: report on an advisory visit to Malawi. British Geological Survey Report. WD/OS/80/20.
BGS (British Geological Survey) (2004). Groundwater quality: Malawi. Natural Environmental Research Council (NERC). Available on http://www.wateraid.org [accessed in September 2013].
Bloomfield JP (2002). Trends in average and extreme groundwater levels in the Chalk. British Geological Survey/National Environmental Research Council.
Carter GS, Bennett JD (1973). The geology and mineral resources of Malawi. Bull. Geol. Surv. Malawi 6.
Chilton PJ, Foster SSD (1995). Hydrological characterization and water-supply potential of basement aquifers in tropical Africa. Hydrogeol. J. 3(1): 36 – 49.
Chilton PJ, Smith-Carington AK(1984). Hydrogeological characterisation and water-supply potential of basement aquifers in tropical Africa. Hydrogeol. J. 3: 36-49.
Chimphamba J, Ngongondo C, Mleta P (2009). Groundwater chemistry of basement aquifers: A case study of Malawi, In: Titus R, Beekman H, Adams S, Strachan L, (Eds.). Basement Aquifers of Southern Africa. WRC Report No. TT 428-09. ISBN 978-1-77005-898-9.
DeGabriele J (2002). Improving community based management of boreholes: A case study from Malawi. Available from: http://www.ies.wisc.edu/ltc/live/basprog9.pdf [Accessed August 2013].
Dzwairo B, Hoko Z, Love D, Guzha E (2006). Assessment of the impacts of pit latrines on groundwater quality in rural areas: a case study from Marondera district, Zimbabwe. Phys. Chem. Earth 31: 779–788.
Foster SSD (1986). Getting to grips with groundwater pollution protection in developing countries. Natural Resources Forum. United Nations, New York.
Fourie AB, van Ryneveld MB (1995). The fate in the subsurface of contaminants associated with on-site sanitation: a review. Water SA 21(2): 101-111.
GOM (2007). National Water Policy, 2nd Edition. Ministry of Irrigation and Water Development. Capital Printing Press. City Centre, Lilongwe, Malawi.
GOM (2011). National State of Environment Report, 2011. Environmental Affairs Department, Lilongwe
GOM (Government of Malawi) (2008). The Second National Communication of the Republic of Malawi to The United Nations Framework Convention on Climate Change. Climate Change Project.
Grimason AM, Morse TD, Beattie TK, Masangwi SJ, Jabu GC, Taulo SC, Lungu KK (2013). Classification and quality of groundwater supplies in the Lower Shire Valley, Malawi – Part 1: Physico-chemical quality of borehole water supplies in Chikhwawa, Malawi. Water SA 39(4): 563 – 572.
Hall AH (2002). Chronic arsenic poisoning. Toxicol. Lett. 128: 69–72.
Jones MJ (1985). The weathering zone aquifers of the basement complex areas of Africa. Q. J. Eng. Geol. 18: 35–46.
Kanyerere T, Levy J, Xu Y, Saka J (2012). Assessment of microbial contamination of groundwater in upper Limphasa River catchment, located in a rural area of northern Malawi. Water SA 38(4): 581–596.
Kundell J (Ed.) (2008). Water profile of Malawi. Food and Agriculture Organisation. http://www.eoearth.org/article/water_profile_of_Malawi [Accessed September 2013].
Kushe JF (2009) Assessment of the chemical quality of groundwater for drinking in Dedza District, Malawi. MSc Thesis. University of Zimbabwe (Available Online). http://ir.uz.ac.zw/jspui/bitstream/10646/1038/1/Thesis%20Kushe%20%20Final%20Thesis%202008-2009.pdf. [Retrieved July 2013].
Lyman GH, Lyman CG, Johnson W (1985). Association of leukemia with radium groundwater contamination. JAMA 254(5): 621-626.
Malawi Government - United Nations Development Programme (1986). National Water Resources Master Plan: Groundwater Resources of Malawi. Department of Water, Ministry of Works and Suppliers. United Nations.
Malawi Standards Board (MSB) (2005a). Malawi standard; borehole and shallow well water quality specification MS 733:2005. Blantyre, Malawi.
Masamba WRL, Sajidu SMI, Thole B, Mwatseteza JF (2005). Water defluoridation using Malawi's locally sourced gypsum. Phys. Chem. Earth Parts A/B/C. 30(11-16): 846 - 849
McCarthy JJ, Canziani OF, Leary NA, Dokken DJ, White KS (Eds.) (2001). Climate change 2001: impacts, adaptation, and vulnerability. Contribution of Working Group II to the Third Assessment Report of the Intergovernmental Panel for Climate Change (IPCC) Report. GRID ARENDAL.
McFarlane MJ, Bowden DJ (1992). Mobilization of aluminium in the weathering profiles of the African Surface in Malawi. Earth Surf. Process. Landforms 17: 789 – 805.
McFarlane MJ, Chilton PJ, Lewis MA (1992). Geomorphological controls on borehole yields: a statistical study in an area of basement rocks in central Malawi. Geological Society, London, Special Publications. 66: 131–154.
McFarlane MJ, Whitlow R (1990). Key factors affecting the initiation and progress of gullying in dambos in parts of Zimbabwe and Malawi. Land Degradation Rehabil. 2: 215 – 235.
Mkandawire PP (2004). Groundwater resources of Malawi, In: Appelgren B (Ed.). Managing Shared Aquifer Resources in Africa. IHP-VI, Series on Groundwater No. 8. UNESCO, France.
Mkandawire T (2008). Quality of groundwater from shallow wells of selected villages in Blantyre District, Malawi. Phys. Chem. Earth. 33: 807–811.
Mleta PWEC (2010). An Analysis of Groundwater in Mchinji District of Central Malawi. MSc Thesis. University of Fort Hare, Zimbabwe (Available online) http://ufh.netd.ac.za/bitstream/10353/366/1/Mleta%20(M%20Sc)%20Geography.pdf [Retrieved August 2013].
Molapo P, Puyo S (2004). Transboundary aquifer management in the context of integrated water resources management in the SADC region, In: Appelgren, B. (Ed.). Managing Shared Aquifer Resources in Africa. IHP-VI, Series on Groundwater No. 8. UNESCO, France.
Monjerezi M, Ngongondo C (2012). Quality of groundwater resources in Chikhwawa, lower Shire valley. Malawi. Water Qual. Exp. Health 4: 39–53
Monjerezi M, Vogt RD Gebru AG, Saka JDK, Aagaard P (2012b). Minor element geochemistry of groundwater from an area with prevailing saline groundwater in Chikhwawa, lower Shire valley (Malawi). Phys. Chem. Earth 50–52: 52–63.
Monjerezi M, Vogt RD, Aagaard P, Saka JDK (2011). Hydro-geochemical processes in an area with saline groundwater in lower Shire River valley, Malawi: An integrated application of hierarchical cluster and principal component analyses. Appl. Geochem. 26: 1399–1413.
Monjerezi M, Vogt RD, Aagaard P, Saka JDK (2012a). The hydro-geochemistry of groundwater resources in an area with prevailing saline groundwater, lower Shire Valley, Malawi. J. Afr. Earth Sci. 68: 67–81.
MSB (2005b). Malawi standard; drinking water – specification. Malawi Standards Board, MS 214:2005, ICS 13.030.40 (first revision).
Msilimba G, Wanda EMM (2013). Microbial and geochemical quality of shallow well water in high-density areas in Mzuzu City in Malawi. J. Phys. Chem. Earth. 66: 173-180
Msonda KWM, Masamba WRL, Fabiano E (2007). A study of fluoride groundwater occurrence in Nathenje Lilongwe Malawi. Phys. Chem. Earth 32: 1178–1184.
Msonda WK (2003). A study of groundwater quality, water defluoridation and impact of dental fluorosis on children in Nathenje, Lilongwe, Malawi. MSc Thesis. University of Malawi. Malawi.
Muneer M, Bhatti IA, Adeel S (2010). Removal of Zn, Pb and Cr in Textile Wastewater Using Rice Husk as a Biosorbent. Asian J. Chem. 22(10): 7453-7459.
National Statistical Office (2002). Malawi Core Welfare Indicators Questionnaire Survey: Report of Survey Results. Zomba, Malawi.
Ngongondo CS (2006). An analysis of long-term rainfall variability, trends and groundwater availability in the Mulunguzi river catchment area, Zomba mountain, Southern Malawi. Quat. Int. 148(1): 45-50.
Palamuleni LG (2002). Effect of sanitation facilities, domestic solid

waste disposal and hygiene practices on water quality in Malawi's urban poor areas: a case study of South Lunzu Township in the city of Blantyre. Phys. Chem. Earth 27: 845–850.

Pedley S, Howard G (1997). The public health implications of microbiological contamination of groundwater. Q. J. Eng. Geol. 30: 179-188.

Pietersen K (2005). Basic human needs and the right to health: contribution of groundwater. Proceedings of Biennial Groundwater Conference 2005. Pretoria. 7-9 March 2005.

Pritchard M, Mkandawire T, Edmondson A, O'Neill JG, Kululanga G (2009). Potential of using plant extracts for purification of shallow well water in Malawi. Phys. Chem. Earth 34: 799–805.

Pritchard M, Mkandawire T, O'Neil JG (2007). Biological, chemical and physical drinking water quality from shallow wells in Malawi: Case study of Blantyre, Chiradzulu and Mulanje. Phys. Chem. Earth 27: 845-850.

Pritchard M, Mkandawire T, O'Neill JP (2008). Assessment of groundwater quality in shallow wells within the southern districts of Malawi. Phys. Chem. Earth 33: 812-823.

Ramoeli P (2002). SADC Protocol on Shared Watercourses: Its History and Current Status, In: Turton AR, Henwood R (Eds.). Hydropolitics in the Developing World: A Southern African Perspective. Pretoria. African Water Issues Research Unit (AWIRU). 105-111.

Robins N, Davies J, Farr J (2013). Groundwater supply and demand from southern Africa's crystalline basement aquifer: evidence from Malawi. Hydrogeol. J. 21: 905–917.

Robins NS, Chilton PJ, Cobbing JE (2007). Adapting existing experience with aquifer vulnerability and groundwater protection for Africa. J. Afr. Earth Sci. 47: 30–38.

Robins NS, Davies J, Farr JL, Callow RC (2006). The changing role of hydrogeology in semi-arid southern and eastern Africa. Hydrogeol. J. 14: 1483–1492.

Rowland HAL, Gault AG, Lythgoe P, Polya DA (2008) Geochemistry of aquifer sediments and arsenic-rich groundwaters from Kandal Province, Cambodia. Appl. Geochem. 23: 3029–3046.

Sajidu SM, Masumbu FFF, Fabiano E, Ngongondo C (2007). Drinking water quality and identification of fluoritic areas in Machinga, Malawi, Malawi. J. Sci Technol. 8: 042 – 056.

Sajidu SMI, Masamba WRL, Henry EMT, Kuyeli SM (2007). Water quality assessment in streams and wastewater treatment plants of Blantyre, Malawi. Phys. Chem. Earth Parts A/B/C. 32(15):1391-1398.

Sajidu SMI, Masamba WRL, Thole B, Mwatseteza JF (2008). Groundwater fluoride levels in villages of Southern Malawi and removal studies using bauxite. Int. J. Phys. Sci. 3(1): 1–11.

Sajidu S, Kayira C, Masamba W, Mwatseteza J (2012). Defluoridation of Groundwater Using Raw Bauxite: Rural Domestic Defluoridation Technology. Environ. Nat. Resour. Res. 2(3).

Sibale FK, Chidothi C, Tsakala E (1998). Oral health status among the 12 and 15 year old school pupils in Machinga District, Malawi. Proc. of the 1st International Chancellor College Research Seminar. pp. 13-18.

Smedley PL, Kinniburgh DG (2002). A review of the source, behaviour and distribution of arsenic in natural waters. Appl. Geochem. 17(5): 517-568.

Smith-Carrington AK, Chilton PJ (1983). Groundwater Resources of Malawi. Department of Lands, Valuation and Water. Lilongwe, Republic of Malawi.

Stoupy O, Sugden S (2003). Halving the number of people without access to safe water by 2015 – a Malawian perspective. Part one and part two. WaterAid. Malawi Country Programme.

Turton A, Godfrey L, Julien F, Hattingh H (2004). Unpacking groundwater governance through the Lens of a trialogue: A Southern African Case Study. In: Appelgren B (Ed.). Managing Shared Aquifer Resources in Africa. IHP-VI, Series on Groundwater No. 8. UNESCO, France.

UN (1989). Malawi: ground water in eastern, central and southern Africa. Natural Resources/Water Series No. 19. United Nations, New York.

von Hellens A (2013). Groundwater quality of Malawi – fluoride and nitrate of the Zomba-Phalombe plain. BSc Thesis. University of Malawi. http://stud.epsilon.slu.se/5651/1/von_hellens_a_130611.pdf

Wanda E, Monjerezi M, Mwatseteza JF, Kazembe LN (2011). Hydrogeochemical appraisal of groundwater quality from weathered basement aquifers in Northern Malawi. Phys. Chem. Earth 36: 1197–1207.

Wanda EMM, Gulula LC, Phiri A (2013). Hydrochemical assessment of groundwater used for irrigation in Rumphi and Karonga districts, Northern Malawi. J. Phys. Chem. Earth 66: 51-59

Wen D, Zhang F, Zhang E, Wang C, Han S, Zheng Y (2013). Arsenic, fluoride and iodine in groundwater of China. J. Geochem. Explor. 135: 1-21.

WHO (2008). Guidelines for drinking-water quality. (Third edition incorporating the first and second addenda). Volume 1, Recommendations. World Health Organization. Geneva. ISBN 9789241547611.

World Bank (Online). http://data.worldbank.org/indicator/SH.H2O.SAFE.RU.ZS [Accessed September 2013].

World Bank (Online). http://data.worldbank.org/indicator/SP.POP.TOTL [Accessed September 2013].

Xie XJ, Wang YX, Su CL, Duan MY (2013). Effects of recharge and discharge on delta H-2 and delta O-18 composition and chloride concentration of high arsenic/ fluoride groundwater from the Datong Basin, Northern China. Water Environ. Res. 85: 113-123.

14

Characterization of groundwater quality in Oran Sebkha basin

BOUALLA Nabila [1], BENZIANE Ahmed[1], MOUSSA Kacem[2] and DERRICH Zoubir[3]

[1]Department of Civil Engineering, Laboratory Materials, Soil and Thermal, 31000 Oran, Algeria.
[2]Department of Earth Sciences, Laboratory Geodynamics Basins and Sediment Budget, 31000 Oran, Algeria.
[3]Department of Chemistry, Laboratory of Physico-Chemical Material/Catalyst/Environment, 31000 Oran, Algeria.

Growing populations and increasing industrialization causes increase in living standard, which result in decrease in the quality of water and may put stresses on natural waters by impairing both the quality of the water and the hydrological budget. This research aimed at determining the origin of the chemical elements of groundwater from the Oran Sebkha basin. It applied the inverse geochemical modeling to derive the sources of variation in the hydrochemistry. Fifty five (55) water samples were selected from different point in Oran Sebkha basin for sampling purpose in July 2011. Physico-chemical parameters such as pH and electric conductivity were measured *in situ*. Moreover, chloride, sulfate, alkalinity, calcium, magnesium, sodium, potassium, were measured in the laboratory. Inverse geochemical models of the statistical groups were developed using PHREEQC to elucidate the chemical reactions controlling water chemistry. The inverse geochemical modeling demonstrated that relatively few phases are required to derive water chemistry in the area. In a broad sense, the reactions responsible for the hydrochemical evolution in the area fall into three categories: (1) dissolution of evaporite minerals; (2) precipitation of carbonate minerals; and (3) weathering reactions of silicate minerals. The high values of the physico-chemical parameters of water obtained in the present study sites indicate a variation in the physico-chemical parameters and demonstrated that relatively few phases are required to derive water chemistry in the area. Range of values were found as pH (5.1-7.6), conductivity (720-15820 µS cm^{-1}), chloride (994-7810 mg l^{-1}), sulfate (6.1-112.4 mg l^{-1}), alkalinity (421-19962 mg l^{-1}), calcium (80-680 mg l^{-1}), magnesium (212.4-4525 mg l^{-1}), sodium (124.2-4687.4 mg l^{-1}) and potassium (0.9-42.5 mg l^{-1}).

Key words: Physico-chemical parameters, water, Sebkha.

INTRODUCTION

The availability of good quality water is an indispensable feature for preventing diseases and improving quality of life (Oluduro and Aderiye, 2007). Natural water contains some types of impurities whose nature and amount vary with source of water (Drever, 1988).

Water quality characteristic of aquatic environment arise from a multitude of physical, chemical and biological interactions (Deuzane, 1979). The water bodies, rivers, lakes, dams and estuaries are continuously subject to dynamic change with respect to the geological age, geochemical characteristics and anthropogenic influences.

This is demonstrated by continuous circulation, transformation and accumulation of energy and matter through the medium of living things and their activities. The dynamic balance in the aquatic ecosystem is upset by human activities, resulting in pollution which is manifested dramatically as fish kill, offensive taste, odor, color and unchecked aquatic weeds. The objective of the monitoring

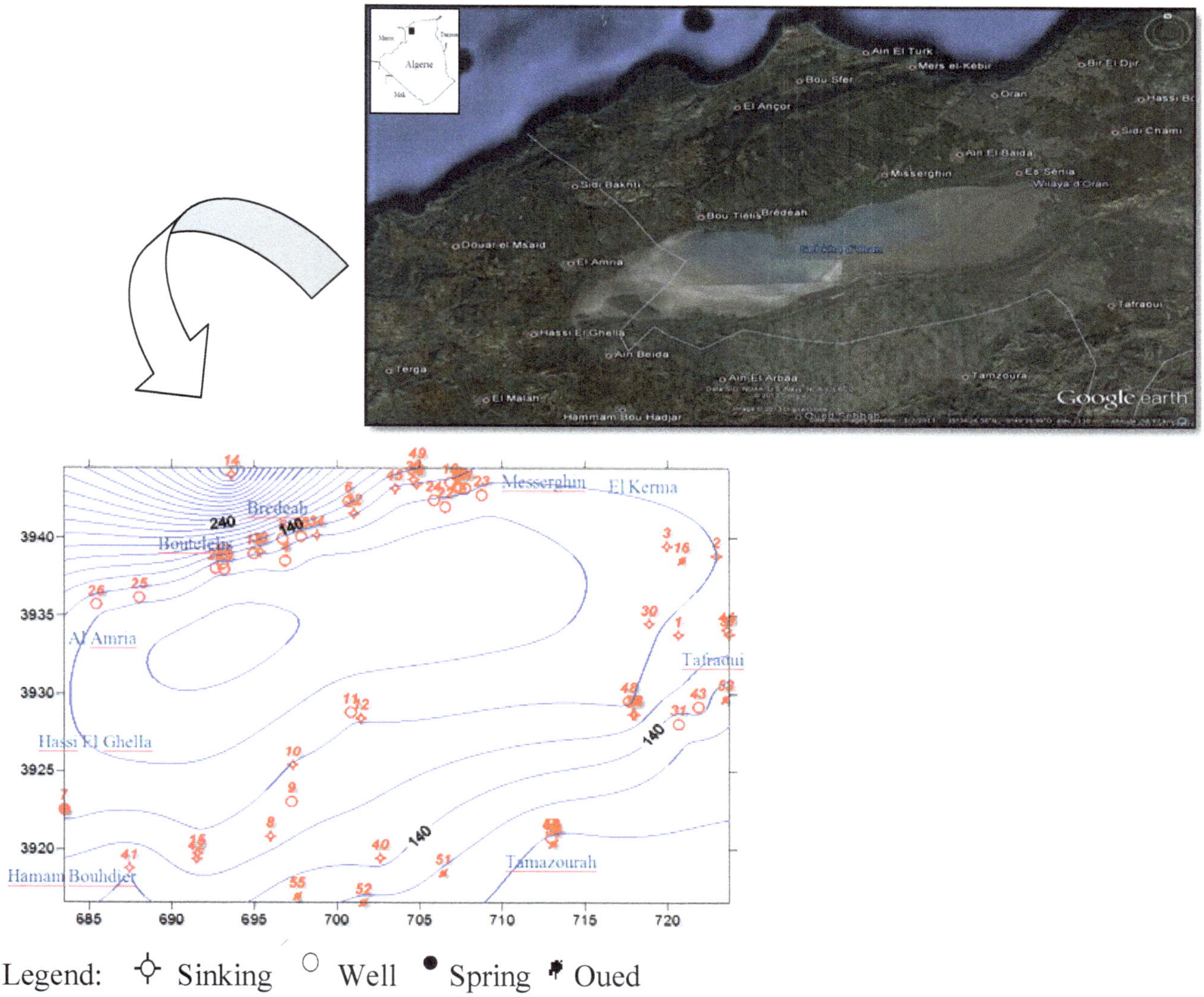

Figure 1. Localization of sampling points in the study area.

studies undertaken for water body is to assess variation in water quality.

The soundness or validity of the results in the inverse modeling depends on a valid conceptualization of the groundwater system, validity of the basic hydrochemical concepts and principles, accuracy of input data into the model, and level of understanding of the geochemical processes in the area (Güler and Thyne, 2004).

Variation in groundwater chemistry is mainly a function of the interaction between the groundwater and the mineral composition of the aquifer materials through which it moves. Hydrochemical processes, including dissolution, precipitation, ion exchange, sorption and desorption, together with the residence time occurring along the flow path, control the variation in chemical composition of groundwater (Apodaca et al., 2002). The zone of study covers: the great Oran Sebkha basin (with an area of 1878 km²); the Sebkha itself (with an area of 298 km²) and the stretches (over 40 km long and 6 to 13 km wide). The zone is located in the Central Coastal Oran basin and is bounded by: the Djebel Murdjadjo (530 m) in the north, the Mount Tessala (1061 m) in the south, the plain oued Tlelat in the east, and oued Mellah in the west (Figure 1).

The integrated geological study (sedimentology, geomorphology, tectonic, hydrogeology) determined factors which conditioned its setting up, since the lower-middle Miocene (Burdigalian-Langhian-Serravallian); that emerged in the beginning of the Tortonian age, then it met, and until the Pleistocene under a marine regime that evolved toward a continental regime since the superior Pleistocene. The southward swing of the northern area (Murdjadjo south side) at the Soltanian age would have been the main reason of the Sebkha of Oran endoreism and its subsequently closing (full) with the setting up of the wind accumulations.

Table 1. The physico-chemical parameters of water samples from Oran Sebkha basin.

Parameter	Average	Minimum	Maximum	Standard deviation	Median
pH	6.9	5.1	7.6	0.4	6.9
EC (µS m^{-1})	5905.3	720	15820	3303.7	5710
Cl^-(mg L^{-1})	2472.7	994	7810	1180.8	2307.5
SO_4^{-2}(mg L^{-1})	26.1	6.1	112.4	20.3	22
HCO_3^-(mg L^{-1})	4764	421	19962	4128	3288
Na^+(mg L^{-1})	863.7	124.2	4687.4	717.7	733.7
K^+(mg L^{-1})	8.6	0.9	42.5	8.4	6.2
Ca^{2+} (mg L^{-1})	304.4	80	680	131.9	280
Mg^{2+} (mg L^{-1})	1258.8	212.4	4525	920.2	956.4

MATERIALS AND METHODS

In order to characterize the groundwater quality in the dry season, a sampling network was chosen to allow the acquisition of representative data on the spatial and temporal variability (Figure 1). This network covers the entire basin upstream to downstream and consists of a set of points of water wells, boreholes and springs.

Water withdrawals were made at low waters. The samples were collected during the period of 04 to 20 July 2011 where comprehensive analyses were performed. The parameters that were measurement are: pH and electrical conductivity (EC) measured in the field using a multi-parameter HANNA (pH meter HANNA instrument HI 9811).

Subsequently, the samples were analyzed in the laboratory for their chemical constituents such as calcium, magnesium, sodium, potassium, chloride, bicarbonate and sulfate. This was achieved using standard methods as suggested by Rodier (1996). Ca^{2+}, Mg^{2+}, HCO_3^- and Cl^- were analyzed by volumetric titrations. Concentrations of Ca^{2+} and Mg^{2+} were estimated titrimetrically using 0.05 N EDTA and those of HCO_3^- and Cl by H_2SO_4 and $AgNO_3$ titration, respectively. Concentrations of Na^+ and K^+ were measured using an Ionogram (Model: Ionogram Easy lyte Na/K/Cl, 800 ml, MEDICA 001384-001 R2 Analyser, 12160/12014-05) and that of sulfate by Spectrometer method (Model: Spectrometer UV Optizen 2120 UV). The results were tabled for interpretation.

In literature, many different methodologies have been applied to study, evaluate and characterize the sources of variation in groundwater geochemistry. Among these methods is the inverse geochemical modeling in PHREEQC (Parkhurst and Appelo, 1999) based on a geochemical mole-balance model, which calculates the phase mole transfers (the moles of minerals and gases that must enter or leave a solution).

RESULTS AND DISCUSSION

The physico-chemical parameters obtained from analysis of water samples from Oran sebkha basin are presented in Table 1. The groundwater samples of the study area have pH values ranging from 5.1 to 7.6, which indicate that the groundwater is slightly alkaline. The electrical conductivity (EC) values ranged from 720 to 15820 µS cm^{-1} which is the characteristic of mixed water (Ca-Mg-Cl-HCO_3). The gradual increase of conductivity denotes a significantly lateral input of wastewater with higher conductivity (Trinh, 2003). The order of abundance of the major cations is Mg ≥ Na ≥ Ca ≥ K and all samples exceeded the desirable limit of Ca, Mg and Cl for drinking water, respectively :75, 50 and 200 mg l^{-1}, but the sulfate concentrations are all below health guidelines (WHO, 1993) (Table 1).

In all water samples, a critical look at the results revealed that the values of these physico-chemical parameters are higher in the basin based on geological position; discharges of industrial wastewater and contamination of the groundwater by seawater were excluded as the source of high concentrations. The quantity of waste in different phases of a natural aquatic system is reflected by the level of hardness, alkalinity, and other physico-chemical parameters. Since the higher values of all these physico-chemical parameters than those obtained could be as a result of washing away of sulfate based fertilizers into the river (Ipinmoroti and Oshodi, 1993; Vogel, 1970).

Inferring the sources of solutes in groundwater using simple massbalance approaches does not provide unequivocal results, a reasonable way of constraining the processes that might influence the major-ion chemistry. The controls on the hydro chemical evolution of groundwater largely depend on the chemistry of the recharging water, water aquifer matrix interaction (cation exchange), or both, as well as groundwater residence time within the aquifer (Garrels and MacKenzie, 1971).

Three general processes that contribute to the generation of solutes in groundwater are: evaporate dissolution, carbonate dissolution and silicate weathering (Garrels and MacKenzie, 1971). The chemistry of the evolving water depends not only on the bulk chemistry of the matrix, but also on the rate of weathering. Meybeck (1987) reported that weathering rates of evaporates and carbonates are up to 80 and ~12 times, respectively, faster than silicate weathering rates. Hence, even relatively minor proportions of carbonates and evaporates can significantly influence water chemistry.

The study of the Ca^{2+}/Mg^{2+} (méq L^{-1}) ratio of groundwater from the area of study suggests the dissolution dolomite present in the alluvium (Figure 2). That is, if the ratio Ca^{2+}/Mg^{2+} = 1, dissolution of dolomite should occur, whereas a higher ratio is indicative of greater calcite con-

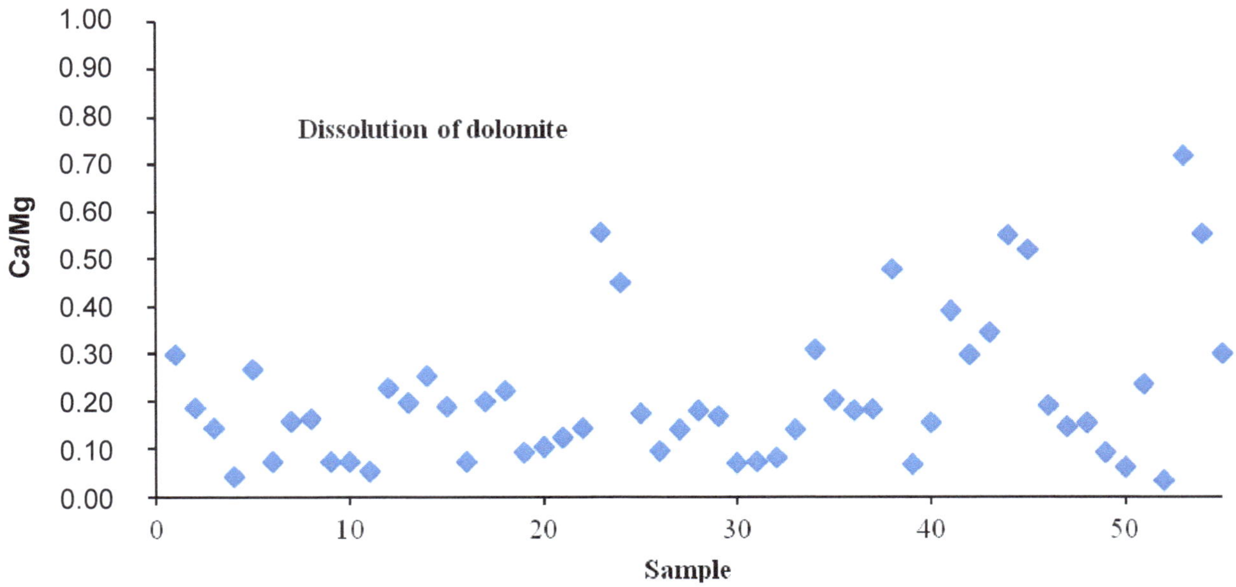

Figure 2. Sample number versus Ca^{2+}/Mg^{2+}.

Table 2. Saturation indices of the water samples from Oran Sebkha basin calculated by PHREEQC.

Phase	Stoichiometry	Average
Anhydrite	$CaSO_4$	-2.19
Aragonite	$CaCO_3$	0.84
Calcite	$CaCO_3$	0.98
Dolomite	$CaMg(CO_3)_2$	2.87
Gypsum	$CaSO_4{:}2H_2O$	-1.98

tribution (Maya and Loucks, 1995). The points are located below the line (Ca^{2+}/Mg^{2+} =1) indicating the dissolution of dolomite. The samples have a ratio between 0.03 and 0.72 indicating the dissolution of dolomite. Dissolved carbonates (calcite and dolomite) occur predominantly in the form of HCO_3^-, due to the pH range. The solubility of calcite and dolomite is largely controlled by CO_2 fugacity and pH, according to the reactions:

$$CaCO_3 + CO_2(g) + H_2O = Ca^{2+} + 2HCO_3^- \quad (1)$$

$$Ca1\text{-}xMgxCO_3 + CO_2\ (g) + H_2O = (1\text{-}x)\ Ca^{2+} + xMg^{2+} + 2HCO_3^- \quad (2)$$

The plot of Ca^{2+}+ Mg^{2+} versus SO_4^{-2} + HCO_3^- will be above the 1:1 line if the dissolutions of calcite, dolomite, anhydrite and gypsum are the dominant reactions in a system (Maya and Loucks, 1995). Ion exchange tends to shift the points to left due to an excess of Ca^{2+}+ Mg^{2+} (Cerling et al., 1989; Fisher and Mulican, 1997; McLean and Jankowski, 2000). If reverse ion exchange is the process, it will shift the points to the right due to a large excess of SO_4^{2-}+ HCO_3^- over Ca^{2+}+ Mg^{2+}. The plot of Ca^{2+}+ Mg^{2+} versus SO_4^{2-}+ HCO_3^- (Figure 2) shows that the all samples are located in the right side due to an excess of SO_4^{-2}+ HCO_3^- (Belkhiri et al., 2010).

A Na (méq L^{-1})-normalized Ca^{2+} (méq L^{-1}) versus HCO_3^- (méq L^{-1}) and plot Mg^{2+} (méq L^{-1}) (Figures 3, 4 and 5) (Gaillardet et al., 1999) shows that the three group samples range from being influenced by silicate weathering to carbonate dissolution. The relation between Na-normalized Ca^{2+}, HCO_3^- and Mg^{2+} shows that although most of the Mg may have been derived from carbonate dissolution, some have a silicate source (Belkhiri et al., 2010).

The saturation indices of some of the common mineral phases is presented in Table 2, which clearly shows that the groundwater is generally supersaturated with respect to most of the carbonate and silicate phases and is most probably responsible for the composition of the groundwater (Belkhiri et al., 2010). The results of saturation calculations show that all the points are under saturated with anhydrite and gypsum, halite mineral phases are minor or absent in the area, suggesting that their soluble components Na^+, Cl^-, Ca^{2+} and SO_4^{2-} concentrations are not limited by mineral equilibrium. In contrast, aragonite, calcite and dolomite reach saturation as groundwater chemistry evolves along the groundwater flow paths. The groundwater in the studied area evolves from less saline Ca-g-HCO_3 water to blended Mg-Ca-HCO_3-Cl water to brackish Mg-Ca-Cl-HCO_3 water along the topographic flow path inverse geochemical modeling along groundwater flow paths indicating the dissolution of evaporite minerals, precipitation of carbonate minerals and weathering reactions of silicate minerals (Table 1 and Figure 6)

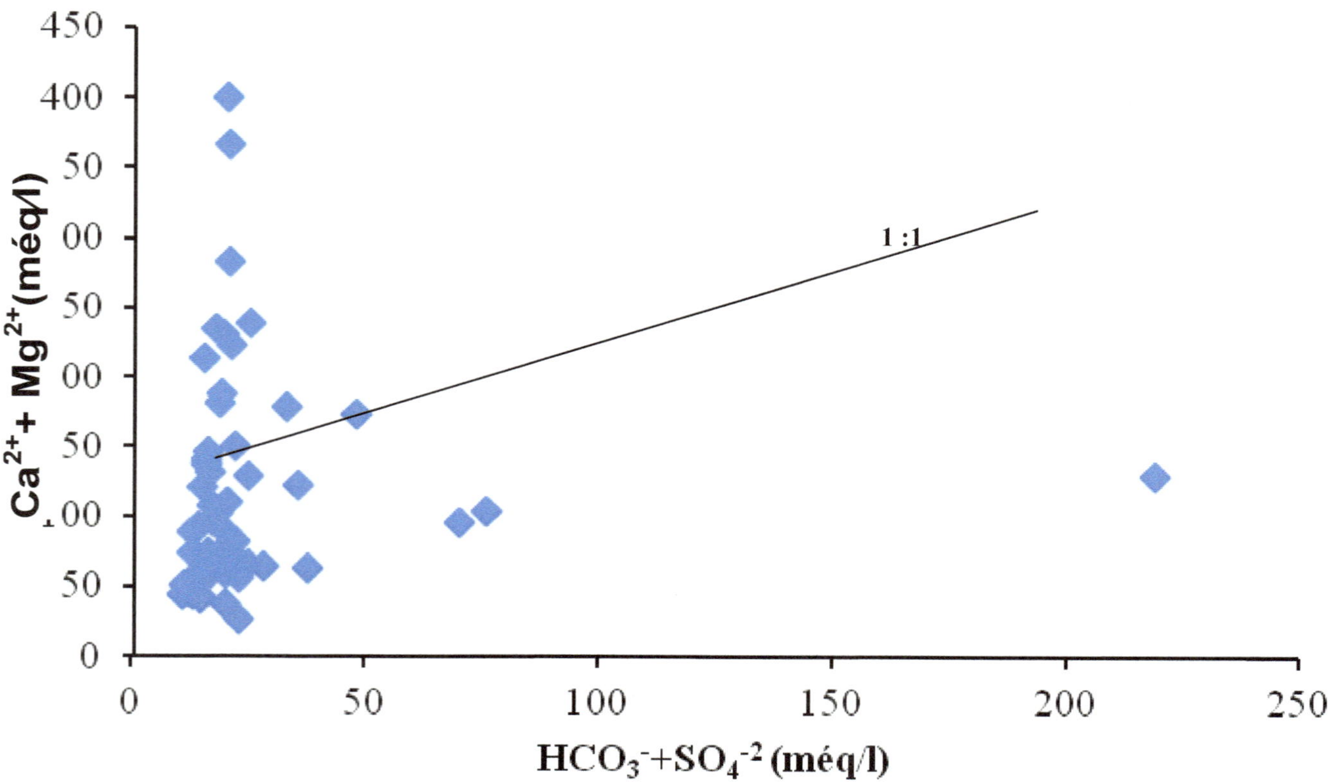

Figure 3. Ca^{2+}+ Mg^{2+}versus SO_4^{-2}+ HCO_3^-.

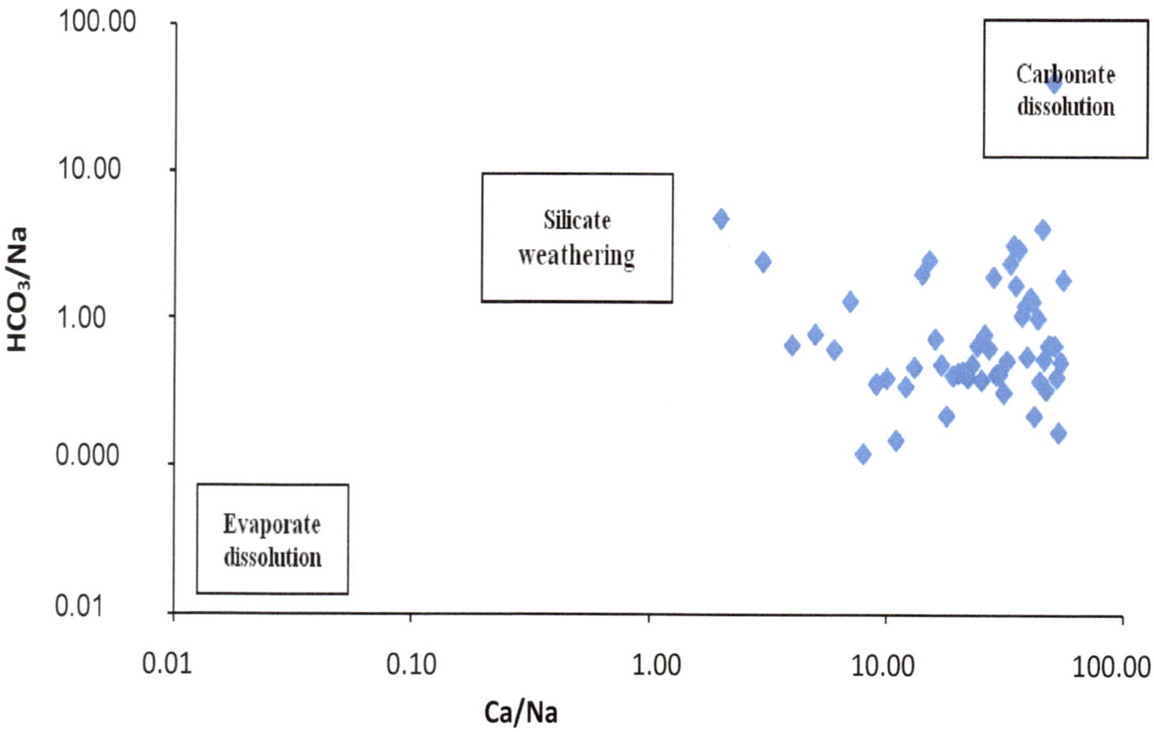

Figure 4. Na-normalized Ca^{2+} and HCO_3^- (HCO_3/ Na) (méq L^{-1}/ méq L^{-1}).

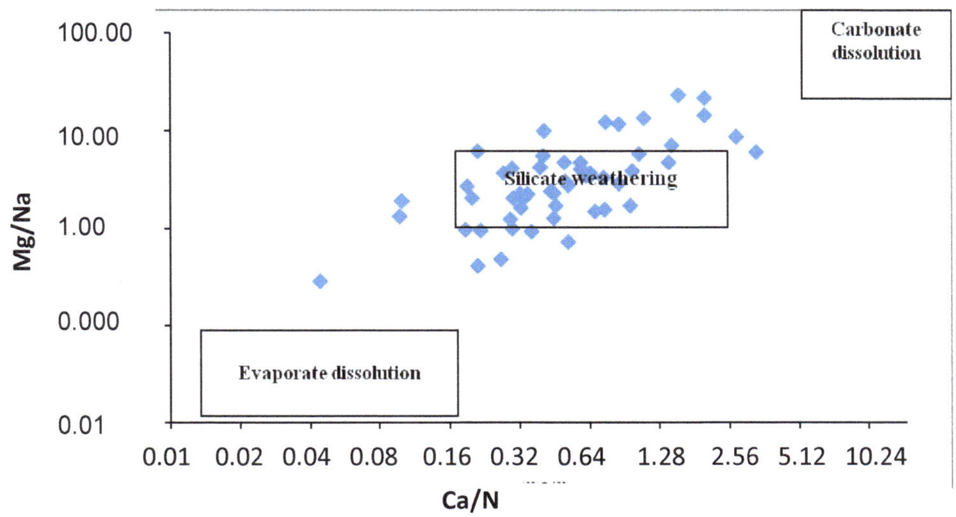

Figure 5. Na-normalized Ca^{2+} and Mg^{2+}.

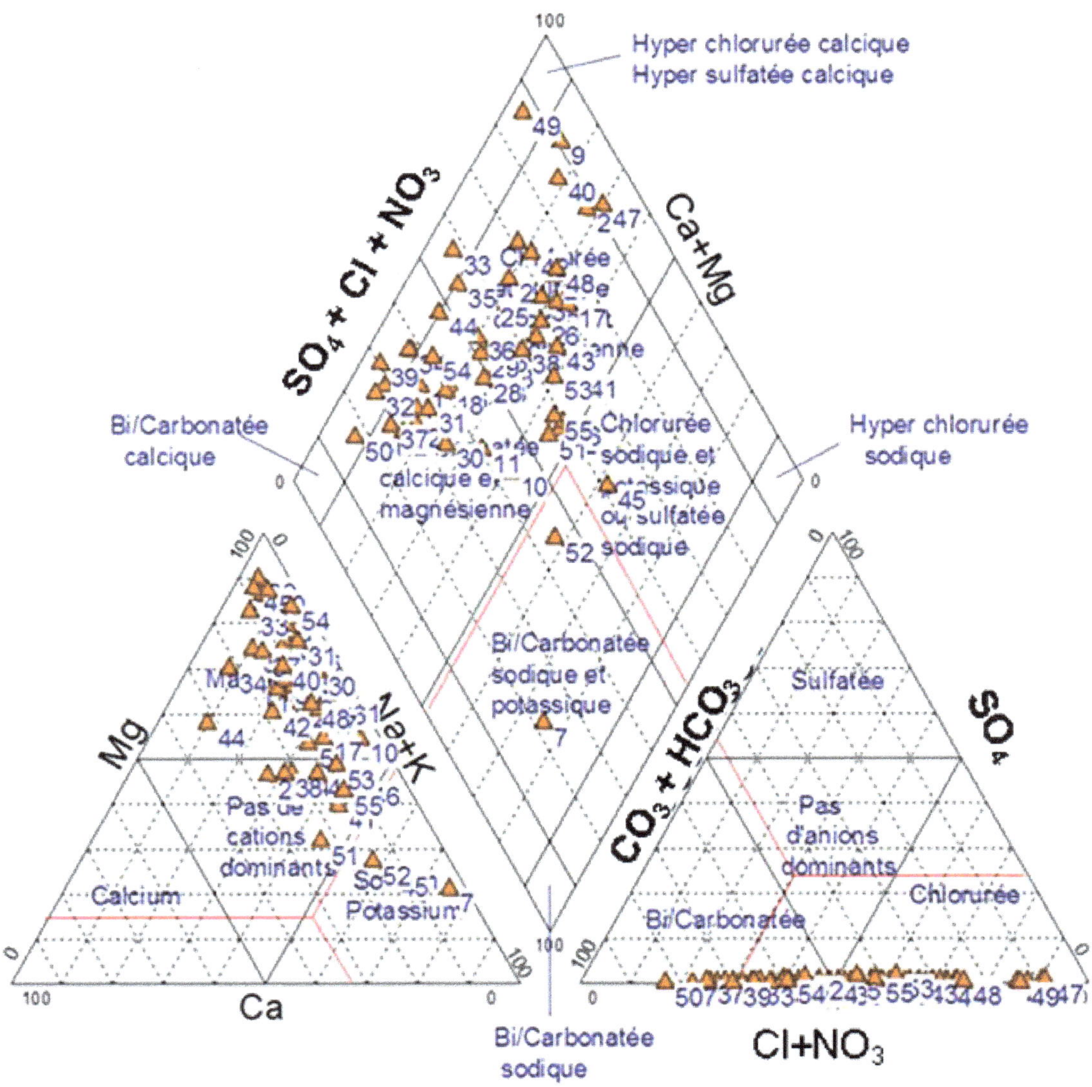

Figure 6. Piper diagram showing groundwater classification (Piper, 1994).

(Belkhiri et al., 2010).

Conclusion

In conclusion, the water quality is a significant aspect of determination of the water body status. The parameters examined in the scope of this thesis were subjected to variety of spatial trends. The results of this study show that the groundwater in the studied area evolves from less saline Ca-Mg-HCO_3 water to blended Mg-Ca-HCO_3-Cl water to brackish Mg-Ca-Cl-HCO_3 water along the topographic flow path. The analysis indicates that the water of Oran sebkha basin is characterized by moderate alkalinity in the investigated period.

In addition to this, it important that bacteriological assessment of water from these different wells be carried out to sure if the water is safe for drinking and other domestic applications.

REFERENCES

Apodaca LE, Jeffrey BB, Michelle CS (2002). Water quality in shallow alluvial aquifers, Upper Colorado River Basin, Colorado, 1997. J. Am. Water Res. Assoc. 38 (1):133-143.

Belkhiri L, Boudoukha A, Mouni L, Baouz L (2010). Application of multivariate statistical methods and inverse geochemical modeling for characterization of groundwater - A case study: Ain Azel plain (Algeria).

Cerling TE, Pederson BL, Damm KLV (1989). Sodium-calcium ion exchange in the weathering of shales: implications for global weathering budgets. Geology 17:552-554.

Deuzuan J (1979). Handbook of drinking water quality. Indiana Univ. Press. pp. 3-17.

Drever JI (1988). The Geochemistry of Natural Waters. Prentice-Hall, Upper Saddle River, NJ.

Fisher RS, Mulican WFIII (1997). Hydrochemical evolution of sodium-sulfate and sodium-chloride groundwater beneath the Northern Chihuahuan desert, Trans- Pecos, Rexas, *USA*. Hydrogeol. J. 10(4):455-474.

Gaillardet J, Dupre B, Louvat P, Allegre CJ (1999). Global silicate weathering and CO_2 consumption rates deduced from the chemistry of large rivers. Chem. Geol. 159:3-30.

Garrels RM, MacKenzie FT (1971). Evolution of Sedimentary Rocks. Norton, New York.

Güler C, Thyne GD (2004). Hydrologic and geologic factors controlling surface and Groundwater chemistry in Indian wells-Owens Valley area, southeastern California, USA. J. Hydrol. 285:177-198.

Ipinmoroti K, Oshodi O (1993). Determination of Trace Metals in fish, associated Wanted and Soil Sediments fresh fish ponds. Discovery innovates 5:138.

Maya AL, Loucks MD (1995). Solute and isotopic geochemistry and groundwater flow in the Central Wasatch Range, Utah. J. Hydrol. 172:31-59.

McLean W, Jankowski J (2000). Groundwater quality and sustainability in an alluvial aquifer, Australia. In: Sililo, et al. (Ed.), Proc. XXX IAH congress on Groundwater: Past Achievements and Future Challenges. Cape Town South Africa 26th November-1st December 2000. AA Balkema, Rotterdam, Brookfield.

Meybeck M (1987). Global chemical weathering from surficial rocks estimated from river dissolved loads. Am. J. Sci. 287:401-428.

Oluduro AO, Adewoye BI (2007). Efficiency of moringa Oleifera Sead extract on the microflora of surface and ground water. J. Plant Sci. 6:453-438.

Parkhurst DL, Appelo CAJ (1999). User's Guide to PHREEQC (Version 2)-A Computer Program for Speciation, Batch-Reaction, One-Dimensional Transport, and Inverse Geochemical Calculations.

Piper AM (1994). A graphic procedure in geochemical interpretation of water analyses. Am. Geophys. Union Trans. 25 :914-923.

Rodier J (1996). L'analyse de l'eau, eaux naturelles, eaux résiduaires, eau de mer. 8e édition Dunod, Paris.

Trinh AD (2003). Étude de la qualité des eaux d'un hydro système fluvial urbain autour de Hanoi (Vietnam); Suivi expérimental et modélisation. Thèse de doctorat de Université Joseph Fourier - Grenoble 1 (France) et Centre National des Sciences et de la Technologie (Vietnam).

United States Geological Survey, Water Resources Investigations Report 99-4259, Washington, DC. p. 326.

Vogel I (1970). A text book of inorganic analysis published by Harley and Sons, 4th Ed. p.148.

WHO (1993). Guideline for drinking water quality 2nd Ed. Recommendation. World Health Organization general 1:30-113.

15

Heavy metals accumulation in edible part of vegetables irrigated with untreated municipal wastewater in tropical savannah zone, Nigeria

Hassana I. Mustapha[1] and Omotayo B. Adeboye[2]

[1]Department of Agricultural and Bioresources Engineering, School of Engineering and Engineering Technology, Minna, Nigeria.

[2]Department of Agricultural and Environment Engineering, Obafemi Awolowo University, Ile-Ife, Osun State, Nigeria.

In this present study, the quality of municipal wastewater used for irrigation of spinach was investigated for its heavy metal build-up. The municipal wastewater used for irrigation and the irrigated spinach samples were collected and analyzed for their heavy metal concentrations. The results indicate that the municipal wastewater used was contaminated with copper (1.90 mg/l), lead (0.09 mg/l) and iron (25 mg/l) and the municipal-irrigated spinach was contaminated with manganese (95 mg/kg) and cadmium (0.03 mg/kg). The results of these investigations were compared with World Health Organization (WHO) and Food and Agriculture Organization (FAO) heavy metal standards for irrigation water quality and permissible levels of metals in food and water. It revealed that the heavy metal concentrations were above the recommended threshold limits. High concentrations of these metals are very detrimental to the health of the inhabitants and crop consumers. Regular monitoring for safe practice is strongly recommended in order to avert terminal diseases in the area.

Key words: Domestic wastewater, irrigation, heavy metals, soil, vegetables.

INTRODUCTION

Agriculture uses about 70% of water withdrawals, therefore, it is expected that in times and regions of water scarcity, farmers would turn to domestic or urban waste water as a water source (FAO, 2010). In recent times, farmers use wastewater to irrigate their crops. This wastewater contains large amount of organic materials, some inorganic elements and/or substantial amount of toxic heavy metals (Zavadil, 2009; Arora et al., 2008; Lone et al., 2003), non essential heavy metals which when present in large amount could be transferred to animal and human beings through food chain (Lone et al 2003). The presence of these substances are harmful to human health because of the non biodegradable nature of heavy metals and their potential to accumulate in different parts of the body (Lawal and Audu, 2011; Arora et al., 2008). Similarly, toxic metals may be absorbed by vegetables through several processes and finally enter the food chain at high concentrations which are capable

Table 1. Heavy metals in soil (mg/Kg) at Soje.

Parameter	Mn	Mg	Cu	Zn	Pb	Fe	Cd
Plot average	3.20	6.85	31.00	25.46	10.00	60.96	0.05
Before irrigation	2.01	10.93	26.32	22.00	6.25	33.10	0.04
Control	1.98	10.95	26.85	22.79	8.97	40.00	0.01
Max. limits	-	-	60.50	370.50	2129.0	-	2.25

of causing serious health risks to consumers (Kihampa et al., 2011). Also, accumulation of highly toxic metals (Cd) even at low concentrations in food and water leads to the buildup of cadmium in kidneys and may likely lead to kidney diseases (Latif, 2009). Other effects of metal toxicity may include damage or reduce mental and central nervous function, lower energy levels, and damage to blood composition, lungs, kidneys, liver and other vital organs (Kihampa et al., 2011). Indeed, it was estimated that 10% of the world's population consumes foods irrigated with wastewater (Finley, 2008). However, these factors: climate, atmospheric deposition, the nature of the soil on which the plants are grown and the degree of maturity of the plants at the time of harvesting can influence the concentration of heavy metals on and with plants (Farooq et al., 2008, Muchuweti et al., 2005). Similarly, Nazif et al. (2006) explained that the fate and effects of pollutants discharged into a particular water body will depend not only on the amount of polluting substances emitted but also on the hydrological, physical, chemical and biology conditions characterizing the water body concerned.

Farmers in Soje took advantage of the continuous flow of water at their vicinity, and were unaware of the quality and the implications of using contaminated water for crop cultivation. Thus, this study aimed at quantifying the content and quality of heavy metals in domestic wastewater used for irrigation of vegetables in Soje, in the soil and edible part of spinach.

METHODOLOGY

Study area

The study site selected for this study is the Soje irrigation scheme. Soje irrigation scheme is located in Soje, few kilometers from Minna railway station, Minna, Niger state, Nigeria. Niger state is situated at the North Central part of Nigeria. It lies in the savannah zone of the tropics between latitude 8°10'N and 11°30' N and longitude 3°30' E and 7°30' E. The study area has two main seasons, the rainy and the dry (harmattan). The rainy season begins in April and ends in October while the harmattan starts in November and ends in March. Soje has an average annual rainfall of 103.3 mm, average annual temperature of 22.5°C and average annual relative humidity of 50.3%. The source of water for Soje irrigation scheme is the wastewaters from Minna Township. The water flows through unlined channels. The farmers take advantage of it continuous flow for irrigation. Investigations revealed that the irrigation scheme was set up by farmers in the area and the area is about 7.5 ha.

Collection of samples

Three plots sizes of 1.803 x 2.44 m were marked out for the purpose of this study. Spinach (*Spinacia oleracea*) was planted on the marked out plots during the dry season of 2010. Soil, wastewater and vegetable samples were sampled.

Composite soil samples made up of five soil samples per plots were collected from a depth of 0 to 25 cm before planting and after the growing period from two irrigated plots and a control plot. The soil samples were air-dried, and crushed to pass a 2 mm mesh sieve. 0.5 g of the finely ground soil samples were digested using 2 ml technical grade trioxonitrate (v) (HNO_3) in beakers at 95°C for 1 h after which hydrogen peroxide (H_2O_2) was added. After cooling, the samples were decanted and diluted with MilliQ water to 10 ml mark for analyses.

The plastic bottles for the wastewater collection were thoroughly washed with detergent and rinsed with tap water before taking the bottles to the study site for sample collection. The bottles were rinsed with the waste water samples before collection. Wastewater samples used for the irrigation of the spinach were collected in 2 l plastic bottles from the Soje unlined channel and transported to the laboratory for heavy metal analysis. These samples were collected in three replicates.

The edible part of the spinach were randomly selected and collected from the plots. The collected samples were stored in labeled polythene bags and then transported to the laboratory for preparation and treatment for analysis. A total of 90 samples were collected from the two plots and the control. The samples were washed with distilled water to eliminate suspended particles. The leafy stalks were removed from the samples; they were sliced and dried on a sheet of paper to eliminate excess moisture. The dried samples were weighed and oven dried at 60°C to a constant weight. The oven dried samples were ground in a mortar until it passed through a 60 mm mesh sieve. A mixture of 2 HNO_3 to 1 $HClO_4$ in a conical flask was used for wet digestion for about 2-3 h on a sand bath, 10 ml of HCL was then added after which the digested samples were filtered with a 0.45 μm pore size cellulose nitrate membrane filter paper. The filtered samples were made up to 100 ml with distilled water and stored for analysis using atomic absorption spectrophotometer.

RESULTS AND DISCUSSION

Spinach was cultivated in soil samples with heavy metal as presented on Table 1. Generally, agricultural soils have low background levels of heavy metals; contaminations are mainly through fertilizer application, irrigation with partially treated or untreated sewage. This was reported by Farooq et al. (2008). The result shows that the soil in Soje is not polluted with Mn, Mg, Cu, Zn, Pb, Fe and Cd. There is need to protect the soil from contamination through regular monitoring.

Table 2. Average concentration of elements in domestic wastewater (mg/l).

Parameter	Wastewater	WHO limit
pH	6.40	6.5-8.5
Mn	0.44	-
Mg	0.22	30
Cu	1.90	1.0
B	1.66	-
Zn	1.28	3
Pb	0.09	0.01
Fe	25.0	0.3

Table 3. Heavy metal concentrations in spinach.

Parameter	Mn	Mg	Cu	Cd	Pb	Fe
Irrigated plot 1	1.22	95	1.45	0.033	0.060	3.93
Irrigated plot 2	1.22	95	2.00	0.030	0.030	3.99
Mean concentration	1.22	95	1.73	0.032	0.045	3.96
Control plot	0.64	97	1.09	0.017	0.020	2.21
WHO&FAO*	-	-	40	0.2	0.3	-
USDA*	-	79	-	-	-	2.71
WHO**	6.61		10.00	0.02	2.0	150.0

Permissible levels in food; **Critical level of metal ions in edible portion of vegetables (Lone et al., 2003).

Analysis of waste water

The average concentration of heavy metal content of domestic wastewater used for irrigation purposes at Soje irrigation scheme is shown in Table 2. The pH of the domestic waste water has a low pH, which is not desirable. Low pH value could decrease the solubility of certain essential elements such as selenium and at the same time increase the solubility of many other elements such as Fe, Al, B, Cu, Cd, Hg and Mn (Akan et al., 2008). The concentrations of the heavy metals in the domestic effluent are in the following order of decreasing magnitude Fe > Cu > B > Zn > Mn > Mg > Pb. The concentrations of Fe, Cu, B and Pb were all above the safe limit for WHO standards for domestic wastewater quality for irrigation. Although, Pb is the lowest in concentration of 0.09 mg/l, but it is about the most toxic metals found in aquatic ecosystem. Boron is essential for plant growth in small quantity; the wastewater contained high levels of boron and the toxicity could be observed on older plants by yellowing or drying of the leaf tissue at the tips and edges (Abbott and Hasnip, 1997).

Metal concentration in spinach

The heavy metal concentrations in spinach are shown in Table 3. The concentrations of heavy metals in edible part of spinach vary from metal to metal. The trend of accumulation in the spinach show an order of decreasing magnitude from Mg to Cd (Mg > Fe > Cu > Mn > Pb > Cd). Magnesium had the highest content in spinach sample with cadmium being the lowest of all the metals analyzed. Vegetables are known to naturally accumulate heavy metals from either wastewater or dumpsite (Arora et al., 2008). The Mg content in the spinach far exceeded the permissible levels in food according to the USDA standards. It was also observed that spinach has a good potential for Fe bioaccumulation. Cu, Cd and Pb concentrations were all below the permissible levels allowed in food. Lone et al. (2003) reported that long term exposure to Cd in food and water leads to build up of Cd in kidney causing kidney problems.

Conclusion

Routine monitoring of heavy metal contents of treated, partially treated or untreated domestic wastewater used for irrigation of edible crops are very vital to ensure safe practice, safety to both the health of humans and animals in particular and the environment in general. The quality of soil can also be changed by the application of sewage to farm lands. This study reveals that domestic wastewater depended upon by farmers at Soje for irrigation of their crops is contaminated with lead, copper and iron. Consumption of these food crops can lead to serious health issues such as cancer. The spinach samples were highly contaminated with magnesium. Excess of magnesium is unlikely to inhibit crop growth though it causes deficiency of calcium and potassium which are essential nutrients of plant (Abbott and Hasnip, 1997).

Conflict of Interests

The author(s) have not declared any conflict of interests.

REFERENCES

Abbott CL, Hasnip NJ (1997). The Safe Use of Marginal Quality Water in Agriculture: A Guide for the Water Resources Planner, Report OD 140 (TDR Project R 6570), DFID (Department for International Development), HR Wallingford.

Akan JC, Abdulrahman FI, Damari GA, Ogunbuaja VO (2008). Physicochemical determination of pollutants in wastewater and vegetable samples along the Jakara wastewatewater channelin Kano metropolis, Kano State, Nigeria. Eur. J. Sci. Res. 23 (1):122-133

Arora M, Kiran B, Rani S, Rani A, Kaur B, Mittal N (2008). Heavy metal accumulation in vegetables irrigated with water from different sources. Food Chem. 111:811-815.

Farooq M, Anwar F, Rashid U (2008). Appraisal of heavy metal contents in different vegetables grown in the vicinity of an industrial area. Pak. J. Bot. 40(5):2099-2106.

Finley S (2008). Reuse of domestic grey water for irrigation of food crops, unpublished thesis submitted to McGill University in partial fulfillment of the requirements of the Master of Science degree.

Food and Agriculture Organization of the United Nations (FAO) (2010). The wealth of waste: The economics of wastewater use in agriculture. FAO Water Reports 35:29:1-142.

Kihampa C, Mwegoha JSW, Shemdoe SR (2011). Heavy metals concentrations in vegetables grown in the vicinity of the closed dumpsite. Int. J. Environ. Sci. 2(2):889-895.

Latif MI (2009). Quantitative Assessment of Heavy Metals in Soils and vegetables irrigation with sewage in Rawalpindi area, unpublished thesis submitted in partial fulfillment of the requirements for the degree of Doctor of Philosophy in Soil Science, department of Soil Science and Water conservation, Pir Mehr Ali Shah, Arid Agriculture university Rawalpindi, Pakistan.

Lawal AO, Audu AA (2011). Analysis of heavy metals found in vegetables from some cultivated gardens in the Kano metropolis, Nigeria. J. Environ. Chem. Ecotoxicol. 36(6):142-148.

Lone MI, Saleem S, Mahmood T, Saifullah K, Hussain G (2003). Heavy metal contents of vegetables irrigated by sewage/tubewell water. Int. J. Agric. Biol. 5(4):533-535.

Muchuweti M, Birkett JW, Chinyanga E, Zvauya R, Scrimshaw MD, Lester JN (2005). Heavy metal content of vegetables irrigated with mixtures of wastewater and sewage sludge in Zimbabwe: implications for human health. Agric. Ecosyst. Environ. 112(1):41-48.

Nazif W, Perveen S, Shah AS (2006). Evaluation of irrigation water for heavy metals of Akbarpura Area. J. Agric. Biol. Sci. 1(1):51-54.

Zavadil J (2009). The effect of municipal wastewater irrigation on the yield and quality of vegetables and crops. Soil Water Res. 4 (3):91-103.

Assessment of changes in drinking water quality during distribution: A case study of Area 25 Township in Lilongwe, Malawi

I. B. M. Kosamu[1] , S. C. Gama[2], M. Tsakama[1], B. U. G. Mughogho[1] and C. Tenthani[1]

[1]The Polytechnic, University of Malawi, Private Bag 303, Blantyre 3, Malawi.
[2]COM/MLWTB Laboratory, College of Medicine, Private Bag 360, Blantyre 3, Malawi.

The quality of drinking water at the point of delivery to the consumer is crucial in safeguarding people's health. This study assesses changes in drinking water quality during distribution at Area 25 Township in Lilongwe, Malawi. Water samples were collected from the exit point of the treatment plant, storage tank and taps at consumers' households. All samples were tested using standard procedures for pH, turbidity, feacal coliforms, manganese, lead, zinc and residual chlorine. One-way ANOVA showed some significant water quality changes during distribution but the average values fell within World Health Organisation (WHO) and Malawi Bureau of standards (MBS) allowable levels for pH, turbidity, feacal coliforms, manganese, lead, zinc and residual chlorine (p < 0.05). Tap water at Area 25 Township is generally safe for human consumption.

Key words: Drinking water, distribution system, biochemical parameters, human health.

INTRODUCTION

Tap water is generally perceived by consumers as being safe and worth paying for. In most developing countries, like Malawi, drinking water distribution networks are made of galvanized iron pipes which are susceptible to internal corrosion and subsequent leaching of iron and other contaminants into water. There are also many other physical, biological and chemical processes that may affect the quality of water in a distribution system. Materials used in the distribution networks may therefore serve as a potential source of chemical contamination (WHO, 2011). Water may also be biologically contaminated through the development of biofilms along the interior walls of distribution systems and external entry through cracks and faults. The distribution pipes are mostly laid underground and there is little or no maintenance; a fact that makes them a potential health hazard to water consumers (Mendels, 1998). Leakages in distribution systems contribute to sediment infiltration and studies have shown that the key factors contributing to drinking water contamination in distribution systems are age of the pipes, extent of internal deposition, plus hydraulic issues such as flow rate, residence time and pressure (Napacho and Manyele, 2010). The end result is that consumers may receive water that satisfies the organoleptics (colour, odour and taste) but whose overall quality is bad. Most consumers have a tendency to believe that if the quality of water entering a distribution system is high then the quality will still remain good at the point of delivery. The intrusion of pathogens into drinking water in distribution systems, for example, may lead to outbreaks of diarrhoeal diseases. Ashbolt (2004) reported that worldwide the consumption of poor quality water and lack of proper sanitation and hygiene are responsible for an estimated1.7 million deaths annually.

Kalua and Chipeta (2005) reported that in Malawi, nearly 50% of all illnesses are related to waterborne diseases. The ever increasing reliance on tap water and the fast growth of populations in urban centres in Malawi and other sub-Saharan Africa therefore poses a new challenge for water distribution managers.

Lilongwe Water Board (LWB) is responsible for water treatment and distribution of potable water in the city of Lilongwe. Established in 1947, the Board abstracts raw water from Lilongwe River which originates from Dzalanyama Ranges. There are two dams, Kamuzu Dam I and Kamuzu Dam II which are constructed along the river. The catchment area is approximately 1,870 square kilometers. Kamuzu Dam I was constructed in 1966 and has a storage capacity of 4.5 million cubic meters. Kamuzu Dam II was constructed in 1989 with an initial storage capacity of 9.2 million cubic meters. The dam was rehabilitated and raised in 1999 thus increasing the storage capacity from 9.2 million to 19.8 million cubic meters and now acts as a balancing reservoir and its outflow goes directly into Kamuzu Dam I. Water flows by gravity down to the abstraction point, about 20 km downstream (Lilongwe Water Board, 2011).

The Board has two main Treatment Plants (TW I and TW II) with a combined design capacity of 95,000 cubic meters per day. The treatment process involves the following major steps: screening; coagulation and flocculation; sedimentation; filtration and chlorination. There are about 40,000 metered customers which also include about 650 water Kiosks within the City (Lilongwe Water Board, 2011).

Water is only treated at the Lilongwe Water Board treatment plant before it is distributed by underground pipes to all the residential, industrial and institutional areas. Area 25 (2278.276 hectares) is found in the northern part of Lilongwe city, some 25 km from the treatment plant. Although, there have been no reports on serious contamination from consumers, systematic, independent assessment is always a good practice for quality control. The aim of this study was therefore to determine the extent of water quality changes from the treatment plant through the distribution network to consumers.

MATERIALS AND METHODS

Study area

Lilongwe City (13°59′S 33°47′E) is located in the central region of Malawi (Figure 1). It was declared Malawi's capital city in 1975 leading to rapid population growth from 102,924 in 1977 to over 900,000 people in 2009. The Board currently serves about 63% of the population (LWB, 2011). The city is characterised by an altitude of 1,036 m, an average annual temperature of 20°C, and a mean annual precipitation of 750 mm. Rain usually falls between November and March but sometimes extends to April. The seasonal variation in the rainfall patterns has effects on water quality entering a treatment plant and it would be scientifically prudent to sample at different times of the year during both dry and wet season. However, due to some constraints, the data presented in this paper only reflects the dry season.

Sampling design

Water samples were collected from the exit point of the treatment plant, storage tank and taps at eight randomly chosen consumer households. Sampling was done three times over a period of five months (July to November 2011). The samples were collected in prewashed (with detergent, dilute HNO_3, doubly de-ionized distilled water, respectively) double caped polyethylene bottles. In the field, the sampling bottles and caps were rinsed three times with the water to be sampled prior to sampling. Most of the samples were obtained directly from the tap after allowing the water to run for at least 5 minu so as to stabilise the variation in temperature. Then, the samples were acidified to 1% with nitric acid and stored at 4°C before analysis to minimise changes in physicochemical characteristics of the metals (Tuzen and Soylak, 2006).

Sample and data analysis

The standard analytical methods that were used for determination of biochemical parameters of water were from American Public Health Association series of Standard Methods of Examination of Water and Effluent (APHA, 1998). Samples were analysed within 2 h after collection. Results of laboratory analysis were subjected to data analysis using SPSS, version 16. To analyse changes in the levels of pH, turbidity, feacal coliforms, manganese, lead, zinc and residual chlorine, one-way Analysis of Variance (ANOVA) at 0.05 significance level was employed.

RESULTS AND DISCUSSION

A summary of the results is presented in the following set of tables which also compare each of the parameters with the acceptable levels as stipulated by the World Health Organisation (WHO, 2004) and Malawi Bureau of Standards (Malawi Bureau of Standards, 2005).

Manganese levels, 0.03 ± 0.002 to 0.05 ± 0.005 (Table 1), showed significant variation between the treatment point and consumer points (α = 000). Grazuleviciene and Balcius (2009) also found that the concentration of manganese varies with increase of the distance from the water treatment plant. The concentration of lead, 0.37 ± 0.061 to 0.56 ± 0.053 (Table 2) and zinc, 1.04 ± 0.071 to 1.83 ± 0.306 (Table 3) did not show significant variation from the treatment point to consumers. The levels of manganese, lead and zinc in the distribution system were found to be within the maximum accaptable levels for drinking water as stipulated by WHO and MBS. It has to be noted however that in addition to WHO and MBS guidelines which are specifically related to human health, some water quality regulators such as the United States Environmental Protection Agency (US EPA, 2012) also impose an aesthetic limit of 0.05 mg/L for manganese which is lower than the health guideline to avoid "dirty water" problems. Zinc has an aesthetic guideline of 5 mg/L relating to taste, as no adverse health effects are expected from any level likely to be encountered in water supplies.

Residual chlorine levels (Table 4) showed decreasing concentration with increasing distance from the treatment point. The concentration ranged from 2.17 ± 0.083 to 0.17 ± 0.083 mg/L at treatment point and furthest, consumer, respectively. This trend is normal and could consumer,

Figure 1. Map of Malawi showing the position of the city of Lilongwe.

Table 1. Manganese levels (given as means and standard errors of the means) compared to WHO and MBS standards.

Sample station	Mn in water (mg/l)	Maximum Acceptable Levels (MAL) (mg/l)	
		WHO*	MBS*
Treatment Plant	0.02 ± 0.002	0.40	0.10
Storage Tank	0.02 ± 0.001	0.40	0.10
Consumer 1	0.02 ± 0.002	0.40	0.10
Consumer 2	0.04 ± 0.006	0.40	0.10
Consumer 3	0.05 ± 0.007	0.40	0.10
Consumer 4	0.04 ± 0.006	0.40	0.10
Consumer 5	0.03 ± 0.004	0.40	0.10
Consumer 6	0.04 ± 0.004	0.40	0.10
Consumer 7	0.05 ± 0.006	0.40	0.10
Consumer 8	0.05 ± 0.005	0.40	0.10

*WHO and MBS standards were not exceeded in all cases.

consumer, respectively. This trend is normal and could be attributed to the fact that chlorine is still effective or being used in the distribution system. The mean levels of chlorine reaching the farthest consumer tap are at minimum levels, five times less than at the treatment point but still meet the acceptable WHO and MBS standards.

Turbidity levels (Table 5) did not show any significant variation ($\alpha = 0.214$) between the treatment plant and consumer points with values ranging from 1.19 ± 0.075 to 1.32 ± 0.155 NTU. This is the normal trend for treated water if there is no pollution along the distribution system. Pollution could be in the form of soil particles, organic matter and even heavy metal entry into the distribution system causing the color of water to significantly change along the distribution system. When water is not turbid, it is not evidence enough to rule out any microbial or chemical contamination but the aesthetic quality is acceptable to most consumers.

The water pH (Table 6) ranged from a mean of 7 to 8 in the distribution system. This is within the WHO standard of drinking water quality. However, the levels from the treatment point to the consumer showed a significant dif-

Table 2. Lead levels (given as means and standard errors of the means) as compared to WHO and MBS standards.

Sample Station	Pb in water (µg/l)	Maximum Acceptable Levels (MAL) (µg/l)	
		WHO*	MBS*
Treatment Plant	0.37 ± 0.061	10.0	5.00
Storage Tank	0.71 ± 0.033	10.0	5.00
Consumer 1	0.50 ± 0.021	10.0	5.00
Consumer 2	0.51 ± 0.023	10.0	5.00
Consumer 3	0.08 ± 0.037	10.0	5.00
Consumer 4	0.67 ± 0.028	10.0	5.00
Consumer 5	0.22 ± 0.011	10.0	5.00
Consumer 6	0.33 ± 0.082	10.0	5.00
Consumer 7	0.56 ± 0.053	10.0	5.00
Consumer 8	0.11 ± 0.038	10.0	5.00

*WHO and MBS standards were not exceeded in all cases.

Table 3. Zinc levels (given as means and standard errors of the means) as compared to WHO and MBS standards.

Sample Station	Zn in water (mg/l)	WHO* and MBS* maximum acceptable levels (MAL) (mg/l) **NGL
Treatment Plant	1.04 ± 0.071	3.00
Storage Tank	1.16 ± 0.051	3.00
Consumer 1	1.19 ± 0.058	3.00
Consumer 2	1.29 ± 0.051	3.00
Consumer 3	1.25 ± 0.073	3.00
Consumer 4	1.23 ± 0.033	3.00
Consumer 5	1.83 ± 0.306	3.00
Consumer 6	1.27 ± 0.034	3.00
Consumer 7	1.31 ± 0.036	3.00
Consumer 8	1.28 ± 0.055	3.00

*WHO and MBS acceptable levels were not exceeded in all cases. **NGL = No guidelines are given but levels above 3 mg/l are considered unacceptable.

Table 4. Residual chlorine levels (given as means and standard errors of the means) as compared to WHO and MBS standards.

Sample Station	Residual chlorine in water (mg/l)	WHO* and MBS* acceptable minimum range at delivery point (mg/l)
Treatment Plant	2.17 ± 0.083	5.0 mg/l
Storage Tank	1.00 ± 0.000	0.2 -1.0
Consumer 1	0.98 ± 0.022	0.2 -1.0
Consumer 2	0.86 ± 0.073	0.2 -1.0
Consumer 3	0.68 ± 0.081	0.2 -1.0
Consumer 4	0.49 ± 0.130	0.2 -1.0
Consumer 5	0.40 ± 0.109	0.2 -1.0
Consumer 6	0.40 ± 0.055	0.2 -1.0
Consumer 7	0.37 ± 0.047	0.2 -1.0
Consumer 8	0.20 ± 0.083	0.2 -1.0

*WHO and MBS acceptable standard were met in all cases.

Table 5. Turbidity levels (given as means and standard errors of the means) as compared to WHO and MBS standards.

Sample Station	Turbidity (NTU)	WHO* and MBS* maximum acceptable Levels (MAL) (NTU)
Treatment Plant	1.19 ± 0.075	5.00
Storage Tank	0.97 ± 0.087	5.00
Consumer 1	1.03 ± 0.097	5.00
Consumer 2	1.24 ± 0.050	5.00
Consumer 3	1.09 ± 0.072	5.00
Consumer 4	1.11 ± 0.061	5.00
Consumer 5	1.08 ± 0.036	5.00
Consumer 6	1.32 ± 0.155	5.00
Consumer 7	1.11 ± 0.082	5.00
Consumer 8	1.22 ± 0.130	5.00

*WHO and MBS acceptable levels were not exceeded in all cases.

Table 6. pH values in water (given as means and standard errors of the means) as compared to WHO and MBS standards.

Sample Station	pH	Acceptable range	
		WHO*	MBS*
Treatment Plant	7.4 ± 0.1	6.5 – 8.5	5.0 – 9.5
Storage Tank	7.1. ± 0.3	6.5 – 8.5	5.0 – 9.5
Consumer 1	7.5 ± 0.1	6.5 – 8.5	5.0 – 9.5
Consumer 2	7.2 ± 0.2	6.5 – 8.5	5.0 – 9.5
Consumer 3	7.6 ± 0.3	6.5 – 8.5	5.0 – 9.5
Consumer 4	7.7 ± 0.1	6.5 – 8.5	5.0 – 9.5
Consumer 5	7.5 ± 0.2	6.5 – 8.5	5.0 – 9.5
Consumer 6	7.4 ± 0.1	6.5 – 8.5	5.0 – 9.5
Consumer 7	7.7 ± 0.2	6.5 – 8.5	5.0 – 9.5
Consumer 8	6.9. ± 0.1	6.5 – 8.5	5.0 – 9.5

*WHO and MBS acceptable levels were not exceeded in all cases.

Table 7. Feacal coliform levels in water as compared to WHO and MBS standards.

Sample Station	Feacal coliform in water (cfu/100 mL)	Maximum Acceptable Levels (MAL) (cfu/100 mL)	
		WHO*	MBS*
Treatment Plant	0	0	0
Storage Tank	0	0	0
Consumer 1	0	0	0
Consumer 2	0	0	0
Consumer 3	0	0	0
Consumer 4	0	0	0
Consumer 5	0	0	0
Consumer 6	0	0	0
Consumer 7	0	0	0
Consumer 8	0	0	0

*WHO and MBS acceptable levels were not exceeded in all cases.

ference ($\alpha = 0.003$). This might be attributed to leaching of salts from the distribution pipes (Schock, 1999; Agatemor and Okolo, 2007).

There was no variation in the bacteriological quality of water (Table 7) between the treatment point (no colonies/ml) and consumer point (no colonies/ml). This signifies that there was no feacal contamination between the treatment point and consumer point to cause any risk of water borne diseases. The WHO and MBS standards for drinking water standard require that there should be no fecal coliforms in drinking water and that if any are found, an intervention should be instigated to resolve the problem.

It should also be noted that samples were only taken in the dry season when source water is be expected to have low turbidity and little variation in quality. Depending on the efficiency of treatment, some changes in the water quality leaving the treatment plant might be expected in wet weather when some parameters such as turbidity and manganese levels in the source water may increase.

Conclusion and recommendation

Based on laboratory assessment, tap water provided by Lilongwe Water Board to residents of Area 25 Township in Malawi is safe for human consumption with respect to residual chlorine, manganese, pH, zinc, turbidity, lead and fecal coliform bacteria as the levels met both the WHO and MBS standards for drinking water quality. However, as the data presented in this study only reflects the dry season, it would be expected that turbidity and manganese levels are higher in wet weather and therefore warranting further research

REFERENCES

Agatemor C, Okolo PO (2007). University of Benin water supply system: microbiological and physico-chemical assessments. Environmentalist 27:227–239.

American Public Health Association (APHA) (1998). American Water Works Association and Water PollutionControl Federation, Standard methods of the examination of water/wastewater, 20th Edition. APHA, AWWA, and WPCF, New York.

Ashbolt N J (2004). Microbial contamination of drinking water and disease outcomes in developing regions. Toxicology 198(1):229-23.

Gražulevičienė R, Balčius G (2009). Assessment of iron and Manganese Concentration Changes in Kaunas City Drinking Water Distribution System. Environ. Res. Engin. Manag. 50(4):37-43

Kalua PWR, Chipeta WPC (2005). 'A Situation Analysis of Water Sector in Malawi.' A paper presented at the workshop on Situation Analysis of Water Sector in Malawi.

Lilongwe Water Board (2011). Madzi Bulletin. http://www.lwb.mw/index.php?option=com_content&view=article&id=101&Itemid=87

Malawi Bureau of Standards (2005). Drinking water specifications, 1st revision, MS 214:2005.

Mendels P (1998). Dealing With Disaster and a Fragile Infrastructure. New York Times, January 12.

Napacho ZA, Manyele SV (2010). Quality assessment of drinking water in Temeke District (part II): Characterization of chemical parameters. Afri. J. Environ. Sci. Technol.4(11):775-789.

Schock MR (1999) Internal corrosion and deposition control. In Association AWW (ed) Water quality and treatment: a handbook of community water supplies. McGraw-Hill pp. 1–17.

Tuzen M, Soylak M (2006). Evaluation of Metal Levels of Drinking Waters from the Tokat-Black Sea Region of Turkey. Polish J. Environ. 15(6):915-919.

US EPA (2012). Secondary Drinking Water Regulations: Guidance for Nuisance Chemicals. http://water.epa.gov/drink/contaminants/secondarystandards.cfm.

WHO (2004). Guidelines for drinking water quality, 3rd edition. World Health Organization, Geneva.

WHO (2011). Manganese in drinking water. Background document for development of WHO guidelines for drinking water quality. WHO/SDE/WSH/03.04/104/REV/1. http/www.lwb.mw/online water monitoring.htm.

Dimensions in rural water coverage and access in Akwa Ibom State, Nigeria

Jacob Atser[1] and Peter U. Udoh[2]

[1]Department of Urban and Regional Planning Faculty of Environmental Studies, University of Uyo, PMB 1017, Uyo, Nigeria.
[2]Department of Architecture, Faculty of Environmental Studies, University of Uyo, PMB 1017, Uyo, Nigeria.

This study examined the levels of rural water access and coverage in Akwa Ibom State, Nigeria. In Akwa Ibom State, there are two statutory bodies involved in water supply: Akwa Ibom State Water Company Limited and Akwa Ibom State Rural Water Supply and Sanitation Agency. Apart from these two statutory bodies, the federal government, United Nations Development Programme (UNDP), Exxon Mobil, and other foreign bodies are involved in rural water supply. This study explores the contribution of these bodies towards improved rural water coverage and access using scientific approach. Data on mini water schemes and hand pump boreholes in rural areas in the state were obtained from official records and utilized for analysis. The result indicated inequity in the location of projects. The distribution however does not conform to the clearly spelt out criteria for sitting of water schemes in the rural communities. With this distribution gap created by such chaotic pattern, it was apt to examine the level of access which also revealed that access to safe water in the state is grossly inadequate. At the state level, the coverage and access levels of 37.69 and 33.99% were observed. Four spatial factors were examined to ascertain their influence on rural water supply using correlation analysis and result singled out rural population to be highly significant correlated with the number of safe water points(r=0.678). This implies that about 46% of variance in safe water points in rural areas of the state is explained by rural population factor. It can be inferred that poor coverage and access to safe water facilities provide clear indicators of the state of poverty in rural areas of Akwa Ibom State and thus, call for urgent attention.

Key words: Water coverage, water access, rural areas, trends, dimensions, Nigeria.

INTRODUCTION

Water is one of the essential needs of man. Securing access to potable water supply is a central issue of concern not only in urban areas but much more in rural areas. The importance of water supply for domestic uses cannot be compromised not only because of its social and economic values, but also because water based sources of livelihoods have become critical to the survival and health of most rural households, providing valuable contributions to rural livelihoods (Bain et al., 2014). Water is therefore a very strategic socio-economic asset especially in poor economies where wealth and survival are measured by the level of an access to water. Water is an important element in the varied and complex social relations of production within which conflict between individuals and communities are bred. Therefore no nation, city or rural area can survive as an entity without access to sustainable water supply. Access to water supply is therefore one of the key factors that enhance the wellbeing of the rural households. Inaccessibility and unequal access to safe water supply can constrain the inclusiveness of growth and thus result in low standard of living among the rural people (Yange et al., 2013). In circumstances where access is denied, communities' as well as individuals standard of living and productivity becomes drastically reduced.

However, despite the critical role of water for social and economic development, only about 89% of the population had access to improved/safe drinking water supply in the year 2012 (WHO/UNICEF, 2014). This implies that that nearly one in every ten people worldwide have access to improved drinking water and as a result, millions of people, mostly children, still die yearly from diseases mainly associated with lack of access to safe water source (Bain et al., 2014). One of the millennium development goals (MDGs) targets is that by 2015 the proportion of people without sustainable access to safe drinking water should have been reduced by 50%. Perhaps the most frightening is the UNWWD Report (2006) which predicted that by 2050, at least one in every four people are likely to live in a country affected by chronic or recurring lack of access to safe water source. This calls for immediate and sustaining action amongst our policy makers and various stakeholders involved especially in rural water supply if we are to avert this menace.

Rural water coverage and access are terms that have been used interchangeably. However, the same term may have different meanings for different practitioners (Ross and Bostoen, 2010). According to WHO/UNICEF (2004), water coverage refers to the proportion of the population using improved sources of drinking water. It is based on the principle that an improved source of water is designed to deliver water to a certain number of people. Rural water coverage is often calculated by multiplying the number of each safe water points by the number of people who should be served by those water points. However, coverage may not give accurate estimate of access due to functionality and distance to the water source(s). For example, it could be assumed that water point can serve a particular number of people but the number actually having access could be very different (IIDL, 2008).

The Ministry of Local Government and Housing in Zambia; defined access to water in rural areas based on the ability of people to collect at least a minimum of 25 liters of water per person per day for domestic purposes all year round, and also walk less than 500 meters to the water point (Village Water, 2010). This could be a good definition; however, it is important to note the length of time spent queuing at the water point. The Mozambique definition, for example, refers to a 30 min round trip including going to the water point, queuing, fetching the water, and returning home (Ross and Bostoen, 2010). However, WHO/UNICEF (2004) broadly defined rural water access as the availability of at least 20 L per person per day from an improved source within one kilometer of user's dwelling. The standard for Nigeria according to FGN (2000) is that access to rural water supply should guarantee minimum level of service, serving about 250-500 persons per point. WHO/UNICEF (2004) estimated that in Africa, the rural areas enjoy between 25-50% of rural water coverage. WHO/UNICEF (2014) estimates in 2012 indicate that Sub-Saharan Africa achieved 53% of safe water access while the rural water courage for Nigeria in the same year was 49%. It was against the background of poor water coverage that the federal government in the year 2000 came up with the following targets while emphasizing on the provision of sufficient potable water to all Nigerians in an affordable and sustainable way:

i) Improving service initial coverage of 40 to 60% by year 2003;
ii) Extension of service coverage to 80% of population by the year 2007;
iii) Extension of service coverage to 100% of the population in the year 2011; and
iv) Sustain 100% full coverage of water supply and wastewater service for the growing population beyond the year 2011.

Improved water sources are defined in terms of the types of technology and levels of services that are more likely to provide safe water than unimproved technologies. Improved water sources include household connection, public standpipes, boreholes, protected dug wells, protected spring and rainwater harvesting. Unimproved sources are unprotected wells, unprotected springs, vendor provided water, and bottle water (WHOSIS, 2008). It is emphasized here that potable/wholesome water is the same thing as safe water source as long as its availability is enough to guarantee water demand. The more reason bottled water is not considered as safe source unless there are other improved sources to compliment it. This research adopts safe water source instead of potable/wholesome to be in line with the current trend in public water supply.

Equitable and sustainable distribution of water supply facilities in our rural communities is very fundamental for guaranteeing rural water coverage and access. Population of a place is very fundamental when issue of water coverage is to be addressed whereas water access looks beyond population issue, where the water points are actually sited is also of paramount importance. This implies that for equitable distribution of safe water to our rural populace, coverage and access should be seen to satisfy these tenets (Atser, 2012). Compromise to these tenets could likely result in under utilization, mismanagement and neglect of the facility. Nigeria has a water policy and rural water programs should be implemented in line to guarantee right to both coverage and access of safe water delivery especially in the rural communities of Akwa Ibom State. This research is concerned with investigating the extent to which water coverage and access is guaranteed in Akwa Ibom State.

Improving water coverage and access is not only imperative but more importantly, it is within the tenets of sustainable development, which according to the WCED (1987), is development that meets the needs of the present generation without compromising the ability of future generation to meet their own needs and is widely understood as either intergenerational equity principle or intra-generational equity principle. The overall aim is to eliminate poverty which Millennium Development Goals intend to achieve through equitable distribution of amenities such as right access to safe water. On this basis, sustainability has been added to equity in distribution (Haughton, 1998). Furthermore, Haughton while examining equity did not only based on how environmental "disamenities" such as pollution are distributed but also on environmental assets, in particular, the issue of inequitable access to environmental resources, such as water, remarked that many water developments failed to satisfy the basic distribution equity and environmental justice tenet; and that no groups, particularly the disadvantaged, should be made worse off in absolute or relative terms because of water policies. This issue has been taken care in the water policy of Nigeria and there is no discrimination between the rich and poor in rural and urban areas of the country. This is also linked to the research because the degree of success or failure in guaranteeing safe and sustainable water coverage and access to a large extent depends fully on the understanding of emphasis placed in this concept, which are some of the salient points the Nigeria water policy is anchored on.

The stratification of equity into horizontal and vertical dimensions still relates to two broad issues namely: the universality of needs (horizontal equity) and special or targeted supplies (vertical equity). For horizontal equity, everyone needs a particular basic necessity at some point where as vertical equity relates to targeting a specific supply to the needs of a special group such as the poor. The distinction between these situations still turns on the interpretation of need. In line with the benchmarks in Article 2 and 25 of the Universal Declaration of Human Right of 1992, access to domestic water supply services, of an acceptable level, is vital to human well being and dignity and is widely recognized as a human need and, therefore, a basic human right (Anan, 2003). It is therefore essential that coverage be universal. Contaminated water jeopardizes both the physical and social health of the people. It is an affront to human dignity. Once again, this recent definition of access to water as basic human need and right emphasized the responsibilities of governments and international community in the water sector to come up with a sustainable framework on water coverage and access without negating the tenet of distributional equity.

MATERIALS AND METHODS

Akwa Ibom State is the study area. It is one of the oil rich states in the Niger Delta Region of Nigeria. Located in the southeastern coast of Nigeria, Akwa Ibom State was created on September 23, 1987 from the former Cross River State of Nigeria. The State is wedged in between Rivers, Abia and Cross river States and the Republic of Cameroon to the Southwest, North, East and Southeast respectively while the Bight of Bonny bordered the State to the South. It lies between latitudes 4°32' N and 5°32' N; and longitudes 7°28' E and 8° 25' E. According to NPC (2007), Akwa Ibom State has a total land area of 6,187 km^2, which represents 0.67% of the total land mass of Nigeria. The State has 31 Local Government Areas with Uyo, Eket, Ikot Ekpene, Abak, Etinan, and Oron being the most developed urban centres (Table 1). According to the most recent National Population Census conducted in 2006, Akwa Ibom State had a total population of 3,920,208 persons out of which 87.89% constituted rural population while 12.11% formed the urban population (NPC2007). The most striking characteristic of the population of Akwa Ibom State is its crude density. When compared with other states in the south and southeastern parts of the country, the area is one of most densely settled state. Apart from Imo and Anambra States, Akwa Ibom State is the third most densely populated state with densities as high as 634 persons per square kilometer in Nigeria (NPC 2007).

In Akwa Ibom State, there are two statutory bodies involved in water supply. The first is Akwa Ibom State Water Company Limited which is charged with the responsibility of urban and peri-urban water supply while the second, Akwa Ibom State Rural Water Supply and Sanitation Agency has the mandate of rural water supply. Apart from these two statutory bodies set up by Akwa Ibom State Government, the Federal Ministry of Agriculture also intervenes in rural water supply in the state directly or through Cross River Basin Development Authority. There is also foreign collaboration with UNICEF and UNDP on rural water supply. Multinational oil companies like Exxon Mobil, Elf and Shell Petroleum Development are also involved in rural water supply in the state. Sometimes Local Government Administration extends their statutory function of maintaining the water schemes in their domains to direct execution of water supply projects. The research examined the dimensions in rural water schemes in the state in terms of coverage and access particularly that the level of coverage does not necessarily translates to the same level of access unless it is matched with population. Data on the total length of all the roads in the state and total area of all the 31 Local Government Areas in the state were obtained from the State Ministry of Works and Transport. Data on poverty index were obtained from the Akwa Ibom State Ministry of Economic Planning and Development. The data on rural population was projections from the 2006 census results. To compute water coverage, the following formula was used:

$$Pc = (ps / pp) \, 100\%$$

Where pc =%age water coverage, ps = Population served (500 as maximum number of persons to be served per one safe water point multiplied by the number of safe water points) and pp = projected population to year 2013.

For rural water access, different formulae were applied in computing the percentage rural water access for each LGA. The

Table 1. Population distributions in urban and rural Akwa Ibom State in 2013.

Local Government Area	Urban	Semi-Urban	Rural	Total
Abak	61826	18398	111528	191752
Eastern Obolo			24509	24509
Eket	64790	37962	42797	145549
Esit-Eket		32528	38739	71267
Essien Udim		18557	211865	229423
Etim Ekpo		17948	105642	123590
Etinan	32359	38357	88104	158720
Ibeno		20924	51956	72880
Ibesikpo Asutan		8875	143333	152208
Ibiono-Ibom			182264	182264
Ika			79294	79294
Ikono			162012	162012
Ikot-Abasi		23309	93234	116543
Ikot-Ekpene	100297	58971	12165	171433
Ini			125608	125608
Itu		13589	127327	140916
Mbo			118578	118578
Mkpat-Enin		18240	165219	183459
Nsit-Atai			78965	78965
Nsit-Ibom		32194	79808	112002
Nsit-Ubium		31852	98219	130071
Obot-Akara			114155	114155
Okobo		47091	75241	122332
Onna		108703	90457	199178
Oron	63819	14727	19637	98183
Oruk Anam			223276	223276
Udung Uko			40813	40813
Ukanafun		25699	12923	38622
Uruan		59070	81719	140789
Urue Offong/Oruko			54150	54150
Uyo	151566		71275	222841
State Total (%)	474657 (11.17)	626994 (15.55)	2924812(72.66)	4026463 (100.0)

Source: National Population Census 2006 projected to 2013.

basic assumption is that each safe water point is expected to service a maximum of 500 people within maximum distance of 250 m radius (FGN, 2000).

Ppc - Pp/Nc
Where Ppc = Average community population in a LGA; Pp = LGA population; Nc = Number of communities per LGA.

Wprc = Ppc/Tpwp
Where Wprc = Required water points per community in LGA; Tpwp= Threshold population for water point (500 persons)

Wprl = Wprc x Nc
Where Wprl = Required water points in LGA

Percentage access = (NSWP/WPrl) x 100

Where NSWP =Total number of water points found in LGAs

Pearson correlation statistics was employed to investigate the relationships between spatial influences of the number of safe water points among the 31 Local Government Areas in the State using version 17 of Statistical Package for Service Solutions (SPSS) software.

RESULTS AND DISCUSSION

Table 2 shows the rural water coverage in the 11 Local Government Areas (LGAs) with its total population as rural and while Table 3 shows the access status in the same 11 LGAs adjudged as entirely rural. To analyze the rural water coverage in each of the LGAs, water per capita in each of the LGAs was calculated based on Nigeria water demand standard. This is based on the fact that each water point should be able to serve a maximum of 500 persons per day and delivers at least 30 liters of

Table 2. Analysis of rural water coverage for eleven rural LGAs in Akwa Ibom State.

LGA	Population	HPB	MWS	NSWP	Pop. served	% coverage
Eastern Obolo	24509	4	20	24	12000	48.96
Ibiono Ibom	182264	4	63	67	33500	18.38
Ika	79294	3	80	83	41500	52.34
Ikono	162012	5	84	89	44500	27.47
Ini	125608	8	70	78	39000	31.05
Mbo	18578	7	40	47	23500	19.82
Nsit Atai	78965	11	60	71	35500	44.96
Obot Akara	114155	17	70	87	43500	38.12
Oruk Anam	223276	19	70	89	44500	20.00
Udung Uko	40813	5	50	55	27500	67.38
Ureoffiong/Oruko	54150	9	50	59	29500	54.48

HPB= Hand pump borehole, MWS= Mini water schemes, NSWP= Number of safe water points.

Table 3. Percentage rural water access analysis for eleven rural LGAs of Akwa Ibom State.

NAME OF LGA	Pop	NC	Ppc	WPrc	WPrl	NSWP	% access
Eastern Obolo	24509	17	1442	3	51	24	47.06
Ibiono Ibom	182264	161	1132	3	483	67	13.87
Ika	79294	54	1468	3	164	83	50.61
Ikono	162012	79	2051	5	395	89	22.53
Ini	125608	79	1590	4	316	78	24.68
Mbo	118578	78	1520	4	312	47	15.06
Nsit Atai	78965	49	1612	4	196	71	36.22
Obot Akara	114155	61	1871	4	244	87	35.66
Oruk Anam	223276	108	2067	5	540	89	16.48
Udung Uko	40813	25	1633	4	100	55	55.00
Urue Offong Oruko	54150	37	1464	3	111	59	53.15

water per day per person (FGN, 2000). The maximum of 500 persons to be served by one safe water point was then multiplied by the total number of safe water points in each of the LGAs to arrive at estimated water coverage. This implies the expected population that water supply could meet their water demand given the right access. The percentage of this was then found with respect to the entire population of the LGA to give percentage coverage for coverage analysis and the result is in Table 2.

For rural water access analysis, the eleven LGAs adjudged by National Population Census of 2006 to be 100% rural were first considered. The idea of averaging the community approximate population was to enable application of 250 – 500 persons service standard per safe water point so as to arrive at required number of points for the respective communities which is to be used in calculating the number of points required in a particular LGA. With this the percentage access was then calculated by finding the number of water points within a community against the number of community in the LGA. Table 3 shows the analysis of percentage rural water access in the eleven rural LGAs.

There is an indication of inequity in the location of projects. The distribution however does not conform to the FGN (2000) which clearly spelt out criteria for sitting of water schemes in our rural communities. With this distribution gap created by such chaotic pattern, it is apt to say there is urgent need to correct the lopsidedness being observed. Similarly, the study also revealed that safe coverage in the state is grossly inadequate. Based on FGN (2000) which states that Nigeria should attend 100% full coverage of safe water supply for the growing population beyond 2011, many LGAs are still not able to meet the target. Even among the LGAs that are classified as urban, the attainment of 100% full coverage is still not achievable. There are 18 LGAs with coverage range of 18 – 40% with Ibiono Ibom LGA being the least with coverage level of 18.38%. Few LGAs that had coverage levels above 70% are those LGAs having peri-urban and urban centres and these are Eket, Ikot Ekpene and Ukanafun. Probably all the water points could not be said to belong entirely to the rural areas and vice versa. A total of 12 LGAs had coverage levels of between 41 to 69% while only 3 LGAs had coverage levels of 70% and

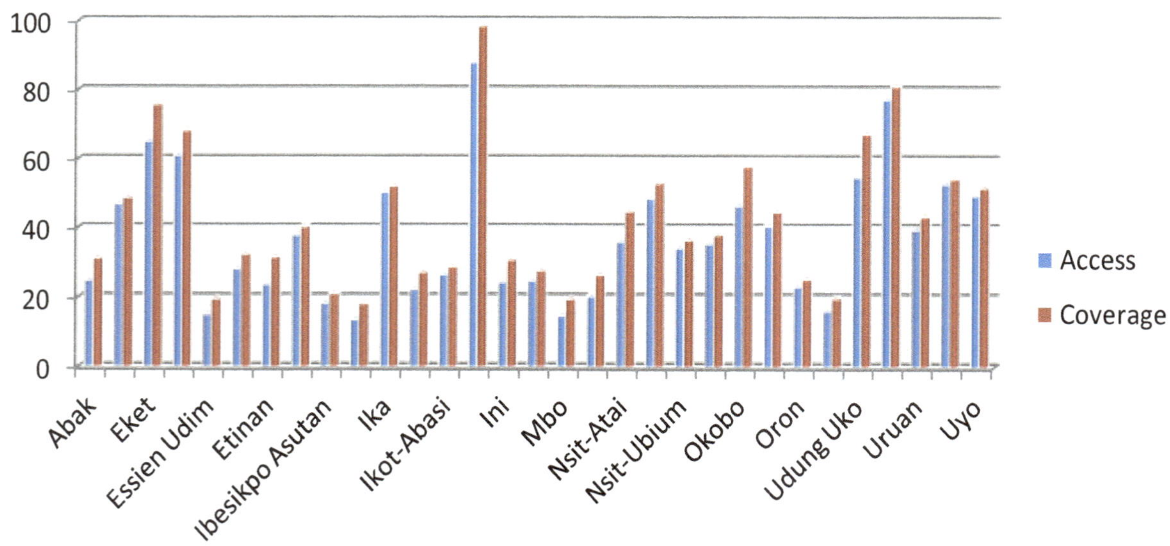

Figure 1. Rural water coverage and access relationship among the LGAs.

above (Figure 1). Table 4 shows the dimensions in rural water coverage and access for the entire state. The data in Table 4 were utilized to examine the relationship between rural water access and coverage as illustrated graphically in Figure 1. The R-square coefficient value of 0.979 indicates that about 98% of access level relate to coverage levels. This implies that improving coverage levels would indirectly impact positively on access to safe water points. Water coverage has spatial connotation and therefore could improve access. The result of poor access revealed in this study is a good pointer to policy makers and various stakeholders to reflect on the design and implementation of mini water schemes in the state with emphasis given to water distribution.

Based on FGN (2000) rural water access benchmark no LGA could rightly be said to have access to safe water. Ikot Ekpene had what could be regarded as the highest precent age water access of 88.20% and closely followed by Ukanafun (77.51%) while Ibiono Ibom had the least access level (13.87%). A comparison of water access with coverage even revealed more water access problem among the LGAs. In ideal situation and given the right framework, water generated should be equal to water access (FGN, 2000). That is to say that the volume of water generated must be distributed proportionately in accordance with the standard stipulated by FGN (2000) before coverage could be equated to access. A situation where many inhabitants and communities have to travel long distances (more than 30 min) as well as queue for water, irrespective of the number of water points within the water headwork does not add any special advantage but rather creates access problem. For a community to have access to water, it therefore means that the households within the community should travel a maximum distance of 250m to fetch water and each water points should not serve more than 500 persons per day (FGN, 2000). Thus, if the number of water points were said to be evenly distributed across the community such that no individual would travel more than 250 m, then such area would have met the required access target.

In Table 5, four spatial factors were examined to investigate their influence on number of safe water points among the LGAs. The choice of these factors was based on their relative importance. Road infrastructure is the artery of major developments, that is, road infrastructure is preceded by other developments. This implies that where a road goes, development follows. Mini water schemes are provide using heavy motorized driller equipment and as such the more motorable the LGA is in terms of road network, the more the potentials for the provision of mini water schemes. All the four independent variables are surrogate to rural development. The poverty index as well as rural population is good indicators of the level of water accessibility. It is expected that poverty incidence is low among households and communities that have improved access to water supply while areas with high rural population should have more safe water points than those with less population. However, the result as presented in Table 6 and Figure 2 shows that only the rural population factor highly significant correlated with the number of safe water points in the state ($r = 0.678$, $p < 0.001$), implying that about 46% of variance in number of safe water points is explained by rural population. The area (0.402) and road (0.280) factors though positively related, had weak influences on water supply going by their R-square values of 16.2 and 8.0% respectively. The poverty index factor however shows negative symptoms (-0.130) of relationships which is expected. The higher the index factor, the lesser the number of safe water points. This relationship is however very weak accounting

Table 4. Water supply coverage and access in rural Akwa Ibom State.

LGAs	Rural pop	HPB	MWS	NSWP	%access	% coverage
Abak	111528	10	60	70	25.11	31.40
Eastern Obolo	24509	4	20	24	47.06	48.96
Eket	42797	10	55	65	65.30	75.94
Esit-Eket	38739	3	50	53	61.22	68.40
Essien Udim	211865	12	72	84	15.44	19.82
Etim Ekpo	105642	9	60	69	28.55	32.66
Etinan	88104	12	44	56	24.07	31.78
Ibeno	51956	3	39	42	38.11	40.42
Ibesikpo Asutan	143333	6	55	61	18.66	21.27
Ibiono-Ibom	182264	4	63	67	13.87	18.38
Ika	79294	3	80	83	50.61	52.34
Ikono	162012	5	84	89	22.53	27.47
Ikot-Abasi	93234	14	40	54	26.82	29.00
Ikot-Ekpene	12165	5	19	24	88.20	98.64
Ini	125608	8	70	78	24.68	31.05
Itu	127327	1	70	71	25.11	28.00
Mbo	118578	7	40	47	15.06	19.82
Mkpat-Enin	165219	10	78	88	20.54	26.63
Nsit-Atai	78965	11	60	71	36.22	44.96
Nsit-Ibom	79808	5	80	85	48.80	53.25
Nsit-Ubium	98219	10	62	72	34.44	36.65
Obot-Akara	114155	17	70	87	35.66	38.12
Okobo	75241	7	80	87	46.71	58.00
Onna	90457	15	66	81	40.66	44.77
Oron	19637	1	10	11	23.41	25.46
Oruk Anam	223276	19	70	89	16.48	20.00
Udung Uko	40813	5	50	55	55.00	67.38
Ukanafun	12923	8	13	21	77.51	81.25
Uruan	81719	1	70	71	39.67	43.44
Urue Offong/Oruko	54150	9	50	59	53.15	54.48
Uyo	71275	4	70	74	49.71	51.91
State Total (%)	2924812(72.66)	238	1750	1988	33.99	37.69

for only 2% of the variance in the number of safe water points across the state.

RECOMMENDATIONS AND CONCLUSION

There is a need for the statutory government organs to interface with stakeholders involved in rural water programs to maintain a robust spatial database which amongst other things should include hydrologic parameters of all their water programs. This will aid in design, costing and execution of rural water projects. The redesign of the mini water schemes should be looked into. Concentration of the water points at the headwork does not help to address the issue of water access. Steps should be taken to reduce the widening gap between water points away from immediate confines of the headwork but within the limit of hydraulic head. By the year 2015, the target year of MDGs to increase access of population by 50% to safe water, Akwa Ibom State rural population will still be left behind. The result of this research has revealed very low access in the rural communities of Akwa Ibom State. There is urgent need for Akwa Ibom State Government, at least for interim measures, to harness other sources of water to complement groundwater. Effort should be made to explore the possibility of rainwater harvesting and utilization. Some parts of China with annual rainfall less than 500mm have utilized rainwater potentials to balance her water needs, so the lesson from other developing countries' experiences and international collaboration are very important (Yang et al., 2013). Water quality in our numerous streams should be looked into and improve upon as the case may be to serve as a stop gap in those

Table 5. Underlying factors for rural water supply in Akwa Ibom State.

L.G.A	Total Length (Km)*	Area (Km^2)*	Poverty index**	Rural population***	NSWP
Abak	152.4	252	83.64	111528	70
Eastern Obolo	141.0	117	50.82	24509	24
Eket	218.5	175	78.31	42797	65
Esit Eket	41.0	164	49.74	38739	53
Essien Udim	364.0	295	56.67	211865	84
Etim Ekpo	189.9	235	59.58	105642	69
Etinan	214.2	182	30.81	88104	56
Ibeno	19.0	248	53.37	51956	42
Ibesikpo	501.0	191	75.13	143333	61
Ibiono Ibom	273	333	26.88	182264	67
Ika	97.2	68	45.93	79294	83
Ikono	150.3	390	43.98	162012	89
Ikot Abasi	318.6	335	45.33	93234	54
Ikot Ekpene	216.9	115	36.59	12165	24
Ini	129.0	320	41.21	125608	78
Itu	193.0	273	39.79	127327	71
Mbo	83.6	335	56.18	118578	47
Mkpat Enin	393.0	332	45.61	165219	88
Nsit Atai	174.0	101	77.04	78965	71
Nsit Ibom	194.1	109	69.35	79808	85
Nsit Ubium	202.5	243	45.23	98219	72
Obot Akara	75.0	227	59.38	114155	87
Okobo	77.5	360	46.48	75241	87
Onna	291.3	174	45.70	90457	81
Oron	59.8	96	66.67	19637	11
Oruk Anam	449.9	512	30.96	223276	89
Udung Uko	113.5	64	90.36	40813	55
Ukanafun	223.4	254	67.20	12923	21
Uruan	356.0	422	71.36	81719	71
Urue Offong	121.6	118	77.89	54150	59
Uyo	253.9	249	68.75	71275	74

Sources: *Atser (2012), **Akwa Ibom State Ministry of Economic Planning and Development (2013); ***National Population Census 2006 projected to 2013.

Table 6. Correlations.

Parameter		Road	Area	Poverty index	Rural Pop
NSWP	Pearson Correlation	0.280	0.402*	-0.130	0.678**
	Sig. (1-tailed)	0.064	0.012	0.244	0.000

*Correlation is significant at the 0.05 level (1-tailed). **. Correlation is significant at the 0.01 level (1-tailed).

communities that are in dying need of safe water source (Bain et al., 2012). In conclusion therefore, sustainable rural water coverage and access for Akwa Ibom State cannot be overemphasized. Thus for rural water coverage and access to be sustainable, creation and maintenance of efficient and up-to-date database containing vital information is essential. Also, rural communities in the state are still far from attaining the

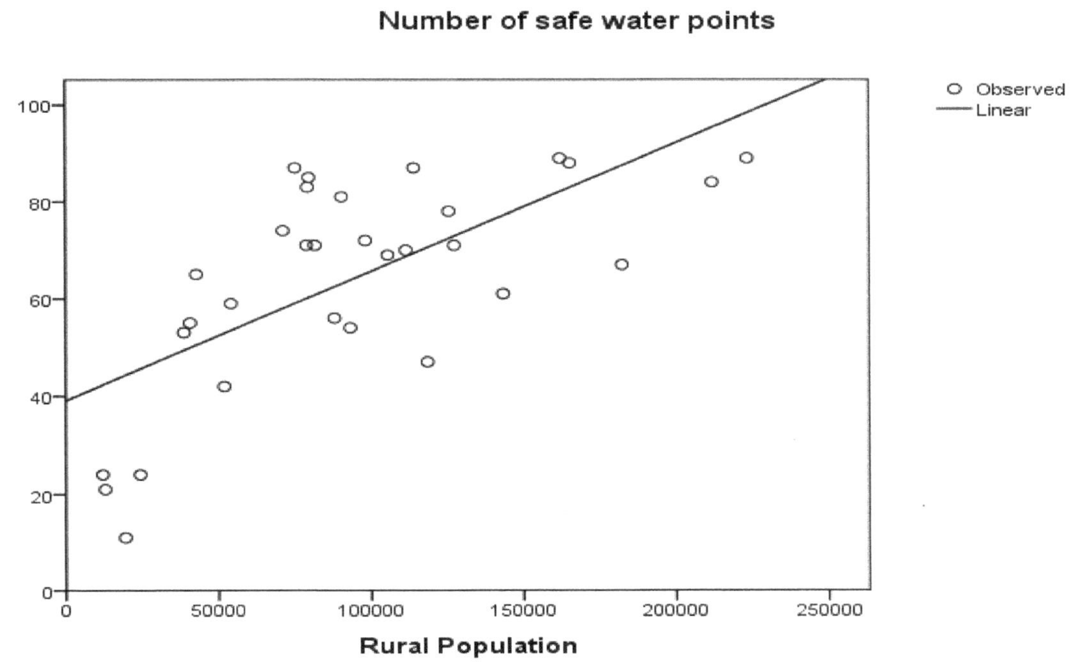

Figure 2. Relationship between rural population and safe water points.

100% coverage target as specified in Nigeria water policy. In some areas, more safe water points are generated or covered but only to be poorly distributed or accessed.

REFERENCES

Anan I (2003). Technical background paper on water, sanitation and health focus on International changes for achieving the water related targets of MDGs in Asia. UN Water Resources section, Environmental and Sustainable development division.

Atser J (2012). Spatial patterns of social infrastructure in Nigeria. LAP LAMBERT Academic Publishing, Av Akademikerverlag GmbH & Co. KG

Bain R, Gundry S, Wright JA, Yang H, Pedley S, Bartram J (2012). Accounting for water quality in MDG monitoring: lessons from five countries. Bulletin of the World Health Organization. 90(3):228-235.

Bain R, Cronk R, Hossain R, Wright J, Yang H, Slaymaker T, Bartram J (2014). Faecal contamination of drinking-water in low- and middle-income countries: A systematic review and meta-analysis. *PLoS Medicine*. 11; e1001644.

Bain R, Cronk R, Hossain R, Bonjour S, Onda K, Wright J, Yang H, Slaymaker T, Hunter P, Pruss-Ustun A, Bartram J (2014). Global assessment of exposure to faecal contamination through drinking water based on a systematic review. Trop. Med. Int. Health. 19(8):917-927.

FGN (2000). National water supply and sanitation policy. www.nwri.gov.ng/userfiles/national water policy.

Haughton G (1998). Geographical Equity and Regional Resource Management: Water Management in Southern California. Environ. Planning and Design. 25:297.

IIDL (2008). National urban water reform project on National water supply and sanitation baseline survey (south-south zone). Federal Ministry of Agriculture and Water Resources Report, Abuja.

NPC (1998).1991 National Population Census of the Federal Republic of Nigeria: Analytical Report at the National Level". National Population Commission (NPC), Abuja, p8.

NPC (2007). 2006 National Population Census Result: Legal Notice on the Details of the Breakdown of the National and State Provisional Totals". National Population Commission. *Printed and published by the* Federal Government Printer, Lagos, Nigeria. FGP71/52007/2500(OL24).

Ross I, Bostoen (2010).Data reconciliation in South Africa: Looking at monitoring approaches in water and sanitation. A Water Aid Regional workshop in collaboration with WHO/UNICEF.

UNWWD(2006).Water: a shared responsibility. www.unesco.org/water/wwap/wwd2.

Village Water (2010).Sustainable Water and Sanitation in Zambia. http://www.villagewater.org/node/6

Water Aid (2006). Effective Financing of Local government to provide water and sanitation services. www.wateraid.org/other/startdownload.

WCED (1987).*Our common future*. World Commission on Environment and Development. OUP, Oxford.

WHO/UNICEF (2004). Joint monitoring programme for meeting the MDG drinking water and sanitation target: a midterm assessment report. http://www.who.int/water/sanitation_health/monitoring/jmp2004/index.

WHOSIS (2008). Access to improved drinking water sources and improved sanitation.www.who.int/../2wst.

Yang H, Flower RJ, Thompson JR (2013). Sustaining China's water resource. Science. 339:141-141

Yang H, Bain RES, Bartram J, Gundry S, Pedley S, Wright JA (2013). Water safety and inequality in access to drinking-water between rich and poor households. Environ. Sci. Technol. 47(3):1222-1230.

18

Climate change, environmental security and displacement in Nigeria: Experience from the Niger Delta Flood Disaster, 2012

Luke Amadi and C. U. Mac Ogonor

Department of Political Science and Administrative Studies, University of Port Harcourt, Nigeria.

The coastal regions of Africa are prone to series of environmental disasters arising from vulnerability of climate change. The October 7th and November 3rd, 2012 coastal floods in the Niger Delta region Nigeria, provides an evidence of the persistence and inevitability of climate change vulnerability which has been an issue of global concern with potential for havoc on human existence including environmental security, displacement and their far reaching consequences. Using primary and secondary data sources, the paper foreshadows the imminent dangers of climate change vulnerability. It deployed a participatory methodology through focused group discussions (FGDs), questionnaires and oral interview guide as primary data sources. While secondary data sources included relevant authoritative reports from National Emergency Management Agency (NEMA), UNDP, UNEP, newspapers, magazines and documents published by governmental and Non-Governmental Organizations (NGOs). The sampling technique was largely purposive due in part to the sensitivity of the issues investigated. Two open-ended questionnaires were used to elicit two types of information on coastal flood, environmental security and displacement within the purposively selected areas of study namely; Bayelsa, Delta and Rivers States. The findings suggest that the 2012 flooding negatively affected the region with evidence of displacement, out migration, impoverishment, food production decline, etc. The paper made some policy recommendations on mitigation of climate change vulnerability.

Key words: Climate change, environmental security, development, Niger Delta, Nigeria.

INTRODUCTION

The African coastal zone consists of a narrow, low-lying coastal belt. It also includes the continental shelf and coasts of 32 mainland countries. It is composed of a variety of ecosystems, including barrier/lagoons, deltas, mountains, wetlands, mangroves, coral reefs and shelf zones. These ecosystems vary in width from a few hundred meters (in the Red Sea area) to more than 100 km, especially in the Niger and Nile deltas. In west Africa (Mauritania to Namibia), the coastal zone spans a broad range of habitats and biota and includes the pristine islands of

Bijagos Archipelago; the offshore island nations of Cape Verde and São Tomé and Principe; and the remote central Atlantic islands of San Helena and Ascension (IPCC, 2007).

A large percentage of West Africa's urban population lives in coastal cities. In Nigeria, for example, about 20 million people (22.6% of the national population) live along the coastal zone; about 4.5 million Senegalese (66.6% of the national population) live in the Dakar coastal area. About 90% of the industries in Senegal are located within the Dakar coastal zone. In Ghana, Benin, Togo, Sierra Leone and Nigeria, most of the economic activities that form the backbone of the national economies are located within the coastal zone. Coastal areas also form the food basket of the region. Offshore and inshore areas, as well as estuaries and lagoons, support artisanal and industrial fisheries accounting for more than 75% of fishery landings in the region (IPCC, 2007).

Similarly, the coastal zone of East Africa, including coastal wetlands, extends from Sudan to South Africa and includes the near-shore islands off the coast of Tanzania and Mozambique and the oceanic islands of Madagascar, the Seychelles, Comoros, Mauritius and Reunion. The desert margins of the Red Sea feature some of the richest coral reefs in the world. Coral reefs further south, extending from Kenya to the Tropic of Capricorn, are well distributed around most of the oceanic islands. They buffer the coastline against the impact of wave breakers and the full force of storms and cyclones. Many principal east African cities are located inland. Despite their low densities, however, coastal cities like Dar es Salaam and Mombasa are experiencing annual population growth of 6.75 and 5%, respectively (World Bank, 1995a). Coastal tourism and fisheries represent large inputs into the GNP of east African states (IPCC, 2007).

In recent times, environment and security discourse have emerged as a central policy concern for countries across the world. This has been under consideration since the 1980s mainly by two groups: (1) the environmental policy community, addressing the security implications of environmental change and security, and (2) the security community, looking at new definitions of national security, particularly in the post-Cold War era. With the surge in environment and security challenges, the General Assembly officially introduced the concept of security and environment at its 42nd session.

The 90s and 2000s revealed increasing vulnerability of climate change such as recent environmental disasters namely; the Japanese Okushiri, Hokkaidō tsunami which struck Okushiri Island of Hokkaidō on July 12, 1993, the 2004 Indian Ocean tsunami with over 230,000 people killed in 14 countries bordering the Indian Ocean, the US Atlantic hurricanes (Andrew, 1992; Katrina, 2005; Irene, 2011) Sandy flood, 2012 and the Haiti earthquake; In West Africa, the most recent severe floods in the coastal regions of Nigeria on 2nd November 2012, etc (Amadi, 2012).

There are several perspectives in recent literature on environmental security discourse (Mathews, 1989; Homer Dixon, 1992; Awosika et al., 1992). The 1990s witnessed a paradigm shift to human security (UNDP, 1994; Klare, 1996).

Adibe (1994) had examined the change, security had assumed in recent literature encompassing "security expansionism". Suhrke (1995) argued on "environment change, migration and conflict". Her treatise is premised on terminological shifts in line with social changes and transformations in environment and conflict discourse.

Homer-Dixon (1991) identified "gaps" among the poor countries to meet environmental demands which could result conflicts unlike rich countries. He observed that the rich countries have the potential for a widening gap between demands on the state and its financial ability to meet these demands- a gap which he argued could lead to internal conflict between competing ethnic groups or significant out-migration to countries better able to cope with environmental stresses.

Choucri (1994) argued that environmental degradation forces people to move, sometimes across borders and more assuredly to impinge on and ultimately challenge those (host) populations, thus becoming a key element of conflict.

Suhrke (1995) identified four principal ways pressure on the environment has affected security of the state. They are: (1) environmental degradation can cause health hazards or jeopardize the economic livelihood of a significant part of the population; (2) intensified competition for declining or degraded resources can create conflicts within or among States: these in turn can generate regional instabilities that affect nations further afield; (3) environmental degradation may force people to migrate, thereby creating conflict over scarce resources in the receiving areas; and (4) the existentialist argument advanced by a school of environmentalists and "ecological economists" to the effect that an environmental resource has an intrinsic value regardless of its being consumed (even in the form of being seen), and hence its loss is a matter of security (Suhrke, 1996).

Kaplan (1994) re-echoed the persistent environmental degradation and deforestation in West Africa. He writes that in Sierra Leone, Guinea, Ivory Coast and Ghana, most of the primary rain forest and the secondary bush is being destroyed at an alarming rate. He reports that he "saw convoys of trucks bearing majestic hardwood trunks to coastal ports". (p.48)

In 1987, the World Commission on Environment and Development linked security with environment as foreshadowed in the Brundtland Report:

> Humankind faces two great threats. The first is that of a nuclear exchange. Let us hope that it remains a diminishing prospect for the future. The second is

that of environmental ruin world-wide and far from being a prospect for the future, it is a fact right now (Bruntland, 1987).

These terms are marginalized in policy discourse or at the best superficially addressed in the poor countries. Thus, environmental security and climate change issues seem not to have been prioritized in development discourse especially in the coastal regions of Africa such as the Niger Delta region in Nigeria.

Issues of displacement in the region now take various forms such as in and out migration, flood induced displacement, acid rain and oil spill, to crisis induced displacement, etc.

Ibeanu (1998) observed the continued marginalization of displaced persons in Nigeria; "the proclivity among the displaced is rarely to contest their predicament. Only in a few celebrated cases, like that of the Bakolori peasants who were displaced by a dam in the 1980s and the Ogoni whose livelihoods are threatened by oil exploration, have internally displaced people organised and effectively put their condition on the political agenda. The lack of organization among displaced people generally serves to keep their plight concealed".

In 1999 was the adoption of the United Nations Framework Convention on Climate Change (UNFCC) which was aimed to develop a global framework on climate change. The coastal areas in Africa are fraught with environmental challenges and flood related disasters. The East African floods of 1998 and the Mozambique floods in early 2000 and 2001 caused considerable damage to property and infrastructure. The major infrastructure damage was road and rail network damage. Communications among human settlements in Kenya, Uganda, Rwanda and Tanzania were seriously disrupted, impeding movement of goods and persons in the region (AEO, 2002).

Aspects of coastal erosion have been examined in recent literature (Ibe and Quelennac, 1989; Awosika et al., 1992; Dennis et al., 1995; French et al., 1995). For instance, the occurrence of coastal erosion has been reported in the Niger Delta by Okon and Egbon (1999). The report of Udofa and Fajemirokun (1978) showed a rise in sea level along Nigerian coastal water. They conducted a mechanical analysis of tide data from 1960 - 1970 and reported mean sea level rise to be 0.462m above zero level of the tide gauge which have been of increasing concern in the region.

In 2007, the Fourth Assessment Report of the Intergovernmental Panel on Climate Change (IPCC) predicted that rising global temperatures will contribute to an upsurge in severe storms, floods, droughts, glacier melt and sea level rise. In vulnerable areas of the developing world, extreme weather is expected to intensify pressures on land and water resources, disrupt agricultural production, and threaten food security. In the wake of the IPCC report, a number of policy studies concluded that there is a strong likelihood that the natural hazards and environmental stresses associated with climate change will trigger or amplify conflict, especially in vulnerable or unstable areas of the developing world (CNA Corporation, 2007; Campbell et al., 2007; Smith and Vivekananda, 2007; Fingar, 2008; UN, 2009).

Among the projected scenarios were severe resource scarcity, dramatic increases in internal and external migration, disease outbreaks and a host of destabilizing social and political effects (Campbell and Weitz, 2008). The CNA Corporation envisioned a confluence of factors that might overwhelm weak or flawed systems of governance and public institutions, setting the stage for "internal conflicts, extremism and movement toward increased authoritarianism and radical ideologies" (CNA Corporation, 2007.)

The Nigerian Environmental Study/Action Team (NEST), reported that sea-level rise and repeated ocean surges will not only worsen the problems of coastal erosion that are already a menace in the Niger Delta, the associated inundation will increase problems of floods, intrusion of sea-water into fresh water sources and ecosystems destroying such stabilizing system as mangrove, and affecting agriculture, fisheries and general livelihoods (NEST, 2004).

The Niger Delta is the product of both fluvial and marine sediment build-up since the upper Cretaceous period, some 50 million years ago. Over time, up to 12,000 m of shallow marine sediments and deltaic sediments have accumulated, contributed mainly by the Niger River and its tributaries (UNEP, 2011).

The coastal area comprises three vegetation zones: (i) beach ridge zone, (ii) saltwater zone and (iii) freshwater zone. The beach ridge zone is vegetated by mangroves on the tidal flats and by swamptrees, palms and shrubs on the sandy ridges. The saltwater zone is mainly vegetated by red mangrove (*Rhizophora mangle*). The coastal plain and freshwater zone is vegetated by forest tree species and oil palm. The Niger River floodplains are covered by rainforest trees, oil palm, raffia palms, shrubs, lianas, ferns, floating grasses and reeds (UNEP, 2011).

The coastal regions are at the most prone to these threats as they repeatedly experience intense environmental stress with minimal response from the government and the international community. There are several causes of environmental insecurity such as the depletion of the ozone layer, pollution, deforestation, acid rain, erosion, flood, sea level rise, gas flaring, etc.

This study focuses on the recent Niger Delta coastal flood, where no fewer than three million persons were displaced and a few others died. The study and its case scenario are linked to a broader elucidation of Nigeria's commitment to global environmental conventions. This is significant in an era of post Rio +20 world environmental summit of 2012. We argue that climate change vulnerability remains an issue of urgent policy attention among the poor coastal regions.

In this context, we seek to examine the recent environmentally induced travail of the Niger Delta people namely; the October and November 2012 coastal floods. This is important in view of the reoccurring incidence of coastal erosion, acid rains, oil minority politics, ozone layer depletion, oil spill, environmental pollution, sea level rise and ocean surge within the volatile region. The region is a minority area in Nigeria which implies that they are numerically less as compared to the major ethnic groups. This study significantly examines how the federal government is responding to the problems of these groups. This is linked to a brief elucidation of Nigeria's commitment to some global environmental protocols and how these have impacted awareness on climate change vulnerability. The essay argues that policy discourse to improve the lots of the poor coastal regions is important.

MATERIALS AND METHODS

This study provides a case analysis of displacement arising from coastal erosion, the victims and implications for policy discourse. It adopted a participatory methodology through Focused Group Discussions (FGD), Questionnaires and Oral interviews as primary data sources to evaluate the level of environmental damage and security threats arising from the October and November 2012 flood disaster.

Participatory methodology is suitable, being a direct and all inclusive approach, the victims were reached for interviews and on the ground analysis of the problems of the study. According to Chambers (2010), current development thinking and practice have diverged into two clusters, with procedures associated with the paradigm of things imposed by powerful actors and organizations in tension and contradiction with participatory methodologies (PMs) associated with the paradigm of people. PMs that combine methods have proved increasingly versatile and adaptable to contexts and purposes. For our purpose we explore the "paradigm of people".

Secondary data sources included authoritative reports from National Emergency Management Agency (NEMA), UNEP reports, newspapers, magazines, reports and documents published by government and non-governmental organizations. The sampling technique was largely purposive due in part to the sensitivity of the issues investigated.

Two open-ended questionnaires were used to elicit two types of information on climate change vulnerability, displacement and insecurity. The first, the general sample (GS) questionnaire elicited general information about climate change in the Niger Delta and was directed at community members, youths, women and elders. A total of 255 questionnaires were received from those distributed, (80 in Bayelsa, 100 in Delta and 75 in Rivers). The second questionnaire (key informant sample) sought in-depth, broad and sensitive information from more informed citizens of the region such as teachers, civil servants, environmental engineers, literate farmers, etc.

A total of 55 questionnaires were received: 20 in Bayelsa, 20 in Delta and 15 in Rivers State). The overall return rate of distributed instruments for data collection revealed 88% response. Oral interviews were also conducted. The objective was to elicit detailed and informed information from rural women, heads of households, fishermen and farmers who could not write. Beyond the interviews were FGDs, three each were conducted in Bayelsa (Yenagoa), Delta (Asaba) and Rivers (Port Harcourt) with sample sizes of 12, 8 and 8, respectively. The aim was to further retest the earlier data for validity.

Nigeria and commitment to some global environmental summits and protocols

Nigeria is committed to several multilateral environmental agreements (MEAS), summits and protocols. The Stockholm conference of 1972 became a spring board on which subsequent global environmental summits were organized. This include the 1992 Rio earth summit, the UN –Rio +5 by group of experts in 2007, in 2002 the Johannesburg Plan of Implementation and the Rio +20 in 2012.

The international community has negotiated and brought into effect various Multilateral Environmental Agreements (MEAs), including the Rio Conventions: the UNFCCC, UNCCD, CBD developed at the Rio Earth Summit in 1992. These conventions are also related to other MEAs that have been agreed over the years including the CITES and Ramsar Convention on Wetlands. Nigeria is party to all these MEAs listed above. However, like most of the developing world, progress in implementing the obligations of these agreements has been limited, as has been identified in the reports that the country has been submitting to the conferences that these conventions hold regularly.

Ramsar, Convention on Wetlands 1971

The Convention on Wetlands, commonly referred to as the Ramsar Convention was signed in the city of Ramsar, Iran in 1971. The purpose of this convention is to conserve and promote wise use of the wetlands. Furthermore, the convention aims to raise awareness among the nations about the ecosystem services that wetlands provide including biodiversity aspects of habitats to a wide variety of plants and animals (Final Act, 1971).

Kyoto Protocol to the United Nations Framework Convention on Climate Change

The Kyoto Protocol was adopted in Kyoto, Japan, on 11 December 1997 and entered into force on 16 February 2005. The detailed rules for the implementation of the protocol were adopted at COP 7 in Marrakesh, Morocco, in 2001, and are referred to as the "Marrakesh Accords." Its first commitment period started in 2008 and ended in 2012. The objective of the protocol is that each party included in Annex I, in achieving its quantified emission limitation and reduction commitments under Article 3, in order to promote sustainable development, shall:

(a) Implement and/or further elaborate policies and measures in accordance with its national circumstances, such as:
(i) Enhancement of energy efficiency in relevant sectors of the national economy;
(ii) Protection and enhancement of sinks and reservoirs of greenhouse gases not controlled by the Montreal Protocol, taking into account its commitments under relevant international environmental agreements; promotion of sustainable forest management practices, afforestation and reforestation;
(iii) Promotion of sustainable forms of agriculture in light of climate change considerations;
(iv) Research on, and promotion, development and increased use of, new and renewable forms of energy, of carbon dioxide sequestration technologies and of advanced and innovative environmentally sound technologies;
(v) Progressive reduction or phasing out of market imperfections, fiscal incentives, tax and duty exemptions and subsidies in all greenhouse gas emitting sectors that run counter to the objective of the convention and application of market instruments;
(vi) Encouragement of appropriate reforms in relevant sectors aimed at promoting policies and measures which limit or reduce emissions of greenhouse gases not controlled by the Montreal

Protocol;
vii) Measures to limit and/or reduce emissions of greenhouse gases not controlled by the Montreal Protocol in the transport sector;
(viii) Limitation and/or reduction of methane emissions through recovery and use in waste management, as well as in the production, transport and distribution of energy (Kyoto Protocol, 1998).
(b) Cooperate with other such parties to enhance the individual and combined effectiveness of their policies and measures adopted under this Article, pursuant to Article 4, paragraph 2 (e) (i), of the Convention. To this end, these parties shall take steps to share their experience and exchange information on such policies and measures, including developing ways of improving their comparability, transparency and effectiveness. The Conference of the Parties serving as the meeting of the Parties to this Protocol shall, at its first session or as soon as practicable thereafter, consider ways to facilitate such cooperation, taking into account all relevant information (Kyoto Protocol, 1998).

In Doha, Qatar, on 8 December 2012, the "Doha Amendment to the Kyoto Protocol" was adopted. The amendment includes:
1. New commitments for Annex I Parties to the Kyoto Protocol who agreed to take on commitments in a second commitment period from 1 January 2013 to 31 December 2020;
2. A revised list of greenhouse gases (GHG) to be reported on by Parties in the second commitment period; and
3. Amendments to several articles of the Kyoto Protocol which specifically referenced issues pertaining to the first commitment period and which needed to be updated for the second commitment period.

On 21 December 2012, the amendment was circulated by the Secretary-General of the United Nations, acting in his capacity as Depositary, to all Parties to the Kyoto Protocol in accordance with Articles 20 and 21 of the Protocol.

During the first commitment period, 37 industrialized countries and the European Community was committed to reduce GHG emissions to an average of five percent against 1990 levels. During the second commitment period, Parties committed to reduce GHG emissions by at least 18% below 1990 levels in the eight-year period from 2013 to 2020; however, the composition of Parties in the second commitment period is different from the first (Doha Amendment Report, 2012).

United Nations Convention on Biological Diversity 1992

The main objectives of this convention as indicated in Article 1, are the conservation of biological diversity, the sustainable use of its components and the fair and equitable sharing of benefits arising out of the utilization of genetic resources, including by appropriate access to genetic resources and by appropriate transfer of relevant technologies, taking into account all rights over those resources and to technologies, and by appropriate funding (UN, 1992).

United Nations Framework Convention on Climate Change 1999

The objective of the Convention on Climate Change is to achieve, in accordance with the relevant provisions of the convention, stabilization of greenhouse gas concentrations in the atmosphere at the level that would prevent dangerous anthropogenic interference with the climate system. Such a level should be achieved within a timeframe sufficient to allow ecosystems to adapt naturally to climate change, to ensure that food production is not threatened and to enable economic development to proceed in a sustainable manner (UNFCCC, 1999).

Within African sub region, there are regional environmental summits. The first ever regional comprehensive report on the state of Africa's environment-*Africa Environment Outlook* (AEO),which was specifically requested by African Ministerial Conference on Environment (AMCEN), in collaboration with United Nations Environmental Programme (UNEP), reinforces environmental breakdown as it traces environment and development trends since the 1972 United Nations Conference on the Human Environment and provided a comprehensive analysis of status and trends of the environment in Africa which has been abysmal.

Also, the First Africa Drylands Week and World Day to Combat Desertification which recently held in Dakar, Senegal, from June 10-17, 2011 was aimed to Combat desertification. Overall deforestation has declined globally, but persists in Africa and South America, according to the FAO's 2010 Global Forests Resource Assessment. The pressure on arid zone forests and the rangelands that protect them may increase, especially in the tropical and subtropical regions, from two opposing forces, according to the assessment (Wolf, 2011).

RESULTS AND DISCUSSION

Dynamics of the October and November 2012 flooding

The data collated from the study including Focused Group Discussions (FGD) questionnaires and oral interviews provided objective significant relationship between flooding, environmental security and displacement.

Similarly, a recent empirical study in this direction equally corroborates this result (Amadi, 2013). On October 7, 2012 a flood disaster affected communities lying on the path of the River Niger and River Benue, this included the people of the Niger Delta area as the River Niger overflew and had disruptive effects on lives and properties (Figure 1). Data collated from the study revealed 95% damage on farms lands and animals, while several houses submerged and many people were rendered homeless especially in the Delta and Edo states axis. This phenomenal occurrence was historic and significant to the growing discourse on vulnerability of climate change as corroborated in a similar study (Amadi, 2013; NEMA, 2012).

In Rivers State, a subsequent occurrence was witnessed the Orashi river a tributary of river Niger overflew on November 3, 2012. Our data confirmed that "about three persons: an infant and two women died", several others displaced, as "no fewer than 24 communities and several houses were submerged" (Agbu, 2012). Most affected included Udoda, Igovia, Kunsha, Ikodi Town, Mbiama, Akinima, Okolobiama, Okparaki, Okarki town, Akiogbologbo, Agbo, Ishiaye, Akioiso, Isusu, Isua, Odqwa, Edagberi and Betterland as well as Oshiebele, Oshi, Odiereke-Ubie, Enito, Akara-mini, all in Ahoada-West local government area of the state.

One of the displaced flood victims in Omoku, the headquarters of Ogba/Egbema Local Government Area of Rivers State, Mr. Ikediukwu Adiele, lamented that people in his village lost many property and many of them are now homeless. "For three days now, we have been wading through the water to pick some valuable items from our flooded homes. As you can see, we are now using canoe to move around (Agbu, 2012). Several

Figure 1. Map of Niger Delta showing the major cities in the region. Source: National Bureau on Statistics.

dimensions of hazards and challenges were experienced by the affected communities.

"Also, areas badly hit by the disaster included, Okwuzi, Ase-Azaga, Isukwa, Odugiri, Agwe, Onikwu, Ugbaja, Isala, Ogbe-Ogene, Utu, Adiawai, Obiofu, Kreigani, Obrikom, Idu and Ebocha as well as Umu-Onyema quarters in Ndoni, all in Ogba/Egbema/Ndoni local government area of the state. Unfortunately, the Rivers State government, the authorities of two affected local government councils as well as the National Emergency Management Agency (NEMA) was not in the know of the flooding in the communities until the Senate Committee on Environment and Ecology raised the alarm" (Daily News, 2012).

While families parked out of the affected communities in the area, farmers in the area began early harvesting of their root crops, especially cassava and yam so as to avoid losing them to the flood. (Amdi, 2013). An indigene of Edageberi community, Bonny Otatie Ulolo, said although, people of the area usually experience minor flooding every November as a result of over-flow of the Orashi River, the present situation was unprecedented in the history of the communities (Leadership Newspaper, 2012).

According to some victims, "the intensity of the disaster has never been experienced in recent times in the area as floods usually come and go but this persisted magnificiently" (Leadership Newspaper, 2012).

The Rivers State Commissioner for Special Duties, Chief Emeka Nwogu, in company of the chairman of the council, Hon. Raymond Nwokeocha has visited the affected communities to ascertain the level of damage. Nwogu, who was deeply touched by the plight of the victims, expressed sympathy over the incident, describing it as a great loss to the state and appealed to those that are yet to evacuate to do so to avoid more loss of lives (Leadership Newspaper, 2012).

Indigenes of the affected communities blamed the cause of flood on the over flow of the River Niger, explaining that the last time they experienced such flooding was in 1994 and 1998 respectively and described this year's disaster as exceptional (Leadership Newspaper ,2012).

To Emenike Umesi, the South-South co-ordinator of NEMA, the agency had been aware of the flood situation in the affected communities in Ahoada-West and Ogba/Egbema/Ndoni Local Government Area of the state and had been working on the affected areas. He said they had concluded assessment of the affected communities to determine the level of assistance that could be given to the victims, adding that the agency identified a camp for victims but it was swept off by flood (Leadership Newspaper, 2012).

In Delta State, communities have been sacked, farmland destroyed from Asaba, the Delta State capital to Oko, down to Utchi to Aboh, Abari and Bomadi, residents were displaced even as over 1,000 houses are believed to have collapsed. "We thought it will not come to some of us who were some distances away from the river. "We kept vigil all night and it was when the tide increased that we decided to move our things upland in the open.

"It is amazing to us that the water is increasing this time

of the year", Mr. Ifeanyi Okafor, a tenant at OgbeOfu/Jarret area of Asaba, said (Leadership Newspaper, 2012). "Areas that were spared at the initial time, have now been submerged. For instance, the former secretary to the then Bendel State government and Iyase of Asaba, Chief Patrick Onyeobi was hit severely as his palace compound and others were submerged. Before now, the shoreline around his house by the River Niger was very secured and no one ever thought the vexatious Niger could have been so offended to rise to such a level as to sack the former SSG and many others in Delta Capital" (The Tribune, 2012).

One of the earlier sufferers of the flood in Asaba was the Nigeria Immigration service office and the tourist Ganda hotel by the bridge head. The riverine people braved the odds, paddling canoes to their homes, retrieved their property and ferrying them to the highways which have become home for them (The Tribune, 2012). The Asaba experience speaks for several communities now devastated by the ravaging flood especially in Aniocha North, Ndokwa land, Isoko, Patani, Bomadi and Burutu communities amongst others. The case of Ewulu community in Aniocha South Council is Pathetic in the sense that over 100 houses, were pulled down, and both elderly and the sick ones trapped as water from River Umuoni, a tributary of the Niger overflowed its bank, The Ase River, which traversed the Ndokwa nation was also a channel through which the rising Niger flood unleashed more troubles on inhabitants (The Tribune, 2012).

Scores of displaced farmers in Ndokwa East of Delta are now seeking refuge on the upland of Ashaka and Utagba-Ogbe, Kwale towns following the washing away of their communities, farmland and crops by flood from the River Niger and its tributary, Ase Creek (The Tribune, 2012).

A visit to the area revealed that the water rose in fury and submerged farmlands and crops on the alluvial rich islands and adjoining farms on both sides of the River Niger thereby sacking the farmers and destroying their fish ponds, homes and markets. "We have never had it this bad in a life time, government and its agencies should come to our rescue" (The Tribune, 2012). Communities most affected along the River Niger are, Aballa-Oshimili, Utchi Communities of Okwumedo, Umuochi, Owelle, Obalu and Obeche, Okpai, Abalagada, Aboh, Abuato, Ugbene, Agwe-Iyom, Ise-Onokpo, Onuobiuku, Umu-Ugbome, Umu-Uti, Afiankwo, Umuolu, Adiai, Utuoku, Oworubia, Wari-Irri, and Onyah. Other communities affected included Aballa-Obodo, Aballa-Uno, Inyi communities of (Umu-Inyagbo, Obeche, Umu-Agwuyam, Isiolu, Umuoga, Ezinyi, Utuke, Ude, Ogigogwe, Ezeagba), Umu-Eche, Ogwasi, Umugwo, Umuazu, Ozala, all in Onuaboh. The flood also ravaged Akarai, Azagba, Ekpe, Ibedeni, Osafu, Ase, Asaba-Ase, Onogbokor, Iyede-Ame and Anyama (Kiagbodo, 2012). "Chief Nnamdi Olise-Atuoku and Mr. Raphael Nwammana who are now seeking refuge at Ashaka and Kwale respectively said that yam, cassava, plantain, vegetables, garden eggs and pepper among others were completely destroyed in their farms" (The Tribune, 2012).

Camps were opended in Asaba, Tuomo in Burutu Bomadi, Okwagbe in Ughelli South, Patani and Aboh among other places to take in the displaced persons. On a daily basis, the camps received more victims, even in Asaba, another camp was opened at I.C.E to take in people from Utchi and Oko communities to compliment the St Patrick College camps for hundreds of displaced persons in Asaba (The Tribune, 2012).

According to the Delta State Governor, Emmanuel Uduaghan, "After the aerial overview, I started the ground assessment by vehicle and boat. The situation is pathetic. My first encounter with the victims was at Oko communities in Oshimili South Local Government Area of the state. The three communities of Oko Amakom, Oko Ogbele and Oko Anala were flooded and the villagers gathered as refugees by the roadside. I had to talk to them that we will open camps to offer them temporary accommodation from the rage of nature, but some of them were reluctant to come to the camp, one of which was already opened at Saint Patrick's College, Asaba" (The Tribune, 2012).

"The displaced persons occupied half of the Benin-Asaba-Onitsha expressway and it was obvious that there would not be movement from either Asaba to Onitsha or vice versa if the situation was not taken care of" (The Tribune, 2012).

According to a rural farmer, "our farms are badly damaged we are faced with dual challenges of hunger and housing" (Ifejika, 2012).

The situation in Oguta was no less distressing several victims of the flood in Oguta and Ohaji/Egbema Local Government Areas of Imo State were equally lamenting. About eight communities were affected in Egbema. According to the director of administration of the Ohaji/Egbema LGA, Barrister Alloy Obinna, "eight communities were affected in the Egbema area alone. Our fish ponds, farm lands, plantain, cassava, economic trees and crops are now under water. Over 123 farm settlements and satellite towns are under water and about 8,000 persons have no homes now", he said (The Vanguard, 2012).

In Edo State, the Edo State Deputy Governor, Dr. Pius Odubure counted the intensity of the disaster in the communities in the area (Daily News, 2012).

In Cross-River State, Mr Fabian Okpa, the Special Adviser to Gov. Liyel Imoke on Special Duties, recounted that, " nine of the 18 local government areas of the state were affected" (Daily News, 2012).

'In an interview, Obi Emmanuel Obiechina, a leader of the Aika community in Ndokwa East local government area, said his residence was submerged by the flood. In addition, he stated that the lives of members of his community were shattered, with most of them refugees in various communities in Delta and Anambra States. Mr. Obiechina had taken refuge in a hotel in Asaba' (Daily

News, 2012). 'The traditional ruler said that he refused to send his people to camps for displaced people set up by the state government because those in the camps were being dehumanized. He blamed the excessive flooding on dereliction of duty by the government'. "We have lived in that community for more than 500 years and we have not had this kind of disaster before. It's the opening of the dams that have now brought this flooding on us. We are so devastated and everything that the Aika people have– land, crops, houses have been washed away" (Daily News, 2012).

Till date, rehabilitating the impacted areas have been a challenge. Awosika et al. (1992) study foreshadowed this concern as they observed that, "In Nigeria, a potentially massive "environmental refugee" migration will occur. For a 1-m rise, more than 3 million people are at risk, based on the present population. The estimated number of people that would be displaced ranges from 740,000 for a 0.2m rise to 3.7 million for a 1m rise and 10 million for a 2m rise".

Also, the presence of crude oil and its exploitation means further environmental degradation and hazards for the region. These are threats to environmental sustainability.

Experts have noted that, riverine flooding, is often a function of precipitation and water runoff volumes within the watershed of the stream or river; also there is the Coastal flooding, which is typically a result of storm surge, wind-drive waves, and heavy rainfall (Awosika et al., 1992). They contend that storm surges may overrun barrier islands and push seawater up coastal rivers and inlets, blocking the downstream flow of inland runoff. Also, there is the Urban flooding, which occurs where there has been development within stream flood plains, these are indicators that the Niger Delta region like most African coastal areas are volatile and prone to environmental hazards.

The growing discussion is that several coastal communities are prone to dangers of climate change vulnerabilities in several disruptive manners.

Challenges and effects of the flooding

There are divergent challenges of the effects including peasant food production decline in the region (Amadi, 2013), hunger, displacement and migration.

The United Nations says Nigeria will need $38 million (about N5.7 billion) in emergency aid to help 2.1 million people uprooted from their homes by flooding (Channels Report, 2012). According to Jens Laerke the spokesperson for the UN's Office for the Coordination of Humanitarian Affairs, explaining the aid plan said, "the plan includes help with food, water, shelter and schools mainly in farming and fishing communities along the Niger River" (Channels Report, 2012).

There are several dimensions to the effects of the flood disaster. A close observer on the Delta State incident recounts: At least five people, including two children and a traditional ruler, have died as a result of continued massive flooding that has ravaged parts of Delta State in Nigeria's oil-rich Niger Delta. In addition, the floods have submerged a multi-billion naira Okpai Independent Power Plant (IPP). Large parts of Ndokwa East local government area are also under water". The still rising flood has affected parts of Asaba, the state capital, as well as such communities as Ovrode, Ofagbe, Okpe-Isoko, Lagos Iyede, Igeh, Ikpidelrri, Ivrogno, Onogboko, Itebioge, Iyede-Ame and Azagba. Other flooded areas include Otoka-Ekegbresi, Egbeme, Okrama-Oyede, Warri, Iwrie Ogbokor, Ekpe, Asafo, Umeh, Aviara, Uzere, Asaba-Ase, Aboh, Kwale and Ashaka. Isoko South and North and Ndokwa East local government areas are the most devastated. In Aboh community, the flooding led to the death of a local monarch and two children. Three other persons reportedly lost their lives at the relief camp in Ivrogbo, Isoko South council area. Survivors of the flood were then relocated to St. Michael's College in Oleh, the headquarters of the local government area. The flood wreaked havoc on farms, schools, courts, health centers, markets and electricity installations. People now use canoes as the only means of getting about in the flooded communities (Sahara Reporters, 2012).

According to the National Secretary of the Ijaw National Congress, Mr. Robinson Esite, "some of the residents of Patani who struggled to get some valuables from the flooded community were just standing by the road side not knowing what to do. He said, "I can tell you that there is a serious humanitarian crisis in this part of the Niger Delta now. The town of Patani, Adagbabiri and Bomadi have been sacked by flood. "In Patani, the flood is such that no building is spared and the people of Patani are in a terrible state. Many of them are just standing by the road side with the few properties they were able to rescue from the flood (Sahara Reporters, 2012).

Also, a Niger Delta activist from the town Mr. Presidor Ghomorai, said, "Patani is gone, it is a hopeless situation, only the youths are left there. And as it is, the flood is threatening Bomadi seriously" (Sahara Reporters, 2012).

"In Bayelsa, for instance, much of government business has been limited to managing the flood fall-outs. Even the Creek Haven, Yenagoa, the seat of the state government is not on safe grounds. The state has been marooned by flowing water for close to a month now into an island" (Sahara Reporters, 2012).

In Yenogoa (capital of Bayelsa State) where the ex-militants were undergoing a rehabilitation training, the spokesperson for the presidential aid on amnesty, expressed concern over the magnitude of destruction inflicted on Niger Delta by the recent flood, saying; "the greatest challenge facing the region would be how to manage the disaster caused by the deluge". Coordinator of the Presidential Amnesty Programme and Special Adviser to the President on Niger Delta Affairs, Mr. Kingsley Kuku, in Yenagoa while presenting three truckloads of relief materials to the state government for victims of the flood, said the aftermath of the disaster would be more challenging than the disaster itself. Represented by Mr. Tawari Dortimi, Kuku called on all the affected states to map out strategies to contend with the challenges ahead (In NEWS, 2012). The ex-militants undergoing rehabilitation donated some money from their allowances to alleviate the sufferings in the affected areas.

In his remarks, the Deputy Governor, Rear Admiral John Jonah (rtd), said the government had held series of meetings to develop post disaster management, saying that government was also worried about transporting relief materials to displaced persons in remote areas. "Our post flood management strategies will include policies in agriculture. We are thinking of how to get our food back to avoid food shortage. We are brainstorming to know how to reduce the pains", he said (In NEWS, 2012).

Other areas affected included the Niger Delta University at the Wilberforce Island in Amassoma. The flood destroyed several offices, lecture halls, hostels, students properties- including credentials, textbooks/materials, vehicles, farmlands, etc .In a statement by the Registrar of the institution Mr. Tonbra Morris-Odubo a notice of indefinite suspension of resumption date for the 2012/2013 academic session' supposedly fixed for 8th October 2012 was made (In NEWS, 2012). There was fear of outbreak of epidemic in the flood-ravaged Obuwari settlement, equally referred to as Mile I, in Sagbama, headquarters of Sagbama Local Government Area in Bayelsa. About 300 buildings, including public and private schools in the area, were submerged.

Equally affected are those living along the Rivers Niger and Benue and with the volume of water from the two biggest rivers confluencing in Lokoja, Kogi State, if the water rose above sea level toward the south down to the creek mostly of Delta and Bayelsa State. Communities on both sides of the river Niger as well as those along the distributional became victims (In NEWS, 2012).

In Rivers State, the impact is mainly in the areas traversed by the Orashi and Sombreiro Rivers in the west senatorial district. Specifically, Ahoada West and East, Ogba/Egbema/Ndoni and Abua/Odua councils were in the bull's eye. From Akabuka and westward through Oboburu, Obagi, Edu, Obigbo and Etu, down to Ohali, the road had become part of the Orashi River. Eastward to Erema, Ibewa and Ituli, there is respite, thereby forcing a drift in that direct from the western end. Both sides, demographically, constitute Egi clan, which is part of Ogba Kingdom, where about 90% of the upstream operations of French oil giant, Total, is concentrated (The Guardian, 2012).

Chibuzor Ugwoha, the immediate past managing director of the Niger Delta Development Commission (NDDC), whose father's house in Erema has since been taken over by distant relations and even unknown persons sacked by the flood in the west end of the Egi kingdom, said: "I have never seen a thing like this since I was born. In situations like this, you do not ask questions or even probe the clans and identities of persons flooding in. The first and most logical thing to do is to let them in and then do whatever is possible to offer them immediate hope" (The Guardian, 2012).

Several heart-ripping stories came from the impacted communities several animals including pythons,antelopes and grasscutters were seen floating" (Wokocha, 2012). Ojumite, who claimed he had remained separated from his children and wife since the flood began, explained: "Our people are not conversant with canoes; we cannot swim, we conduct all our businesses on land and so what we are currently witnessing is like asking us to adopt new ways of life after centuries of existence" (The Guardian, 2012). That helplessness was most tragically expressed in the event involving one Odoka Benson from Obagi a rural community in Rivers State. He had returned to Obagi from Omoku, where he was taking refuge. At sunset, he was ready to go. As if actually bent on departing finally, he turned down entreaties to stay the night anyhow in the flooded town, because it had become late and the boat operators had closed business for the day (The Guardian, 2012). He reportedly got to the anchorage, forced out a boat and boarded, in company of his pregnant wife and daughter of one Pastor Blessing Dikogwo, their two children, a lady and himself (The Guardian, 2012). Apparently, Benson might have reasoned that the business of moving canoes on water was all mechanical without any form of mental calculations. But midway or midstream, because the Akabuka-Obagi road had also become a stream, 'the captain' got what he did not bargain for. He was up against the forceful tide of the water flowing across the road (Sahara Reporters, 2012).

It became an ill-fated voyage, as everyone on board, except himself, was swept away by the tide. The lady on board was the elder sister of Wisdom Nwoko-Omere, who is among persons awarded scholarship by the NDDC to pursue post-graduate studies in universities abroad (Sahara Reporters, 2012).

'One Azubike Nwaoga returned from Babcock University in Ogun State, where he is studying hard to become a better priest of the Seventh Day Adventist Church, to begin from the scratch'. 'The flood overwhelmed his home while away in school and the folks he left behind reported that all his belongings went with the water, except his motor-bike. He took solace in the Bible: "In all things, we give thanks to God" (Sahara Reporters, 2012). Ojumlte could not also do much when his house was overrun in Ahoada. "The only thing I could take out of my house is my certificate," he bemoaned. His ponds were flooded and the fishes in them spilled into the wild.' The collective loss is equally enormous (Sahara Reporters, 2012).

'Many of the social facilities provided by Total as part of its Corporate Social Responsibility (CSR) to its host communities are endangered. For instance, the health centre in Ogborgu stood submerged as at last week. The water scheme at Akabuka was itself under water and it was not likely it was still giving good water to the community' (Sahara Reporters, 2012).

'The grand benefactor itself, Total, is battling for a lifeline. It has been forced to temporarily closed shop in Ogba Kingdom. Its External Communications manager, Charles Ebereonwu, neither confirmed nor denied outrightly the claim that the oil company had shut down operations. He only said: "We are talking about lives here and I think life is more important than production" (Sahara Reporters, 2012).

'In real terms, this means that about 40,000 barrels of crude, representing Total's contribution, have been cut back from the 2.6 million barrels per day (bpd) national output. One environmentalist in Port Harcourt added a frightening dimension'. He said: "Total's waste pits and other disposal sites have been flooded. Chemicals and other harmful substances that the company uses in its operations have all been washed into flooded homes and farmlands. "We cannot measure the extent of impact until a proper assessment is conducted" (Sahara Reporters, 2012).

'Reacting, Ebereonwu sounded more like a lawyer in court, saying whoever that was asserting should offer proofs'. He said: "This is a nationwide tragedy, not restricted to Total, and so whoever that is alleging pollution should provide the evidence" (Sahara Reporters, 2012).

'All the same, the company has been part of the coordinated efforts by the Rivers State government at rehabilitating the victims, donating materials to the various relief camps in the area. Unlike fire, water can hardly be contained. It dissipates on its own and it is not different in the current situation' (Sahara Reporters, 2012).

With several persons displaced and living in temporary camps, their daily subsistence has been a challenge. An observer on the allocation of the relief materials provided by the Niger Delta Development Commission (NDDC) said, "We are not satisfied based on what the masses are facing in the hands of those feeding us. They are mismanaging the relief materials that are given to them; even though government supply those things in quantity; they economise them," one of the victims complained. The representative of the Agency, Prof. Aminigo advised the people to always cooperate with those managing affairs in the camps, noting that the materials donated by the agency would go a long way to ameliorate their plight. The materials were provided in most affected communities in the various Niger Delta states' (Sahara Reporters, 2012).

"Flooding in the oil rich Niger Delta, has disrupted oil production to the tune of around 500,000 barrels per day (bpd)– more than a fifth of nation's oil output according to the Department of Petroleum Resources" (Sahara Reporters, 2012).

Managing the issue has been a complicated affair for both the victims and the governments of the region. Most of the camps established for the Internally Displaced Persons (IDPs), could hardly contain them, others are not in good habitable conditions. The temporary camp has been described as worse and could lead to outbreak of diseases as most of them are over-stretched.

Recommendations: Mitigation strategies

Climate change has been an issue of global concern. Flood control in the region needs collaborative efforts including the cooperation of government, policy makers, international and local communities, enlightenment programmes and capacity building through environmental education involving NGOs, CBOs and the mass media, could reassert the level of awareness of the poor inhabitants of the coastal regions to understand the enormity of the challenges posed by climate change.

Gender and environmental security should be given novel policy priority especially among the volatile and poor coastal regions. According to the Rio 2012 report, "the implementation of sustainable development will depend on the active engagement of both the public and the private sectors. We recognize that the active participation of the private sector can contribute to the achievement of sustainable development, including through the important tool of public-private partnerships. We support national regulatory and policy frameworks that enable business and industry to advance sustainable development initiatives, taking into account the importance of corporate social responsibility. We call on the private sector to engage in responsible business practices, such as those promoted by the United Nations Global Compact" (Rio, 2012).

Importantly, key mitigation strategies could involve mainstreaming climate change adaptation into development planning at all levels of government including local, state and federal. Novel environmental policies should as a matter of expediency evolve participatory strategies to interface with the rural populace on climate change issues such as Rural Climate Change Monitoring Forum (RCCMF). This could be linked directly to State Emergency Management Agency as focal points to relate climate change issues in the rural areas.

Adaptation to climate change and concomitant adverse effects will involve an understanding of climate change parameters and dynamics, including monitoring and data analysis of climate change parameters. This strategy should lead to an African Climate Change Scenario (ACCS), upon which countries can base their adaptation options. Existing scenarios and adaptation measures for climate change and sea-level rise are built around Western experiences (IPCC, 2007).

Within an integrated approach, there is a great opportunity to anticipate problems associated with sea-level rise rather than simply reaction to change as, or after, it occurs (Nicholls and Leatherman, 1994).

Furthermore, a well-planned response that seeks to anticipate the physical impacts of sea-level rise in a timely fashion will minimize unwise decisions and result in lower costs for reactive responses such as protection (Nicholls and Leatherman, 1995). Anticipatory responses include urban growth planning, building setbacks, wetland preservation and mitigation, public awareness, and integrated coastal zone management (IPCC, 2007).

Policies and regulations concerning the use of the coastal zone for any form of human activity should include consideration of sea-level rise. Physical planning and building-control measures and regulations should be instituted and implemented. Allocation of land for any economically useful purpose in areas likely to be flooded or inundated should be avoided. The public should be informed of the risk of living in coastal and lowland areas that are threatened by sea-level rise. Timely public education on erosion, sea-level rise, and flooding risks could be a cost-effective means of reducing future expenditures. Where coastal infrastructures such as roads, fish land, and curing plants are approved and must be constructed, the authorities and owners of these infrastructures should make sure that marginal increases in the height of the structures are included to offset sea-level rise (Smith and Lenhart, 1996) and other related phenomena. People located in high-risk areas should be offered incentives to relocate out of these areas. Setbacks could be used as buffer zones to allow sea level to rise without threatening coastal development. French et al. (1995) recommended incorporating buffer zones between the shore and new coastal development in Nigeria (IPCC, 2007).

There is urgent need of bringing all actors in the coastal areas together to address coastal-zone problems. The program should consist of a set of principles and plans to guide the use of coastal land and resources for conservation, recreation and sustainable.

Environmental activism and awareness will ameliorate unsustainable environmental consumption by all stakeholders including the MNCs operating in the region.

Conclusion

Climate change and global warming are now scientifically established facts. Climate change is a massive threat to human development and in some places it is already undermining the achievement of the Millennium Development Goals (MDGs) and the international community's efforts to reduce extreme poverty. The issue of climate change is global. Nigeria and indeed Africa is not excluded from its threats.

Our study had the aim of understanding environmental security in the coastal regions of Nigeria; how disasters such as flooding and its effects have impacted the region, the level of response by the state and the international community. The results derived from this study show that majority of the people in the area have little knowledge of climate change vulnerability, frequency in occurrence of the incident and the causes. Again, most flood victims do not get compensation or relief during flood disaster, majority of those we interviewed who are artisans and peasants, with occupation such as fishing, subsistence agriculture, expressed dissatisfaction with the level of response by the government and policy makers to their plights.

Climate change will exacerbate existing physical, ecological/biological and socioeconomic stresses on the African coastal zone. Most existing studies focus on the extent to which rising sea level could inundate and erode low-lying areas or increase flooding caused by storm surges and intense rainstorms. The coastal nations of west and central Africa (Senegal, The Gambia, Sierra Leone, Nigeria, Cameroon, Gabon, Angola) have low-lying lagoonal coasts that are susceptible to erosion and hence are threatened by sea-level rise, particularly because most of the countries in this area have major and rapidly expanding cities on the coast (IPCC, 2007).

Africa's west coast often is buffeted by storm surges and currently is at risk of erosion, inundation, and extreme storm events. Inundation could be a significant concern (Awosika et al., 1992; Dennis et al., 1995; French et al., 1995; ICST, 1996; Jallow et al., 1996; IPCC, 2007). Major cities such as Banjul (Jallow et al., 1996), Abidjan, Tabaou, Grand Bassam, Sassandra, San Pedro (ICST, 1996), Lagos and Port Harcourt (Awosika et al., 1992)- all situated at sea level would be very vulnerable. Finally, tidal waves, storm surges, and hazards also may increase and may modify littoral transport (Allersman and Tilsmans, 1993 cited in IPCC, 2007).

Carbon dioxide emissions in the area are among the highest in the world. Some 45.8 billion kilowatts of heat are discharged into the atmosphere of the Niger Delta from flaring 1.8 billion cubic feet of gas every day. It has been reported that gas flaring has raised temperatures of the region and rendered many areas uninhabitable.

This study attempted to examine the scenarios of the climate change in the Niger Delta region, its vulnerability and impacts and provided possible mitigations to stem the tide of global warming and environmental degradation in the region.

Finally, the study concludes that climate change is largely though not exclusive attribute of anthropogenic activities of man and could be checked only if these activities are reduced.

REFERENCES

Adibe E (1994). "Weak States and The Emerging Taxonomy of Security in World Politics", FUTURES, June 4961-505.

Agbu I (2012). "Wiping Tears of Flood in Niger Delta" Nigeria Masterweb Daily News Posted 02/11/12 available at www.nigeriamasterweb.com/.../wiping-tears-of-flood-in-niger-delta.

Amadi L (2013). Climate Change, Peasantry and Rural Food Production

Decline in the Niger Delta Region: A case of the 2012 flood disaster. J. Agric. Crop Res. Vol. 1(6), pp. xxx-xxx . Research Paper.

Amadi L (2012b). Africa and Sustainable Environmental Health: Challenges and Policy Implications.International J. Health, Safety Environ. Manag. (IJHSEMR) 1(7):26-44

Amadi L (2012). Talking Right, Walking Wrong: Global Environmental Summits, Unsustainable Consumption and Environmental Engineering .Paper presented at the 3rd International Chemical and Environmental Conference (ICEEC 2012) 21-23 Dec Kuala Lumpur, Malaysia.

Awosika L, French G, Nicholls R, Ibe C (1992).The impacts of sea level rise on the coastline of Nigeria (O'Callahan, J. (ed.). In: Global Climate Change and the Rising Challenge of the Sea. Proceedings of the IPCC Workshop at Margarita Island, Venezuela, 9-13 March 1992. National Oceanic and Atmospheric Administration, Silver Spring, MD, USA, 690 pp.

Campbell K , Weitz R (2008). The clear implications of global climate change. In Climatic cataclysm:The foreign policy and national security implications of climate change, ed. Kurt M. Campbell, pp.213–223. Washington, D.C: Brookings Institution Press.

Channels Television Report (2012) "Flooding: UN says Nigeria will need $38 million in aid" Published, Fri, Nov 9th. available *www. Channelstv. com/home/tag/un)*

Choucri N Ed (1992). "Environment, Development and International Assistance:CrucialLinkages"In: Resolving Third World Conflicts Challenges for a New Era,edSherylJ.Brown and Kimber M Schraub.Washington DC United States Institute of Peace press. 89-118. Responses. Cambridge, Mass.: MIT Press.

Dennis K, Niang-Diop I, Nicholls R (1995). "Sea level rise and Senegal: potential impacts and consequences". J. Coastal Res., special issue 14:243-261.

Doha Amendment to the Kyoto Protocol (2012). Available athttp://unfcc.int/Kyoto_protocol/doha amendment/items/7363php

Final Act (1971) .Ramser Convention on Wetland ,Iran.

French G, Awosika L, Ibe C (1995): Sea level rise and Nigeria: potential impacts and consequences. J. Coastal Res., special issue 14:224-242.

Homer- Dixon T (1991). "On the Threshold: Environmental Changes as Causes of Acute Conflict" International Security 16(2)76-116

Ibeanu O (1998). Exiles In Their Own Home: InternalPopulation Displacement In Nigeria African. J. Political Sci. Vol. 3 No. 2. 80-97.

Ibe A, Quelennac R, (1989). Methodology for assessment and control of coastal erosion in west Africa and central Africa. UNEP Regional Sea Reports and Studies No. 107. United Nations Environment Programme, New York, NY, USA

Interview with Mr Innocent Kiagbodo a rural farmer in Bomadi,Delta State.10/1/2013.

Interview with a rural farmer MrIfejikaOgbu in Asaba 12/1/2013

Interview with Ijaw youth leader comrade joseph Mayomi john in YenagoaBayelsa State.20/2/2013.

Interview with comrade John EziefunaAsaba Delta State.3/2/2013

Interview with Bar (Mrs)Wokochaof Ogba /Egbeman /Ndoni LGA, Rivers State in Port harcourt Rivers State.9/2/2013.

IPCC (2007). "Climate Change 2007: The Regional Impacts of ClimateChange"Contribution of Working Group I to the Third Assessment Report of the Intergovernmental Panel on Climate Change). Cambridge University Press, Cambridge and New York, 572.

Kaplan R (1994). "The Coming Anarchy How security crime, overpopulation, tribalism and disease are rapidly destroying the social fabric of our planet," The Atlantic Monthly:44-76.

Klare M (1996). "Redefining Security: The New Global Schisms". Current History, November. 32-37. Nigerian Environmental Study/Action Team (NEST, 2004) Report.

Okoh R, Egbon P (1999). Fiscal Federalism and Revenue Allocation. The Poverty of the Niger-Delta in Aigbokhan B.E(eds): Fiscal federalism and Nigeria's Economic Development Selected Papers of the 1999 Annual Conference of the Nigerian Economic Society.NES, Ibadan.

Onukwugha A "Rivers Flood: Tear Flows Freely In 24 Communities" Leadership Newspaper Mon, 08/10/2012 Available at http. www. leadership. ng/ nga/ anayo_ onukwugha/index.html.

Suhrke A (1996). "Environmental change, migration and conflict: A lethal feedback dynamics? in Chester A. Crocker et al. (eds.): Managing Global Chaos: International Cooperation in Issues of Displacement. Washington D.C: Carnegie Endowment for International Peace. 113-127.

Sahara Reporters "Delta *Flood*: *Power* Plant Submerged As Monarch, 5 Others Die" Available at www.saharareporters.com.Accessed 2/12/2012.

Udofa I, Fajemirokun F (1978). On a height datum for Nigeria. In: Proceedings of International Symposium on Geodetic Measurements and Computations. Zaria, Nigeria: Ahmadu Bello University, P. 12.

UNEP, Africa Environmental Outlook2(2002) Our Environment Our Wealth.Available www.unep.org/DEWA/Africa/docs/EAO2-OurEnvironourwealth.Accessed10/10/2013.

UNEP (2011). Environmental Assessment Of Ogoniland ,Oxford.

United Nations(1994). Framework Convention on Climate Change. (UNFCCC).

United Nations, (1998).Kyoto Protocol the United Nations FrameworkConvention on Climate Change (UNFCCC).

United Nations (1999). Framework Convention on Climate Change(UNFCCC).

United Nations (1992). Convention on Biological Diversity.

United Nations (2012). Report of the United Nations Conference on Sustainable DevelopmentRio de Janeiro, Brazil 20–22 June.

United Nations Environment Program (UNEP) (2002). Africa Environment Outlook, past, present and future perspectives, 2002.Oxford Press.

United Nations (1987). World Commission on Environment and Development, Oxford.

United Nations (1971). Final Act of the International Conference on the Conservation of Wetlands and Waterfowl.

Wolf J (2011). "Africa Addresses Desertification and Biodiversity Loss 2011" IDN-InDepthNews Available at http: //www .indepthnews. net/area2 php? key= PEhttp ://www. indepthnews. net/area2.php?key=PE. Accessed 12/8/2012.

Biodegradation of petroleum oil by fungi isolated from *Treculia africana* (Dec'ne) seeds in Nigeria

Adekunle, A. A. and Adeniyi, A. O.

Department of Botany and Microbiology, University of Lagos, Yaba, Lagos, Nigeria.

Petroleum crude oil biodegrading fungi were isolated from *Treculia africana* seeds in the presence and absence of petroleum fumes. An assessment of the relative ability of each fungus to biodegrade petroleum crude oil, kerosene, diesel, unspent engine oil, spent engine oil and extracted oil from *T. africana* seeds on minimal salt solution was investigated using changes in optical density read on a spectrophotometer and gas chromatographic analyses. Ten fungi were isolated from *T. africana* seeds in the presence and absence of petroleum fumes. These included one species each of *Mucor*, *Paecilomyces*, *Rhizopus* and *Syncephalastrum*, four species of *Aspergillus* and two species of *Penicillium*. It was evident that the fungi used in this research work were capable of biodegrading the petroleum and extracted *T. africana* seed oil hydrocarbon, though at different rates. *Rhizopus* had the highest degrading ability in kerosene, unspent engine oil, crude oil and the extracted oil from the seed, while *Penicillium pinophyllum* had the lowest ability to degrade the oil. The gas chromatogram (GC) showed that *Paecilomyces* biodegraded the hydrocarbons in the crude oil compared to the control (crude without fungi) using up some carbon atoms (C_{12}-C_{24}) after the 40 days of incubation, suggesting n-alkane biodegradation. Also the GC analysis of the seed oil of *T. africana*, after 40 days of incubation, showed a reduction in the seed oil hydrocarbons, removing C_{10}- C_{15}.

Key words: Hydrocarbon utilization, *Treculia Africana*, seeds, petroleum crude oil, fungi.

INTRODUCTION

Various microorganisms have been reported to possess the capability for utilizing hydrocarbons as their source of carbon and energy (Atlas, 1981). The ability to degrade petroleum hydrocarbons is not restricted to a few microbial genera. A diverse group of bacteria and fungi have been shown to have this ability. In a review made by Zobell (1946), more than 100 species representing 30 microbial genera have been shown to be capable of utilizing hydrocarbons. Recent studies continue to expand the list of microbial species, which have been demonstrated to be capable of degrading petroleum hydrocarbons (Nwachukwu, 2000). Microorganisms available for bioremediation include a range of bacteria like Arthrobacter, *Pseudomonas*, *Flavobacterium* and fungi such as *Penicillium, Cladosporium,*

Aspergillus etc, isolated from water (freshwater, brackish and marine) or soil (Atlas and Bartha, 1972).

Treculia africana (Dec'ne) belongs to the family Moraceae. It contains many seeds, which are buried in the spongy pulp of the fruit. *T. africana* is a riverine fruit tree of the tropical African rain forest. The seeds are extracted after macerating the fruit in water. Analysis of the hexane extracts of *T. africana* seed indicate that it contains a stearine solid fat fraction, resembling that of palm-kernel oil, and an oleine fraction with a composition similar to that of cotton seed oil (Burkill, 1997).

Microorganisms used for biodegradation have been isolated from the soil or water (Amund et al, 1978; and Salminen et al., 2004). No report has been on the use of fungi isolated from *T. africana* seeds in the biodegradation of crude petroleum. The aim of this paper was to isolate pathogenic fungi from *T. africana* seeds. These pathogenic fungi were then used to biodegrade petroleum hydrocarbon to assess the potential ability of the fungi to biodegrade the hydrocarbons using optical density and gas chromatographic methods.

MATERIALS AND METHODS

Collection of seeds

African bread fruit (*T. africana*) seeds were collected from three different markets, Agege and Oyingbo markets in Lagos state, and Ogere market in Ogun state, Nigeria (Long 8° N Lat 4°E). Diseased seeds were separated from the healthy seeds. The visually diseased seeds were used in this experiment. The seeds were sampled from the markets monthly for 9 months. About 1000 seeds were collected from each market at every sampling period. The various oil used here were: crude oil, spent engine oil, unspent engine oil, kerosene, diesel and extracted oil from the seeds of *T. africana*.

Isolation, identification and screening for hydrocarbon degrading fungi

Sixty diseased seeds of *T. africana* were surface sterilized by soaking them in a solution of common bleach (sodium hypochloride) and sterilized distilled water in a ratio of 3:2, for 2 min and then rinsed with three changes of sterilized distilled water. To isolate fungi from the seeds under petroleum crude' oil fumes, the modified methods of Amund et al. (1987) was adopted. Sixteen filter papers (Whatman No. 1001125) were sterilized in the oven at 40°C for 15 min. Eight of the dried filter paper was dipped in 250 ml petroleum crude oil contained in a 500 ml beaker for about 15 s with the help of a sterile forcep, and drained. The petroleum crude oil was obtained from Shell Escravos Port Harcourt, Nigeria. Each of the eight crude oil treated filter papers was placed on the cover of 8 Petri dishes containing solidified potato dextrose agar (PDA) and sterilized diseased seeds (seven seeds per plate) under sterile conditions. The aim of the petroleum fumes was to supply the fungi with hydrocarbons through vapour transfer from petroleum fumes. Another set of eight Petri dishes with the PDA and *T. africana* surface sterilized seeds, were without the 'oiled' filter papers, but contained the 8 non-oil treated filter papers and they served as control. All the plates were incubated at room temperature (28 -31°C) in the incubator, and observed daily for fungal growth. This process was carried out monthly for the 9 months sampling period. The developing fungal colonies were sub cultured aseptically into fresh PDA plates to get pure culture of isolates. A part of each pure culture was then aseptically transferred into sterilized PDA slants, which was previously prepared in the 14 ml McCartney bottles, and served as stock culture.

To identify the fungi, light microscopic examination was carried out and cultural characteristics such as colour of the fungal colony, number of days taken for the fungi to reach maximum growth or diameter (9 cm) of the Petri dish, and texture of the fungal growth were noted. The morphological and cultural features of each fungus were compared with descriptions given by Talbot (1971); Deacon (1980); Domoschet et al. (1980) and Bryce (1992) for identify-cation. Some mycologist (Prof N.U Uma) within the department of Botany and Microbiology, University of Lagos, was consulted for confirmatory identification of the fungi.

Extraction of oil

The method of extraction of oil of *T. africana* seeds was adopted from the oil extraction methods of Egan et al. (1981). The visually healthy seeds were ground using a ceramic pestle and mortar before blending in an electric blender. An amount of 200 g of the ground seed was packed into the extraction thimble before covering with a small ball of cotton wool. The thimble was inserted in a quick fit plain body soxhlet extractor. Petroleum ether in the quantity of 200 ml (60 - 80°C) was poured in a 250 ml round-bottom flask of known weight, which was connected to the extractor and refluxed on an electric thermal heater for 5 h. The other was then collected in the plain body extractor and then separated from the flask that contained the oil. The flask containing the oil was then heated in an oven at 103°C for 30 min. It was cooled and weighed to get the final weight. The percentage oil content of the sample was calculated using the ratio of the amount of oil produced to the weight of sample used expressed as a percentage. The procedure was repeated until at least 250 ml of the oil was extracted from the seed.

Confirmatory test for hydrogen utilization potential of the fungi

The estimation of hydrocarbon utilizers were obtained using enrichment procedure as described by Nwachukwu (2000). A minimal salt solution (MSS) containing 2.0 g of Na_2HPO_4, 0.17 g of K_2SO_4, and 0.10 g of $MgSO_4.7H_2O$, 4 g of NH_4NO_3 and 0.53 g K_2SO_4 dissolved in 1000 ml distilled water and sterilized in an autoclave. Fifty-six test tubes were sterilized, plugged with cotton wrapped in aluminum foil, and placed on test-tube racks. There were 7 test-tube racks, containing eight test-tube each. In each test-tubes 10 ml of the minimal salt solution (MSS) was added and each of 6 racks had either 2 ml of petroleum crude oil, diesel, kerosene, spent engine oil, unspent engine oil or extracted oil from *T. africana* seed. The seventh rack served as control, it had only the MSS in its test-tubes. Six fungi (choice was based on their growth from seed in the petroleum fumes), were inoculated in each test-tube in the rack. The last tube (seventh tube) on each rack served as a second control, and was not inoculated with any fungus. Each of the test tubes was plugged with sterilized cotton wool wrapped with aluminum foil to ensure maximum aeration and prevent cross contamination. All the test tubes were then inoculated at the room temperature in an incubator for 40 days. Constant shaking of the test-tubes was ensured to facilitate oil/cell phase contact. The ability to degrade petroleum crude, diesel, kerosene, spent engine oil, unspent engine oil and extracted oil from *T. africana* seed (based on growth of the organism on the MSS medium) were measured every 5 days using the visual method based on the turbidity of the MSS. The turbidity was measured using the spectrophotometer at a wavelength of 530 and 620 nm. This experiment was repeated twice. Results were statistically analyzed using T-test, Anova (F-test) and Duncan multiple range test as described by Parker (1979). The percentage contribution of each fungus in biodegrading the various oil

Table 1. Fungi Isolated from diseased seeds of *Treculia africana* in the presence and absence of petroleum fumes.

Treculia africana **seeds without petroleum fumes**	***Treculia africana*** **seeds with petroleum fumes**
Aspergillus flavus	*Aspergillus flavus*
Aspergillus niger	*Aspergillus niger*
Aspergillus japonicus	*Aspergillus wentii*
Aspergillus wentii	*Paecilomyces* sp
Mucor sp	*Penicillium chrysogenum*
Paecilomyces sp	*Penicillium pinophyllum*
Penicillium chrysogenum	*Rhizopus* sp
Penicillium pinophyllum	
Rhizopus sp	
Syncephalastrum sp	

(kerosene, crude oil, diesel, unspent engine oil, spent engine oil and extracted oil from seed) after the 40th day of incubation was calculated. To determine the percentage contribution of each fungus in each oil, after the 40th day, the optical density (OD) of each fungus was divided by the total OD of the various fungi in particular oil and expressed as a percentage.

Gas chromatographic (GC) analysis of some oil samples

Gas chromatographic analysis were carried out to assess and further confirm the ability of each fungi, *Paecilomyces* and *Aspergillus niger* isolated from *T. africana* seeds to biodegrade the hydrocarbons. After incubating for 40 days, as done above, the degraded hydrocarbons were extracted based on the physical changes and OD that were observed in the test-tubes after the 40 days incubation. These tubes were: (a) MSS + crude oil; (b) MSS + crude oil + *Paecilomyces*; (c) MSS + extracted oil and (d) MSS + extracted oil + *A. niger*.

The method of Song and Bartha (1990), Kampfler and Steoif (1991) and Salminen et al. (2004) was used for extraction of samples and GC analysis. For the extraction process about 20 ml hexane was used. Each sample was poured into a separating funnel and 20 ml of hexane was added and shaken well and the different oil collected. Column was prepared to get pure extracts. This was done using a 150 ml burette; the base of the burette was blocked with cotton wool to provide a base for the silca gel. About 2 0ml of hexane was poured into a beaker and about 5 g of Na_2CO_3 was added, which was the drying agent. The drying agent was poured into the burette. The silca gel helped to remove impurities.

The extract was poured into the burette and 20 ml hexane was added. The extract diffused down the column and was collected in a 14 ml McCartney bottle. All traces of hexane in the extract were allowed to evaporate by leaving the McCartney bottle opened, and the final extract was used subsequently. Gas chromatographic analysis was carried out using Perkin Elmer Auto-system GC equipped with flame ionization detector. A 30 m- fused capillary column with internal diameter of 0.25 nm and 0.25 m thickness was used, and the peak areas were analyzed with a SRT model 203 peak simple chromatography Data system. About 1-2 ml of extracted sample was injected. The column temperature was 60°C for 2 min to 300°C programmed at a rate increase of 120°C/min. Nitrogen was used as carrier gas with pressure of 30 ml. Hydrogen and air flow rates were 30 ml/min respectively.

RESULTS AND DISCUSSION

Table 1 shows the fungal species isolated from diseased seeds of *T. africana* in the presence and absence of petroleum fumes. Ten fungal species were isolated which included one species each of *Mucor, Paecilomyces, Rhizopus* and *Syncephalastrum* , four species of *Aspergillus* and two species of *Penicillium*.. More fungal species were isolated from the diseased seeds incubated without petroleum fumes. The growth pattern of fungi in the MSS and oil shows that the growth of each fungus had different maximum growth peaks (Figures 1 and 2), providing a fluctuation in the growth pattern of the fungi in the oil media. The growth pattern thus shows that the utilization of the different hydrocarbon used varied widely among the fungi. *Rhizopus* had the highest percentage contribution in the biodegradation of 33.13% in unspent engine oil while *P. pinophylum* had the least of 6.80% in kerosene (Table 2). Generally, *Rhizopus* ranked highest in the degradation of oil in all the oil used except diesel and spent engine oil (Table 3). The least in ranking in the biodegradation was *P. pinophylum* for all the oil used except in the spent engine oil as shown in Table 3.

The results of this work indicate that many of the fungal species isolated from *T. africana* seeds in the presence and absence of petroleum fumes were capable of biodegrading petroleum hydrocarbon. This *T. africana* seeds may be added to the known sources of hydrocarbon degrading fungi. The results further prove that fungi could also play a role in surviving in hydrocarbon rich environment, supporting the reports of Plante-Cunny (1993). An interesting observation in this study was the growth of each fungus in the presence of oil compared to when oil was absent. This probably means the fungi used the oil for its growth. Shaw (1995) found that microorganism breakdown hydrocarbons and use the energy to synthesize cellular components. After being completely broken down the reaction releases $C0_2$, H_20 and energy used to create cellular biomass (Keeler, 1996). It is evident from the results obtained that the fungi were more active in the extracted oil than in other oil (Figure 2). This may be due to the fact that the fungi were isolated from the *T. africana* seed where the oil was extracted, and these fungi have being adapted to using the seed oil hydrocarbon as their source of carbon. Also the

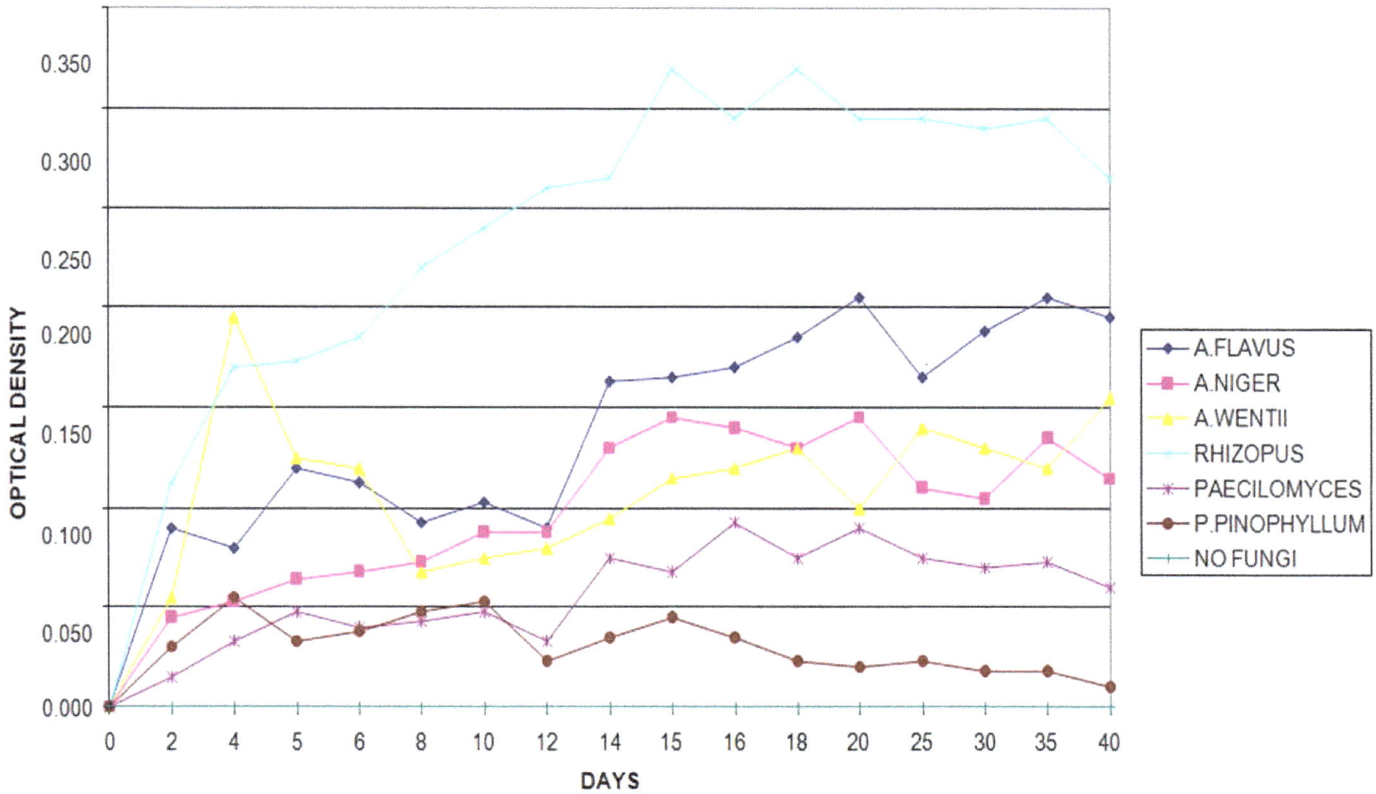

Figure 1. The growth pattern of fungi in minimal salt solution and unspent engine oil using 530 nm.

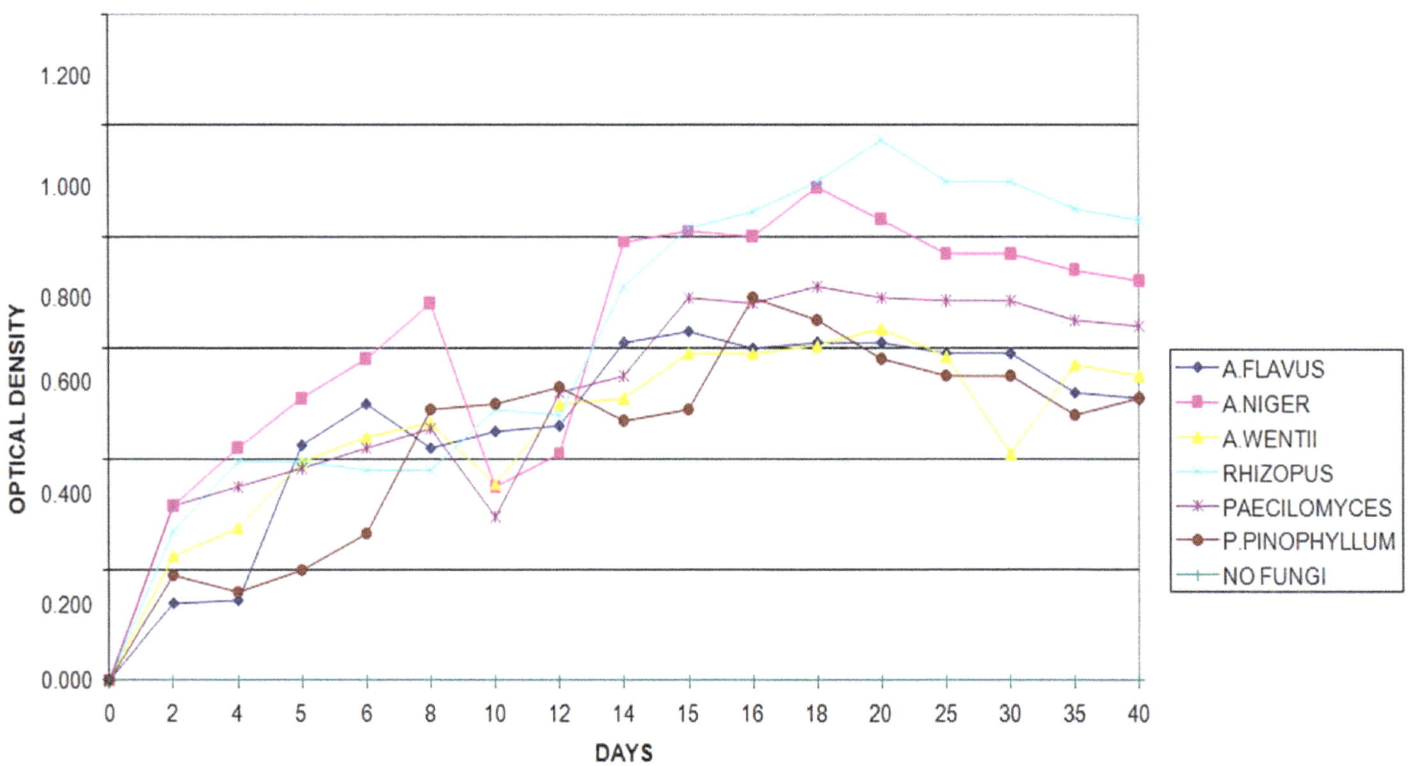

Figure 2. The growth pattern of fungi in minimal salt solution and extracted oil from *Treculia africana* seed using 530. nm

Table 2. Percentage contribution of Biodegradation of each fungus in the different oil after 40 days incubation.

Oil	Percentage contribution of biodegradation of each fungus (%)					
	Aspergillus flavus	***Aspergillus niger***	***Aspergillus wentii***	***Rhizoipus***	***Paecilomyces***	***Penicillium pinophyllum***
Kerosene	20. 87	14.08	16.51	27.19	14.56	6.80
Diesel	17.83	20.93	11.63	18.61	20.61	10.85
Unspent engine oil	24.38	14.38	19.38	33.31	7.50	1.25
Spent engine oil	14.63	18.58	21.74	19.37	11.46	14.23
Crude oil	17.65	18.70	17.85	19.26	14.16	12.45
Extracted oil from the *Treculia africana* seed	13.56	19.15	14.63	22.08	17.02	13.56

Table 3. The order of biodegradation of the various fungi on the different oil after 40 days of incubation.

Oil	***Aspergillus flavus***	***Aspergillus niger***	***Aspergillus wentii***	***Paecilomyces sp***	***Penicillium pinophyllum***	***Rhizoipus sp.***
Kerosene	2*	5	3	4	6	1
Diesel	4	1	5	2	6	3
Unspent engine oil	2	4	3	5	6	1
Spent engine oil	4	3	1	6	5	2
Crude oil	4	2	3	5	6	1
Extracted oil from the *T. africana* seed	5	4	4	3	6	1

* Ranking 1 to 6 represents the order of biodegradation from the highest to the lowest.

various fungi were less active in kerosene compared to other oil, which is probably due to the hydrocarbon present in kerosene. Amanchukwu et al. (1989) observed that most microorganisms find it difficult biodegrading kerosene, attributing this to its type of hydrocarbon chain. Increased turbidity and emulsification of the oil was observed during the course of these investigations. Emulsification is a known part of hydrocarbon degradation (Geyer, 1980), and might be a probable indication of hydrocarbon utilization by the fungi.

The chromatogram of the MSS and crude oil (control), shows the detection of C_{12} - C_{24}; the peaks for these carbons were obvious (Figure 3). The growth of *Paecilomyces* in MSS and crude oil indicates a complete disappearance of the peaks in C_{12} - C_{24}, after 40 days incubation (Figure 4). It was only C12 peak that was not completely absent but reduced in the chromatogram (Figure 4). Also the chromatogram for the MSS and extracted oil from seed (control), C_{12} - C_{24} were detected and their peaks were obvious (Figure 5). The chromatogram of the growth of Aspergillus *niger in* MSS and the extracted oil from *T. africana* seed after 40 days incubation showed that the peaks of C_{10}-C_{16} were absent (Figure 6). The peak system chromatogram could not detect some of the carbons in the oil artificially inoculated with *A. niger*, and P*aecilomyces* probably due to the biodegradation of the oil by these fungi suggesting n-alkane biodegradation. This observation is similar to the work of Salminen et al. (2004) who worked on the potential for aerobic biodegradation of petroleum hydrocarbons in boreal subsurface, suggesting an n-alkane degradation due to the removal of C_n-C_{15} in their study.

In conclusion, it is evident from the results in this work that all the fungi isolated from *T. africana* seed have the potential to biodegrade petroleum crude oil and petroleum products.

Conflict of Interests

The author(s) have not declared any conflict of interests.

ACKNOWLEDGEMENT

The authors thank the Director of Tudaka environmental Consultants Ltd, Lagos, for allowing the use of the Peak system Elmer Auto-system GC in his establishment.

Lab name: TUDAKA Envir. Cons. Ltd
Client: DAYO
Method: Syringe Injection
Description: FID-CHANNEL 1
Column: Capillary 30m
Carrier: nitrogen at 37 psi
Sample: MSSEXTAN
Comments: TOTAL HYDROCARBON CONTENT CHROMATOGRAM FOR THE GROWTH PATTERN OF *Aspergillus niger* IN MINIMAL SALT SOLUTION AND EXTRACTED OIL. (APPENDIX IV)

Temperature program:

Init temp	Hold	Ramp	Final temp
60.00	2.000	12.000	300.00

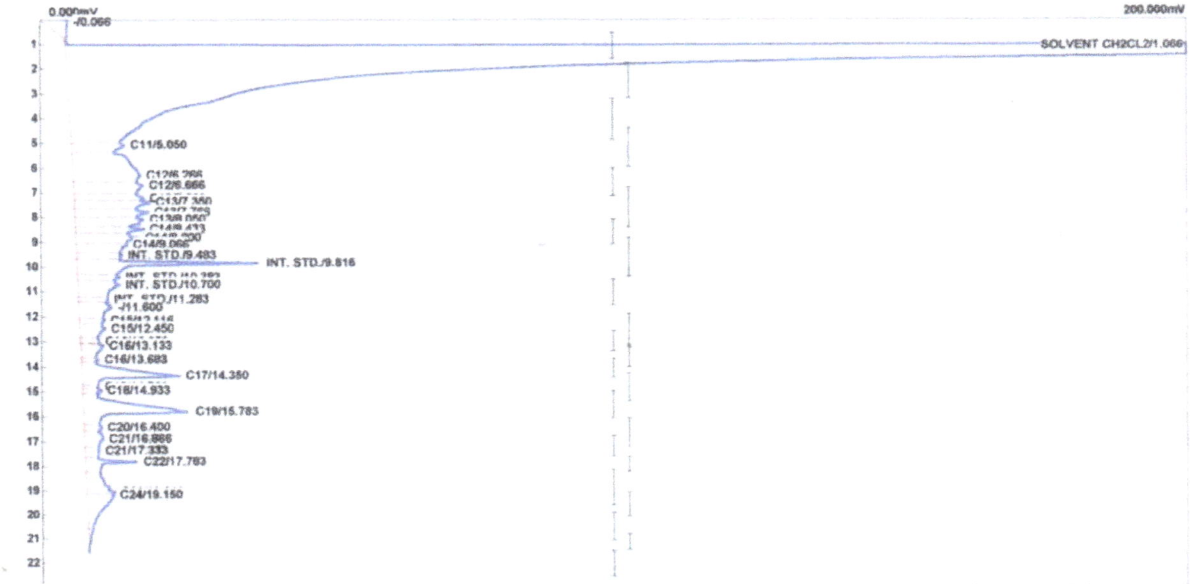

Component	Retention	Area	External	Area %
SOLVENT CH2CL2	1.066	18896.0505	0.0000	75.7608
C11	5.050	197.3560	0.0000	0.7913
C12	6.266	744.8690	0.0000	2.9864
C12	6.666	316.5700	0.0000	1.2692
C13	7.200	183.5590	0.0000	0.7360
C13	7.350	235.2700	0.0000	0.9433
C13	7.766	238.6245	0.0000	0.9567
C13	8.050	249.9315	0.0000	1.0021
C14	8.433	162.5470	0.0000	0.6517
C14	8.800	227.1440	0.0000	0.9107
C14	9.066	186.0145	0.0000	0.7458
INT. STD.	9.483	114.2050	0.0000	0.4579
INT. STD.	9.816	452.5215	0.0000	1.8143
INT. STD.	10.383	88.2150	0.0000	0.3537
INT. STD.	10.700	241.6860	0.0000	0.9690
INT. STD.	11.283	54.5545	0.0000	0.2187
C15	12.116	52.6105	0.0000	0.2109
C15	12.450	122.1715	0.0000	0.4898
C16	12.950	44.2210	0.0000	0.1773
C16	13.133	100.3460	0.0000	0.4023
C16	13.683	24.4640	0.0000	0.0981
C17	14.350	388.4185	0.0000	1.5573
C18	14.766	22.0190	0.0000	0.0883
C18	14.933	31.5875	0.0000	0.1266
C19	15.783	519.9685	0.0000	2.0847
C20	16.400	47.9150	0.0000	0.1921
C21	16.866	107.4710	0.0000	0.4309
C21	17.333	55.8360	0.0000	0.2239
C22	17.783	128.6560	0.0000	0.5158
C23	19.016	178.7360	0.0000	0.7166
C24	19.150	227.2535	0.0000	0.9111
		24640.7920	0.0000	100.0000

Figure 3 :Total hydrocarbon content chromatogram for minimal salt solution (MSS) and crude oil after 40 days incubation; Peaks for C_{12} –C_{24} were obviously detected.

Figure 3. Total hydrocarbon content chromatogram for minimal salt solution (MSS) and crude oil after 40 days incubation; peaks for C_{12}- C_{24} were obviously detected.

Carrier: nitrogen at 37 psi
Sample: MSSEXT
Comments: TOTAL HYDROCARBON CONTENT CHROMATOGRAM FOR MINIMAL SALT SOLUTION AND EXTRACTED OIL. (APPENDIX III)

Temperature program:

Init temp	Hold	Ramp	Final temp
60.00	2.000	12.000	300.00

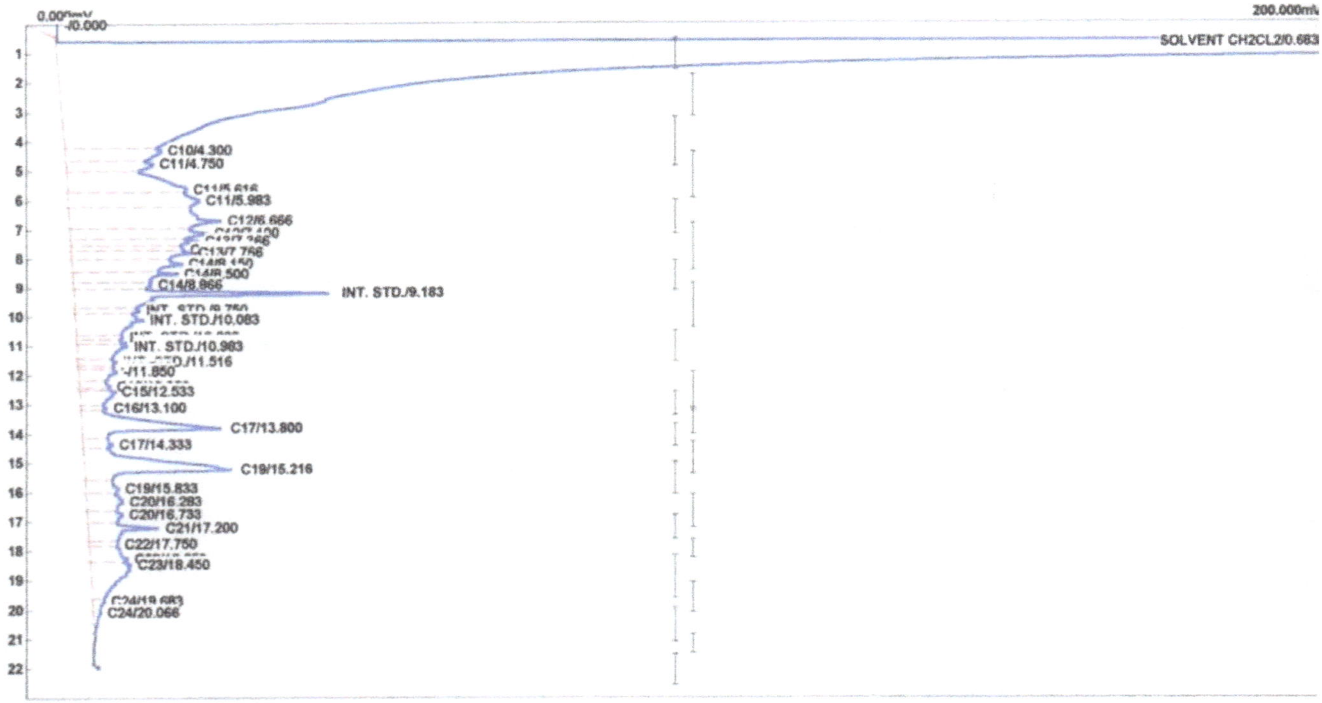

Component	Retention	Area	External	Area %
SOLVENT CH2CL2	0.683	25813.1990	0.0000	73.9408
C10	4.300	359.3530	0.0000	1.0294
C11	4.750	262.5605	0.0000	0.7521
C11	5.616	662.7135	0.0000	1.8983
C11	5.983	574.4450	0.0000	1.6455
C12	6.666	850.8200	0.0000	2.4371
C12	7.100	404.6635	0.0000	1.1591
C13	7.366	253.8000	0.0000	0.7270
C13	7.666	188.5515	0.0000	0.5401
C13	7.766	296.5740	0.0000	0.8495
C14	8.150	369.2975	0.0000	1.0578
C14	8.500	299.8050	0.0000	0.8588
C14	8.866	152.1810	0.0000	0.4359
INT. STD.	9.183	630.2110	0.0000	1.8052
INT. STD.	9.750	114.0360	0.0000	0.3267
INT. STD.	10.083	369.5800	0.0000	1.0586
INT. STD.	10.666	62.6790	0.0000	0.1795
INT. STD.	10.866	71.0390	0.0000	0.2035
INT. STD	10.983	188.2160	0.0000	0.5391
INT. STD.	11.516	77.2240	0.0000	0.2212
C15	12.350	65.8045	0.0000	0.1885
C15	12.533	140.7430	0.0000	0.4032
C16	13.100	33.2115	0.0000	0.0951
C17	13.800	512.0300	0.0000	1.4667
C17	14.333	76.7040	0.0000	0.2197
C19	15.216	653.7195	0.0000	1.8726
C19	15.833	117.5240	0.0000	0.3366
C20	16.283	183.1820	0.0000	0.5247
C20	16.733	121.0160	0.0000	0.3466
C21	17.200	213.0020	0.0000	0.6101
C22	17.750	48.0665	0.0000	0.1377
C22	18.250	153.4695	0.0000	0.4396
C23	18.450	312.2080	0.0000	0.8943
C24	19.683	34.6700	0.0000	0.0993
C24	20.066	20.6500	0.0000	0.0592
		34686.9490	0.0000	100.0000

Figure 4. Total hydrocarbon content chromatogram for the growth of paecilomyces in MSS and crude oil after 40 days incubation; peaks for C_{12}- C_{24} were completely absent except C_{21} peak that was only reduced.

Lab name: TUDAKA Envir. Cons. Ltd.
Client: DAYO
Method: Syringe Injection
Description: FID-CHANNEL 1
Column: Capillary 30m
Carrier: nitrogen at 37 psi
Sample: MSSCR
Comments: TOTAL HYDROCARBON CONTENT CHROMATOGRAM FOR MINIMAL SALT SOLUTION AND CRUDE OIL. (APPENDIX I)

Temperature program:

Init temp	Hold	Ramp	Final temp
60.00	2.000	12.000	300.00

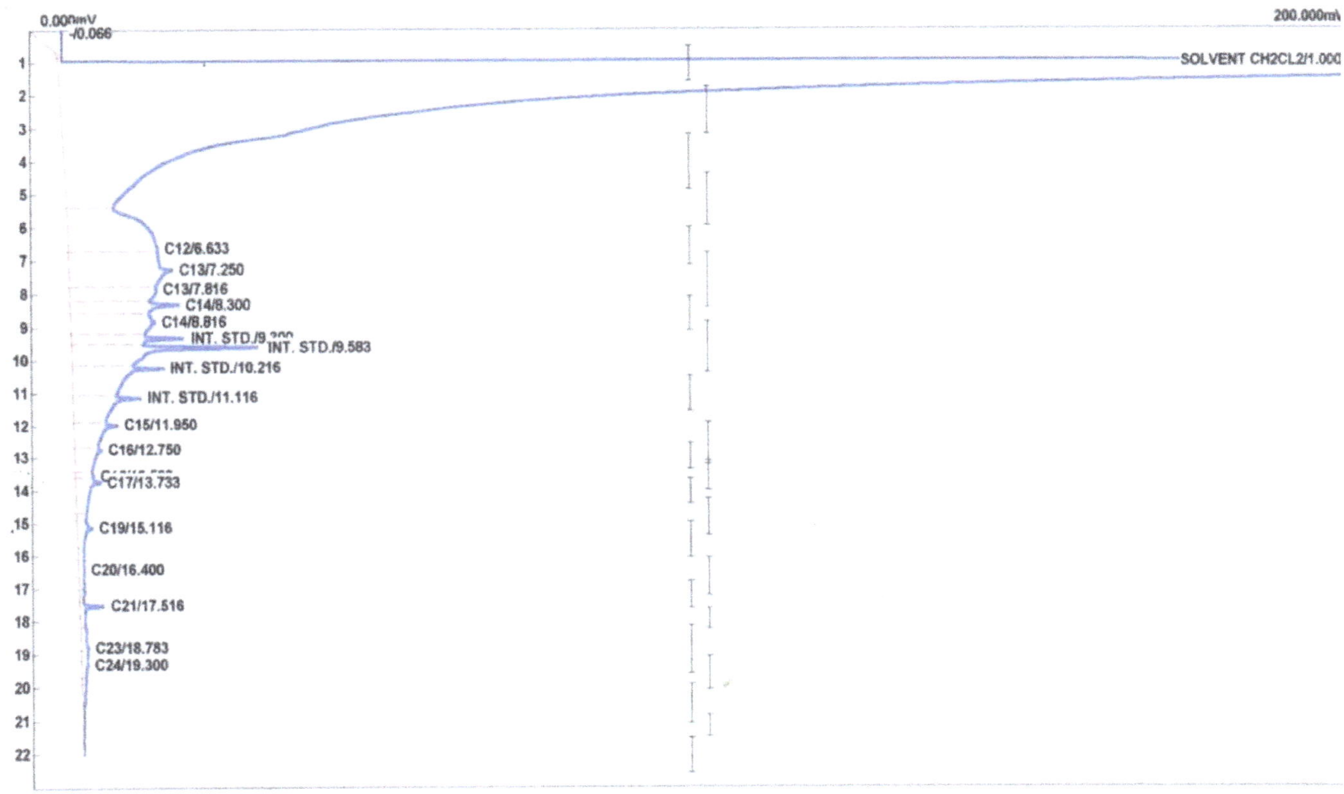

Component	Retention	Area	External	Area %
SOLVENT CH2CL2	1.000	20451.4510	0.0000	79.7118
C12	6.633	921.0840	0.0000	3.5900
C13	7.250	880.8705	0.0000	3.4333
C13	7.816	335.3110	0.0000	1.3069
C14	8.300	293.6030	0.0000	1.1444
C14	8.816	459.5920	0.0000	1.7913
INT. STD.	9.300	239.5140	0.0000	0.9335
INT. STD.	9.583	496.1410	0.0000	1.9338
INT. STD.	10.216	452.0780	0.0000	1.7620
INT. STD.	11.116	308.7980	0.0000	1.2036
C15	11.950	217.9295	0.0000	0.8494
C16	12.750	140.0580	0.0000	0.5459
C16	13.533	26.5730	0.0000	0.1036
C17	13.733	136.7620	0.0000	0.5330
C19	15.116	66.6020	0.0000	0.2596
C20	16.400	22.0500	0.0000	0.0859
C21	17.516	32.3470	0.0000	0.1261
C23	18.783	28.6540	0.0000	0.1117
C24	19.300	24.8020	0.0000	0.0967
		25534.2200	0.0000	100.0000

Figure 5. Total hydrocarbon content chromatogram for minimal salt solution (MSS) and extracted oil after 40 days incubation; peaks for C10-C24 were obviously detected.

Lab name: TUDAKA Envir. Cons. Ltd.
Client: DAYO
Method: Syringe Injection
Description: FID-CHANNEL 1
Column: Capillary 30m
Carrier: nitrogen at 37 psi
Sample: MSSCRPA
Comments: TOTAL HYDROCARBON CONTENT CHROMATOGRAM FOR THE GROWTH PATTERN OF Paccilomyces IN MINIMAL SALT SOLUTION AND CRUDE OIL. (APPENDIX II)

Temperature program:

Init temp	Hold	Ramp	Final temp
60.00	2.000	12.000	300.00

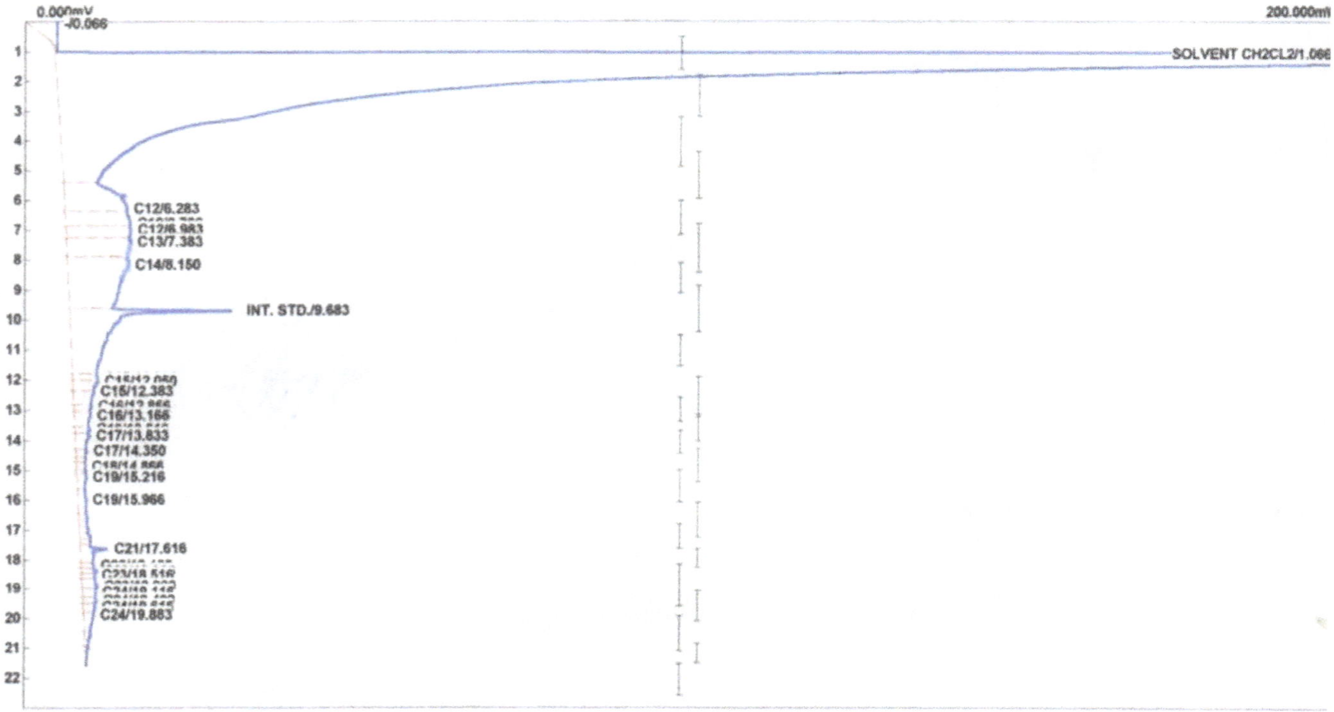

Component	Retention	Area	External	Area %
SOLVENT CH2CL2	1.066	15646.1700	0.0000	79.6692
C12	6.283	474.2340	0.0000	2.4148
C12	6.766	292.6210	0.0000	1.4900
C12	6.983	237.1550	0.0000	1.2076
C13	7.383	363.9930	0.0000	1.8534
C14	8.150	837.2515	0.0000	4.2632
INT. STD.	9.683	822.8200	0.0000	4.1897
C15	11.933	48.4070	0.0000	0.2465
C15	12.050	72.8140	0.0000	0.3708
C15	12.383	77.8290	0.0000	0.3963
C16	12.866	36.1450	0.0000	0.1840
C16	13.166	68.3560	0.0000	0.3481
C16	13.616	27.4260	0.0000	0.1397
C17	13.833	57.4270	0.0000	0.2924
C17	14.350	23.2645	0.0000	0.1185
C18	14.866	22.1100	0.0000	0.1126
C19	15.216	40.6400	0.0000	0.2069
C19	15.966	43.1500	0.0000	0.2197
C21	17.616	79.7630	0.0000	0.4061
C22	18.183	21.0715	0.0000	0.1073
C23	18.366	23.4165	0.0000	0.1192
C23	18.516	20.4620	0.0000	0.1042
C23	18.900	47.3250	0.0000	0.2410
C24	19.116	36.9550	0.0000	0.1882
C24	19.400	27.2670	0.0000	0.1388
C24	19.616	31.3120	0.0000	0.1594
C24	19.883	46.7650	0.0000	0.2381
		19526.1500	0.0000	100.0000

Figure 6. Total hydrocarbon content chromatogram for the growth of *Aspergillus niger* in minimal salt solution (MSS) and extracted oil after 40 days incubation; peaks for C10- C16 were absent or reduced detected.

REFERENCES

Amanchukwu SC, Obafemi A,Okpokwasili AN (1989). Hydrocarbon degradation and utilization by a palm wine yeast isolate. FEMS microbiology 57:151-154.

Amund DO, Adebowale AA, Ugoji EO (1987). Occurrence and characteristics of hydrocarbon utilizing bacteria in Nigerian soils contaminated with spent motor oil. Ind. J. Microbiol. 27:63-87.

Atlas RM (1981). Microbial degradation of petroleum hydrocarbon: An Environment Perspective. Microbiol Rev 45: 180-209.

Atlas RM, Bartha R (1972). Biodegradation of Petroleum in seawater low temperatures. Canadian Journal of Microbiology .22:1851-1855.

Bryce K (1992). The Fifth Kingdom. Mycologue publications, Ontario. 412pp.

Burkill HM (1997). The useful plants of West Tropical Africa. Vol. 4. Royal Botanic gardens, Kew. 960 pp.

Deacon JW (1980). Introduction to modern mycology. Blackwell scientific publications, London. 197 pp.

Domoschet KHW, Gams W, Anderson T (1980). Compendium of Soil fungi. Vol.1. Academic Press, London. 859 pp.

Egan H, Kirk HS, Sawyer R (1981). Pearson's Analysis of Foods. Churchill Living Stone Publishers, London, 596 pp.

Geyer RA (1980). Marine Environmental Pollution. Elsevier Company, New York. 35 pp.

Kampfer P, Steoif M (1991). Microbiological characteristics of a fuel oil contaminated site, including a numerical identification of heterotrophic water and soil bacteria. Microb. Ecol. 21:227-251.

Keeler R (1991). 'Bioremediation' Healing the environment naturally. Research and Development Magazine 2:139-40.

Nwachukwu, S.C.W. (2000). Enhanced rehabilitation of tropical Aquatic Environments polluted with crude petroleum using *Candida utilis*. J. Environ. Biol. 21 (3):241-250.

Parker RE (1979). Introductory Statistics for Biology. 2nd Edition. Edward Arnold, London. 112 pp.

Plante - Cunney MR (1993). Experimental field study of crude oil, drills, cuttings and natural biodeposits on mirco-phyto communities in a Mediterranean Area. Marine Biol. 117:355 - 366.

Salminen, J.M; Tuomi, P.M.; Suortti, P.M and Jorgensen, K.S. (2004). Potential Biodegradation of Petroleum hydrocarbons in boreal subsurface. Biodegradation. 15(1):29-39.

Shaw B (1995). Microbes safety, effective bioremediation oil field pits. Oil and gas J. 4:85-88.

Song, H and Bartha, R. (1990). Effects of jet fuel spills on the microbial community of a soil. Appl. Environ.Microbiol. 56:646-651.

Talbot OHK (1971). An Introduction to Mycology. Leonard Hill, London. 252 pp.

Zobell CE (1946). Action of Microorganisms on hydrocarbons. Bacterial. Rev. 10:1-49.

An assessment of flood vulnerability on physical development along drainage channels in Minna, Niger State, Nigeria

Musa Dalil[1], Nda Hauwa Mohammad[2], Usman Mohammad Yamman[3], Abdul Husaini[4] and Sanni Lekan Mohammed[1]

[1]Department of Urban and Regional Planning, Federal University of Technology, Minna, Niger State, Nigeria.
[2]Government Reserve Area, Minna, Niger State, Nigeria.
[3]Department of Urban and Regional Planning, The Federal Polytechnic, Bida, Niger State, Nigeria.
[4]Niger State Geographic Information Systems, Minna, Nigeria.

The occurrence of floods and its effects on human existence as well as the general environment has unfortunately been on the increase owing to human-induced climate change. Consequently, the vulnerability of the poor and the downtrodden has also increased; therefore the need to embark on sustainable human settlement development as well as awareness creation on the effects of flooding. As a result of this, this paper set out to analyze the causes of flooding in parts of Minna, with a view to providing solutions to forestall its impacts along the channels of River Suka. In order to achieve this, secondary as well as primary sources of data collection such as questionnaire administration and reconnaissance survey were undertaken. The data obtained from the questionnaires were analysed using the descriptive statistical method while the topographic and land use maps of Minna were digitized using the ArcView GIS package, which enabled the mapping of the Flood Risk Vulnerable Areas in Minna, Niger State. The result shows that human activities like construction on the flood plains, poor drainage network and relief of the area are primarily responsible for the perennial floods along the bank of River Suka. To this end, it was recommended that sensitization campaigns should be embarked upon by the government and stakeholders in order to create public awareness to the likely effects of flooding. The paper also recommended the monitoring by authorities, of water levels during the raining season, thereby allowing for the transmission of warning signals to the residents of the flood plains.

Key words: Drainage channel, flood, risk assessment, vulnerability.

INTRODUCTION

Natural disasters have been a regular occurrence globally right from time immemorial; but its frequency and intensity have, in the recent past, increased due to human activities. It is therefore the developmental choices (actions or inactions) made by individuals, communities and nations that pave way for disasters, which are unequally distribute (Ladan, 1998). The negative effects of which are debilitating and multifaceted, ranging from the destruction of the ecosystem, agricultural activities, infrastructural facilities and amenities, injury, illness, death, inhibition of access to education, health services, comfortable housing, drinking water and sanitation, aggravation of poverty and hunger, among others. One of such natural disasters is flooding, and it is usually associated with some social, economic, and environmental consequences (Queensland Government, 2014) that critically threaten the ecosystem and human existence. As a result of this, meeting some of the Millennium Development Goals (MDGs) has extremely been challenged in many communities, especially those of the less developed countries.

Flooding, according to *Geoscience Australia* (2013) can simply be described as "water where it is not wanted". It can also be conceptualised as a situation that results when a part of the earth surface that is usually dry is inundated and covered with water due to high amount of rainfall or the overflowing of a water body. Furthermore, flooding was expounded by *pagasa.dost* (2013) as a "natural hydrological phenomenon and its occurrence is usually the aftermath of metrological events such as seismic activities, astronomically influenced phenomena (high tides coinciding with occurrence of high rainfall), construction of temporary dams, as well as the failure of hydraulic and other control structures." The effects of floods are always debilitating, though their intensity and scope vary depending on terrain, intensity of human activities, quantum of water and the level of preparedness by the stakeholders.

Flooding, especially River flooding are among the most devastating natural disasters in the world, claiming more lives and causing more property damage than any other natural phenomenon (Abubakar, 1993). Though not the leading cause of death in Nigeria, but it affects and displaces more people than any other natural disaster (Usman, 2012). Therefore, there is the need to understand, prevent, prepare for, and mitigate its effects by authorities concerned, especially in developing countries. This has now particularly become rife due to the rapid urbanisation and population growth experienced in developing countries, which often results in increased population concentration in unplanned environment. Consequently, Ishaya et al. (2009) opined that identifying areas that are vulnerable to flooding as well as collecting and analysing information on the "elevation, slope orientation, proximity of built-up areas to drainages, network of drains, presence of buffers, extent of inundation, cultural practices as well as attitudes and perceptions" are the most effective means of ensuring flood preparedness and risk reduction.

Owing to this therefore, there is always the need for the production of maps of flood risk zones with a broad range of functions capable of manipulating both spatial and attribute data (Abah, 2013). Daffi et al. (2014) stated that although "conventional traditional methods can be used for flood hazard assessment, the use of remote sensing and geographic information system techniques have been suggested to provide quick, efficient and effective results as investigated and documented by Balanova and Vassilev (2010), Damayanti (2011), Kafle et al. (2006), Salimi et al. (2008), Ahmed et al. (2010) and others." In view of the foregoing therefore, this paper sets out, with the aid of Geographic Information System (GIS), to map flood prone areas of Minna, as well as analyze the causes of flooding in parts of Minna.

The aim of this study was to analyze the causes of flooding in parts of Minna with a view to providing solutions to developmental challenges along the channels of River Suka. This has been achieved through the following objectives: i) Analyze the causes of flooding along the River channel;ii) Identify the areas at risk in the study area and; and iii) Map the flood prone areas using the GIS technique.

The study area

Minna is located between latitude 9° 36'22"N and longitude 6° 33'l 5"E; and it is situated on a geographical base of undifferentiated basement complex of much gneiss and magnetic. The town is drained by many drainage channels, with the main course of River Chanchaga taking its source from the north central highlands and thereafter, flowing towards the western lowlands before meeting River Kaduna at a point south west of Minna. Its main tributaries include Rivers Wana, Shaho, Godina, and Dunalape, which are flowing from their respective highlands and isolated compounds such as Gwam, Kpewi, Zuru, and Tsauran Nabi hills.

The study area is located in a tropical climate region which is characterized by two seasons in a year, the wet seasons and dry seasons. The annual rainfall received within the region is less than 1000 mm in the wet season and it lasts between May and October with a maximum downpour between the months of July and September. The dry season lasts between the month of October and March. Temperature varies within the region annually, with the dry season having low temperatures because the sun is at the southern hemisphere. Thus minimum temperatures of below 30°C are recorded during the harmattan period, which is late December and January in the following year, and its maximum temperature often do not exceed 42°C. During the wet season, the sun moves northwards from the equator to the tropics of cancer. This results in high temperatures because the sun overheads at noon.

METHODOLOGY

The data used for the study which included the research design and techniques were obtained through both the primary and secondary sources.

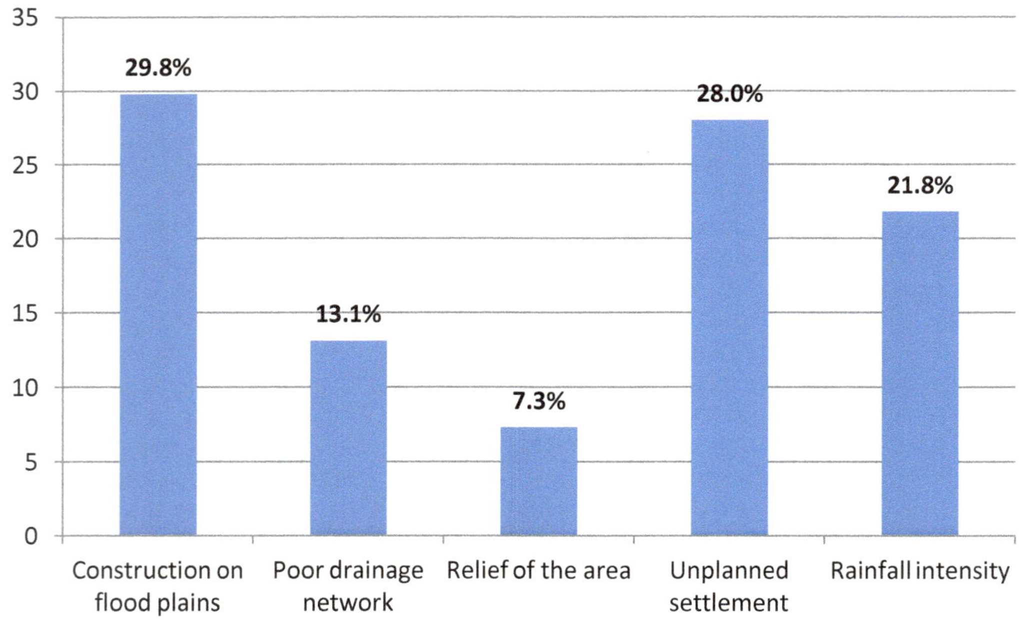

Figure 1. Causes of flooding along River Suka.

Primary and secondary data collection

The primary data used for the study were obtained through questionnaire administration and reconnaissance survey, where the Global Positioning System (GPS) and digital camera were robustly used. The questionnaires sought to feel the pulse of the respondents on the causes of flooding as well as the flood control measures adopted by the residents of the study area, while the reconnaissance survey allowed for the independent assessment of the effects of flooding in the study area, taking snapshots of objects of interest to the study using the digital camera and the delineation of flood plains with the aid of the GPS which aided the production of the flood risk vulnerable areas map of Minna, Niger State whereas the secondary data used were basically obtained from internet materials and journals, as well as the topographic and land use maps of Minna obtained from the Niger State Ministry of Lands and Housing, Minna.

Sampling technique

For the purpose of this study, a total of 350 households were randomly selected (using the random number table) within the areas declared as flood-prone by the Niger State Emergency Management Agency (NSEMA). This meant that all surveyed households were at risk, but due to local topographic effects and construction of drainage channels, not all households had been flooded. Responses from 275 households were filled and returned appropriately, giving an overall response rate of 78.6%.

Data analysis

The data collected from the field was coded and analyzed using the Statistical Package for Social Sciences (SPSS), and presented with the aid of descriptive statistics. Also, in order to create a map of flood risk vulnerable areas of Minna, the ArcView GIS package was used to digitize the topographic and land use maps of Minna. GIS Thereafter, the maps were spatially superimposed with the aid of to create a Flood Risk Vulnerable Areas in Minna, Niger State.

RESULTS AND DISCUSSION

Floods usually occur along the bank of River Suka when Minna records an unprecedented amount of precipitation or when the river is inundated by high amount of water from upstream. But its effects can be mitigated or eliminated through drainage construction, which unfortunately does not cover the entire length of the study area (Figure 2); thus areas without constructed drainages are more prone to flooding and erosion, and they as well, do not have well laid setbacks (Plate 1 and 2) when compared to those with constructed drainages (Plate 3 and 4). Therefore, flooding along parts of the bank of the river is inevitable, and this has been compounded by the reduction in the depth and width of parts of the River course due to sediments deposition resulting from the myriads of human activities taking place along the course of the River. But in order to feel the pulse of the household members residing on the floodable areas of Minna, they were asked to state the single most likely cause of flooding being experienced in their area. As shown in Figure 1, 29.8% of the respondents stated that construction on flood plains are the causes of flooding in Minna, 28.0% adduced it to the problem of unplanned settlements, whereas 7.3% of the respondents stated that relief was responsible for flooding. The usually varying opinions as regards causes of flooding was expounded in Kofo (2012), but the study conversely asked the residents of Lagos metropolis to state as many causes of flooding as possible in their neighbourhoods. The result showed that "torrential rains (94.10%), filled/silted/dirty drainage channels (87.15%), blocked canals (97.55%), inadequate drainage channels (94.30%), non-compliance with Environmental Laws(81.45%), and nature of the physical

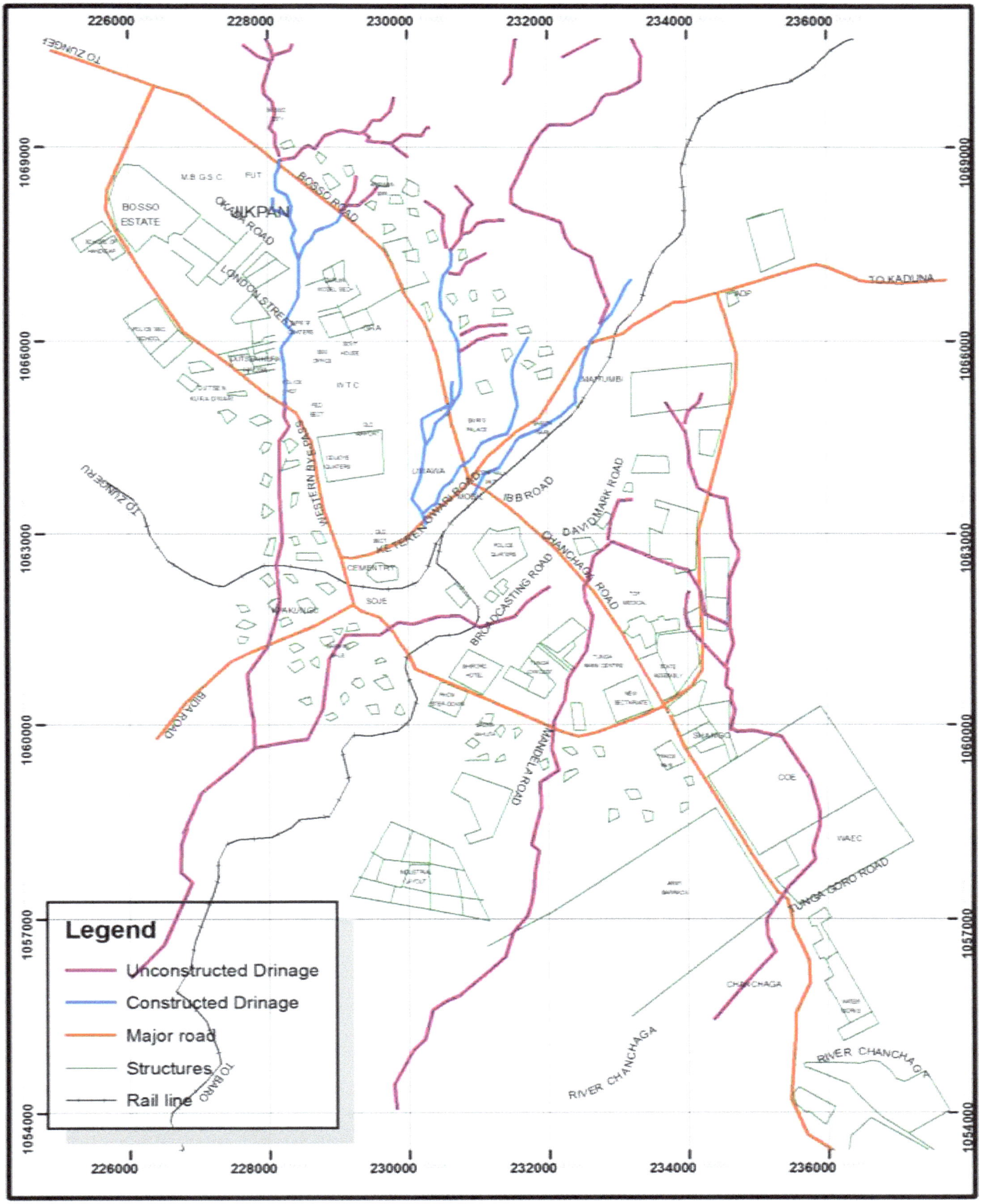

Figure 2. Constructed and unconstructed drainage channels in Minna.

Plate 1. The drainage channel that passes through Kpakungu area of Minna.

Plate 2. Human activities along the drainage channel in Soje area of Minna.

Plate 3. The drainage channel along Sabon Gari area of Minna.

Plate 4. The drainage channel along Kateren Gwari area of Minna.

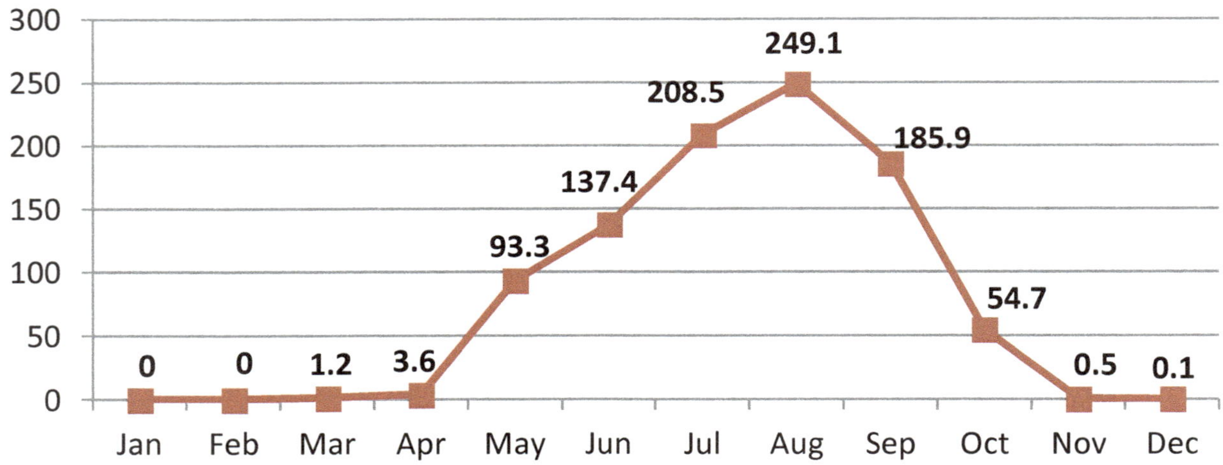

Figure 3. Rainfall pattern in the study area (in mm). Source: Nigerian Metrological Agency.

terrain (90.55%), planlessness (88.95%) and encroachment on drainage channels (90.90%)" were adduced as the causes of flooding in Lagos. Also, the residents of Minna residing along the bank of River Suka also confirmed that there are usually some negative environmental, economic and social problems that results from flooding in the study area. This apprehensive state of affairs was elaborated in Ajayi et al. (2012) which cited different sourcing as stating that over the years, flooding in the watersheds of Ogunpa and Kudeti streams in Ibadan, South West Nigeria has resulted in the loss of hundreds of lives and properties worth millions of Naira.

Rainfall variability of the study area (2002-2011)

The mean monthly discharge of a 10-year period ofMinna was obtained from the Nigerian Metrological Agency in order to analyse the monthly variability of the rainfall pattern. The result of the analysis shows that the study area experiences about 933 mm of rainfall per annum, with about half of that occurring between the months of July and August every year. Consequently, the residents of the flood-prone areas confirmed that the chances of flood occurring are usually higher between the months of July to October, as presented in Figure 3. This is in tandem with that of Makurdi, Benue State, Nigeria as expounded by Ologunorisa and Tersoo (2006); the study pointed that the outcome of seasonal variability of rainfall between 1979 and 2004 shows that the month of August recorded the highest amount of rainfall with 36.43% of the total, June (21.39%), September (20.43%) and July (19.67%).

Identifying flood prone areas using GIS

As earlier mentioned, the GIS technique was adopted in identifying the flood prone areas in Minna. Parts of Soje towards Makera, down to Kpakungu were identified as risk areas, as well as parts of Bosso due to lack of proper drainage channels. Downstream of Keteren Gwari towards River Suka was also identified as risk area due to un-constructed drainage channel and human activities along the River channel. Settlements around new extension of Farm Centre, down to some areas around Mandela road were also identified as risk areas, including parts of Shango and Chanchaga, as presented in Figure 4. Also in a study by Ikusemoran, Anthony and Maryah (2013) which aimed at assessing flood risk and vulnerability of communities in the Benue Floodplains, Adamawa State, Nigeria using the GIS method, the flood vulnerability map was developed through the use of Geo-information techniques which involved the use GPS to capture the studied communities which were consequently linked to a generated digital map of River Benue valley using ArcGIS software to assess each of the communities for flood vulnerability.

Flood control methods adopted by the residents

In the course of the research work, the following were discovered to be the local flood control methods adopted by the residents of the flood-prone areas of Minna: Opening up of new water channels and the construction of drainages; Evacuation of debris from existing drainages and; the use of sand bags to block excess water from reaching residential areas.

Conclusion

The challenges of reducing the risk of floods in the rapidly increasing global urban population in the face of vagaries of the weather due to climate change and dwindling economic fortunes has attracted enormous attention from

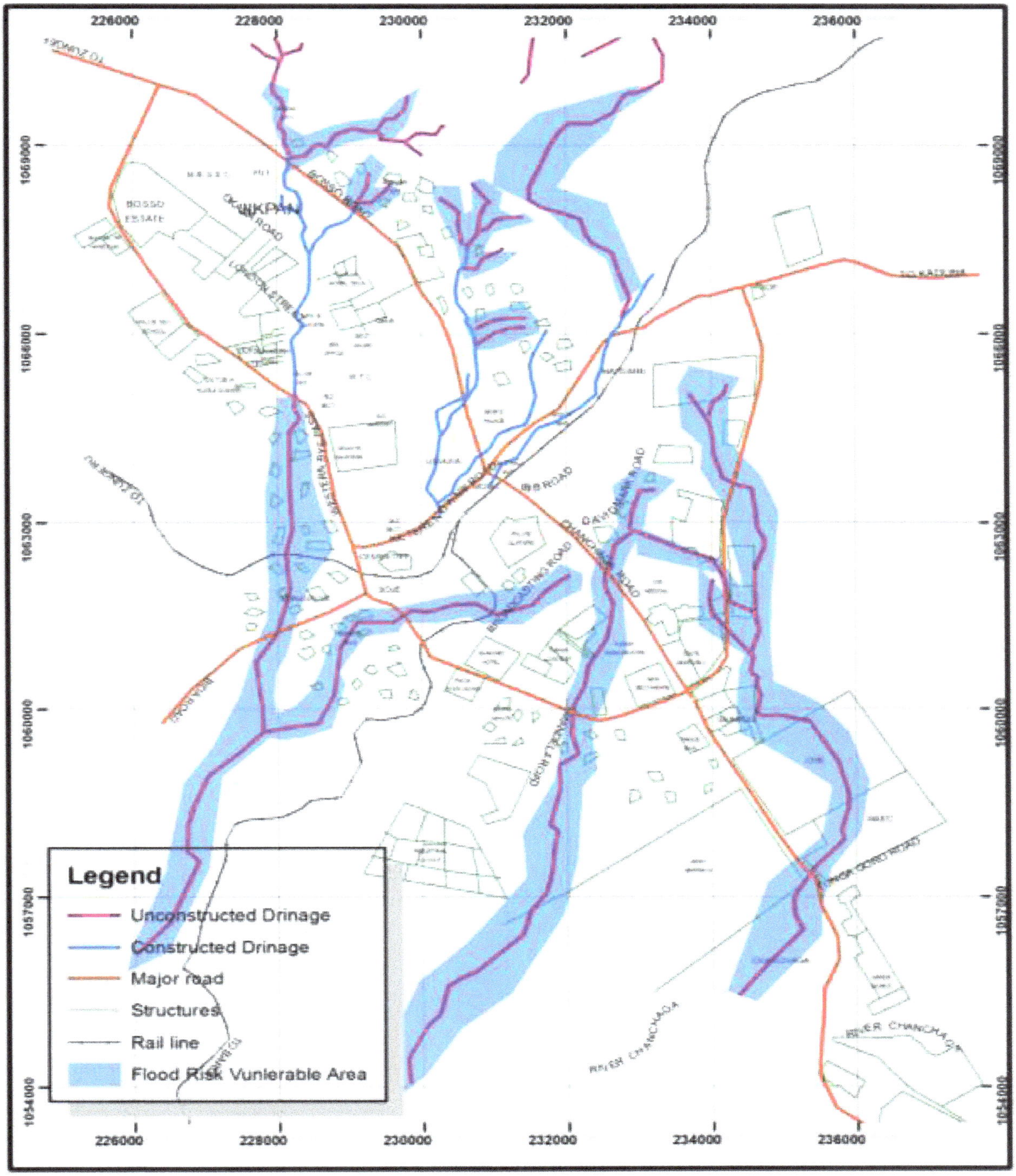

Figure 4. Flood Risk Vulnerable Areas in Minna, Niger State.

the international community, especially as regards achieving the sustainable human settlement development and the MDGs. Therefore, the need to assiduously tackle the problem of flooding through systematic physical planning approach and enlightenment campaigns, as well as the convocation of regular stakeholders' meeting for the purpose of stock taking. This, is believed, would enhance the capacity of communities in preparing for and tackling the likely occurrences of flooding.

Recommendations

In order to effectively ensure that the effects of flooding are less felt in the flood-prone parts of Minna, the following recommendations have been advanced:

1) Embarkment on sensitization campaigns in order to create public awareness on the need to understand, prevent, prepare for, and mitigate the likely effects of flooding;
2) The monitoring by authorities, of water levels during the raining season, thereby allowing for the transmission of flood signals to the residents of the flood plains and;
3) The restriction of human activities along the floodable areas of Minna by the Physical Planning Agency.

REFERENCES

Abah RC (2013). An application of Geographic Information System in mapping flood risk zones in a north central city in Nigeria. Afr. J. Environ. Sci. Technol. 7(6)365-371.

Abubakar AS (1993). A Parametric Approach to Hydro-meteorological Aspect of Flood Forecasting. An M. Tech Thesis submitted to Geography Department, Federal University of Technology, Minna.

Ajayi O, Agbola SB, Olokesusi BF, Wahab B, Taiwo OJ, Gbadegesin M, Taiwo DO, Kolawole O, Muili A, Adeola M A, Olutade OG, Shiji F And Abiola NA (2012). Flood Management in an Urban Setting: A Case Study of Ibadan Metropolis. “Hydrology for Disaster Management Special Publication of the Nigerian Association of Hydrological Sciences, 2012.” Retrieved from: http://www.unaab.edu.ng

Daffi RE, Otun JA, Ismail A (2014). Flood hazard assessment of River Dep floodplains in North-Central Nigeria. Int. J. Water Resour. Environ. Engr. 6(2)67-72.

Geoscience Australia, (2013). What is Flooding? Retrieved from: http://www.ga.gov.au/hazards/flood/flood-basics/what.html

Ishaya S, Ifatimehin OO, Abaje IB (2009). Mapping Flood Vulnerable Areas in a Developing Urban Centre of Nigeria. J. Sust. Dev. Afr (Volume 11, No.4, 2009). ISSN: 1520-5509. Clarion University of Pennsylvania, Clarion, Pennsylvania

Ikusemoran M, Anthony D, Maryah UM (2013). GIS Based Assessment of Flood Risk and Vulnerability of Communities in the Benue Floodplains, Adamawa State, Nigeria. Journal of Geography and Geology; Vol. 5, No. 4; 2013; ISSN 1916-9779 E-ISSN 1916-9787: Published by Canadian Center of Science and Education. doi:10.5539/jgg.v5n4p148. Retrieved from: http://dx.doi.org/10.5539/jgg.v5n4p148

Kofo AA (2012). Global Warming and Challenges of Floods In Lagos Metropolis, Nigeria. Academic Research International. ISSN-L: 2223-9553, ISSN: 2223-9944, Vol. 2, No. 1, January 2012. Retrieved from: www.journals.savap.org.pk

Ladan B (1998). Application of Satellite Remote Sensing in Terrain Analysis and Flood Plain Delineation, a case study of the Sokoto-Rima River System. A Master Degree Thesis submitted to Department of Geography, Federal University of Technology, Minna

Ologunorisa TE, Tersoo T (2006). The Changing Rainfall Pattern and Its Implication for Flood Frequency in Makurdi, Northern Nigeria. J. Appl. Sci. Environ. Mgt. September, 2006 Vol. 10 (3) 97 – 102. JASEM ISSN 1119-8362. Retrieved from: www.bioline.org.br/ja

Pagasa.dost (2013). Causes and Types of Floods. Retrieved from: http://kidlat.dost.gov.ph/genmet/floods/causes_types.html

Queensland Government (2014). What are the Consequences of Floods? Retrieved from: http://www.chiefscientist.qld.gov.au/publications/understanding_floods/consequences.aspx

Usman I (2012). A Survey of the Area Vulnerable to Flood and the Mitigation Strategies Adopted by Akare Community, Niger State, Nigeria. A Master Degree Thesis submitted to the Centre for Disaster Risk Management and Development Studies, Federal University of Technology, Minna.

Adsorption studies of phosphate ions on alginate-calcium carbonate composite beads

Zahid Mahmood[1], Saqib Nasir[1], Nadia Jamil[2], Asma Sheikh[1] and Adnan Akram[1]

[1]Minerals Processing Research Centre, PCSIR Laboratories Complex, Lahore-54600, Pakistan.
[2]College of Earth and Environmental Sciences, University of the Punjab, Lahore-54590, Pakistan.

Alginate-calcium carbonate composite beads was prepared by the sol-gel method and characterized by Fourier transform infra-red spectroscopy (FT-IR) and scanning electron microscope (SEM) instruments. Adsorption potential of phosphate ions have been studied on laboratory scale. The effects of contact time, adsorbent dosage, initial analyte concentration, pH and agitation speed were optimized at 298.15 K (25°C) in a single ion system. The adsorption process was in accordance with Freundlich model and pseudo second order kinetics. Thermodynamic parameters like $\Delta G°$, $\Delta H°$ and $\Delta S°$ have also been determined. Negative values of ΔG establish the favorability of the process. Values of ΔH and ΔS indicate that the process was endothermic and during the adsorption of phosphate ions (PO_4^{-3}), less randomness takes place at the adsorbent-adsorbate interface.

Key words: Alginate-calcium carbonate, composite beads, phosphate ion adsorption.

INTRODUCTION

It is an established fact that waste from industrial and agriculture sectors are the leading sources of contamination in water bodies. It introduces various chemical contaminants into the surface and ground water (Azizullah et al., 2011). Fertilizer industry, laundries and household washings produce waste water enriched in phosphate ions. Although phosphorous is beneficial to human being and agricultural lands, its excessive amount causes the phenomenon of eutrophication in the water bodies. The studies have indicated that effluents from major cities of Pakistan have higher concentrations of pollutants than the permissible limits, that is, National Envrionment Quality Standards (Khan et al., 2007). The removal and rapid refinement of metallic and non-metallic pollutant have become vital for the environmental remediation. Many processes such as chemical precipitation, coagulation, solvent extraction, membrane separation, micro and ultra filtration, forward and reverse osmosis, electro dialysis, phytoremediation and phyto-extraction, ion-exchange and adsorption have been used to cleanse wastewaters (Farooq et al., 2010). Adsorption is the process in which atoms, ions or molecules from an effluent adhere to or accumulate on a suitable interface.

A number of adsorbent variety such as activated carbon, clays, fly ash, agri wastes, biomaterials, polymers and composites have been used as adsorbents (Gotoha

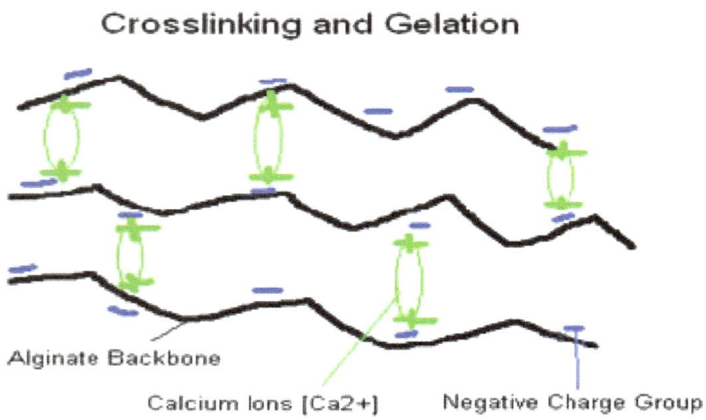

Figure 1. The representation of turning of sodium alginate solution into a gel.

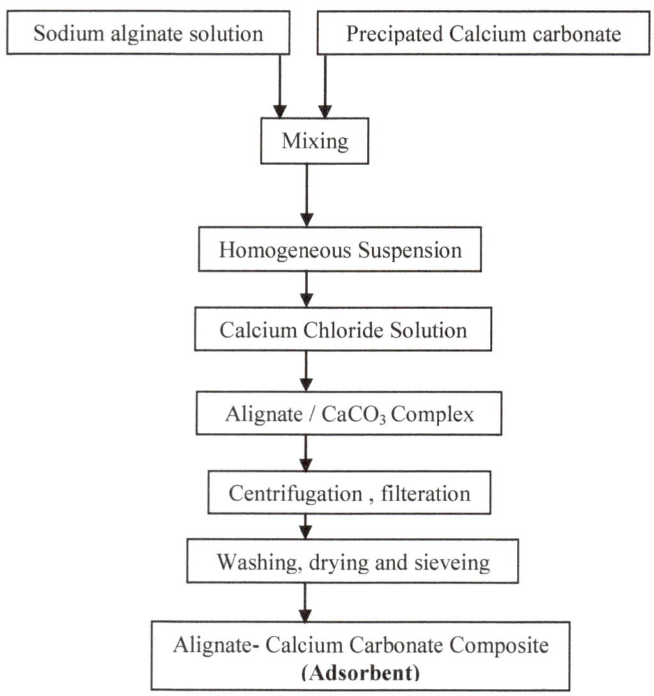

Figure 2a. Flow sheet for adsorbent synthesis.

et al., 2004). The adsorbents used should have some specific properties, such as a high ability to reduce the concentration of pollutant below the acceptable limits, high adsorption capacity, fit to purpose, durable, environmentally friendly and cost effective (Abu Al-Rub et al., 2002). Alginates have been used as low cost adsorbents for waste water purification. These are naturally occurring polysaccharides having D-mannuronic acid and L-guluronic acids. During adsorption process alginates makes a gel and appear to exchange with divalent and trivalent ions (Jorda and Mijangos, 2003).

Mechanical properties of alginates have been improved by combining it with other materials. Sodium alginate has been combined with polyethylene oxide to remove phosphorous from waste water (Lin, 2009-02). Alginate gel beads have also been used effectively for phosphorous removal (Lin et al., 2007-08). Immobilized alginate beads suspended in an aerated activated sludge were used to remove phosphates from the activated sludge (Osee, 1995). Alum sludge combined with alginates has also been used for phosphate removal (Wadood and Sarmad, 2012). A significant increase in the removal of ammonium and soluble phosphorus ions was observed by the co-immobilization of the freshwater micro alga *Chlorella vulgaris* onto alginate beads (de-Bashan, 2002). It has been noted that volatile fatty acid (VFA) content, phosphorus load, pH, COD content, cation concentration and food to microorganism ratio (f/m) influence the biological removal of phosphate ions from wastewaters (Mulkerrinsa et al., 2004). More than 90% phosphate and >50% ammonium ions have been successfully removed during tertiary wastewater treatment using 10% (w/v) modified ca-alginate beads (Cruz et al., 2013).

MATERIALS AND METHODS

Adsorbent preparation

Precipitated calcium carbonate was impregnated on sodium alginate gel to prepare adsorbent in the form of beads in aqueous calcium chloride solution. 5 g of commercial grade sodium alginate was dissolved in 200 mL of distilled water with vigorous stirring and heated at 40°C to get a gel like material (Figure 1). The fine calcium carbonate powder (20 g) was mixed thoroughly under constant mechanical stirring to get the homogeneous mixture. The suspension was injected dropwise into a 1 M solution of calcium chloride with the help of 25 mL syringe. Fine complex beads of alginate-calcium carbonate were obtained. The beads were separated from the solution through centrifuge process followed by filtration and drying for overnight at (70 ± 2°C). The dried mass of beads was sieved and used as an adsorbent. The flow sheet of the process is demonstrated in Figure 2a.

Adsorbent characterization

The characterization of adsorbent material was made by Fourier transform infra-red spectroscopy (FT-IR) and scanning electron microscope (SEM) techniques.

FT-IR spectra of alginate-calcium carbonate composite shown in Figure 2b was recorded in the range of 4,000-400 cm^{-1}. The spectra exhibits several peaks with variable intensities merged to form a broad band in the functional group region between 3,200-3,600 cm^{-1} associated with stretching vibrations of free or bonded -OH groups along with presence of inter and intra-molecular hydrogen bonding between these hydroxyl groups (Bojić et al., 2013). The distorted peaks of medium intensity corresponds to the asymmetric stretching vibrations of alkyl chains. The relatively distorted small bands at 1,634, 1,465, 1,109 and 874 cm^{-1} are assigned to the stretching vibrations of the functional groups like C=O, -COOH and -OH (Kirova et al., 2012). It can be inferred from the FT-IR results that a strong bonding has occurred between functional groups of alginate and host material, that is, calcium

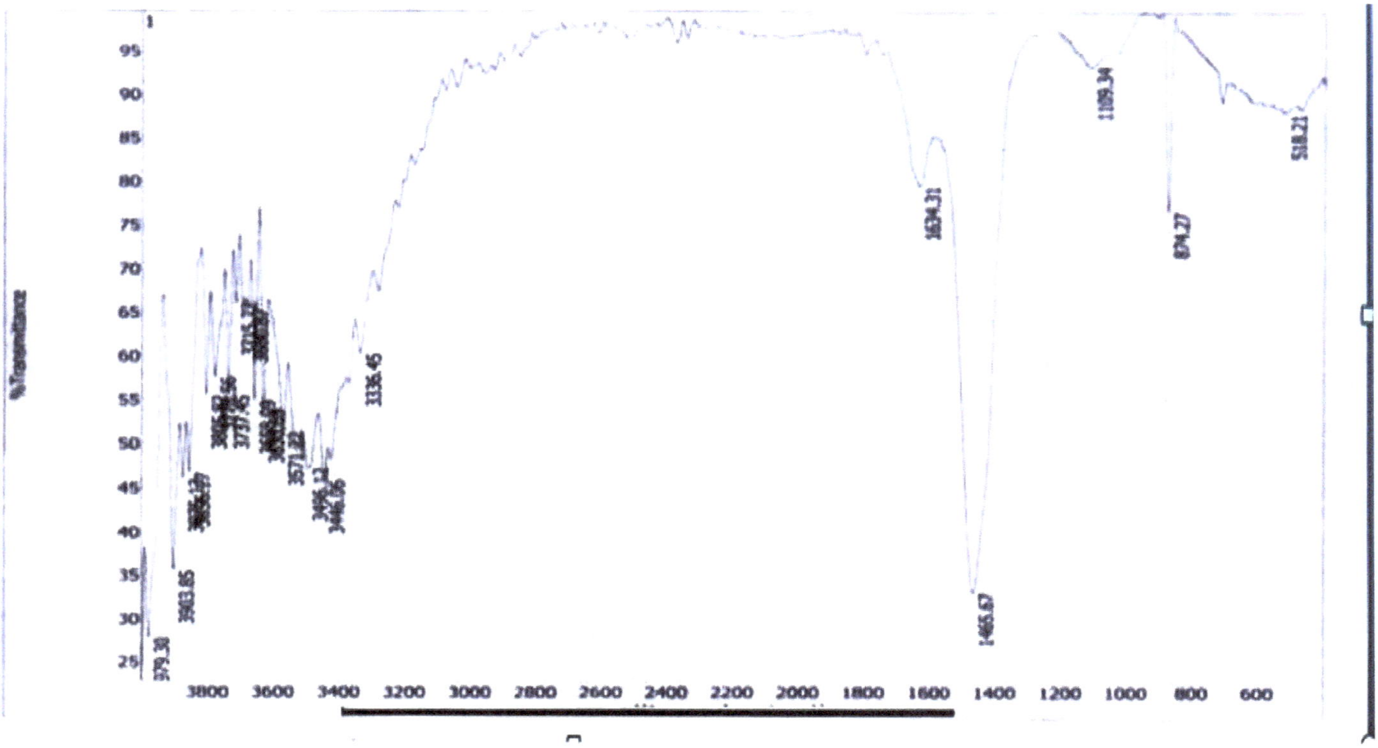

Figure 2b. FT-IR spectra of alginate-$CaCO_3$.

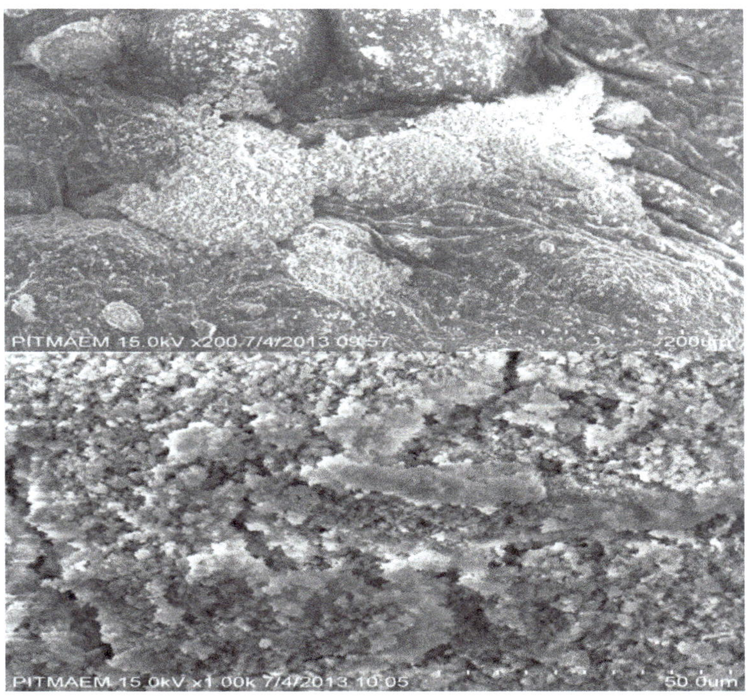

Figure 2c. SEM image of alginate-$CaCO_3$.

carbonate.

The morphology of the adsorbent beads was observed by scanning electron microscope. SEM images from different areas of adsorbent material were taken at 15.0 kv with varying resolutions (Figure 2c). The SEM images indicate the formation of a composite material along with sites responsible for adsorp-

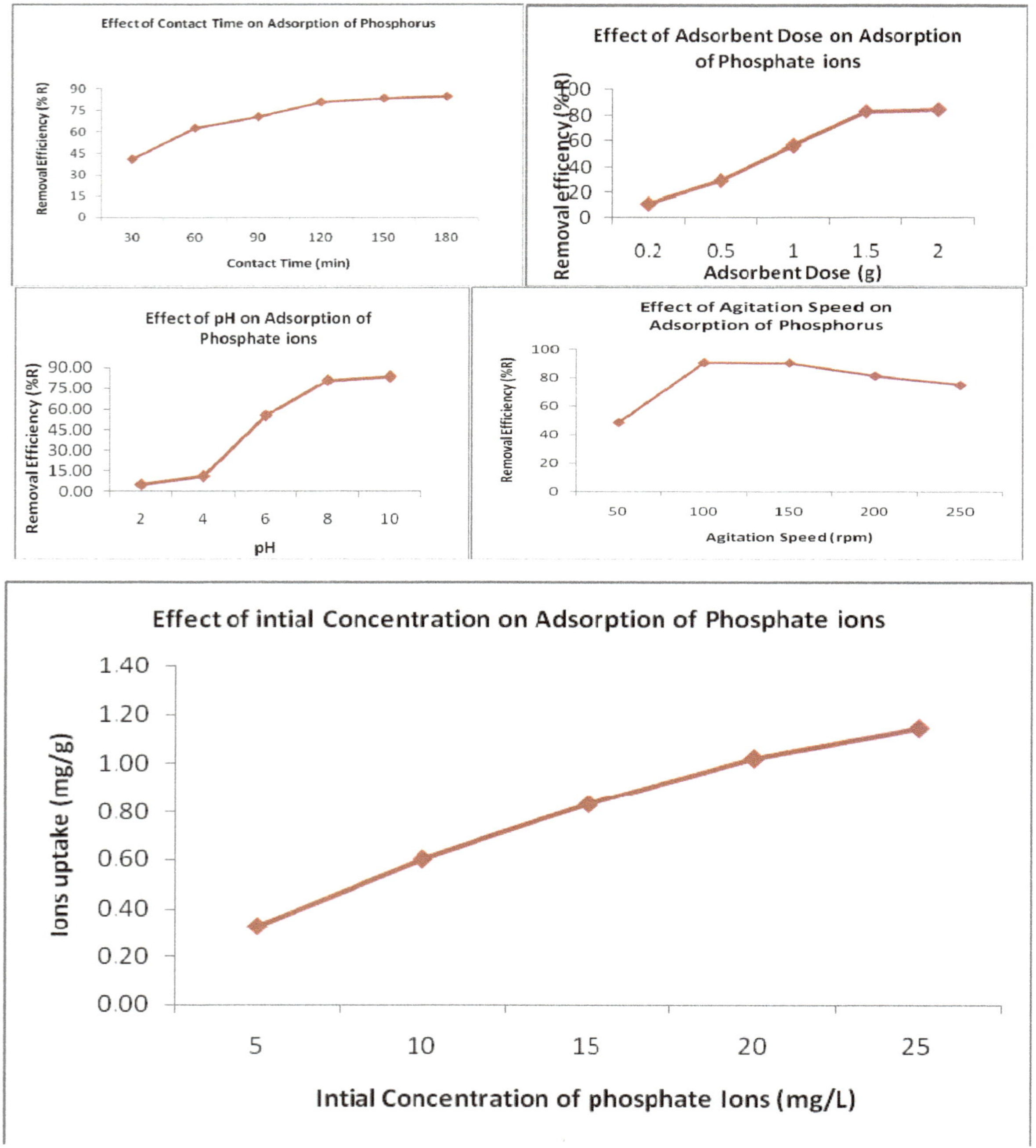

Figure 3. Optimization of parameters: **(a)** Effect of contact time, **(b)** effect of adsorbent dose, **(c)** effect of pH, (d) effect of agitation speed, **(e)** effect of initial metal ion concentration.

tion at the absorbent interface.

Batch experiments

Batch experiments were conducted to remove phosphate ions from the synthetic wastewater. The parameters like adsorbent dose, pH, contact time, shaking speed and initial concentration were optimized at 298.15 K (25°C). The residual concentration of phosphate ions were measured by colorimetric method on spectronic 20 at 460 nm.

RESULTS AND DISCUSSION

Process optimization

Factors affecting adsorption of phosphate ions on alginate-calcium carbonate complex beads were optimized by batch experiments. Figure 3a shows the graphical

Table 1. Optimized parameters for phosphate ions adsorption.

Parameter	Contact time (min)	Adsorbent dose (g)	pH	Agitation speed (rpm)	Initial concentration (mg/L)	Temperature (K)
Cadmium	150	1.5	10	100	10	298.15

Table 2. Langmuir constants calculated from Langmuir isotherm.

Adsorbate	Slope	Intercept	q_{max} (mg/g)	b	R^2
PO_4^{-3}	8.423	-3.245	0.12	-2.60	0.86

Table 3. Freundlich constants calculated from Freundlich isotherm.

Adsorbate	Intercept Log K_f	Slope (1/n)	K_f	n	R^2
PO_4^{-3}	-0.219	0.300	0.60	3.33	0.95

representation of contact time studies taken during batch experiments. The equilibrium was established after 150 min with 83.2% removal efficiency.

The effect of adsorbent dose on the adsorption of phosphates was determined by keeping other parameters constant. It has been observed that adsorption increased with increase in adsorbent dose up to certain point and then it remains constant. The increase in solid-to-liquid ratio develops some additional number of sites thus making adsorption process more efficient. Figure 3b indicate that 1.5 g adsorbent dose was sufficient to remove maximum adsorbed ions.

The effects of pH change on adsorption of phosphate ions was investigated and presented in Figure 3c with maximum removal efficiency noted at pH 10. It has been concluded from the study that the adsorbent behaviour is better under alkaline conditions, that is, at higher pH values. The effect of shaking speed is plotted in Figure 3d, at 100 rpm shaken speed maximum removal was noted. Figure 3e indicates the effect of initial concentration on adsorption capacity of phosphate ions. It was noted that adsorption was higher at lower concentrations. Optimized conditions extracted from batch experiments have been summarized in Table 1.

Adsorption models

Adsorption models provide fundamental physicochemical data for evaluating the applicability of the adsorption process as a unit operation (Vadivelan and Kumar, 2005). For modeling of the phosphate ions from wastewater commonly used, Langmuir and Freundlich models were applied. The Langmuir isotherm (Graph 4a) was obtained by the plotting the data between functions C_f (mg/L) and q (mg/g). The values of Langmuir constants "q_{max}" and "b" were calculated from slope and intercept of the isotherm (Table 2). The Langmuir constant "b" is the parameter which gives affinity of a species towards adsorption (Atkins, 1994). The value of q_{max} (0.12 mg/g) calculated from this model does not agreed with the experimentally determined value (0.62 mg/g). Furthermore, the value of R^2 is far away from unity. It is therefore concluded from observations that Langmuir model does not describe the adsorption of phosphate ions (Malakootian et al., 2009; Lin et al., 2009-02).

Freundlich isotherm (Figure 4b) was constructed by plotting log q vs. log C_f. The value of Freundlich constant, K_f" and Freundlich exponent, n" (Table 3) were calculated from slope and intercept of the corresponding isotherm. It is an established fact that, $1<n<10$ is an indication of the favorability of the adsorption process (Harikumar and Litty, 2012).

The magnitude of the Freundlich constant K_f less than unity signs shows that the adsorption process under study was viable (Schwarzenbach et al., 2003). The trend of straight line (R^2 = 0.95) also set the applicability of Freundlich model to the experimental data.

Adsorption kinetics

Pseudo first order kinetic equaiton is stated as:

$$\ln (q_e - q_t) = \ln q_e - K_1 t \tag{1}$$

Pseudo Second Order Rate Kinetics equation is demonstrated as:

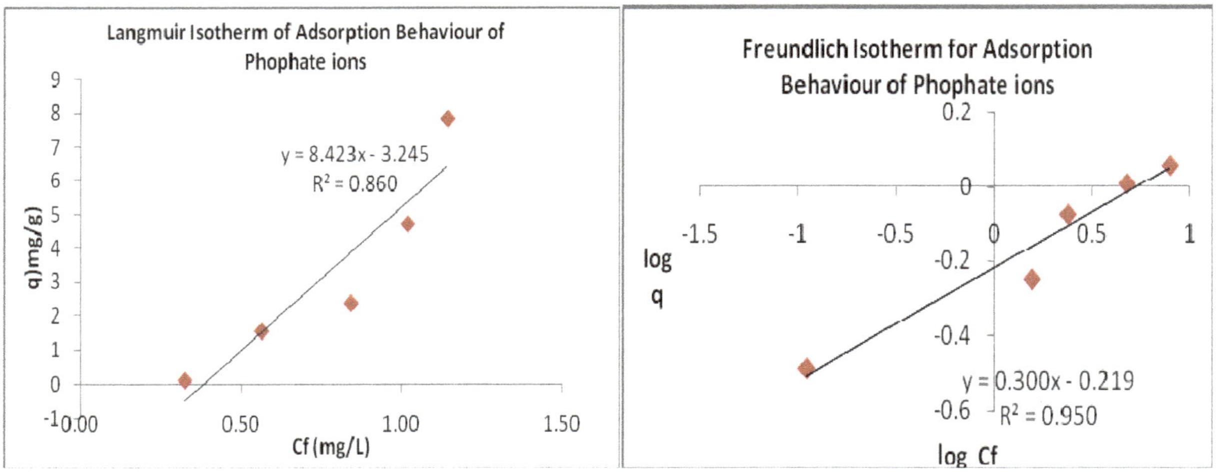

a b

Figure 4. Langmuir isotherm (a) Isotehrm adsorption behaviour of phosphate ions. (4b) Freundlich Isotherm adsorption behaviour of phosphorus.

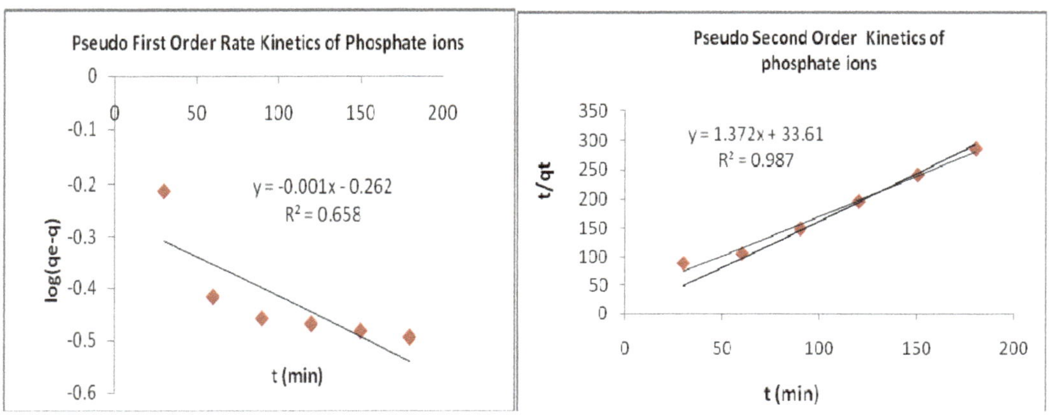

Figure 5. Kineteic isotherms.

Table 4. Rate constants calculated from pseudo first order kinetics isotherm.

Adsorbate	Intercept	Slope	$q_{e\ A.log}$ (Intercept)	$K1_{2.303}$ (Slope)	R^2
PO_4^{-3}	-0.262	-0.001	0.547	-0.0023	0.658

Table 5. Rate constants calculated from pseudo second order kinetics isotherm.

Adsorbate	Intercept	Slope	$q_e^{1/(Slope)}$	$K2^{1/(intercept)(qe)2}$	R^2
PO_4^{-3}	33.61	1.372	0.72	0.0298	0.987

$$t/qt = 1/k2qe^2 + t/qe \quad (2)$$

From the results of Figure 5, it is clear that pseudo second order kinetic model (Ho model) is in good compatibility (R^2 = 0.987) to the process as copmared to the pseudo first order kinetic model (R^2 = 0.658) (Table 4). Furthermore, the theoretically calculated value of q_e (0.72 mg/g) from pseudo second order reaction (Table 5)

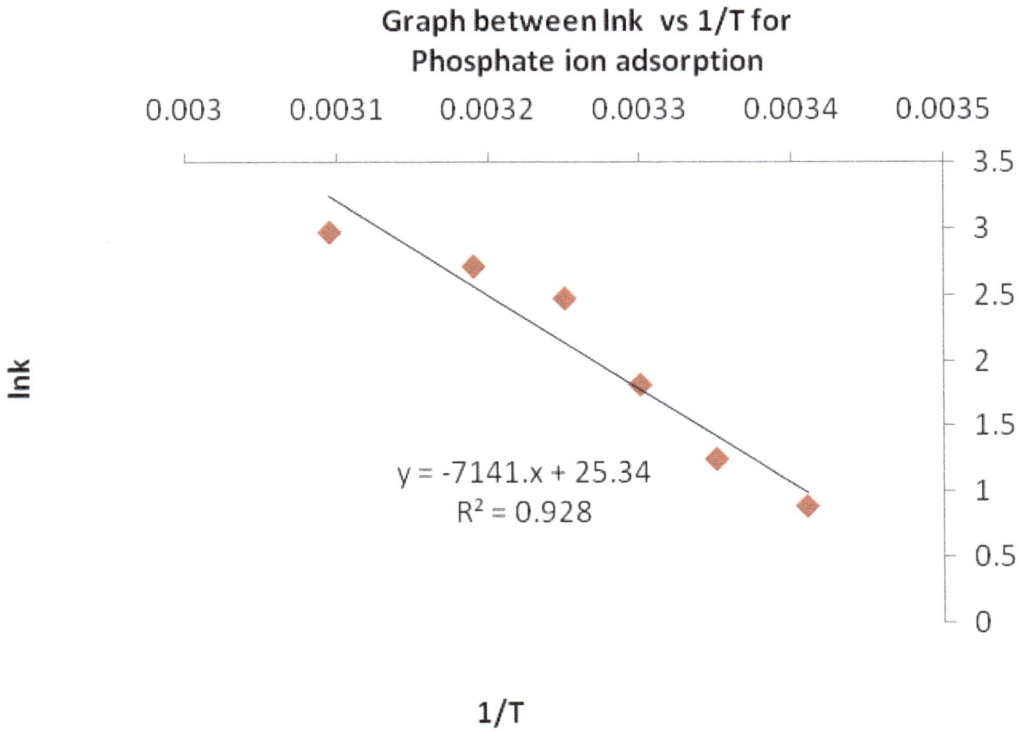

Figure 6. Graph of lnk versus 1/T for phosphate ions adsorption.

Table 6. Thermodynamic constants for phosphates adsorption at various temperatures.

Temperature (k)	ΔG (kJ/mol)	Slope (-ΔH/R)	Intercept (ΔS/R)	ΔS^0 (KJ/mol k)	ΔH^0 (KJ/mol)
293	-2.463				
298	-3.518				
303	-4.573	-7141.26	25.34	0.211	59.36
308	-5.628				
313	-6.683				
318	-7.738				

corresponds to the experimentally determined value of q_e (0.62 mg/g). These facts suggest that adsorption of phosphate ions follow the pseudo second order reaction, which was the result of chemisorption, that is, chemical adsorption (Bernard et al., 2013).

Thermodynamic parameters

Thermodynamic parameters, the standard enthalpy change (ΔH°), standard entropy change (ΔS°), and standard Gibbs energy (ΔG^0) of adsorption were determined. Figure 6 was plotted between lnk and 1/T. ΔG confirmed that the process was spontaneous and feasible (Ho and McKay, 2009). The values of ΔS and Δ H were calculated from the slope and intercept of the graph (Table 6). The value of ΔH indicates that the process is endothermic, lower value of ΔS for phosphate ion adsorption indicates a very less randomness in the sorption process at the adsorbent-adsorbate interface.

Conclusions

The developed absorbent material can effectively be used for the removal of phosphate ions from waste-water. The recovery of phosphorous is a significant factor regarding the tertiary treatment of waste-water. It is inferred from this research work that alginates modified with calcium carbonate could be an effective alternative material to recover phosphate ions from contaminated water. This investigation may be extended for the confiscation of organic pollutants and hazardous metals from wastewater.

Conflict of interest

The authors did not declare any conflict of interest.

REFERENCES

Abu Al-Rub FA, Kandah M, Al-Dabaybeh N (2002). Nickel removal from aqueous solutions using sheep manure wastes. Eng. Life Sci. 2:111-116.

Azizullah A, Khattak MNK, Richer P, Hader DP (2011). Water pollution in Pakistan and its impact on public health - A Review. Environ. Int. 37(2):479-497.

Bernard E, Jimoh, A, Odigure JO (2013). Heavy metals removal from industrial wastewater by activated carbon prepared from coconut shell. Res. J. Chem. Sci. 3:3-9.

Cruz I, Bashan Y, Hernández-Carmona G, de-Bashan LE (2013). Biological deterioration of alginate beads containing immobilized microalgae and bacteria during tertiary wastewater treatment. App. Microbio. Biotechnol. 97(3):9847-9858.

de-Bashan LE, M oreno M , Hernandez JP, Bashan Y (2002). Removal of ammonium and phosphorus ions from synthetic wastewater by the microalgae Chlorella vulgaris coimmobilized in alginate beads with the microalgae growth-promoting bacterium Azospirillum Brasilense. Wat. Res. 36:2941-2948.

Farooq U, Kozinski JA, Khan M A, Athar M (2010). Biosorption of heavy metal ions using wheat based biosorbent; a review of the recent literature. Bioresource Tech.101:5043-5053.

Gotoha T, Matsushimab K, Kikuchia KI(2004). Preparation of alginate–chitosan hybrid gel beads and adsorption of divalent metal ions. Chemosphere 55:135-140.

Ho YS, M cKay G(2009). The kinet ics of sorption of basic dyes from sphagnum. moss peat, The Canad.J. Chem. Eng. 76:822-827.

Harikumar P.S., Litty J (2012). Kinetic and thermodynamics studies of AS (III) adsorption onto iron nanoparticles entrapped ca-alginate beads. Int. J. Plant Animal Env. Sci. 2:159-156.

Jorda Y, M ijangos F(2003). Phenol adsorption in immobilized activated carbon wit h aliginate gels. Sep.Sci.Technol.38:1851-1867.

Khan MJ, Bhatti AU, Hussain S, Wasiullah (2007). Heavy metal contamination of soil and vegetables with industrial effluents from sugar mill and tanneries. Soil Envron. 26(2):139-145.

Kirova G, Velkova Z, Gochev V(2012).Copper (II) removal by heat inactivated Streptomyces fradiae biomass: Surface chemistry characterization of the biosorbent. J. BioSci. Biotech. 77-82.

Yong-bo L (2009). Research on Removal of Phosphorus in Waste Water by Polyethylene Oxide-Sodium Alginate Gel Beads [J]. Environmental Protection Science. 2:005.

Yongbo L, Dan Z, Shaochen S, Tijiu C (2007). Adsorption Performance of Alginate Gel Beads for Phosphorus in Aqueous Solution. Journal of Northeast Forestry University. 8:023.

Malakootian M, Nouri J, Hossaini H (2009). Removal of heavy metals from paint industry wastewater using Leca as an available adsorbent; Int. J. Environ. Sci. Technol. 6:183-190.

Mulkerrinsa D, Dobsona ADW, Colleranb E(2004). Parameters affecting biological phosphate removal from wastewaters. Environ. Int. 30:249-259.

Osee M uyima NY, Thomas E Cloete. (1995) Immobilization of acinetobact e johnsonii cells within alginat e beads. Water SA 21 (1995):239-244.

Schwarzenbach RP, Gschwend PM, Imboden DM (2003). Environmental Organic Chemistry, 2nd edit ion, Amazon, USA.

Vadivelan V, Kumar KV (2005). Equalibrium kinetics, mechanism and process deisgn for the soprtion of methylene blue onto rice husk. J. Colloid Interf. Sci. 286:90-100.

Wadood TM, Sarmad AR (2012). Phosphorus Removal from wastewater using oven- dried alum sludge. Int. J. Chem. Eng. 1:1-11.

Seasonal variation of meteorological factors on air parameters and the impact of gas flaring on air quality of some cities in Niger Delta (Ibeno and its environs)

Asuoha, Adaku N.[1] and Osu Charles I.[2]

[1]Safety Health and Environmental Department, Mobil Producing Nigeria, Nigeria.
[2]Department of Pure and Industrial Chemistry, University of Port Harcourt, P.M.B 5323 Choba, Port Harcourt, Rivers State, Nigeria.

The impacts of gas flaring on meteorological factors at Ibeno, Eket, Onna, Esit Eket and Umudike - Nigeria were investigated by measuring air quality parameters. The results show that the mean concentration of air parameters value were below Federal Environmental Protection Agency (FEPA) and United States Environmental Protection Agency (USEPA) National air quality standards with exception of carbon monoxide which exceeded the limit of 35 ppm in March at Ibeno. Concentration of air parameters at Umudike showed a similar trend to that of study locations at Eket, Ibeno, Esit Eket and Onna. Air quality parameters (Cl^-, SPM and SO_2) were found to have positive correlation with vapour pressure, humidity and rainfall values in the study areas. It was also established that a positive correlation exits between NO_2, H_2S, SO_2, SPM, chloride, carbon monoxide and wind speed relative humidity, temperature and vapour pressure in the study locations. Chloride and nitrate also have a positive relationship with wind speed in some of the study locations.

Key words: Air parameters, gas flaring, meteorological factors.

INTRODUCTION

The atmosphere is a complex dynamic natural gaseous system that is essential to support life on planet Earth. Stratospheric ozone depletion due to air pollution has long been recognized as a threat to human health as well as to the Earth's ecosystems. Indoor air pollution and urban air quality are listed as two of the world worst pollution problems in the 2008 Blacksmith Institute World's Worst Polluted Places report (Anderson, 2005). Nigeria holds the highest record (19.79%) of natural gas flaring globally and is responsible for about 46% of Africa's total gas flared per tonne of oil produced. Until present, there are not less than 123 flaring sites in the region making Nigeria one of the highest emitter of green-house gases in Africa (Tawari and Abowei, 2012).

Air pollution is a mixture of solid particles and gases in the air. Car emissions, chemicals from factories, dust and pollen and mold spores may be suspended as particles. Some air pollutants are poisonous. Inhaling them can increase the chances of having health problems. People with heart or lung disease, older adults and children are at greater risk from air pollution. Air pollution has been identified as one of the most critical environmental problems confronting the Nigeria. Traffic, industry and gas flaring are the major air pollution sources in the region (Tawari and Abowei, 2012). More than two million premature deaths each year can be attributed to the effects of urban outdoor and indoor air pollution and these effects are more prominent in developing countries (Mabahwi et al., 2014). Pollution of air has great influence on living organisms and man in particular. Alarming quantity of gases and smoke are discharged daily into the atmosphere. Globally, the pollutants which account for almost 98% of the total pollution are carbon monoxide 52%, sulphurdioxide 18%, hydrocarbon 2%, particulates 10% and oxides of nitrogen, 6%.

Hydrogen sulphide (H_2S) is emitted from a number of metallurgical, petroleum and petrochemical processes. Fugitive emission of hydrogen sulphide can occur from sour gas wells and certain petrochemical processes. It is a highly toxic gas due to its chemical asphixant characteristics. Despite its strong rotten egg odour, it is often difficult to detect at high concentrations due to rapid olfactory fatigue. Hydrogen sulphide is highly soluble in water and can be easily oxidized to form sulphur dioxide (Aliyu et al., 2013). Sulphur emissions influence the level of acidification of soils and freshwater ecosystems (Aliyu et al., 2013), climate change (Aliyu et al., 2013) and have impacts on human health (Aliyu et al., 2013). SO_2 is oxidized to SO_3 in the atmospheric air by photolytic and catalytic processes involving ozone, oxides of nitrogen (NO_x) and hydrocarbon (HC), giving rise to the formation of photochemical smog. Under normal conditions of the atmosphere, SO_3 reacts with H_2O vapour to produce droplets of H_2SO_4 aerosol which give rise to the so called 'acid rain' causing damage to vegetation and materials. If the concentration of SO_2 is higher along with other gaseous pollutants in the troposphere and continues to accumulate over time, the overall concentration can have a negative effect on health, vegetation and structures (Abdul Raheem et al., 2009)

This study investigate the impact of gas flaring on air quality of some cities in Niger Delta, a case study of Ibeno and its environs in Akwa Ibom State and the seasonal variation of meteorological factors on air parameters was also investigated.

MATERIALS AND METHODS

Study area

Ibeno, Eket, Onna and Esit Eket as the main study area are located in the Niger Delta region of Nigeria. Ibeno is located at latitude 04° 33' N and longitude 08° 00' E, Eket is located at latitude 04° 38' N and longitude 07° 56' E, Onna is located at 04° 37' N and longitude 07° 51' E, Esit Eket is located at latitude 04° 39' N and longitude 08° 03' E all in in Akwa Ibom State. This area experiences a tropical climate consisting of rainy season and dry season. Onshore south-westerly winds are experienced throughout the year, due to the maritime location of the study area (Figure 1). Umudike in Abia State was chosen as the control location for this study.

Sample collection

Air parameters were sampled from March 2007 to November 2007 with the months of March, April, October and November representing dry season and May to September representing rainy season. The sampling locations were labeled; Eket, Esit Eket, Ibenno, Onna and Umudike (control sample).

Determination of air quality parameters

Crowcon Gasman Monitors were used for the determination of the ambient concentration of air parameter in each location. It operates by gas diffusion through an air filter into the sensors which is fitted directly under the air filter. The concentration of air parameters is then displayed on the output meter. The instrument is calibrated before use and was moved from one site to the other between the hours of 8 am to 12 pm.

Determination of meteorological factors

Meteorological data for the sampling location were collected in Umudike, Eket, Ibeno, Onna and Esit Eket from January, 2006 to July 2007 using standard meteorological instruments.

RESULTS AND DISCUSSION

Mean concentration of H_2S ranges from 0.15 ± 0.05 - 0.40±0.12 ppm, Eket; 0.15 ± 0.02 - 0.50 ± 0.03 ppm, Esit eket; 0.15 ± 0.08 - 0.40 ± 0.12 ppm, Ibeno; 0.10 ± 0.02 - 0.25 ± 0.11 ppm, Onna and 0.15±0.01 - 0.35 ± 0.10 ppm in Umudike (Table 2). H_2S were present in all the study locations at a lower quantity. This shows that there are other contributory factors to emission of H_2S such as fumes from automobile exhaust and power generators. Results obtained was lower than results (with the range from 0.18 - 0.76) obtained by Aliyu et al. (2013).

Mean concentration of NO_2 ranges from 0.20±0.02 – 0.60±0.03 ppm, Eket; 0.10±0.01 - 0.60 ± 0.11 ppm, Esit Eket; 0.20±0.04 - 0.65±0.06 ppm, Ibeno; 0.23±0.01 - 0.50±0.14 ppm, Onna and 0.28±0.08 - 1.00±0.04 ppm in Umudike (Table 1). NO_2 is recorded in all study locations. Higher concentration in Umudike may be due to the use of fertilizer in agriculture while that of Ibeno, Eket, Onna and Esit Eket may be as a result of sea spray.

Mean concentration of SO_2 in the study locations shows the same trend of being higher during dry season and lower concentration during rainy season. This is due to dilution effect of rain on the atmosphere. Its values ranges from 0.95 ± 0.06 – 2.00 ± 0.11 ppm, Eket; 1.00 ± 0.10 - 3.00 ± 0.07 ppm, Ibeno; 0.95 ± 0.04 – 2.00 ±0.12 ppm, Esit; 1.00 ± 0.02 – 2.00 ± 0.01 ppm, Onna and 1.10 ± 0.11 – 2.40 ± 0.01 ppm, Umudike (Table 3). The results

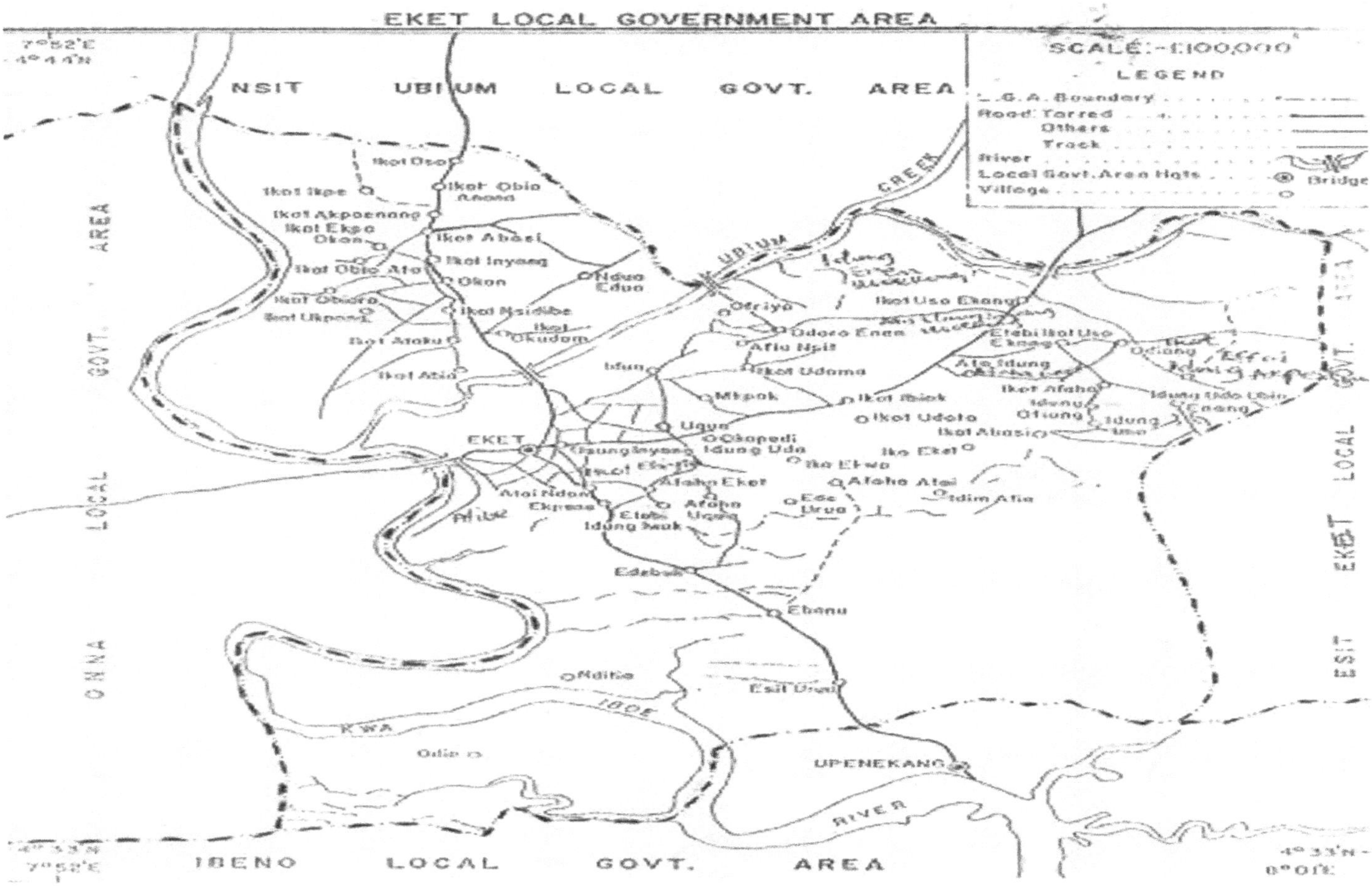

Figure 1. MAP showing sampling locations in Akwa Ibom and Abia State.

Table 1. Mean concentration ± standard deviation (ppm) of NO_2 in the studied area.

Period	Eket	Esit Eket	Ibeno	Onna	Umudike
March	0.40±0.01	0.40±0.06	0.40±0.11	0.45±0.12	0.30±0.01
April	0.40±0.02	0.35±0.09	0.30±0.09	0.40±0.04	0.40±0.11
May	0.30±0.04	0.30±0.03	0.30±0.07	0.30±0.08	0.30±0.06
June	0.20±0.02	0.10±0.01	0.20±0.04	0.25±0.02	0.28±0.08
July	0.25±0.02	0.10±0.01	0.20±0.04	0.23±0.01	0.30±0.02
August	0.40±0.01	0.20±0.02	0.60±0.01	0.30±0.05	0.55±0.01
October	0.60±0.03	0.50±0.08	0.65±0.06	0.50±0.14	1.00±0.04
November	0.60±0.03	0.60±0.11	0.60±0.10	0.40±0.11	0.80±0.02

obtained in the dry season (March, April, October and November) were higher than that of rainy season (May to August). Higher values during dry season may be attributed to vehicular emissions, refuse dump and other combustion activities within the area. SO_2 is known as an acid precursor and was identified in all samples studied. The concentration of SO_2 is lower than ranges of 3.21 – 5.18, 7.4 – 15.5 and 16 – 64 ppm reported by Ettouney et al. (2010) and Kalabokas et al. (1999), respectively but was higher than range of 0.03 – 0.09 ppm reported in Kano metropolis, Nigeria (Okunola et al., 2012) and range of 0.022–0.087 ppm reported by Aliyu et al. (2013). Comparing data with the FEPA standard level of 0.06 ppm, the results obtained from this study was found to be higher.

Mean concentration of carbon monoxide (CO) ranges from 15.00±0.02 – 33.40 ± 0.15 ppm, Eket; 17.00 ± 0.01 – 33.10 ± 0.11 ppm, Esit Eket; 12.50 ± 0.02 – 36.80 ±0.12 ppm, Ibeno; 16.50 ± 0.01 – 30.90 ± 0.06 ppm,

Table 2. Mean concentration ± standard deviation (ppm) of H_2S in the studied area.

Period	Eket	Esit Eket	Ibeno	Onna	Umudike
March	0.30±0.11	0.20±0.02	0.28±0.05	0.20±0.04	0.35±0.10
April	0.30±0.14	0.15±0.02	0.22±0.02	0.15±0.01	0.20±0.08
May	0.15±0.05	0.20±0.01	0.15±0.08	0.10±0.02	0.15±0.01
June	0.20±0.02	0.20±0.02	0.20±0.01	0.20±0.06	0.18±0.04
July	0.22±0.01	0.30±0.01	0.18±0.11	0.17±0.08	0.20±0.04
August	0.40±0.12	0.50±0.03	0.25±0.04	0.10±0.01	0.23±0.07
October	0.30±0.08	0.20±0.01	0.30±0.08	0.25±0.11	0.30±0.09
November	0.30±0.04	0.30±0.01	0.40±0.12	0.10±0.02	0.32±0.11

Table 3. Mean concentration ± standard deviation (ppm) of SO_2 in the studied area.

Period	Eket	Esit Eket	Ibeno	Onna	Umudike
March	2.00±0.01	2.00±0.02	2.00±0.04	2.00±0.02	2.20±0.02
April	1.80±0.08	1.00±0.02	3.00±0.01	2.00±0.01	2.00±0.05
May	1.10±0.02	2.00±0.01	2.00±0.08	1.00±0.02	2.00±0.14
June	1.00±0.05	0.95±0.04	1.00±0.10	1.00±0.02	1.10±0.11
July	0.95±0.06	1.00±0.01	1.20±0.11	1.20±0.01	1.30±0.12
August	1.00±0.09	2.00±0.12	2.00±0.09	1.50±0.01	1.50±0.08
October	2.00±0.03	1.00±0.08	3.00±0.07	2.00±0.03	2.00±0.04
November	2.00±0.11	1.00±0.06	3.00±0.04	2.00±0.01	2.40±0.01

Table 4. Mean concentration ± standard deviation (ppm) of CO in the studied area.

Period	Eket	Esit Eket	Ibeno	Onna	Umudike
March	27.00±0.01	33.10±0.11	36.80±0.12	30.90±0.06	13.20±0.01
April	33.40±0.15	25.50±0.09	33.50±0.08	23.00±0.09	12.50±0.01
May	20.00±0.08	25.20±0.07	28.50±0.04	21.00±0.03	8.00±0.03
June	19.70±0.06	25.00±0.04	13.40±0.01	18.70±0.11	7.00±0.01
July	15.00±0.02	23.50±0.04	12.50±0.02	16.50±0.01	7.20±0.02
August	21.00±0.02	17.00±0.01	18.00±0.05	19.20±0.08	7.50±0.01
October	24.30±0.01	27.00±0.08	28.80±0.14	21.00±0.02	15.90±0.02
November	25.20±0.04	28.00±0.10	30.00±0.11	24.00±0.05	14.00±0.02

Onna and 7.00 ± 0.01 – 15.90 ± 0.02 ppm, Umudike (Table 4). It shows the same seasonal pattern with other air pollutants measured in the study. Higher concentration of carbon monoxide may be from vehicular emission and gas flaring at Ibeno, Eket, Esit Eket and Onna. Umudike recorded lower concentration of carbon monoxide than Ibeno, Eket, Esit Eket and Onna. Carbon monoxide was higher than USEPA national ambient temperature in March 2007 at Ibeno which may be due to higher industrial activities in this area. Barman et al. (2008) conducted a study on the ambient air quality of the city Lucknow, India during Diwali festival. On Diwali day, 24-h average concentration of SO_2 was found to be 139.1 µg/m^3 (0.05 ppm) and this concentration was found to be higher at 1.95 and 6.59 times higher when compared with the respective concentration of Pre Diwali and normal day. On Diwali night (12-h) mean level of SO_2 was 205.4 µg/m^3 (0.074ppm) which was 2.82 times higher than the daytime concentration.

Mean concentration of HCN is lower at Umudike than Eket, Esit Eket, Ibeno and Onna. It ranges from 1.00±0.03 – 8.00±0.06 ppm, Eket; 1.00 ± 0.08 – 8.00 ± 0.06 ppm, Esit Eket; 1.00 ± 0.01 – 6.00 ± 0.03ppm, Ibeno; 1.50 ± 0.05 – 5.00 ± 0.08ppm, Onna and 1.00 ±

Table 5. Mean concentration ± standard deviation of SPM/(mg/m^3) in the studied area.

Period	Eket	Esit Eket	Ibeno	Onna	Umudike
March	0.18± 0.01	0.24±0.02	0.35± 0.01	0.28±0.01	0.10±0.01
April	0.15±0.02	0.13±0.02	0.35±0.11	0.25±0.06	0.23±0.09
May	0.13±0.04	0.20± 0.01	0.24±0.07	0.23±0.06	0.15±0.01
June	1.80±0.09	0.21±0.03	0.24± 0.04	2.10±0.02	0.18±0.02
July	1.50±0.06	0.20±0.05	0.26±0.03	1.70±0.08	0.20± 0.01
August	0.22±0.04	0.24± 0.01	0.32± 0.01	0.26±0.09	0.20±0.04
October	0.28±0.01	0.30±0.08	0.49±0.03	0.28±0.11	0.25±0.06
November	0.28±0.04	0.50±0.08	0.50±0.02	0.22±0.01	0.22±0.01

Table 6. Mean concentration ± standard deviation (ppm) of NH_3 in the studied area.

Period	Eket	Esit Eket	Ibeno	Onna	Umudike
March	1.20±0.06	1.00±0.12	1.00±0.02	1.30±0.06	0.92±0.04
April	1.30±0.07	1.00±0.08	1.00±0.08	1.10±0.09	0.85±0.01
May	0.98±0.13	0.98±0.07	0.90±0.04	1.00±0.02	0.75±0.02
June	0.92±0.04	0.95±0.09	0.88±0.11	0.80±0.11	0.70±0.02
July	0.85±0.01	0.93±0.04	0.90±0.05	0.86±0.05	0.66±0.01
August	0.94±0.02	0.98±0.08	0.95±0.14	0.85±0.03	0.68±0.04
October	1.10±0.03	1.10±0.01	1.08±0.03	0.99±0.08	0.98±0.05
November	1.40±0.02	1.30±0.01	1.20±0.06	1.00±0.01	1.10±0.01

Table 7. Mean concentration ± standard deviation (ppm) of HCN in the studied area.

Period	Eket	Esit Eket	Ibeno	Onna	Umudike
March	1.20±0.01	1.50±0.12	1.00±0.01	2.00±0.09	1.00±0.06
April	1.30±0.01	1.00±0.08	1.00±0.15	3.50±0.11	3.00±0.09
May	1.00±0.03	1.50±0.04	2.00±0.08	1.50±0.05	1.00±0.03
June	3.00±0.02	3.00±0.01	3.00±0.04	3.00±0.07	2.00±0.05
July	2.70±0.02	2.50±0.14	3.60±0.06	3.10±0.01	1.90±0.02
August	4.50±0.01	8.00±0.06	6.00±0.03	5.00±0.08	2.00±0.05
October	2.50±0.04	2.50±0.02	3.00±0.02	3.00±0.01	2.50±0.01
November	2.70±0.01	3.00±0.05	4.00±0.01	3.00±0.04	2.50±0.11

0.06 – 3.00 ± 0.09ppm, Umudike (Table 7). Higher concentrations were recorded in Esit Eket, Ibeno, Onna and Eket in the month of August. This may attributed to the sea spray.

Mean concentration of ammonia (NH_4) ranges from 0.85 ± 0.01 – 1.40 ± 0.20 ppm, Eket; 0.93 ± 0.04 – 1.30 ± 0.01 ppm, Esit Eket; 0.88 ± 0.11 – 1.20 ± 0.06 ppm, Ibeno; 0.80 ± 0.11 – 1.30 ± 0.06 ppm, Onna and 0.66 ± 0.01- 1.10 ± 0.01 ppm, Umudike (Table 6). Concentrations of ammonia were recorded in all location with higher volume during dry season and lower volume during rainy season. The presence of ammonia in the study locations may be linked to the use of fertilizers during agricultural activities in Umudike and sea spray for Eket, Ibeno, Onna and Esit Eket.

Mean concentration of SPM in the study area ranges from 0.13 ± 0.04 – 1.80 ± 0.09 mg/m^3, Eket; 0.13 ± 0.02 – 0.50 ± 0.08 mg/m^3, Esit Eket; 0.24 ± 0.07 – 0.50 ± 0.08 mg/m^3, Ibeno; 0.22 ± 0.02 – 2.10 ± 0.03 mg/m^3, Onna; and 0.10 ± 0.01 – 0.25 ± 0.06 mg/m^3, Umudike (Table 5). These results are below FEPA limits. Presence of SPM may be from dust particle; gas flaring and vehicular emission in Ibeno, Eket, Onna and Esit Eket and it may be due vehicular emissions or bush burning at Umudike.

Table 8. Mean concentration ± standard deviation (ppm) of Cl^- in the studied area.

Period	Eket	Esit Eket	Ibeno	Onna	Umudike
March	0.10±0.06	0.10±0.01	0.30±0.11	0.30±0.01	0.50±0.12
April	0.10±0.09	0.10±0.12	0.70±0.08	0.20±0.01	0.50±0.08
May	0.05±0.03	0.10±0.08	0.20±0.07	0.20±0.03	0.30±0.04
June	0.08±0.11	0.05±0.06	0.20±0.04	0.20±0.01	0.23±0.01
July	0.10±0.01	0.10±0.02	0.25±0.04	0.21±0.02	0.21±0.02
August	0.20±0.08	0.20±0.02	0.40±0.01	0.40±0.01	0.20±0.05
October	0.30±0.02	0.20±0.01	0.30±0.08	0.30±0.02	0.30±0.14
November	0.30±0.05	0.25±0.04	0.30±0.10	0.30±0.02	0.35±0.11

Table 9. Model summary for vapour pressure.

Station	NO_2	H_2S	SO_2	HCN	SPM	C^{l-}	CO	NH_3
Umudike	+	-	-	NE	-	+	+	-
Eket	NE	-	-	-	-	+	+	-
Ibeno	-	+	-	-	-	+	+	NE
Onna	NE	-	+	-	+	+	-	-
Esit Eket	NE	-	+	-	+	+	-	-

R^2 = 1.00, R = 1.00, dependent variable: vapour pressure, NE = no effect.

Table 10. Model summary for relative humidity.

Station	NO_2	H_2S	SO_2	HCN	SPM	Cl^-	CO	NH_3
Umudike	+	+	+	NE	+	-	+	-
Eket	NE	+	-	-	+	+	+	-
Ibeno	+	+	+	-	-	-	-	NE
Onna	NE	-	+	-	+	+	-	-
Esit Eket	NE	-	-	+	+	+	+	-

R^2 = 1.00, R = 1.00, dependent variable: humidity, NE = no effect.

Fine particles in the atmosphere provide suitable loci for various atmospheric conditions to occur. Water forms fine films which dissolves or absorbs various contaminants presence in the atmosphere which reacts with each other. When two or more than two particles fuse, higher and heavier particles are formed. Being heavier they tend to drift down and cause dry precipitation. The entire content of the atmosphere can be wash down by rain or dew, thus acid rain is caused. Results obtained were found to be higher than concentration of PM10 ranging from 70.00 – 326.75 and 41.61 – 157.65 µg/m^3 in dry and wet seasons, respectively (Babatunde et al., 2013).

Mean concentration of Cl^- in the study area ranges from 0.05±0.03 - 0.30±0.05 ppm, Eket; 0.05±0.06 - 0.25±0.04 ppm, Esit Eket; 0.20±0.04 - 0.70±0.08 ppm, Ibeno; 0.20±0.01 - 0.40±0.01 ppm, Onna and 0.20±0.05 - 0.50 ± 0.12 ppm in Umudike (Table 8). Chloride was generally high in the month of March, April, October and November while lower concentrations were recorded in the month of May – August. The source of chloride in the study area may be attributed to influence of marine environment while further study need to be carried out to determine the source of chloride at Umudike.

The regression analyses on seasonal variation of meteorological factors on air parameters (Tables 9 to 13) were calculated. The model revealed that the correlation between of the dependent variable vapour pressure (Table 9) in Umudike and the independent variables of NO_2, H_2S, Cl^-, SO_2, NH_3, SPM and CO is 1.00. This revealed that a strong positive relationship exist between the dependent variable and the independent variables. The R^2 value of 1.00 indicates that degree of variance between the dependent and the independent variables is 100%. The result shows that the coefficients of NO_2, Cl^-, CO are positive; this means that they are positively related to vapour pressure. As such, when the amounts of NO_2,

Table 11. Model summary for temperature.

Station	NO_2	H_2S	SO_2	HCN	SPM	Cl^-	CO	NH_3
Umudike	-	+	+	NE	-	+	+	+
Eket	NE	-	+	+	-	-	-	+
Ibeno	+	+	+	-	-	-	-	NE
Onna	NE	-	+	-	+	+	-	-
Esit Eket	NE	-	-	+	+	+	+	-

R^2 = 1.00, R = 1.00, dependent variable: temperature, NE = no effect.

Cl^-, CO in the air increase, the vapour pressure of the area also increases. The coefficient of H_2S, SO_2, NH_3, SPM are negative showing a decrease in vapour pressure as their volume increases. In Eket, the result indicates that the correlation between the dependent variable of vapour pressure and the independent variables of H_2S, Cl^-, SO_2, HCN, NH_3, SPM and CO is 1.00. The R^2 value of 1.00 indicates that degree of variance between the dependent and the independent variables is 100%. It therefore implies that changes in vapour pressure in the study area are highly accounted for by H_2S, Cl^-, SO_2, HCN, NH_3, SPM and CO. The result shows that the coefficient of H_2S, SO_2, HCN, NH_3, SPM are negative; this means that they are negatively related to vapour pressure. As such when their volume of in the air increases the vapour pressure will drop.

The coefficient of Cl^- and CO is positive, indicating that a positive relationship exists between them and vapour pressure. Their increase brings about an increase in vapour pressure of the area. In Ibeno, It could be observed that the R-value is 1.00. It shows that there exists a positive correlation between the dependent variable of vapour pressure and the independent variables of NO_2, H_2S Cl^-, SO_2, SPM, HCN and CO. Also the R^2 value of 1.00 revealed that, the percentage of variance of the dependent variable accounted for the independent variable which is 100%. It implies that vapour pressure in Ibeno is 100% caused by NO_2, H_2S, Cl^-, SO_2, SPM, HCN and CO. The coefficients of NO_2, Cl^-, SO_2, SPM, HCN are negative showing that they are negatively related to vapour pressure in this research area while H_2S, Cl^-, CO are positively related to vapour pressure. It means that the vapour pressure of the area increases as H_2S, Cl^-, CO in the air increases in this location. In Esit Eket, the results revealed that the R-value is 1.00. It shows that there exists a positive correlation between the dependent variable of vapour pressure and the independent variables of H_2S Cl^-, SO_2 HCN, NH_3 SPM and CO. Also, the R^2 value of 1.00 revealed that, the percentage of variance of the dependent variable accounted for by the independent variable is 100%. It implies that vapour pressure in Esit Eket is 100% caused by H_2S Cl^-, SO_2, HCN, NH_3 SPM and CO. The coefficient of H_2S, HCN, NH_3 and CO are negative, this imply that there is a negative relationship between them and vapour pressure (Table 4 and 6). It shows that when their volume in air increases the tendency for vapour pressure to increase is low. The coefficient of Cl^- , SO_2, SPM are positive showing that vapour pressure will likely increase as Cl^-, SO_2, SPM in air in this study location increases.

The result in Table 11 revealed that the correlation between of the dependent variable of temperature and the independent variables of NO_2, H_2S, Cl^-, SO_2, NH_3, SPM and CO is 1.00. In Umudike, the R^2 value of 100 indicates that degree of variance between the dependent and the independent variables is 100%. This implies that a temperature change in this area is highly associated with NO_2, H_2S, Cl^-, SO_2, NH_3 SPM and CO. From the result, it could also be seen that the coefficient of NO_2 and SPM is negative showing that temperature changes in Umuahia will increase when the volume of NO_2 in the atmosphere reduces. The coefficients of H_2S, Cl^-, SO_2, NH_3 and CO are positive indicating a positive relationship between them and temperature. It therefore implies that when the volume of H_2S, Cl^-, SO_2, NH_3 and CO increases, the temperature of the area under study will also increase. In Eket, the result in Table 11 shows that the correlation between of the dependent variable of temperature and the independent variables of H_2S, Cl^-, SO_2, HCN, NH_3, SPM and CO is 1.00. The R^2 value of 100 indicates that degree of variance between the dependent and the independent variables is 100%. This implies that a temperature change in this area is highly associated with H_2S, Cl^-, SO_2, HCN, NH_3 SPM and CO. The coefficients of H_2S, Cl^-, SPM and CO are negative, indicating a negative relationship between them and temperature. The coefficient of SO_2, HCN and NH_3 are positive indicating that a positive relationship exists between their presence in air and the temperature changes in Eket. Increase in the volume of these substances in air will certainly increase the temperature of the area under study. In Ibeno, It could be observed from Table 11 that the R-value is 1.00. It shows that there exists a positive correlation between the dependent variable of temperature and the independent variables of NO_2, H_2S Cl^-, SO_2, SPM, HCN, CO. Also, the R^2 value of 1.00 revealed that, the percentage of variance of the dependent variable accounted for by the independent variables is 100%. It implies that temperature in Ibeno is 100% influenced by NO_2, H_2S, Cl^-, SO_2, SPM, HCN and CO. The coefficients of NO_2, SO_2 and H_2S are positive

Table 12. Model summary for wind speed.

Station	NO_2	H_2S	SO_2	HCN	SPM	Cl^-	CO	NH_3
Umudike	+	-	-	NE	-	+	-	+
Eket	NE	-	+	+	-	-	+	-
Ibeno	+	+	-	-	-	+	-	NE

R^2 = 1.00, R = 1.00, dependent variable: wind speed, NE = no effect.

Table 13. Model summary for rainfall.

Station	NO_2	H_2S	SO_2	HCN	SPM	Cl^-	CO	NH_3
Umudike	+	-	-	NE	-	-	+	+
Eket	NE	-	+	+	-	-	+	-
Ibeno	+	+	+	-	-	-	-	NE
Onna	NE	+	+	+	-	+	-	-
Esit Eket	NE	-	-	+	+	+	+	-

R^2 = 1.00, R = 1.00, Dependent variable: rainfall, NE = no effect.

indicating that there is a positive relationship between NO_2, SO_2, H_2S and temperature. It shows that when the volume of NO_2, SO_2 and H_2S in air increases the temperature of the study location will increase. The coefficient of Cl^-, SPM, HCN, CO, are negative showing an inverse relationship between them and temperature. In Table 11, the result in Onna shows that the R-value is 1.00. It shows that there exists a positive correlation between the dependent variable of temperature and the independent variables of H_2S Cl^-, SO_2 HCN, NH_3 SPM and CO. Also, the R^2 value of 1.00 revealed that, the percentage of variance of the dependent variable accounted for by the independent variable is 100%. It implies that a temperature change in Onna is 100% influenced by H_2S Cl^-, SO_2 HCN, NH_3 SPM and CO. The coefficients of H_2S, HCN and CO are positive indicating that there is a positive relationship between then and temperature. It shows that when the volume of H_2S, HCN and CO in air increases, temperature will increase. The coefficient of Cl^- SO_2, NH_3, SPM are negative showing that temperature will likely increase as their volume in air decreases. In Table 11, the result of Esit Eket revealed that the R-value is 1.00. This shows that there exists a positive correlation between the dependent variable of temperature and the independent variables of H_2S, SO_2, Cl^-, NH_3 HCN, SPM and CO. Also, the R^2 value of 1.00 revealed that, the percentage of variance of the dependent variable accounted for by the independent variable is 100%. It implies that temperature in Esit Eket is 100% caused by H_2S, SO_2, Cl^-, NH_3 HCN, SPM and CO. The coefficients of H_2S, Cl^-, NH_3 HCN and CO are negative indicating that there is a negative relationship between them and temperature in this area. It shows that when their volume of in air increases, the temperature of the area will decrease. The coefficient of SO_2 and SPM is positive, showing that temperatures will likely increase as their volume in air increases.

For humidity, the result in Table 10 for Umudike revealed that the correlation between of the dependent variable of humidity and the independent variables of H_2S, Cl^-, SO_2, HCN, SPM and CO is 1. This shows a strong positive correlation between the dependent variable and the independent variables. The R^2 value of 1.00 indicates that degree of variance between the dependent and the independent variable is 100%. The coefficients of NO_2, H_2S, SO_2, SPM and CO in this equation are positive, showing that there exist a positive relationship between them and humidity. It implies that when their volume increases, humidity will also increase. Other parameters have negative effect on humidity in Umudike as seen in Table 13. In Eket, the R coefficient of 1.00 shows that the correlation between the dependent variable of humidity and the impendent variables of H_2S, Cl^-, SO_2, HCN, NH_3, SPM and CO is one (1). This implies a strong positive relationship between the dependent variable and the independent variables. The R^2 coefficient of 1.00 revealed that proportion of variance between the dependent variable and the independent variable is 100%. It implies that humidity changes in the study are is highly caused by H_2S, Cl^-, SO_2, HCN, NH_3, SPM and CO. From the equation, it could be observed that the coefficient of H_2S Cl^-, SPM and CO is positive which imply that when their volume of air increases, the humidity of the area will certainly increase. The coefficient SO_2, NH_3, and HCN are all negative, indicating a negative relationship between them and humidity in the research area. In Ibeno, It could be observed from Table 10 that the R-value is 1.00. It shows that there exists a positive correlation between the dependent variable of humidity and the independent variables of NO_2 H_2S Cl^-, SO_2, SPM, HCN and CO. Also, the R^2 value of 1.00 revealed that, the percentage of variance of the dependent variable accounted for by the independent variable is 100%. It implies that humidity in Ibeno is 100% caused by NO_2 H_2S Cl^-, SO_2, SPM, HCN, CO. The coefficients of NO_2, H_2S and SO_2 are positive, indicating that there is a positive relationship between them and humidity. It shows that when the volume of NO_2, H_2S and SO_2 in air increases, the humidity will increase. The coefficient of HCN, SPM, CO and Cl^-, are negative showing an inverse relationship between them and humidity. In Onna, it could be observed that the R-value is 1.00. It shows that there exists a positive correlation between the dependent variable of humidity and the independent variables of H_2S, Cl^-, SO_2, HCN, NH_3, SPM and CO. Also, the R^2 value of 1.00 revealed that, the percentage of variance of the dependent variable accounted for by the independent variable is 100%. It implies that humidity in Onna is 100% influenced by H_2S, Cl^-, SO_2, HCN, NH_3 SPM and CO. The coefficients of H_2S, HCN, NH_3 and CO are negative indicating that there is a negative relationship between them and humidity in this study area. The coefficient of Cl^-, SO_2 and SPM are positive showing that humidity will likely increase as there is an increase in their volume in air. In Table 10, the result of Esit Eket revealed that the R-value is 1.00. This shows that there exists a positive correlation between the dependent variable of humidity and the independent variables of H_2S, Cl^-, SO_2, HCN, NH_3 SPM and CO. Also, the R^2 value of 1.00 revealed that, the percentage of variance of the dependent variable accounted for by the independent variable is 100%. The coefficient of H_2S, SO_2, NH_3 in this equation are negative showing that there exists a negative relationship between them and humidity in the research area. The coefficients of Cl^-, HCN, SPM and CO are positive showing that humidity will likely increase as their volume increases in the air.

From Table 12, it could be seen that our R value is 1.00 for wind speed in Umudike. This indicates that the correlation between the dependent variable of wind speed and the independent variables of NO_2, H_2S Cl^-, SO_2, NH_3, SPM and CO is 1.00. Also, the R^2 of 1.00 shows that the independent variables of NO_2, H_2S, Cl^-, SO_2, SPM, HCN and CO account for 100% of any variation in wind speed changes in the study area. The coefficients of NO_2, Cl^- and NH_3 are positive, this implies that wind speed in this area will increases as the volume of these compounds increases in air. The coefficient of SO_2, SPM, H_2S and CO are negative showing that wind speed of the area will increases as the volume of SO_2, SPM, Cl^- and CO decrease from atmosphere. In Eket, from Table 12, it could be seen that our R value is 1.00. This indicates that the correlation between the dependent variable of wind speed and the independent variables of H_2S, Cl^-, SO_2, HCN, NH_3, SPM, and CO is 1.00. Also the R^2 of 1.00 shows that the independent variables of H_2S, Cl^-, SO_2, SPM, HCN and CO account for 100% of any variation in wind speed changes in the study area. The coefficients of SO_2, HCN and CO are positive; this implies that wind speed in this area will increase as the volume of these compounds increases in air. The coefficient of H_2S, SPM, Cl^- and NH_3 are negative showing that wind speed of the area will increases as the volume of H_2S, SPM, Cl^- and NH_3 decrease from atmosphere. In Ibeno, from Table 12, it could be seen that our R value is 1.00. This indicates that the correlation between the dependent variable of wind speed and the independent variables of NO_2, H_2S Cl^-, SO_2, SPM, HCN and CO is 1.00. Also, the R^2 of 1.00 shows that the independent variables of NO_2, H_2S Cl^-, SO_2, SPM, HCN and CO account for 100% of any variation in wind speed changes in the study area. The coefficients of NO_2, H_2S and Cl are positive, this implies that wind speed in this area will increases as the volume of these compounds increases in air. The coefficient of SO_2, SPM, HCN and CO are negative showing that wind speed of the area will decreases as the volume of SO_2, SPM, HCN and CO increases in the atmosphere.

From Table 13, the result shows that the correlation between the dependent variable of rainfall and the independent variables of NO_2, H_2S, Cl^-, SO_2, NH_3, SPM and CO is 1.00 in Umudike. The R^2 value of 1.00 indicates that degree of variance between the dependent and the independent variable is 100%. This implies that the amount of rainfall in the area is seriously determined by the quantity of NO_2, H_2S, Cl^-, SO_2, NH_3, SPM and CO present in the atmosphere. From the result, it could also be seen that the coefficient of NO_2 is positive showing that as the volume of nitrogen dioxide (NO_2) increases in air, the probability that rainfall will increase is very high. Also, from Table 13, the coefficient of H_2S is negative revealing that rainfall in Umudike is negatively associated with the volume of H_2S present in the atmosphere. It therefore means that as the amount of H_2S increases in the atmosphere, the tendency for the rain to fall is low. Also, the coefficient of Cl^- in this equation is negative, indicating that, an inverse relationship exists between the presence of Cl^- in air and the amount of rainfall in the study area. It therefore implies that when the volume of Cl^- increases in the atmosphere, the rainfall of the area under study will decreases. The coefficient of SO_2 is negative showing a negative relationship between the presence of SO_2 in air and the rainfall in Umudike. It could also be seen from the equation that the coefficient of NH_3 is positive indicating a positive relationship between NH_3 and rainfall of the area under study. This shows that when the volume of NH_3 increases in air, the likely-hood for the area to experience rainfall reduces. The coefficient of SPM in this equation is negative, implying that there exist a negative relationship between SPM and rainfall. It shows that when the volume of SPM increases, the tendency for rain to fall in the area under study decreases. The last variable in this equation that affect rainfall is CO. The coefficient of CO in this equation is positive indicating a positive relationship between the presence of CO in air and rainfall. From the result, it could be seen that rainfall in the study area is highly associated with the amount the volume of CO present in air which means that as CO increases rainfall will also increase. In Table 13, the result in Eket revealed that the correlation between of the dependent variable rainfall and the independent variables of H_2S, Cl^-, SO_2, HCN, NH_3, SPM and CO is 1.00. The R^2 value of 1.00 indicates that degree of variance between the dependent and the independent variables is 100%. The result shows that the coefficient of H_2S, Cl^- and NH_3 is negative; this means that their increase does not cause rainfall in the study area. The coefficients of SO_2, HCN and CO are positive, indicating that a positive relationship exists between them and the rainfall in Eket.

It could be observed from Table 13 that the R-value is 1.00 in Ibeno. It shows that there exists a positive correlation between the dependent variable of rainfall and the independent variables of NO_2, H_2S Cl^-, SO_2, SPM, HCN, CO. Also, the R^2 value of 1.00 revealed that, the percentage of variance of the dependent variable accounted for by the independent variable is 100%. The coefficient of NO_2, H_2S, SO_2 and Cl^- are positive indicating that there is a positive relationship between them and rainfall in Ibeno. It shows that when the volume of NO_2, H_2S, SO, and Cl^- in air increases, the probability that rain will fall in this area also increased. It implies that rainfall in Ibeno is 100% caused by H_2S Cl^-, SO_2 and NO_2. From Table 13, it could be observed that the R-value is 1.00 in Onna. It shows that there exists a positive correlation between the dependent variable of rainfall and the independent variables of H_2S Cl^-, SO_2 HCN, NH_3 SPM and CO. Also, the R^2 value of 1.00 revealed that, the percentage of variance of the dependent variable accounted for by the independent variable is 100%. It implies that rainfall in Onna is 100% influenced of by the present H_2S, Cl^-, SO_2 and HCN in air. The coefficient of NH_3, SPM, CO, CO_2 in this equation is negative showing that there exists a negative relationship between them and rainfall in the research area. In Esit Eket, the result in Table 13 revealed that the R-value is 1.00. This shows that there exists a positive correlation between the dependent variable of rainfall and the independent variables of H_2S, SO_2, Cl^-, HCN, NH_3, SPM and CO. Also, the R^2 value of 1.00 revealed that, the percentage of variance of the dependent variable accounted for by the independent variable is 100%. The coefficients of Cl-, HCN, SPM and CO are positive showing that rainfalls will likely increase as there is increases in its volume in air. It implies that rainfall in Esit Eket is 100% caused by Cl^-, HCN, SPM and CO. The coefficient of H_2S, SO2, NH3 are negative indicating that there is a negative relationship between them and rainfall. It shows that when their volume in air increases, it does not increase rainfall in this location.

Conclusion

Mean concentration of air parameters studied were below FEPA and USEPA National air quality standards with exception of carbon monoxide which exceeds the limit of 35 ppm in March at Ibeno. It was also established that a positive correlation exits between NO_2, H_2S, SO_2, SPM, chloride, carbon monoxide, and wind speed relative humidity, temperature and vapour pressure of the study locations. Chloride and nitrate also have a positive relation-ship with wind speed in some of the study locations.

Conflict of Interest

The authors did not declare any conflict of interest.

REFERENCES

Anderson I (2005). Niger River Basin: A Vision for Sustainable Development. The World Bank, pp: 131.

Abdul Raheem AMO, FA Adekola, IB Obioh (2009). Monitoring Of Sulfur Dioxide In The Guinea Savanna Zone Of Nigeria: Implications Of The Atmospheric Photochemistry. Bull. Chem. Soc. Ethiop. 2009, 23(3):001-008.

Ettouney RS, Zaki JG, El-Rifiai MA, Ettouney HM (2010), An Assessment of the Air Pollution Data from Two Monitoring Stations in Kuwait. Toxicol. Environ. Chem. 92(4):655-668.

Kalabokas PO, Viras LG, Repapis C (1999). "Analysis of the 11 – years record (1987 - 1997) of Air Pollution Measurements in Athens, Greece, Part I: Primary Air Pollutants", Global Nest Journal. 1(3):157-167.

Okunola, OJ, Uzairu A, Gimba CE, Ndukwe GI (2012). Assessment of Gaseous Pollutant along High Traffic Roads in Kano, Nigeria. Int. J. Environ. Sust. 1(1):1-15.

Aliyu YA, Musa IJ, Youngu TT (2013). Appraisal of Sulphur Contaminants from Transportation in Urban Zaria, Nigeria. Academic Journal of Interdisciplinary Studies MCSER Publishing, Rome-Italy. 2(10):155-163.

Tawari CC, Abowei JFN (2012). Air Pollution in the Niger Delta Area of Nigeria. Int. J. Fish. Aquatic Sci. 1(2):94-117.

Mabahwi Nurul Ashikin, Oliver Ling Hoon Leh, Dasimah Omar (2014). Procedia - Social and Behavioral Sciences. 1- 5

Babatunde Saheed Bada1, Kofoworola Amudat Olatunde, Adeola Oluwajana (2013). Air quality assessment in the vicinity of cement company. Int. Res. J. Nat. Sci. 1(2):34-42, June 2013.

Review of methods for modelling forest fire risk and hazard

I. Yakubu[1] , D. Mireku-Gyimah[1] and A. A. Duker[2]

[1]University of Mines and Technology, Tarkwa, Ghana.
[2]Kwame Nkrumah University of Science and Technology, Kumasi, Ghana.

Forest fires (wildfires) have become a major concern for several environmental experts. Assessment of fire effects at local scale is increasingly considered a critical aspect of ecosystem functioning, since fire plays a crucial role in vegetation composition, biodiversity, soil erosion and the hydrological cycle. At global scale, fire is the most generalized means of transforming tropical forest in agricultural areas, and it has severe impacts on global atmospheric chemistry. Fire is a natural factor in many climates with high levels of vegetation stress. However, changes in traditional land use such as hunting, charcoal production, inefficient logging practices and rural abandonment patterns, which have been identified as major causes of wild fires, have recently modified the incidence of fire. Several assessment techniques and methods have been developed to help model and evaluate forest fire risk and hazard. There is the need to identify a method or combination of methods to help model forest fire risk and hazard to enable the sustainability of the natural resources. In this paper, the various methods are reviewed in order to enhance the use of appropriate method(s) for forest fire risk and hazard management. From the review and deductions of the methods, it was concluded that spatial multi-criteria modelling and evaluation (SMCME) of fire risk and hazard is preferred. It was also deduced that combination of SMCME with other methods has proven to be more efficient and effective when compared with the use of individual methods.

Key words: Forest fires, risk, hazard, management.

INTRODUCTION

Wildfires are inevitable companions of forests and foresters across the world and its spread revolves around four main factors: (i) the state and nature of the fuel, that is, proportion of live or dead vegetation, compactness, morphology, species, density, stratification and moisture content (ii) the physical environment, that is, weather conditions and topography (iii) causal factors (human-or natural-related) and (iv) means of prevention and suppression. Fire hazard is defined by both (i) and (ii) and has two types of variations: a spatial and long-term one, related to fuel types and topography and a temporal and short-term one, related to fuel moisture content and weather conditions. Fire risk accounts for (iii) and (iv) (Blanc et al., 1987; Chuvieco and Martin, 1994). Wildfires are considered as a serious problem that distresses many terrestrial ecosystems in the Earth system and causes economic damage to people such as missing income relative to the land use, destruction and loss of property, damages to agriculture, and loss of biodiversity. It is also one of the most important parts of land degradation that is caused by deforestation and desertification (Hernandez-Leal et al., 2006).

Stolle and Lambin (2003) noted that flammable fuel depends on climatic conditions, soil, vegetation and previous fire events. The ignition source is natural (for example lightening) or anthropogenic. If the ignition source is anthro-pogenic, it can be caused deliberately (as part of land management) or accidently through negligence. Preventing a small fraction of these fires would account for significant savings in the natural and human resources. Apart from preventive measures, early detection and suppression of fires is the only way to minimise the damage and casualties. Systems for early detection of forest fires have evolved over the past decades based on advances in related technologies. Wildfire is a paradox, it kills plants and animals and can cause wide-ranging damages to the ecosystem. On the other hand, it can be very beneficial in terms of nutrient recycling and forest regeneration. In some areas, natural wildfires have historically adapted with ecologically positive effects. Other ecosystems are susceptible to severe damages, causing a local extinction of species or considerable changes in ecosystem functions (e.g. soil, hydrology). Integrated modelling approaches could provide helpful insights into wildfire-environmental interactions. Globally, the majority of wildfires are caused by human activities in a direct or indirect form. An anthropogenic influenced wildfire regime (frequency, distribution) will potentially affect human activities. This inter-relationship between humans and wildfires has initiated many scientific studies.

Millington et al. (2008) and Schweitzer and Priess (2010) mentioned in their study (the presentation of an agent-based approach: Simulated land-use management influencing wildfire risk that only a few models exist which consider human activities and the interactions with vegetation-wildfire dynamics. In the paper, the authors presented preliminary results from a modelling approach which captures the wildfire behaviour in Northern Mongolia. The approach aimed at analysing impacts of wildfires on the socio-environment, including feedbacks related to carbon dynamics, biomass availability (in forests and grasslands) and the effects on land use. Therefore newly developed wildfire module on the basis of a well-established wildfire model linked to dynamic land-use model, which integrates new model capabilities of simulating wildfire spread and intensity would be appropriate.

Wildfires play an important role in terrestrial ecosystems, global biogeochemical cycles and climate. They are biological filter, regulator (Bowman et al., 2009; Jin, 2010) and global vegetation consumer (Bond and Keeley, 2005; Jin, 2010). Wildfires influence ecosystems directly by disturbing competition relations between and within species and by accelerating the carbon cycle, nutrient cycle, hydrological cycle and energy cycle (Thonicke et al., 2001; Jin, 2010). They also affect ecosystems indirectly by changing climate. Wildfires favour plants with distinct reproductive and survival strategies in different fire regimes.

Bowman et al. (2009), Bond and Keely (2005) and Jin (2010) argue that fire is another important determinant besides climate in shaping the global biomes distributions. Especially, wildfires have reduced the potential coverage of forest and facilitated the expansion of fire-dependent grassland and shrub land. Their simulations show that forest would at least double in extent in the absence of fire. Wildfires accelerate the natural carbon cycle of primary production and respiration. Regions that have long served as carbon sinks may suddenly become sources of carbon emission due to fires (van der Werf et al., 2004; Jin, 2010). There is, however, the need to detect wild fires and suppress them.

Fire detection and monitoring

Traditional ground-based visual detection methods are not always appropriate for offering reliable information on fire location, size and intensity due to the small field of view and often difficult terrain. Remote sensing has proven to be a valuable data source in different phases of fire management both before (prevention) and after the fire (damage assessment). Remote sensing observation has significant advantages over conventional fire detection and fire monitoring methods because of its repetitive and consistent coverage over large areas of land (Martin et al., 1999). Fire produces four forms of signal that are easily observed from space (Robinson, 1991; Martin et al., 1999). These are direct radiation from active fires (heat and light), smoke, post-fire char, and altered vegetative structure (scar). Fire detection from satellite images initially focused on analysing the first type of signal (Martin et al., 1999). There are a number of satellites and aircraft-borne remote sensing systems which can contribute to fire monitoring from space, including NOAA-AVHRR, Landsat- TM and MSS, SPOT, GOES, DMSP, ERS-ATSR, and JERS. The temporal, spectral and spatial characteristics of these instruments provide a wide range of sensing capabilities and some of them have been shown to be well adapted to fire detection application. NOAA-AVHRR and GOES have provided long-term operational systems, allowing low cost direct reception and near real-time fire information (Martin et al., 1999). The usefulness of operational near real-time fire detection from space is obviously very much dependent on observation frequency. Meteorological satellites are more appropriate because of their high repetition coverage. The geostationary GOES satellite series offer images every 30 min but only covers the American continents. The polar orbiting NOAA-AVHRR series acquire images over the same area every 12 h for the same satellite, but cover the entire world (Martin et al.,

1999). Therefore, NOAA-AVHRR has been used most extensively for detecting and monitoring forest fires. Temporal resolution of AVHRR data may also be used to follow the spatial evolution of large fires, providing significant information for fire behaviour modelling. AVHRR images can provide valuable information because of the possibility of monitoring fire growth at least every 6 h (when using two NOAA satellites, morning and afternoon). Coarse spatial resolution of AVHRR data restrict this potential to large fires, whose size and duration are enough to be followed in time series of AVHRR image data (Martin et al., 1999).

The applications of GIS to fire risk modelling have considered a wide range of hazard variables, depending on the specific characteristics of fire events in the different test sites. Nevertheless it can be summarised into several important variables, such as topography (elevation, slope, aspect and illumination), vegetation (fuel type, moisture content), weather patterns (temperature, relative humidity, wind and precipitation), accessibility to roads and camping sites, land property type, distance to cities, soils, fire history and water availability.

Thus, this paper seeks to review the various methods that are being used to model and evaluate forest fire risk and hazard in order to enhance the appropriate adoption of method(s) for effective and efficient prevention, control and complete elimination of wildfires in the fragile ecosystem.

METHODS FOR FIRE RISK AND HAZARD MODELLING

To model and evaluate fire risk and hazard, there is the need for proper fire risk assessment. The following sections look at factors influencing fire behaviour, the risk assessment methods, modelling of fire risk and hazard methods.

Factors influencing fire behaviour

The factors influencing fire behaviour can either be natural or man-induced. Fire behaviour is a descriptive term used to designate what fire does and how it behaves. It estimates what a fire will do and relates to intensity, flame and rate of spread of specific fire. A product of environmental factors which interact with each other includes fuel, topography, weather and fire. The intensity and speed with which a fire travels depends on the amount and arrangement of the fine dead fuel, moisture content of the dead fuel, wind speed near the flaming zone, terrain and slope (Gould, 2005). The behaviour of a spreading fire is determined by factors such as weather, topography, fuel quantity and fuel moisture content. Countryman (1972) in Pyne et al. (1996) presented the concept of the fire environment- the surrounding conditions, influences, and modifying forces that determine the behaviour of a fire. Topography, fuel, weather and the fire itself are the interacting influences that make up the fire environment. This is illustrated as a fire environment triangle with the fire in the centre.

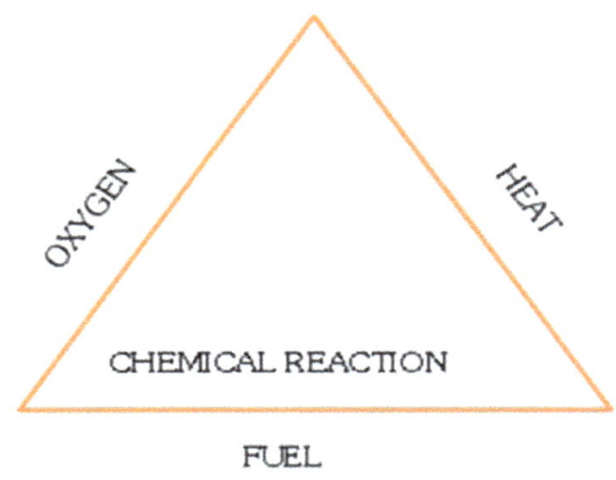

Figure 1. Fire triangle.

The changing states of each of the environmental components; fuel, topography and weather and their interaction with each other as well as the fire itself determine the characteristics and behaviour of a fire at any given moment. Changes in fire behaviour in space and time occur in relation to changes in the environmental components. From a wildland fire standpoint, topography does not vary with time, but can vary greatly in space. The fuel component varies in both space and time. Weather is the most variable component, changing rapidly in both space and time (Pyne et al., 1996). Figure 1 is the fire triangle. The fire triangle or combustion triangle is a simple model for understanding the necessary ingredients for most fires. The triangle illustrates the three elements a fire needs to ignite: heat, fuel and an oxidising agent (usually oxygen). A fire naturally occurs when the elements are present and combined in the right mixture. A fire can be prevented or extinguished by removing any one of the elements in the fire triangle.

Natural factors for fire ignition

The following are present details of some natural factors that contribute to fire ignition and propagation. The natural factors are topography, vegetation, fuel as well as weather.

Topography

Topography includes the elements of slope steepness, aspect, elevation and configuration of the land. Variations in topography can cause dramatic changes in fire behaviour as a fire progress over the terrain. Although topography may not change in time, it affects the way in which fuel and weather change. The fire environment triangle symbolises this interaction among the elements. Topography modifies general weather patterns, producing localised weather conditions that in turn affect fuel type and moisture content (Vadrevu et al., 2009; Pyne et al., 1996). For the past several years, fire behaviour models have incorporated the interaction of fire spread with fuels, weather and terrain (Vadrevu et al., 2009; Albini, 1976; Rothermel, 1983). The effect of terrain attributes on forest survival following wildfire has been assessed by Kushla and Ripple (1997) and others. Four different topographic parameters explained below, as causative factors of fires were elevation, slope, aspect, and a compound topographic index (Vadrevu et al., 2009).

Elevation

Elevation is considered as an influencing factor of forest fire, because elevation relates to precipitation and temperature. In general, with the increasing of elevation, precipitation usually increases. Therefore, the probability of fire is less in areas of higher elevation. In some specific situations however, precipitation does not follow this rule. A similar use of elevation factor for forest fire estimation is performed by Chuvieco and Congalton (1989). For temperature, higher elevation leads to lower temperature, which also means there will be lower probability for fire to appear in higher elevation area. Both precipitation and temperature are influenced by elevation and the effects on the forest fire are the same, so they can be considered together.

Elevation above sea level influences general climate and thereby affects fuel availability. Length of fire season and fuel vary with elevation due to differences in amount of precipitation received, snow melt dates, green up and curing dates (Pyne et al., 1996). Elevation is an important physiographic factor that is related to wind behaviour and hence affects fire proneness (Rothermel, 1983). Fire travels most rapidly up-slope and least rapidly down-slope.

Aspect

Aspect is the direction a slope is facing. Aspect affects fire behaviour through variations in the amount of solar radiation and wind that different aspects receive. Generally, in the northern hemisphere, south and southwest aspects are most favourable for fire to start and spread (Pyne et al., 1996). These aspects receive more sunshine and therefore have lower humidity and higher fuel temperatures (Pyne et al., 1996). Solar radiation intensity is greatest when the slope is perpendicular to the sun angle. In the northern hemisphere, fuels on slopes with an easterly aspect will dry out earlier in the day, but may not become as dry as those on slopes with a westerly aspect (Pyne et al., 1996). It can be deduced that the slope which faces the wind direction is easier to cause raging fire. A north-facing slope also receives less sunlight than a south facing slope. Thus, Southern aspects receive more direct heat from the sun, drying both the soil and the vegetation.

Slope

Slope is an extremely important factor among topographic factors. Slope has a large effect on the speed of fire when it is spreading. Kushla and Ripple (1997), mention that fire always spreads faster up-slope than down-slope. Slope steepness also affects the radiation intensity and fuel moisture. The slopes where the fuel will be the driest vary with time of year, time of day, and latitude. Thus, as a fire moves over the landscape its behaviour can be expected to change with time of day and topographic characteristics because of the variations brought about by the different amounts and intensity of the solar radiation received (Pyne et al., 1996). Slope significantly influences the forward rate of spread of surface fires by modifying the degree of preheating of the unburnt fuel immediately in front of the flames. In a head fire, this is achieved, as with wind; by changing the flames to a very acute angle and with slopes exceeding 15 - 20°C, the flame propagation process involves almost continuous flame contact (Trollope et al., 2002). Conversely a down slope decreases the rate of spread of surface head fires (Trollope et al., 2002). Steep slopes increase the speed of fire a lot, because convective preheating and ignition rate are more effective. In other words, larger slope of the terrain will lead to larger probability of causing fire.

Vegetation and fuel

Vegetation type has a strong relationship with the forest fire risk. Different types of vegetation have different kinds of combustibility. Generally, coniferous forest has a higher probability for fire risk than deciduous forest, because coniferous trees contain less water and higher oiliness (Li, 1998).

Fuel is a critical leg in both of the fire triangles: fuel, oxygen, and heat of the fire fundamental triangle; as well as fuel, topography and weather of the fire environment triangle. Fuel does not cause fire, but it certainly changes the character of a fire, affecting the ease of ignition as well as fire size and intensity (Pyne et al., 1996). Fuel can be described in terms of both fuel state and fuel type.

Fuel state refers to the moisture content of the fuel and whether it is alive or dead. Fuel type is a description of the fuel itself. The description of fuel type includes physical properties of fuel, fuel component and fuel complexes. Fuel properties that affect the way the material burns include quantity, size, compactness and arrangement. Fuel components, which are related to the way vegetation grows may be specified as ground, surface and crown fuel as well as grass, litter, brush, or over story. Fuel complexes, which are associations of components include grass and timber with grass and litter understory (Pyne et al., 1996). Moisture content, expressed as a fraction, is the mass of water held by unit mass of oven dry fuel and is determined primarily by fuel type and weather. It may also be expressed as a percentage of the fuel oven dry weight. Fuel moisture is normally expressed on a dry matter basis and is a critical factor in determining the intensity of a fire because it affects the ease of ignition, the quantity of fuel consumed and the combustion rate of the different types of fuel. The most important influence of fuel moisture on fire behaviour is the smothering effect of the water vapour released from the burning fuel. It reduces the amount of oxygen in the immediate proximity of the burning plant material thus decreasing the rate of combustion. Fuel load is regarded as one of the most important factors influencing fire behaviour because the total amount of heat energy available for release during a fire is related to the quantity of fuel. Assuming a constant heat yield, the intensity of a fire is directly proportional to the amount of fuel available for combustion at any given rate of spread of the fire front.

Weather

Forest fires are strongly linked to weather and climate (Flannigan and Wotton, 2001). Weather is one of the most important factors affecting the behaviour of a fire. The most important components of weather affecting the behaviour of a fire are air temperature, relative humidity and wind speed.

Air temperature plays an important role in fire behaviour. Its direct effect is to influence the temperature of the fuel and therefore the quantity of heat energy required to raise it to its ignition point. Air temperature also has indirect effects via its influence on the relative humidity of the atmosphere and moisture losses by evaporation. Research in South Africa indicated that air temperature had a highly significant positive effect on the intensity of fires in African grasslands and savannas. The relative humidity of the atmosphere influences the moisture content of the fuel when it is fully cured.

It is positively correlated with fuel moisture and therefore plays an important role in controlling the flammability of fine fuels. The combustion rate of a fire is positively influenced by the rate of oxygen supply to the fire (Brown and Davis, 1973; Trollope et al., 2002) hence the effect of wind speed on fire behaviour. Wind also causes the angle of the flames to become more acute. With increased wind velocities, the flames are forced into the unburned material ahead of the fire front resulting in more efficient preheating of the fuel and greater rates of spread in surface head fires (Luke and McArthur, 1978; Cheney, 1981; Trollope et al., 2002). It was stated that increased wind speeds cause greater rates of spread and therefore more intense fires (Brown and Davis, 1973; Luke and McArthur, 1978; Trollope et al., 2002). However, flame height does not necessarily increase with increased wind speeds because these cause the flames to assume a more acute angle and this may prevent the ignition of aerial fuels.

Human factors for fire ignition

There is no fire without a cause. The factors necessary for fires to occur are presence of flammable fuel and an ignition source. The ignition source of fires can be natural causes (for example lightening) or human causes (anthropogenic) (Stolle and Lambin, 2003). In general, natural causes do not seem to be of great interest for the wildfire research community. Most of the fires are caused by human activities and that is where most of the research emphasis is laid on (Stolle and Lambin, 2003). Forest fires, for example, in Ghana are usually human-induced. Human causes of fire can be categorised as direct cause of setting up fires as a method or tool for land clearing (Tomich et al., 1998; Applegate et al., 2001; Santoso, 2006), traditional beliefs and indirect cause of fire by human activities that favour the occurrences and potentially increase the risk of fires such as logging, road development, resettlement, etc. (Santoso, 2006). In the Indonesian context, development can be often equated to opening access to large tracts of sparsely populated forest land for other uses (Bowen et al., 2000). Initially, primary forests are humid, have a closed canopy with little undergrowth and have a low fire risk. However, where roads penetrate primary forest to aid logging, humidity become lower, wind speeds increase, and there is always a ready supply of drier fuel available, therefore the risk to fire is considerably increased (Nicolas and Beebe, 1999). The agro-industrial crop arrives and takes over the area once the forests have been logged. Ideally the new estate crop is established at the start of the next rainy season to allow it to gain a rapid control of the site and suppress weed growth. But often far greater areas are cleared at one time than planted. In either case, the open ground encourages the rapid colonization by herbs. The area then becomes highly vulnerable to repeat fires in the next dry period (Bowen et al., 2000).

In most parts of Africa, the environment is undergoing rapid desertification due to rampant bush burning for farming and hunting. The effects of these activities on the ecosystem cannot be overemphasised. Bush burning activities generally continue to reduce the flora and fauna around the globe to levels that are destructive to biological diversity. The principal elements of the environment such as vegetation and water have been severely impacted. This has significant implication for farmers' income and food security of the family. It is therefore imperative to consider the risks involved in such bush burning.

Risk assessment

Fire risk assessment should be seen as a specific part of a wider, overall, assessment of the risk to which the ecosystem is exposed and may be part of an overall program of risk reduction. There are three parts to fire risk assessment:

Initial assessment

This involves the identification of the hazards and sizing the risks. After identification of the hazard, one important thing is to decide whether the hazard from fire is important enough to be a source of serious potential harm or in any given situation may cause loss, death, injury or damage. Consideration is made on how likely it is that each hazard could cause harm. This will determine whether or not there is the need to do more to reduce the risk. Even after all precautions have been taken, some risk usually remains. A decision is made for each significant hazard whether the remaining risk requires any control measures.

Risk reduction

Having made the initial assessment there follows the important task of reducing the hazards and risks. It will almost certainly be the case that some reductions may be effected immediately, and these short-term measures would include such things as improving the environmental practices- the management of waste and rubbish, and the implementation of a programme of fire safety training for employees and community members. Other long-term measures would include such things as the installation of a fire suppression system, the change in some negative beliefs and the substitution of hazardous processes and materials with less hazardous ones.

Final risk assessment

When the hazards and risks have been reduced to what, at the time, appears to be an irreducible level, there follows a more rigorous final assessment of the risk. The final assessment will determine the risk categorisation which conventionally will be defined as high, normal or low. Of course in larger premises such as a forest, it will be quite normal to have different risk categories for different parts of the area. The final assessment will have three outcomes: it will determine whether the areas, or parts of it, are to be categorised as being of high, normal, or low risk; this in turn will determine the fire precautionary measures required in the area, and it will be the starting point in the formulation of an emergency plan. In carrying out the risk assessment it will be necessary to have in mind but not limited to the following factors: the living things present in the area, the use to which the area is put, the sources of ignition present, the use of flammable materials, the contents of the area, the structural features of the area, traditional beliefs of the people in the area and fire education level in the area. It is worth mentioning that precautionary measures of people are directly connected to their risk perceptions (Rosenstock et al., 1988). For example; people who expect higher probability of being hurt by fire will tend to take more precautionary measures. Previous studies as captured in Shavit et al. (2013): "The Effect of Fire on Emotions and Risk perception: A field study after the Carmel Forest Fire Disaster" clearly showed factors affecting individuals' risk perception of being hurt by fire.

In large or complex forest area it may simplify the task, and indeed be more appropriate, if all three parts of the exercise are carried out by treating distinct areas such as forest, communities, and water bodies as separate entities.

Risk assessment techniques provide a valuable tool in attempting to categorise the degree and severity of risk to which an organisation, nation or the ecosystem might be liable. While no method is infallible, sensible use of risk assessment and application of the lessons drawn can result in more cost effective introduction of fire protective measures.

Risk assessment methods

There is no single 'correct' way of carrying out risk assessment, there are three methods which might be useful, each of which makes clear what is to be understood by the terms high, normal and low risk. These are:

1. The risk category indicator method: This is a diagnostic method in which the various elements in the area are classified in such a way as to indicate the area in which they are found and should be categorised as being high, normal, or low risk. Elements which may give rise to high risk indicators in the case of forest include: communities; vegetation; wind; topography; road network; and negative traditional beliefs.
2. The risk value matrix method: Unlike the Risk Category Indicator method, this method attempts to put the risk assessment onto a quantitative basis. However, it cannot be strongly stressed that the numbers involved are purely relative, and therefore they have no absolute significance whatsoever. Whilst all risks are made up of two elements- the probability that an event will occur and the consequences of that occurrence, the relative contributions of these two elements to risk may vary considerably. Formula for risk value: Remembering that the two elements of risk are the fire hazard and the fire risk, the risk value is defined by the simple formula:

Risk value = fire hazard value x fire risk value

If the size of the fire hazard and the fire risk is expressed by assigning values to them then, by applying the formula, a number obtain would be a measure of the risk value. The size of the risk value then becomes the basis for categorising the area as being of high, normal or low risk.

Quantifying the fire hazard and the fire risk

This is easily done by: classifying the fire hazards;

Table 1. Classifications of fire risk and hazard.

Fire hazard (description)	Value	Fire risk (description)
Negligible	1	Unlikely
Slight	2	Possible
Moderate	3	Quite possible
Severe	4	Likely
Very Severe	5	Very likely

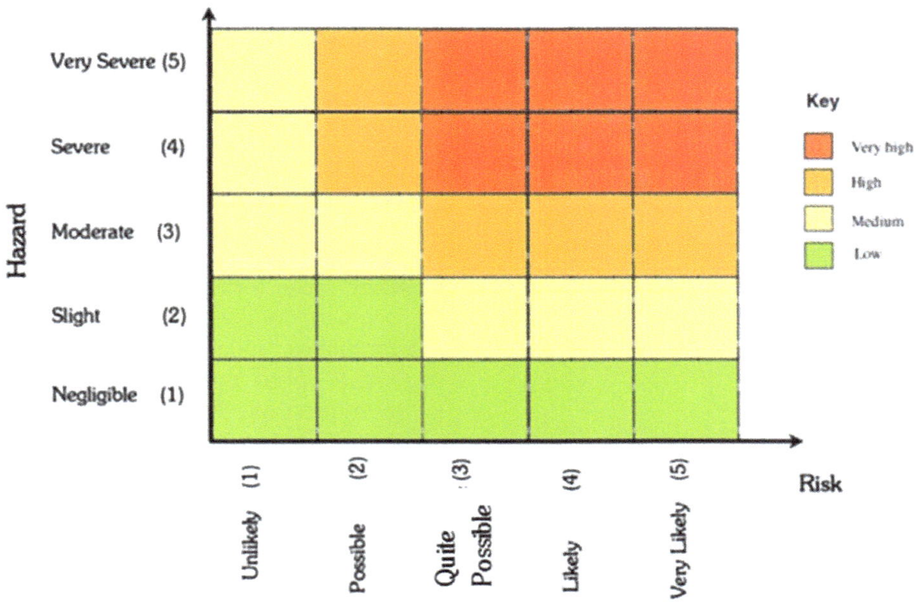

Figure 2. Risk rating matrix (Anon., 2011).

describing them as being between negligible and very severe; and assigning a numerical value to each description. Similarly, it may classify the fire risks by describing them as being between unlikely to very likely, and by assigning a numerical value to each of these descriptions. Table 1 is a classification of fire risk and hazard.

Using the risk value formula for all possible combinations of fire hazard values and fire risk values, a set of twenty five numbers are obtained. The risk values obtained can then be displayed as a two dimensional grid (risk value matrix). Figure 2 shows a risk rating matrix (Anon., 2011). The final task in this method is to decide the ranges of the risk values that will correspond to our three categories of risk. 3. The algorithmic method: An algorithm is a two dimensional diagrammatic representation of the steps to be undertaken in order to make a decision, solve a problem, or carry out a process. In short, it is a flowchart.

METHODS FOR FOREST FIRE RISK MODELLING

Forest fire risk assessment is very important for fire management. It may be considered at different spatial and temporal resolutions: global and local; short term, and long-term fire risk estimation. Global scales can contribute to the establishment of general guidelines for fire management at continental level, while local scales are adapted to specific fire prevention resources of small regions (Chuvieco et al., 1999). Risk should, however, be estimated in order to plan for the necessary resources for fire management.

Short-term fire risk estimation

Short-term estimation of risk is required to take update decisions on fire pre-suppression and suppression activities, which should ideally provide daily estimations of fire risk and it is commonly based on weather data; however, recently satellite information is also being considered (Chuvieco et al., 1999).

The physical basis to estimate fire risk has many similarities in the different ecosystems, the actual formulation varies from one country to another, and therefore a great diversity of indices is available since many of them have been developed primarily for specific geographic area (Marzano et al., 2005). Long term estimation, however, is necessary.

Long-term fire risk estimation

Long-term estimation addresses the general, more permanent, planning of fire fighting resources, which is related to the more structural factors that affect fire ignition or fire propagation, such as topography or terrain characteristics, vegetation structure, human activities or weather patterns. These factors can be considered stable at least during a whole fire season; therefore they do not need to be updated frequently.

Two to five year updates are accurate enough for fire management (Chuvieco et al., 1999). There are some common methods used in long-term fire risk estimation, such as qualitative methods, quantitative methods based on expert knowledge (multi-criteria evaluation techniques), regression techniques (linear regression and logistic regression), and artificial neural networks (Chuvieco et al., 1999; Marzano et al., 2005).

Logistic regression model

One quantitative approach to obtain a fire risk index is to calculate the weights of the different variables using regression analysis techniques such as logistic regression. In the context of fire risk assessment, fire occurrence (usually expressed as number of ignition points/areas or as a proportion of burned area) is the dependent variable, while fire danger variables (slope, fuels, fuel moisture, road network, recreational areas, etc.) are the independent ones. Coefficients of regression become the weights of each danger variable to produce the synthetic fire risk map (Marzano et al., 2005).

Logistic regression is a quite flexible tool, since it accepts the input of a dataset composed of continuous and/or categorical variables as well as non-normally distributed one. Several independent variables can be included in the model. Its main characteristic refers to the binary format of the dependent variable (Chuvieco et al., 1999). Thoha (2006) applied logistic regression on forest fire prediction in peatland areas in Bengkalis, Riau Province.

Multi-criteria evaluation

Multi-criteria evaluation (MCE) is a decision making tool developed for complex multi-criteria problems that include qualitative and/or quantitative aspects of the problem in the decision-making process (Mendoza et al., 1999). The MCE techniques (Chuvieco et al., 1999) may be a good alternative to reduce the subjectivity of this assigning process, since the opinion of experts may be quantitatively assessed. Moreover, each expert's opinion may be weighted according to his/her degree of knowledge in the field of study. The MCE techniques have been used for fire danger mapping, weighting each danger variable after the expert's opinion in two different scenarios (Chuvieco et al., 1999). Multi-criteria analysis can be implemented using analytical hierarchy process (AHP) (Saaty, 1980). The AHP method approaches decision-making by arranging the important components of a problem into a hierarchical structure similar to a family tree. The AHP method reduces complex decisions into a series of simple comparisons, called pairwise comparisons, between elements of the decision hierarchy. By synthesising the results of these comparisons, AHP can provide the best decision and provide a clear rationale for the choice (Mendoza et al., 1999).

Power (2006), stated that "Spatial decision support systems (SDSS) provide a powerful and easy interface to combine cartographic models and other image data to define solutions to unstructured and semi-structured problems. SDSS supports a range of decision-making styles and approaches by generating a series of feasible alternatives through an interactive and recursive process in which decision making proceeds by multiple passes, sometimes involving alternative routes rather than a simple linear path. Examples where SDSS techniques have been used in fire management include Varela et al. (2005), Barrett et al. (1999) and Jones et al. (2004). The basic strategy is to divide the problem into well-defined smaller pieces, analyse each piece separately, and then integrate the pieces logically to produce a solution, following Jankowski's (1995) general framework. Decision criteria are formally evaluated and allocated a score based carry within the decision making process. In many instances this process is achieved using multi-criteria evaluation (MCE). The criteria considered in definition of potential fire hazard are slope, aspect, vegetation and communities".

A comprehensive consideration for fire risk implies taking into account a wide range of variables. A common terminology distinguishes between the concepts of risk associated with the beginning of a fire (fire ignition risk or flammability) and to the spreading of an active fire (fire behaviour risk or fire hazard). In each case, different variables and different risk weights should be considered. However, both approaches require being capable of integrating different spatial variables. GIS provides tools to on expert opinion regarding the weight each criteria will create, transform and combine geo-referenced variables. Therefore, GIS can spatially integrate several hazard variables related to fire risk and provide tools for risk analysis (Chuvieco et al., 1999). The applications of GIS to fire risk modelling have considered a wide range of hazard variables, depending on the specific characteristics of fire events in the different test sites.

Integrated spatial multi-criteria methods for fire risk and hazard modelling

The integrated approach makes extensive use of spatial and non-spatial data for modelling fire risk and hazard.

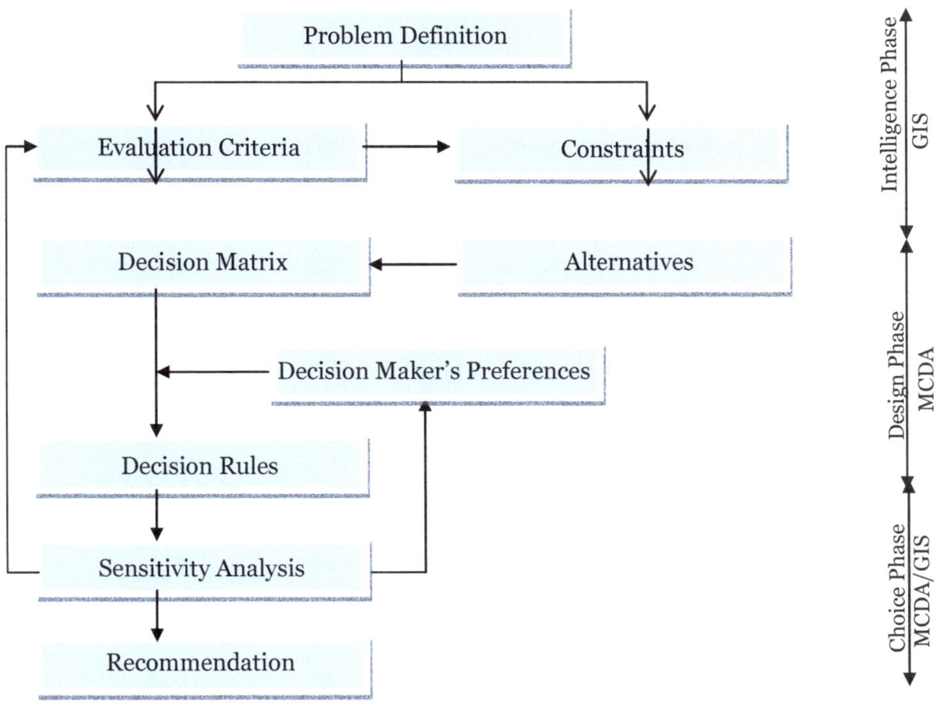

Figure 3. Framework for spatial multi-criteria decision analysis (Malczewski, 1999).

This method combines two or more methods to help in decision analysis of optimal model for fire risk and hazard.

Decision analysis is a set of systematic procedures for analysing complex decision problems. These procedures include dividing the decision problems into smaller and more understandable parts; analysing each part, and integrating the parts in a logical manner to produce a meaningful solution (Figure 3) (Malczewski, 1997). In general, multi-criteria decision analysis (MCDA) problems involve six components (Keeney and Raiffa, 1976; Pitz and McKillip, 1984). These components are as follows:

1. A goal or a set of goals the decision maker wants to achieve;
2. The decision maker or a group of decision makers involved in the decision making process with their preferences with respect to the evaluation criteria;
3. A set of evaluation criteria (objectives and/or physical attributes);
4. The set of decision alternatives;
5. The set of uncontrollable variables or states of nature (decision environment); and
6. The set of outcomes or consequences associated with each alternative attribute pair.

MCDA techniques can be used to identify a single most preferred option, to rank options, to list a limited number of options for subsequent detailed evaluation, or to distinguish acceptable from unacceptable possibilities. There are many MCDA approaches which differ in how they combine and utilise the data. MCDA approaches can be classified on the basis of the major components of multi-criteria decision analysis. Three different classifications can be made. These are:

1. Multi-objective decision making (MODM) versus multi-attribute decision making (MADM);
2. Individual versus group decision maker problems; and
3. Decisions under certainty versus decisions under uncertainty.

The distinction between MADM and MODM is based on the evaluation criteria which are the standards of judgments or rules on which the alternatives are ranked according to their desirability. Criterion is a general term and includes both the concepts of attributes and objectives. An attribute is a measurable quantity whose value reflects the degree to which a particular objective is achieved. An objective is a statement about the desired state of the system under consideration (Chankong and Haimes, 1983). It indicates the directions of improvement of one or more attributes. Objectives are functionality related, or derived from a set of attributes (Malczewski, 1999). There might be formal relationship between objectives and attributes, but usually the relationship is informal. To assign an attribute to a given objective, two properties, which are comprehend-siveness and measurability, should be satisfied. An attribute is comprehensive if its value sufficiently indicates the degree to which the objective is met. And it is measurable if it is reasonably practical to assign a value in a relevant measurement scale. The ratio, interval, ordinal and binary scales are suitable for measuring attributes, whereas nominal scale is not since it does not allow an ordering of the alternatives (Janssen, 1992). MADM problems require that choices be made among alternatives described by their attributes. The set of attributes is given explicitly and multi-attribute problems have a finite set of feasible alternatives. Unlike MADM, MODM problems require that means-ends relationships be specified, since they deal explicitly with the relationship of attributes of alternatives to higher level objectives. MODM involves designing the alternatives and searching for the best decisions among an infinite or very large set of feasible alternatives. Each alternative is defined implicitly in terms of the decision variables and evaluated by means of objective functions (Malczewski, 1997).

Both MADM and MODM problems can be further classified as individual and group decision making depending on the goal-preference structure. If there is a single goal preference, the problem, is considered as individual decision-making regardless of the number of decision makers involved in the process. However, if the individual or interest groups are characterized by different goal preferences, the problem becomes the group decision making (Malczewski, 1997).

The other classification depends on the certainty of the decision. If the decision maker has perfect knowledge of the decision environment and the amount of knowledge available is enough, then the decision is considered as decision under certainty. However, most of the real world decisions involve some aspects that are unknown and difficult to predict. This type of decisions is referred to as decisions under uncertainty. The decisions under uncertainty can be further subdivided into fuzzy and probabilistic decision making (Eastman et al., 1993). The probabilistic decisions are handled by probability theory and statistics; and the outcome of a stochastic event is either true or false. However, if the situation is ambiguous, the problem is structured as the degree of how much an event belongs to a class. This type of problems is handled by fuzzy set theory (Zadeh, 1965).

CONCLUSION

The use of GIS and RS technologies has improved tremen-dously for the collection, availability of data, and integrated management of spatial and non-spatial data for forest fire risk and hazard management.

The fire risk and hazard model generated from individual methods lacks the one stop solution for forest fire management. Hence, the need for the adoption of an integrated spatial multi-criteria approaches to fire risk and hazard modelling and evaluation.

Conflict of interests

The authors declare there is no conflict of interest.

REFERENCES

Albini FA (1976). Estimating Wildfire Behavior and Effects: US Department of Agriculture, Forest Service, Intermountain Forest and Range Experiment Station. Technical Report, Ogden, UT. 92.

Anon. (2011). Fundamentals of Remote Sensing, http://pcmas1.ccrs.nrcan.gc.ca/e.htmhttp://pcmas1.ccrs.nrcan.gc.ca/e. htm, accessed: December 14, 2011, pp.34-52.

Applegate G, Chokkalingam U, Suyanto S (2001). The Underlying Causes and Impacts of Fires in South-East Asia. UnpublishedFinal Report, Center for International Forest Research (CIFOR), Jakarta, Indonesia. p.100.

Barrett TM, Jones JG, Wakimoto RH (1999). Adapting forest planning decision support systems for prescribed fire treatments. Proceedings from the joint fire science conference and workshop. June 15-17, Neuenschwander L. F. and Ryan K. C. (ed), Idaho. pp.12-17.

Bond WJ, Keeley JE (2005). Fire as a Global 'Herbivore: The Ecology and Evolution of Flammable Ecosystems. Trends Ecol. Evol. 20(7): 387-394.

Bowman JS, Balch JK, Artaxo P, Bond WJ, Carlson JM, Cochrane M, D'Antonio CM, DeFries RS (2009). Fire in the Earth System. Science 324:481-484.

Brown AA, Davis KP (1973). Forest Fire: Control and Use, McGraw Hill Book Co., New York. pp. 34-85.

Chankong V, Haimes YY (1983). Multi-Objective Decision Making: Theory and Methodology, North Holland Publishers, New York. pp. 350-400.

Cheney NP (1981). Fire behaviour, In Gill, A. M., Groves, R. H. and Noble, I. R. (eds) Fire and the Australian biota: Australian Acad. Sci. Canberra. pp.151-175.

Chuvieco E, Deshayes M, Stach N, Cocero D, Rian˜o D (1999). Short-term fire risk: Foliage Moisture Content Estimation from Satellite Data. In E, Chuvieco (eds), Remote sensing of large wildfires in the European Mediterranean Basin. Berlin: Springer-Verlag, pp. 17 - 38.

Chuvieco E, Congalton RG (1989). Application of Remote Sensing and Geographic Information System to Forest Fire Hazard Mapping. Remote Sens. Environ. 29:147-159.

Chuvieco E, Martin MP (1994). Global Fire Mapping and Fire Danger Estimation using AVHRR images. Photogramm. Eng. Remote Sens. 60(5):563-570.

Eastman JR, Jin W, Kyem PAK, Toledano J (1993). Raster Procedures for Multi-criteria/Multi-objective Decisions. Photogramm. Eng. Remote Sens. 61(5):539-547.

Flannigan MD, Wotton BM (2001). Connections - Climate/Weather and Area Burned. In: Johnson EA, Miyanishi K (eds) Forest Fires: Behavior and Ecological Effects. Academic Press. pp. 335-357.

Gould J (2005). Fire Danger and Fire Behaviour, Australia Overview International Fire Weather Workshop, Bureau of Meteorology, Melbourne.URL:www.bom.gov.au/bmrc/wefor/projects/fire_wx_workshop_jun_05/08gould.pdf, accessed 24th November, 2010. 4-25.

Hernandez-Leal PA, Arbelo M, Gonzalez-Calvo A (2006). Fire Risk Assessment using Satellite Data. Adv. Space Res. 37: 741-746.

Jankowski P (1995). Integrating Geographical Information Systems and Multiple Criteria Decision-Making Methods. Int. J. Geogr. Inf. Syst. 9(3):251-273.

Janssen R (1992). Multi-objective Decision Support for Environmental Management, Kluwer Academic, Dordrecht. p. 232.

Jin H (2010). Drivers of Global Wildfires-Statistical Analyses Unpublished MSc. *Thesis*, Division of Physical Geography and Ecosystems Analysis, Department of Earth and Ecosystem Sciences. pp. 1-60.

Jones SD, Garvey MF, Hunter GJ (2004). Where's the Fire? Quantifying Uncertainty in a Wildfire Threat Model. Int. J. Wildland Fire 13: 17-25.

Keeney RL, RaiffaH (1976). Decisions with Multiple Objectives: Preferences and value Tradeoffs, Wiley, New York. 1993-569.

Kushla JD, Ripple WJ (1997). The Role of Terrain in a Fire Mosaic of a Temperate Coniferous Forest. For. Ecol. Manage. 95: 97-107.

Li T (1998). Forest Fire Risk influencing Factors and Types. J. Chin. Peoples Armed Police Force Acad. 4 31-33.

Luke RH, McArthur AG, (1978). Bush Fires in Australia. Australian Govt. Pub. Serv., Canberra. 359p.

Malczewski J (1997). Propagation of Errors in Multi-Criteria Location Analysis: A Case Study, In Fandel, G, Gal T. (eds.) Multiple Criteria Decision Making, Springer- Verlag, Berlin. pp. 54-155.

Malczewski J (1999). GIS and Multi-Criteria Analysis, John Wiley and Sons, Canada. 392, 137-269.

Martin MP, Ceccato P, Flasse S, Downey I (1999). Fire Detection and Fire Growth Monitoring using Satellite Data, In: Chuvieco E (ed), Remote Sensing of Large Wildfires in the European Mediterranean Basin, Springer, Berlin, Germany. pp.101-122.

Marzano R, Bovio B, Guglielmet E, Jappiot M, Lampin C, Dauriac F, Deshayes M, Salas J, Aguado I, Martinez J, Molina D, Martinez E, Carrega P, Fox D, BeroloW, Martin P, Gomes I, Conese C, Bonora L, Karteris M, Gitas I, Mallinis G, Giannakopoulos V (2005). Common Methods for Mapping the Wildland Fire Danger, EUFIRELAB. URL: http://eufirelab.org. pp.4-38.

Mendoza GA, Macoun P, Prabhu R, Sukadri D, Purnomo H, Hartanto H (1999). Guidelines for Applying Multi-Criteria Analysis to the Assessment of Criteria and Indicators", Criteria and Indicators Toolbox Series (9), CIFOR, Bogor, Indonesia. pp.15-70.

Millington J, Romero-Calcerrada R, Wainwright J, Perry G (2008). An Agent-Based Model of Mediterranean Agricultural Land-Use/Cover Change for examining wildfire Risk. J. Artif. Soc. Soc. Simul. 11(4):4.

Nicolas MVJ, Beebe GS (1999). Fire Management in the Logging Concessions and Plantation Forests of Indonesia, Unpublished Forest Fire Prevention and Control Project (South Sumatra) and Integrated Forest Fire Management Project (East Kalimantan), European Union, GTZ and Ministry of Forestry and Estate Crops, Jakarta, Indonesia. pp.10-65.

Pitz GF, McKillip J (1984). Decision Analysis for Program Evaluations, Thousand Oaks, CA: Sage Publications. pp. 237 - 287.

Power CJ (2006). A Spatial Decision Support System For Mapping Bushfire Hazard Potential Using Remotely Sensed Data, Bushfire Conference, Life In A Fire-Prone Environment: Translating Science Into Practice 6 - 9th June 2006, *Brisbane*. 34:1- 6.

Pyne SJ, Andrews PL, Laven RD (1996). Introduction to Wildland Fire, John Wileys and Sons Inc. 1:221-227.

Robinson JM (1991). Problems in Global Fire Evaluation: Is Remote Sensing the Solution? In Levine JS (ed.), Global Biomass Burning: Atmospheric, Climatic and Biosheric Implications, MIIT Press, Cambridge, MA. pp.67-73.

Rosenstock IM, Strecher VJ, Becker MH (1988). Social Learning Theory and the Health Belief Model. Health Educ. Behav. 15(2): 175-183.

Rothermel RC (1983). How to Predict the Spread and Intensity of Forest and Range Fire, USDA Forest Service, Inter-mountain Forest and Range Experiment Station. p. 40.

Saaty TL (1980). The Analytic Hierarchy Process, New York, McGraw-Hill. pp. 23-157.

Santoso H (2006). Adaptation to Recurrence Forest Fires and their Risks under the Influence of Climate Change and Climate Variability,16th Asia Pacific Seminar on Climate Change, 5 - 8 September, Jakarta, Indonesia.

Schweitzer C, Priess JA (2010). Linking Wildfire Behaviour and Land-Use Modelling in Northern Mongolia, International Environmental Modelling and Software Society (iEMSs) 2010 International Congress on Environmental Modelling and Software Modelling for Environment's Sake, 5th Biennial Meeting, Ottawa, Canada.

Shavit T, Shahrabani S, Benzoin U, Rosenboim M (2013).The Effect of Fire on Emotions and Risk Perceptions: A Field Study after the Carmel Forest Fire Disaster. J. Environ. Psychol. 36:129-135.

Stolle F, Lambin EF (2003). Inter-Provincial and Inter-Annual Differences in the Causes of Land-Use Fires in Sumatra, Indonesia. Environ. Conserv. 30(4):375-387.

Thonicke K, Venevsky S, Sitch S, Cramer W (2001). The role of fire disturbance for global vegetation dynamics: coupling fire into a Dynamic Global Vegetation Model. Glob. Ecol. Biogeogr. 10: 661-678.

Tomich TP, Fagi AM, de Foresta H, Mudiyarso D, Stolle F, van Noordwijk M (1998). Indonesia's Fires: Smoke as a Problem, Smoke as a Symptom. Agroforestry Today 10:4 -7.

Trollope WSW, Trollope LA, Hartnett DC (2002). Fire Behaviour a Key Factor in the Fire Ecology of African Grasslands and Savannas In: Forest Fire Research and Wildland Fire Safety Viegas (ed), Millpress, Rotterdam. pp.1-14.

Vadrevu KP, Eaturu A, Badarinath KVS (2009). Fire Risk Evaluation Using Multi-Criteria Analysis-A Case Study. Environ. Monit. Assess. 166(1-4):223-239.

Van der Werf GR, Randerson JT, Collatz GJ, Giglio L, Kasibhatla PS, Arellano Jr, AF, Olsen SC, Kasischke ES(2004). Continental-Scale Partitioning of Fire Emissions During the 1997 to 2001 El Niño/La Niña Period. Science 303:73-75.

Varela J, Arias JE, Sordo I, Tarela A (2005). Multicriteria Decision Analysis for Forest Fire Risk Assessment in Galicia, Spain, In: Proceedings of the 4th International Workshop on Remote Sensing and GIS applications to Forest Fire Management. 1-233.

Zadeh LA (1965). Fuzzy Sets. Information and Content 8(3):338-353.

Indications of the changing nature of rainfall in Ethiopia: The example of the 1st decade of 21st century

Lemma Bekele

Geography Department, Kotebe Univesity College, Addis Ababa, Ethiopia.

Climate change was defined as a statistically significant variation in the mean state of climate. Accordingly, with regards to rainfall, a working hypothesis that reads as: there is a statistically significant change (increase/decrease) in the mean annual and seasonal rainfall values at weather stations was forwarded for testing. Eight years (2001-2008) annual and seasonal rainfall data of 17 weather stations from most parts of Ethiopia were used for the purpose. Setting the significance level at 0.05, the simple correlation and regression techniques were used to reach results. Most stations from the western part of Ethiopia were seen to exhibit statistically significant increasing trend in annual rainfall receipt. Albeit not significantly, northward and eastward from here, the receipt was also seen tending first to increase then decrease. This tendency was observed to overlap with the traditional distribution of rainfall in the country. The alternating of dry and wet years and the more localizing of wet years than dry years were identified. The four seasons considered were observed to be statistically significant, but increasing and decreasing trends in seasonal rainfall receipt at the respective stations were not significant. Most conspicuously, 94% of the total number of stations considered in this study show increasing trend in autumn rainfall receipt. Based on the findings of the study, some conclusions policy makers may be taking into consideration were finally drawn.

Key words: Annual, Ethiopia, rainfall, seasonal, significant.

INTRODUCTION

The hypothesis

Climatologically speaking, Ethiopia is a world in miniature, in that, dictated by its high amplitude of relief (Mesfin, 1970), the largeness of its size, 1,127,127 Km^2 (US Library of Congress, 2005: 4) and its location in the heart of the Horn of Africa, it enjoys varieties of climate that are comparable to the effects of latitude, so much so that the high latitude equivalents can be found along the uplands and the low latitude equivalents along the lowlands (Daniel, 1977).

This fact lends the country to serve as a laboratory for examining change in climate. As per the series of the syntheses reports of the Intergovernmental Panel on Climate Change (IPCC) the working definition of climate change is put as a statistically significant change in the mean state of climate and/or its variability over prolonged period (WMO-IPCC, 2001, 2007, 2013).

With respect to rainfall, its variability has been a topic of discussion by several research works both from outside Ethiopia (Otun and Adewumi, 2009; Hendrix and Salehyah, 2010; Wolff et al., 2011) and from within Ethiopia (Woldeamlak, 2009; Hongwei et al., 2011; Rientiger et al., 2013).

Otun and Adewumi (2009) looked at rainfall variability in the Sudano-Sahilian regions of Nigeria, and its connection with the sahelian drought. They used daily rainfall series from seven stations that spanned over 30 year climate interval (1940 – 1970 and 1970 – 2000). Their conclusion includes the fact that the drought condition of 1970's might be recurring in the future (Otuned and Adewuni, 2009). Hendrix and Salehyan (2010), established relationship between rainfall variability and socio political conflict in Africa. They particularly found out that dry and wet years are associated with all types of small scale and large scale instances of political conflict (Hendrix and Salehyan, 2010). Based on their evidence from laminated lake sediment record from Southern Kenya Wolff et al. (2011) reported of changes in El-Nino related Variability of East Africa's rainfall during the last three millennia (Wolff et al., 2011)

Studies on rainfall trends and variability made in Ethiopia include Weldeamlak (2009) and Hongwei et al. (2011). Using rainfall data spanning from 1975-2003 for 12 stations from the Amhara region, Woldeamlak (2009) reported, among other things, of lower variability of rainfall in the western part of the region than the eastern. From their trend analysis of 53 years' daily precipitation data at Debremarkos Hongwei et al. (2011) found no increasing trend in the extreme precipitation at Debre Markos. In their assessment of the diurnal cycle of rainfall across the Upper Blue Nile basin Rientjes et al. (2013) revealed the fact that over most parts of the basin rainfall is highest between mid-and-late afternoons. They used seven years (2002-2008) satellite derived precipitation data (Rientjes et al., 2013).

With regards to rainfall again, research works done in Ethiopia in the style amenable to the definition of climate change given above includes Yilma and Zanke (2004), as quoted by Cheung et al. (2008), and Cheung et al. (2008). Using decadal rainfall records of 134 gauging stations with differing study period (ranging between 35 and 43 years) and employing the regression and correlation techniques, Cheung et al. (2008) analyzed trends in seasonal and annual rainfall at national, watershed, and gauging station levels. In their seasonal analysis, they considered two seasons – *kiremt* (July to September) and *belg* (March to May) (Cheung et al., 2008). At the seasonal level, their findings brought in a significant decline in *kiremt* rainfall for some watersheds eg. the southern Blue Nile watershed (Cheung et al., 2008), and at gauging station level – their finding include a significant increase at Hirna and a significant decrease at Robe from the eastern region of the country (Cheung et al., 2008). Yilma and Zanke (2004), as quoted by Cheung et al. (2008) reported of the decline in annual rainfall in southwestern and eastern Ethiopia (Cheung et al., 2008). Much earlier, Conway and Hulmes (1993), as quoted by Yilma (1996), reported of decrease in annual rainfall in north central highlands, later on confirmed by Yilma (1996).

Unlike such a previous work like Cheung et al. (2008) which based its analyses on two seasons, *kiremt* and *belg*, and non-uniform, but large number of years (35 to 43 years), this research work is based on mean annual and four seasons' mean rainfall of 17 weather stations[1] drawn from most parts of Ethiopia (Figure 1), and spread over eight uniform years (2001-2008) almost the first decade of the twenty first century, would endeavor to test one central hypothesis: that there is statistically significant change (increase/decrease) in mean annual and seasonal rainfall values at the stations.

Although this paper has several shortcomings including the briefness of the study period and the unevenness in the distribution of the stations considered, it is my conviction that the paper may serve two purposes, (a) that it may be indicative of the general direction on the changing nature of rainfall in Ethiopia and (b) it may be a modest addition to the related works in Ethiopia and serve as a reference material for future researches.

METHODOLOGY

Data source

All the eight years mean annual and seasonal values of the 17 stations used in this study were based on the monthly rainfall records collected from the National Methodological Agency[2] (NMA). The years and seasons considered in this study were meteorological years and meteorological seasons. A meteorological year comprises four seasons each composed of three months (Ahrens, 1988). In the northern hemisphere the meteorological definition of winter would be December, January and February; spring would be March, April, and May; summer would be June, July, August; and autumn, September, October, and November (Ahrens, 1988). In order to fit to the meteorological definition. the first month of winter, December, considered here, in each case, was the previous year's December. This procedure was strictly observed throughout. Setting up the mean annual and seasonal

[1] Originally the writer intended to consider more than seventeen stations distributed over the northern, southern, eastern, western and central parts of Ethiopia, and a minimum study period of 10 years, complete with twelve calendar months of rainfall data. After excluding station with incomplete data and taking cognizance of their distribution over most parts of the country, the author, finally, concentrated on the seventeen stations with comparable data from 2001 to 2008, nearly the first decade of the twenty first century. Monthly rainfall data in two cases, however, i) Dessie for November and December (2006); and ii) Goba, for January and February (2005), and May (2006) were missing. In all these cases, nevertheless, the missing data were filled with the respective arithmetic average of the immediate proceeding and succeeding months with data.)

[2] It would be appropriate to mention here the fact that most of the data on the monthly rainfall of the seventeen stations were secured from the National Meteorological Agency (NMA) by my students for a course work on GeEs 1012 (Applied climatology). The author also obtained a good stock of data straight from the same agency for cash.

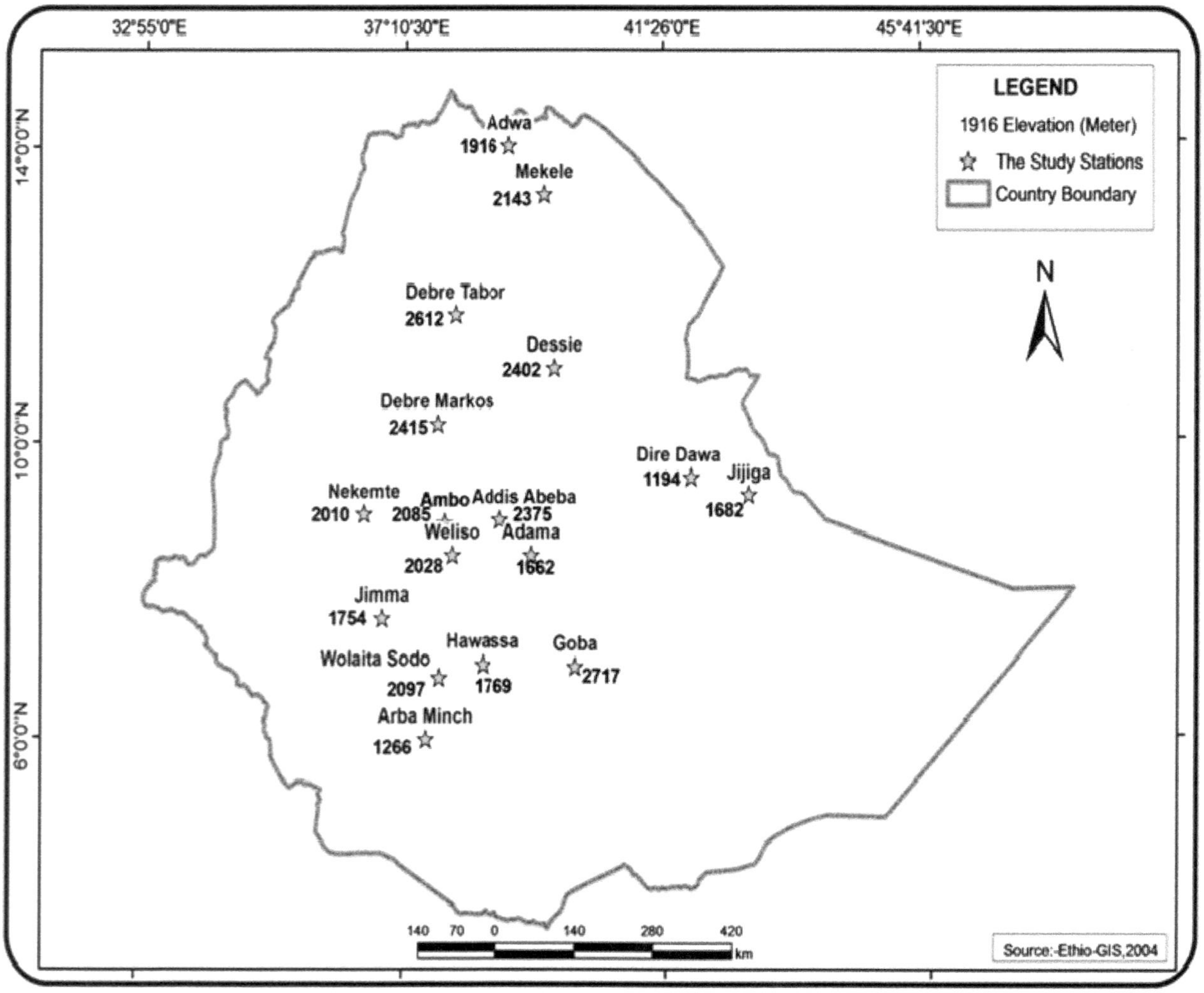

Figure 1. The study stations.

rainfall values as dependent variables: the years and the seasons there in as independent variables; and putting the significance level at 0.05, the regression and correlation techniques (Berenson and Levine, 1983) were used to arrive at the results.

RESULTS AND DISCUSSION

Annual increases and decreases

Table 1 provides a summary of the correlation and regression analyses done on the mean annual rainfall of the fourteen stations. Three distinct groups can be identified. The first group comprises Ambo, Debre Tabor, Nekemte, and Woliso, all but Debre Tabore, from the western part of the Ethiopia, where mean annual rainfall exhibited statistically significant increase over the study period. The second group has Adama, Addis Ababa, Arba Minch, Debre Markos, Goba, Hawassa, Jijiga, Jimma, and Wolaita Sodo – most of which are from the central and southern Ethiopia – which observed increased annual rainfall, albeit not significantly. The third group composed of Dessie, from north east Ethiopia that showed a statistically significant decrease in the mean annual rainfall, plus Adwa and Mekele from northern and Dire Dawa for eastern Ethiopia, which showed decreased rainfall, though not statistically significant again.

It appears, then when, in general annual rainfall receipt in the western part of the country is tending to experience statistically significant increase-northward and eastward from here the tendency is first to continue to increase (in almost all cases not significantly) then decrease (in almost all cases not significantly again). There is the established fact that amount of rainfall in the country is decreasing as one goes northward and eastward from the western part of Ethiopia to modestly increase over the northern and eastern highlands, and decrease thereafter (Kebede, 1964; Mesfin, 1972; Daniel, 1977; Nigtu, 2011), a fact which can also be read from the data on the mean annual rainfall of the 17 stations given in Table 2 and illustrated in Figure 2. Adwa and Mekele (northern most

Table 1. Correlation co-efficient, slope (b_1) decreases/increases and t-test statistics for seventeen stations in Ethiopia (2001 – 2008).

Station	r	Slope (b_1)	I=increased D=decreased	t_{cal}	t_{tab}	R=Reject H_0 A=Accept H_0
Adama	0.500	+3.412	I	1.410	1.9432	A
Addis Ababa	0.261	+1.375	I	0.657	1.9432	A
Adwa	-0.204	-0.979	D	-0.208	1.9432	A
Ambo	0.690	+3.451	I	2.350	1.9432	R
Arba Minch	0.272	+1.645	I	0.689	1.9432	A
Debre Marcos	0.349	+1.040	I	1.196	1.9432	A
Debre Tabor	0.660	+4.400	I	2.200	1.9432	R
Dessie	-0.346	-6.930	D	-2.327	1.9432	R
Dire Dawa	-0.011	-0.043	D	-0.028	1.9432	A
Goba	0.083	+0.587	I	0.206	1.9432	A
Hawassa	0.385	+1.801	I	1.007	1.9432	A
Jijiga	0.176	+0.712	I	0.434	1.9432	A
Jimma	0.106	+0.657	I	0.259	1.9432	A
Mekele	-0.161	-0.820	D	-0.403	1.9432	A
Nekemte	0.835	+7.332	I	3.752	1.9432	R
Wolaita Sodo	0.461	+3.747	I	1.274	1.9432	A
Woliso	0.674	+2.910	I	2.243	1.9432	R

H_1 = There is significant increase/decrease in the mean annual rainfall; H_0 = H_1 is not true. Significance level =0.05. Source: Original data from NMA, compiled by the writer, r = regression.

Table 2. The annual mean, the wettest and driest years for seventeen weather stations is Ethiopia (2001 – 2008).

Station	Annual mean (mm)	Wettest year (W)	Driest year (D)	Difference in N^0 years between W and D
Adama	898.2	2007	2002	5
Addis Ababa	1277.2	2001	2002	1
Adwa	778.4	2001	2002	1
Ambo	990.2	2007	2002	5
Arba Minch	738.5	2007	2002	5
Debre Marcos	1336.9	2006	2002	4
Debre Tabor	1411.1	2006	2002	4
Dessie	1390.0	2003	2008	5
Dire Dawa	602.1	2007	2005	2
Goba	1031.1	2006	2002	4
Hawassa	989.3	2007	2003	4
Jijiga	591.9	2006	2008	2
Jimma	1415.2	2001	2002	1
Mekele	476.7	2006	2004	2
Nekemte	2036.3	2008	2002	6
Wolaita Sodo	1385.0	2007	2002	5
Woliso	1196.6	2006	2001	5

Source: Original data from NMA, compiled by the writer.

stations) reported mean annual rainfall of 778.4 and 476.7 mm, respectively; and Dire Dawa and Jijiga (eastern most stations) revealed mean annual rainfall of 602.1 and 591.9 mm, in that order Nekemte and Jimma (western most stations) presented 2036.3 and 1415.22 mm of annual rainfall, the highest and the second highest, respectively, among the stations considered in this study. The other stations, in general, took the

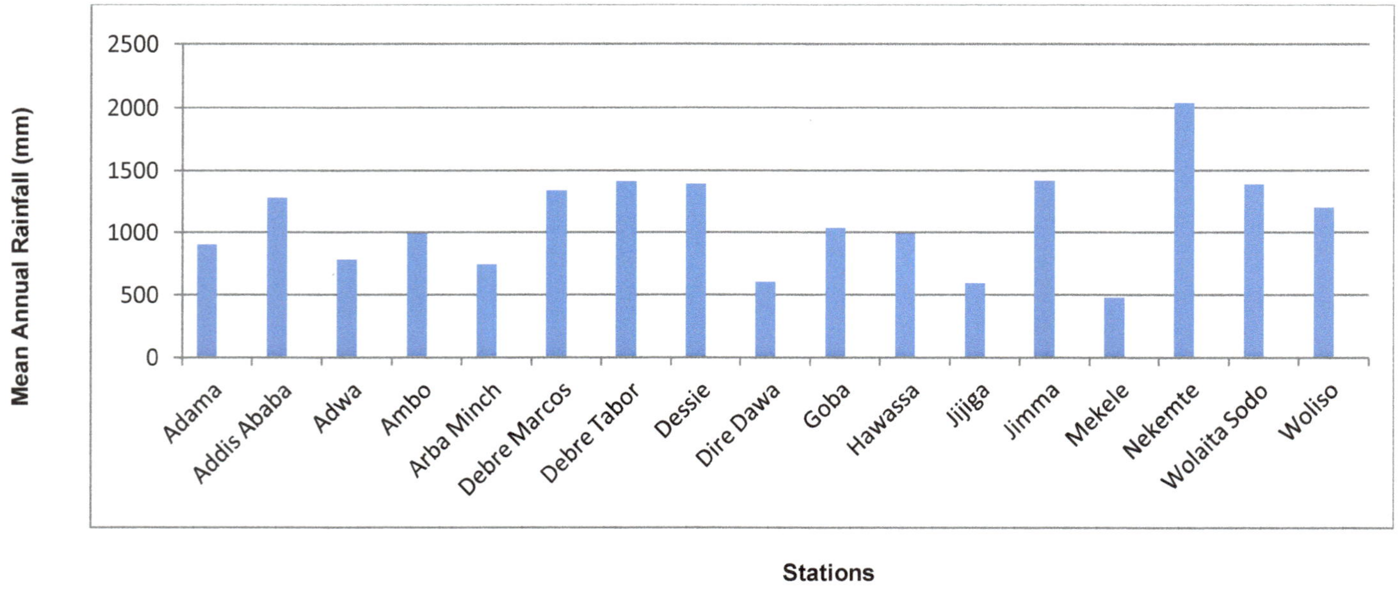

Figure 2. Graphic rrepresentation of the mean annual rainfall of the 17 Stations (2001-2008).

intermediate position between these two extremes. It can be said then that the behaviour of the 17 stations' increases and decreases in the mean annual rainfall amounts observed over the study period, somehow, tends to overlap with the traditional distribution of rainfall in Ethiopia.

Exceptions to this general trend were, however, presented by some stations. For instance, when Jimma from mid-west Ethiopia experienced an increase in the annual rainfall receipt which was not statistically significant; most of the neighbouring stations revealed a statistically significant increase. Likewise, Debre Tabor and Dessie both from northern highlands and at about the same distance from mid-west Ethiopia produced contrasting annual results-statistically significant increase for Debre Tabor and statistically significant decrease for Dessie. Difference in annual values was also observed between the eastern most stations – Dire Dawa, decreasing (not significantly), and Jijiga, increasing (not significantly). These differences may be decidedly explained by local circumstances including elevation factor.

Table 2 also provides a list of the wettest and driest years each station experienced during the study period together with the interval in years between these extreme events. The first feature that may be deciphered from the table would be the fact that wetness tends to be more localized than dryness. This may be learned from the fact that, when in six cases-at Adama, Ambo, Arba Minch, Hawassa, Dire Dawa and Wolaita Sodo or about 35% of the total number of stations, 2007 was the wettest year; and in another six cases-at Debre Marcos; Debre Tabor, Goba, Jijiga, Mekele and Woliso or about 35% of the total again, 2006 was the wettest year; in 11 cases-at Adama, Addis Ababa, Adwa, Ambo, Arba Minch, Debre Marcos, Debre Tabor, Goba, Jimma, Nekemte, and Wolaita Sodo or about 65% of the total number of stations, 2002 was the driest year. The other feature worth mentioning would be the difference in years between the wettest and driest years. The average gap for the 17 stations was put at three and – half years. At some stations, however, that is Addis Ababa, Adwa and Jimma the driest and the wettest years occurred consecutively, while at Nekemte a gap of six years was required.

Seasonal increases and decreases

Table 3 shows the chief points of the statistical analyses done on the seasonal data. Two outstanding features can be learned from the table. First, the compatibility in the pattern of mean annual rainfall increases and decreases with the traditional spatial distribution of rainfall in the country mentioned above, in some way, seems to be repeated here. This is so, for again, mostly the same group of stations with statistically significant annual rainfall increase records, Ambo, Debre Tabor, Nekemte and Woliso emerged with the largest number of seasons with increased rainfall; Three seasons had increase in the case of Ambo, Nekemte and Woliso; and all the four seasons had increase in the case of Debre Tabor. Ambo observed increases in spring (significant), summer (significant) and autumn (significant at 0.10 level); Nekemte in spring (significant), in autumn (significant), and summer (not significant); Woliso in autumn (signify cant at 0.10 level), spring (not significant), and summer

Table 3. Increases and decreases in seasonal rainfall receipt and t-test statistics for seventeen stations in Ethiopia (2001-2008).

Station	Season			
	Winter	Spring	Summer	Autumn
Adama	I(A)	D(A)	I(A)	I(R)
Addis Ababa	I(A)	D(A)	D(A)	I(R)
Adwa	I(A)	I(A)	D(A)	I(R)
Ambo	D(A)	I(A)	I(R)	I(R)*
Arba Minch	D(A)	D(A)	I(A)	I(R)
Debre Marcos	D(A)	I(A)	I(A)	I(R)
Debre Tabor	I(R)	I(R)	I(A)	I(A)
Dessie	I(A)	D(A)	D(A)	I(A)
Dire Dawa	D(A)	I(A)	D(A)	I(A)
Goba	D(A)	I(A)	I(A)	I(A)
Hawassa	D(A)	D(A)	I(A)	I(R)
Jijiga	D(A)	I(A)	D(A)	I(A)
Jimma	I(A)	I(A)	D(A)	I(A)
Mekele	D(A)	I(A)	D(A)	D(A)
Nekemte	D(A)	I(R)	I(A)	I(R)
Wolaita Sodo	D(A)	I(A)	I(A)	I(R)
Woliso	D(A)	I(A)	I(A)	I(R)*

H_1=There is significant increase/decrease in the mean seasonal rainfall; H_2=H_1 is not true. Significance level= 0.05. Source: Original Data from NMA, Compiled by the writer. I = Increased; D = Decreased; A = Accept H_0; R = Reject H_0; * = Reject H_0 at 0.10 significance level.

(not significant). Debre Tabor which can be regarded as a ‘moisture island’ of the northern highlands witnessed increase in winter (significant), spring (significant), summer (not significant), and autumn (not significant). Other stations distributed over most parts of the country also joined the ranks of stations with three seasons of increased rainfall, which includes Adama, Adwa, Debre Markos, Goba, Jimma, and Wolaita Sodo. In contrast to all these Mekele, from Northern Ethiopia, experienced three seasons of decreased rainfall-in summer, autumn and winter (in all cases not statistically significant). The other stations, including Adama and Addis Ababa, from central Ethiopia, Dire Dawa and Jigjiga from Eastern Ethiopia, and Hawassa from south central part of the country, observed two seasons of decreased rainfall.

Indeed, here too, exceptions to the general trend mentioned earlier were observed, locally produced. The comparison of Adwa’s and Mekele’s four seasons increasing and decreasing trends present good example. Adwa, about 100 kilometers north west of Mekele, is at a lower elevation (1916 meters above sea level) than Mekele (2143 meters above sea level), (vide the map attached here with) and still Adwa experienced three seasons of increased rainfall (though not significant in all cases), while Mekele observed three seasons of decreased rainfall (in all the cases not significant again). Besides, Mekele reported of an eight year mean annual rainfall of 476.7 mm, when Adwa reported 778.4 mm (vide Table 2). Facts that exclude elevation factor like geographical position may explain this.

The second feature would be the season to season differences - six stations – Debre Tabor (significant), Adama, Addis Ababa, Adwa, Dessie and Jimma (not significant in the last five), or about 35 percent of the total, witnessed increasing trend in winter rainfall. Seven stations or about 41 percent of the total exhibited decreasing tread in summer. This includes – Addis Ababa, Adwa, Dessie, Jimma and Mekele. (in all cases not significant). Twelve stations or about 71 percent of the total number of stations which includes Deble Tabor (significant), Jijiga (not significant), Dire Dawa (not significant) saw increasing trend in spring rainfall. Lastly and probably most strikingly sixteen stations or about 94 percent of the total showed increasing trend in autumn rainfall. Nine stations, or about 51 percent of these comprising Adama, Addis Ababa, Arba Minch, Hawassa, Debre Markos, Plus Ambo and Woliso (both significant at 0.10 level), exhibited statistically significant increase in autumn.

Conclusion

The eight – year period mean annual and seasonal data of the 17 stations considered in this study supported the rejection and acceptance of the null hypotheses of no significant increase/decrease in the annual and seasonal rainfall. On the annual scale, most stations from western part of Ethiopia show statistically significant increasing trend in annual rainfall. In almost all directions from here the tendency, though inconsistently, was first to increase then to decrease albeit not significantly. This appeared to be in close parallelism with the normal distribution of rainfall in Ethiopia. The alternating recurrence of dry and wet years and the more localizing of wet years than dry years were recognized. At the seasonal level, every station witnessed either statistically significant or not significant increasing and/or decreasing trend in the respective seasonal rainfall amount, the pattern of which, taken as a unit, coincided with the traditional distribution of rainfall in the country. Moreover, more than half of the weather stations considered in this study, observed a statistically significant increase in autumn rainfall receipt.

The following conclusions, which policy makers may take into consideration, are drawn from the findings of the study mentioned above:

a) The establishment of the close parallelism in the behaviour of the increases/decreases in the mean annual and seasonal rainfall values with the traditional distribution of the rainfall in Ethiopian as a fact may persuade one to say that rainfall distribution in the country would continue to be as it was and as it is now with the wet and dry areas growing wetter and drier, respectively.
b) The alternating recurrence of dry and wet years and the localizing of wet years than dry years would mean that the resultants, positive or negative, would be local in scale when it comes to wetness and regional or beyond when it comes to dryness.
c) Autumn comes after summer-the observation of statistically significant increase in autumn rainfall at more than 50% of the total weather stations considered in this study could, therefore, mean (i) the continuation of summer rainfall into autumn when it comes to parts of central, northern and western Ethiopia; and (ii) the further augmentation of autumn rainfall when it comes to the eastern and southern Ethiopia.
d) Finally, it can be said that, climate change as associated with rainfall in Ethiopia would confront decision makers with decrease or increase in annual and seasonal amounts. What needs to be done is not hard to forward – all the possible means should be designed to combat all the possible consequences of the events.

Conflict of interests

The author did not declare any conflict of interest.

ACKNOWLEDGMENTS

First and foremost I express my heartfelt gratitude to the National Meteorological Agency of Ethiopia which directly or indirectly offered me with all the original data used in this paper. I am gratefu l to Dr. Birhan Gessese and Ato Mekuria Delelegne, my colleagues, for their continuous encouragement and constructive suggestions. I also thank Ato Bamlaku Amentie of the Department Of Geography and Environmental Studies (AAU) for the cartographic works of the map attached herewith and Wro Wossene Birhan for typing the manuscript.

REFERENCES

Ahrens D (1988). *Meteorology Today* St. Paul: West Publishing Company.
Berenson M, Levine D (1983). Basic Business Statistics Concepts and Application Englewood Cliffs: Prentice-hall.
Cheung W, Senay G, Singh A (2008). "Trends and spatial distribution of annual and seasonal rainfall in Ethiopia," Int. Metrol. J.
Daniel G (1977). Aspects of climate and Water Budget in Ethiopia Addis Ababa: University press.
Hendrix CS, Salehyan (2010). 'After the rain: Rainfall Variability, Hydro-Meteorological Disasters, and Social Conflict in Africa' in http://ssrn.com/abstract=1641312.
Hongwei S, Yan J, Gebremichael M, Ayalew SM (2011). Trend analysis of extreme precipitation in the Northwestern Highlands of Ethiopia with a case study of Debre Markos. Hydrology and Earth System Sciences. 15(6):1937-1944.
Kebede T (1964). "Rainfall in Ethiopia" Ethiopian Geo. J. 2, 28-36.
Mesfin WM (1970). An Atlas of Ethiopia Asmera: Poligrafico.
Mesfin W M (1972). Ethiopia. Addis Ababa: Berhanena Selam HSI Printing press.
Nigatu M (2011). "Rainfall Intensity Duration Frequency (RIDF) Relationship Under Changing Climate" (Case study on upper Blue Nile River Basin, Ethiopia). Msc Thesis.
Otun JA, Adewumi (2009). 'Rainfall Variablity and Drought Inference in Sudano-Saheliah Region of Nigeria'. An Int. J. Agric. Sci. Technol. 8:2. anaals.edu.hg/index.php/series-B/article/view/429/0.
Rientjes T, Alemseged TH, Ayele AF (2013). 'Diurnal Rainfall Variability Over the Upper Blue Nile Basin; A Remote Sensising Based Approach'. Int. J. Appl. Earth Observation and Geo information. 21:311-326. www.sciencedirect.com.
US Library of Congress (2005). "Ethiopia country profile" :ttp;//covervit.oc.gov/frd/es/profiles/Ethiopia.
Wolff C, Haug GH, Timmermann A, Damsté JSS., Brauer, A., Sigman, DM, Mark AC, Verschuren D (2011). Reduced interannual rainfall variability in East Africa during the last ice age. Science, 333(6043), 743-747.
WMO, IPCC 2001, Climate Change: synthesis Report.
WMO, IPCC 2007, Climate Change: synthesis Report
WMO, IPCC, 2013, Climate Change: synthesis Report
Woldeamlak B (2009). "Rainfall variability and crop production in Ethiopia case study in the Amhara Region" proceedings of the 16th international conference of Ethiopian studies ed. by Ege S, Harald, A, Birhanu T, Shiferaw BT. 823-836.
Yilma S (1996). Stochastic predictions of summer rainfall amounts over northeastern Africa Highlands and over India. Brusels

Gumbel Weibull distribution function for Sahel precipitation modeling and predicting: Case of Mali

Kya Abraham Berthe[1] , Abdramane B[1] and Setephan Reichenbach[2]

[1]Optical Spectroscopy and Sciences of Atmospheres Laboratory (LOSSA).
[2]Department of Computer Science and Engineering, University of Nebraska-Lincoln, Nebraska, United States.

Climate change, trend of precipitation variation is greatly affected by Sahel countries economy in general and Mali in particular, which increased the social instability in recent years. In this paper, we proposed Gumbel Weibull distribution function for modeling and predicting the precipitation of Mali. The methodology is composed of two steps: parameters computations and estimations. We computed the parameters using four computations methods such as: method of moments (MOM), maximum likelihood method (MLM), method of least squares (MLS) and probability weighted moments (PWM). To estimate the best method, firstly we used several good fit tests like: Kolmogorov-Smirnov (Ks), Chi-square, Anderson-Darling and D-index to analyze each method parameters, then the ratio of the standard error to return period for final estimation. For simulation, daily data of the period, 1949 to 2006 provided by Mali Meteorology Department of four localities (Kayes, Koutiala, Mopti and Hombori) was used. Results of simulations were suitable for Anderson-Darling good fit technique and PWM for Koutial, Mopti and Hombori precipitation; and MLS for Kayes precipitation. The plotting of the return period of the precipitations for PWM and MLS in the 1000 years has also confirmed this result.

Key words: Rainfall modeling and predicting, Gumbel and Weibull distribution function, method of moments (MOM), maximum likelihood method (MLM), method of least squares (MLS) and probability weighted moments (PWM) parameters optimization, statistical analysis, good fit test estimation.

INTRODUCTION

Sahel climate is characterized by strong climatic variations of temperature, humidity and precipitation (Foxa and Rockstrom, 2003; Kandji et al., 2006). In Sahel countries generally and Mali in particular, economy depends mainly on agriculture, so these variations and the lack of efficient policy in environment management increase the poverty and affect social instability (food insecurity, migration, social conflicts, etc.).

An efficient management of under and over ground water is a key factor for agriculture sustainability which depends mostly on the precipitation. For farmers, precipitation is the main parameter, because it directly affects their incomes (Udual and Ini, 2012). Researches on modeling, prediction and forecasting climate mitigation in general and precipitations in particular are necessary conditions to increase agriculture production.

Although researches have been done in the field of rainfall modeling and predicting, the complexity of the trend climate mitigation makes it remain a challenge mainly for Sahel countries. Analysis of rainfall data strongly depends on its availability and distribution pattern. It has long been a topic of interest in the fields of meteorology to establish a statistical probability distribution that can provide a good estimation.

Mali climate represent a better sample for Sahel climate variation studies, because of its geographic position. Mali climate is straddling to the sub-tropical band called the Sahel; in which northern parts reach well into the dry Sahara desert, while the southern regions experience a wetter, more tropical, climate. The seasonal rainfall in Mali is controlled by the movement of the tropical rain (also known as the Inter-Tropical Conversion Zone, ITCZ) which oscillate between the northern and southern tropics over the course of a year, and brings rainfall to the southern regions of Mali between June and October, peaking in August. Seasonal temperature variations are large, and differ in their patterns for different parts of the country (Jones et al., 2005).

The analysis and modeling of rainfall data based on time frequency trend is very complex (Phien, 1987; Manik and Datta, 1998; Zhang, 2003; Cousineau et al., 2004). The recent years have seen the increasing use of statistical theory (Vivekanandan, 2013). Gumbel distribution function approach the latest innovation based on statistical theory. This distribution with cumulative probability can be used to estimate the magnitudes of floods with different return periods, because of the simplicity with which the quantiles can be calculated.

FREQUENCY ANALYSIS

Sites description

The four sites have been selected based on the availability of long period data and their position. This climate is mainly characterized by a high variability of inter-annual and inter-decadal times-scales, which can make long-term trend difficult to be identify. The selected four sites represent the complexity of rainfall in Mali.

1. Mopti (14°:406'27N) capital of the fifth administrative region of Mali, limited by Tombouctou in the north and Segou region to the southwest and Burkina Faso to the southeast. The Niger River crosses Mopti, makes this region in terms of water, a well-irrigate; so agriculture is well-developed, with also a particular successful fishing activities and shepherd. Mopti serves as an important commercial crossroads between Mali's north, south and bordering nations. Tourism is also-developed, notably in the cities of Djenne and Mopti.
2. Hombori (15°142'297N) is a small town in the community of Douentza in Mopti region, situated on the national road 15 highway linking Mopti to Gao and Kidal regions. This small commune is the gate of Sahara. The climate in Hombori differs from other sites, the lack of important precipitation activity and continuous aquifer can be tapped by deep wells making the provision of drinking water in the commune difficult.
3. Kayes (4°1126'46N), the main city of the first administrative area of Mali is crossed by Senegal River, the region is bordered by the countries of Mauritania at the north, Senegal at the west, Guinea at south and region of Koulikoro in the east. At the Guinean border, the climate is rather wet, but becomes Sudanian and later Sahelian to the north.
4. Koutiala (12°528'44N) is situated in Sikasso region (southernmost), the third region of Mali. This region receives more rain than any other Malian regions. Koutiala is the heartland of the cotton production in West African regions. The quantity and the quality of cotton produce in the area makes this site the second industrial city of Mali.

Data analysis

To reduce the complexity of analysis, we used regression analysis. Regression analysis is a process of fitting a function to a set of data points. Curve fitting with polynomials is done with polyfit function which uses the least squares method. Polyfit function finds the coefficients of a polynomial representing the data; used them to find new values of the data.

Based on year cumulated precipitation

Figure 1 represent the year-to-year cumulated precipitation, blue curve represented the record data and the red one is the best polynomial fit of the recorded data used polyfit function. It illustrates the downward trend for total precipitation, with minimum in the 1980s followed by small increases in the 1990s.

The downward trend is more important for Mopti, Kayes and Hombori. From 1990, there was a small increase followed by decreases to 2004 except for Kayes (Figure 1c). The decrease of precipitation is observed in all the sites in the period of 1950 to 1980. This decrease is more important for Koutiala and Hombori. A relative increase was observed for the period of 1990 to 2000. The drought (1980's to 1990's) decimated the livestock (20 to 50%), caused scarcities of water and food, and increased immigration and social conflicts between famers and shepherds.

Based on the number of rain days in year

From Figure 2, it is noted that the trend of number of rainfall days is downward for Mopti, Kayes and smaller for Koutaila and Hombori, with a minimum situated in the interval 1980 to1995.

A small increasing of the precipitation (Figure 1) is observed at the end of 1990s for all the sites. The global trend for both curves (Figures 1 and 2) is the reduction of precipitation and the number of precipitation days. Figures 1 and 2 show the variation on trend of the precipitation and the number of rainfall days for 56-year period. A period of particularly high rainfall (cumulate precipitation and number of rain days) occurred in the early 1950s, whilst the early 80s were very dry, causing widespread dryness in Mali and other Sahelian countries. Precipitation has recovered to some extent for all sites since 90s, but the increase has been quickly followed by the decrease in 2000. It is recommended that an efficient underground management policy to compensate this reduction of precipitation should be made.

PROPOSED METHOD: GUMBEL DISTRIBUTION FUNCTION

Since 1970, attention of the research on climate modeling has been focused on the discrete distribution function provides by Lagrangian distributions (Vivekanndan, 2013; Consul et al., 1972; 1973; Heathcote et al., 2004). This distribution function is been found more general in nature and wider in scope. Gumbel distribution is the latest innovations for climate modeling. It can be represented as follows (Consul et al., 1972; Heathcote et al., 2004):

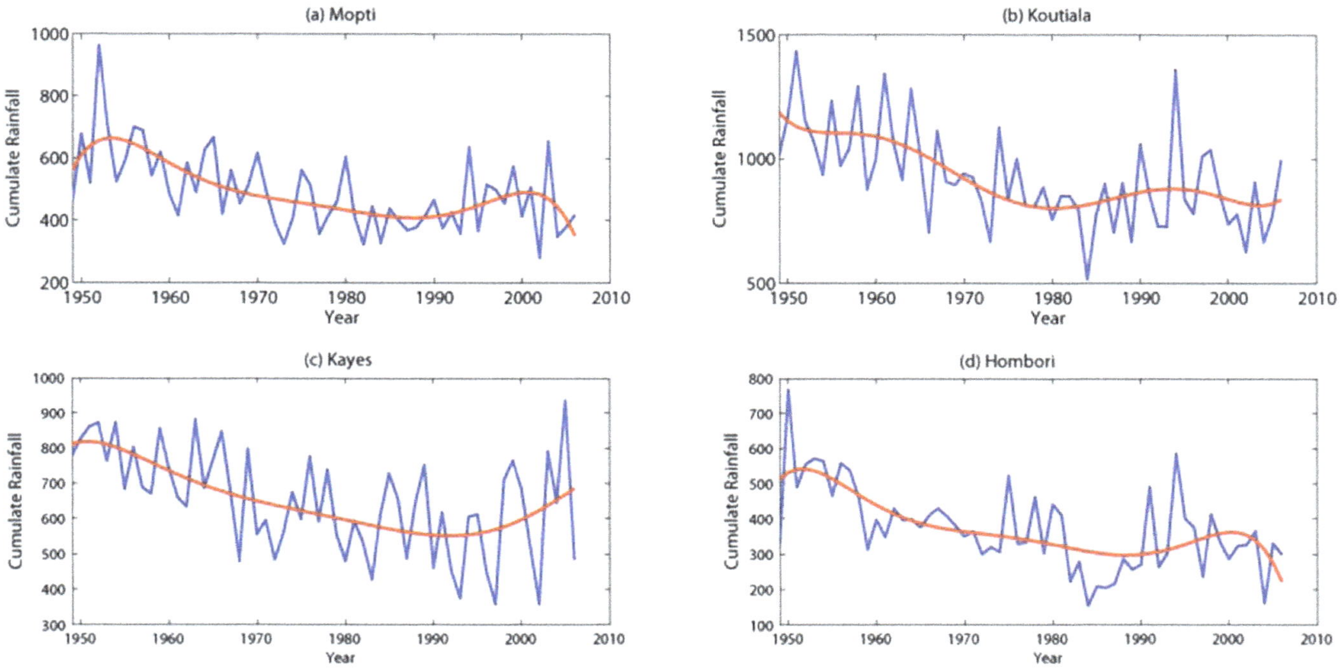

Figure 1. Cumulated year annual precipitation: (a) Mopti, (b) Koutiala,(c) Kayes and (d) Hombori.

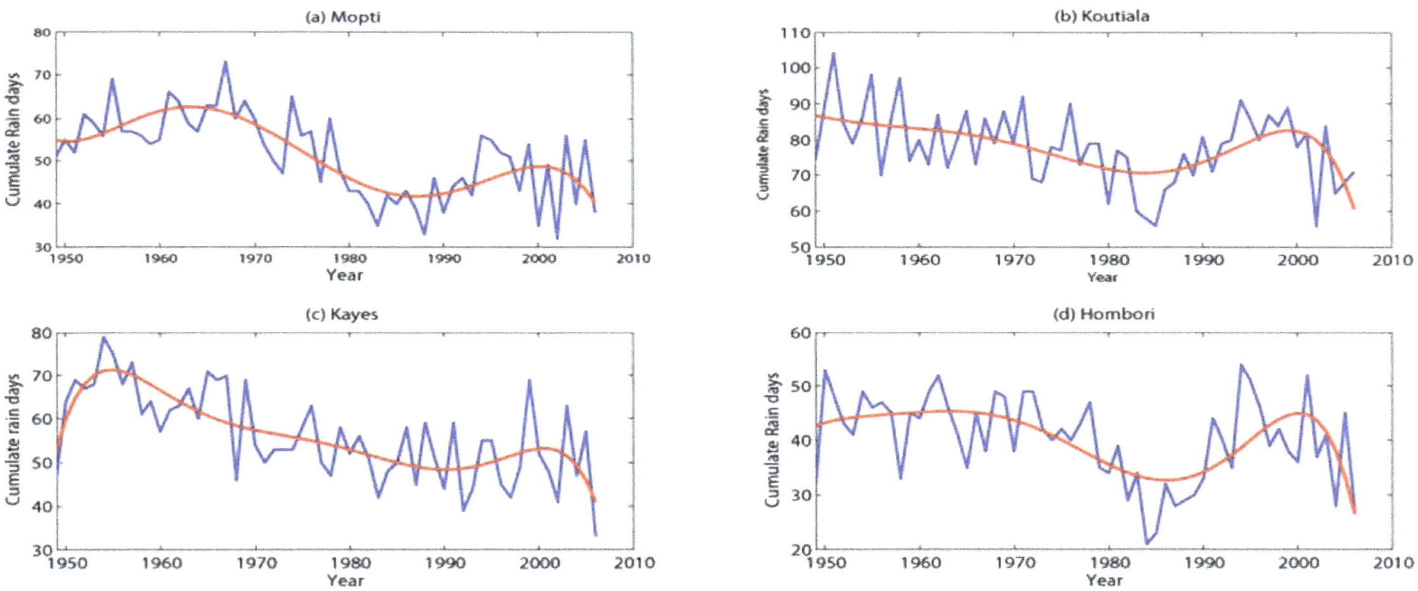

Figure 2. Total rainfall days per year, (a) Mopti, (b) Koutiala, (c) Kayes and (d) Hombori. The blue curve shows year-to-year total rainfall days (recorded rain days) and the red curve represent the recorded rain days using polyfit function.

$$f(x;g,\alpha,\beta,\lambda)=\begin{cases}\left(\alpha\beta g(x)^{\beta-1}+\alpha\lambda g(x)^{\beta}\right)e^{\lambda g(x)}g'(x)\exp\left\{-\alpha g(x)^{\beta}e^{\lambda g(x)}\right\}, & k<x<\infty\\ 0, & \text{elsewhere}\end{cases}$$

Where $g(x)$ is a continuous monotonic increasing function, k is is any positive real number, with $()\alpha,\lambda,\beta,x$ satisfying the conditions $\alpha>0,\lambda>0,\beta\geq 0,x>0$ this condition is also known as Weibull three parameters distribution probability function. Similarly, many distributions can be found by efficient parameters selection (Isaic-Maniu, 1983; Gumbel, 1958; Chausse, 2010).

Parameters estimation

In the wide fields of management, planning, design and prediction parameters play a key role in modeling and predicting. In all this paper, we selected $\lambda = 0$, $\alpha \neq 0$, $\beta \neq 0$. To be specific, some significant parameter estimation methods such as method of moments (MOM), maximum likelihood method (MLM), method of least squares (MLS) and the probability weighted moments (PWM) are used for the determination of Gumbel weibul distribution function parameters.

Method of moments

The method of moment is probably the oldest method for constructing an estimator. The advantage of this method is the simplicity, and the facility of implementation widely used. In MOM α and β are estimated as (Vivekanandan, 2013; Hall, 2006):

$$\alpha = \overline{R} - 0.5772157\beta,\ \beta = 0.60796821\left(\widehat{R}^2 - \overline{R}^2\right)^{0.5}$$

Where, $\widehat{R}^2 = \frac{1}{N}\sum_{i=1}^{N} R_i^2$ and $\overline{R}^2 = \left(\sum_{i=1}^{N} \frac{R_i}{N}\right)^2$ are the mean and standard deviation of the annual recorded data.

Maximum likelihood method

Maximum-likelihood method was first proposed for factor analysis by Bowman et al. (1985). This method consists of setting the Gumbel parameters which maximize the likelihood distribution function by the following formula (Vishwa et al., 2013):

$$\alpha = \beta\left[\ln N - \ln \sum_{i=1}^{N} e^{-\frac{R_i}{\beta}}\right],\ \beta = \overline{R} - \frac{\sum_{i=1}^{N} R_i e^{-(R_i/\beta)}}{\sum_{i=1}^{N} e^{-(R_i/\beta)}}$$

Where, R_i is the recorded annual cumulate rain based on the daily record of ith year and N is the sample size.

Method of least squares

The least-squares method is one of the most important estimation methods, which attempts to obtain such values of the factor loading and the unique variance that minimizes a different loss function. The parameters estimation is computed using the following formula (Vishwa et al., 2013; Manik and Datta, 1998).

$$\alpha = \left(\frac{\sum_{i=1}^{N} R_i}{N}\right) + \frac{\left(\sum_{i=1}^{N} \ln(-\ln(P_i))\right)\beta}{N},\ \beta = \frac{\left(\sum_{i=1}^{N} R_i\right)^2 - \left(N\sum_{i=1}^{N} R_i^2\right)}{\left(N\sum_{i=1}^{N} R_i\left(\ln(-\ln(p_i))\right)\right) - \left(\sum_{i=1}^{N} R_i\right)\left(\sum_{i=1}^{N} \ln(-\ln(p_i))\right)}$$

Where, $p_i = m/(N+1)$ and $\ln(-\ln(p_i))$ defines the cumulative probability of non-exceed for each R_i. Here, m is the rank assigned to each of the data points arranged in ascending order (rank 1 is assigned to the smallest value and N to the largest value).

Probability weighted moments

Probability weighted moment was introduced by Arthur et al. (1979) and Landwehr et al. (1979). This method takes into account the weight moment to determine the parameters computed as (Landwehr et al., 1978; Vishwa, 2013):

$$\alpha = M_{100} - 0.5772157\beta,\ \beta = \left(M_{100} - 2M_{101}\right)/\ln 2$$

Where, $M_{100} = \sum_{i=1}^{N} \frac{R_i}{N}$ and

$$M_{101} = \sum_{i=1}^{N} R_i \times (N-i)/(N(N-1))$$

Where R_i is the recorded annual cumulate record based on the daily record of ith year and N is the sample size.

Good fit techniques

In this section, we used good fit technique to check the compatibility of the methods.

Anderson-Darling test

Anderson-Darling (AD) test was initiated by Andersion and Darling (1954) as an alternative to statistical tests for detecting sample distributions. AD test is non-directional and can be applied to Normal, Weibull, and other types of distributions (Anderson and Darling, 1954; Heatcote, 2004). It is calculated from the following formula:

The A^2 statistic is defined by:

$$A^2 = (-N) - (1/N)\sum_{i=1}^{N}\left\{(2i-1)\ln(z_i) + (2N+1-2i)\ln(1-z_i)\right\}.$$

For a given sample of N values, $z_i = F(R_i)$ for $i = 1, 2, 3, \cdots, N$ and $R_1 < R_2 < \cdots < R_N$

Kolmogorov-Smirnoff test

Kolmogorov-Smirnoff (KS) test was first introduced by Kolmogorov (1933, 1941) and Mezbahur et al. (2006) as a test of the distance or deviation of empirical distributions from a postulated theoretical distribution. KS test for a given theoretical cumulative distribution is defined by:

$$KS = \max_{i=1}^{N}\left(F_e(R_i) - F_D(R_i)\right).$$

Where, $F_e(R_i) = (i-0.35)/N$ the empirical cumulative distribution function (CDF) of is R_i, $F_D(R_i)$ is the computed CDF of R_i. Here, 'i' is the rank assigned to each data point arranged in ascending order (Kolmogorov et al., 1933) (rank 1 is assigned to the smallest value and N to the largest value).

Diagnostic test

D-index statistics is defined by (United States Water Resources

Table 1. Simulation result of the four methods and four good fit techniques.

Site	Good fit test: Chi-square				Good fit test: D-index				Good of fit test: KS				Good fit test: Anderson-Darling			
	MOM	MLM	MLS	PWM	MOM	MLM	MLS	PWM	MOM	MLM	MLS	PWM	MOM	MLM	MLS	PWM
Koutiala	3.6194	1.7325	1.7345	4.4661	2.1571	2.2765	2.2777	3.1857	2.5998	0.1553	0.1562	54.438	0.1137	0.0412	0.0407	0.0102
Mopti	4.5054	2.6747	2.0937	8.8888	1.3228	1.3883	1.4049	1.4958	2.3737	0.2774	0.3378	13.986	0.1408	0.0733	0.0592	0.2481
Hombori	3.75	0.21463	1.3534	5.4947	2.7411	2.7437	2.9617	2.4858	2.4995	11.819	0.6507	49.686	0.0929	0.0279	0.0805	0.41067
Kayes	4.5353	92.624	0.54723	14.471	1.9374	3.3624	3.8472	1.5551	7.1462	805.25	60.2876	193.235	0.1729	0.357	0.0044	0.3809

Council, 1981):

$$D-index=\frac{1}{R}\sum_{i=1}^{6}\left|R_i-R_i^*\right|$$

Where, R, R_i and R_i^* are the average precipitation, the ith highest recorded and estimated annual cumulated records, respectively. The D-index test essentially gives weightage to the upper most six data points rather than the data points at lower levels. The least value D-index is considered as the best method.

Chi-square test

This distribution was introduced by the German statistician Friedrich Robert Helmert in 1876. He used distribution of the sample variance using the following formula.

$$\chi^2=\sum_{i=1}^{N}\frac{\left(R_i-R^*\right)^2}{R^*}$$

Where, R_i, i, N, R^* are the year recorded data, year, total year and the expected rain recorded, respectively

Simulation results

After computing the value of α and β for MOM, MLM, MLS and PWM, we estimated each method using good fit technique, the result of the simulation is presented in Table 1.

From Table 1, Anderson-Darling technique in general provided small values for all the methods then the other good fit test techniques. It may be noted that the good fit test technique of Anderson-Darling is suitable for the four methods: MOM, MLM, MLS and PWM, when compared with the other values given by Chi-square, D-index and Kolmogoro-Sminov test. Anderson-Darling good fit test for all the sites precipitation analysis is suggested. To improve the efficiency of the modeling, we proposed to use the return period and the standard error.

Return period

Estimation of rainfall for a desired return period is a pre-requisite for planning, which can be achieved by probabilistic approach involving fitting of probability distributions to the recorded data which can be done using the return period and the standard error. The return period is defined by Bedient and Huber (1948) as an annual maximum event that has a return period (or recurrence interval) of T years. The return period and the standard error are computed by the following formulas using the parameters α,β from Table 1 to compute the return period.

$$R_T=\alpha+Y_T\beta, SE=\frac{\beta}{\sqrt{N}}(1.1589+0.1919Y_T+1.1Y_T^2)^{0.5}$$

Where, $Y_T=-Ln(-Ln(1-(1/T)))$ and T is the year (period). For the test, we used the lower and upper confidence limits (LCL and UCL) of the estimated rainfall

Table 2. Ratio of standard error to return period of sites for the four methods.

Site\method		MLM	MLS	PWM
Koutiala	Min	2.79	2.78	**1.27**
	Max	7.99	8.5	5.18
Mopti	Min	3.07	3.44	**1.74**
	Max	8.35	8.75	6.27
Hombori	Min	7.82	4.84	**2.64**
	Max	26.03	10.01	7.72
Kayes	Min	**0.47**	3.24	1.19
	Max	2.5	8.57	5

(R_T) at 95% significance level for each return period (T) represented by:

$$LCL=R_T-1.96(SE),\ UCL=R_T+1.96(SE)$$

RESULTS AND DISCUSSION

In order to make the final selected method for each site, we used the Anderson good fit test technique to compute the standard error and the return period, then we introduced the ratio of the standard error to return period for the period; the result of the simulation is presented in Table 2 as estimated in percent.

From Table 2, the minimum ratio (in bold) are

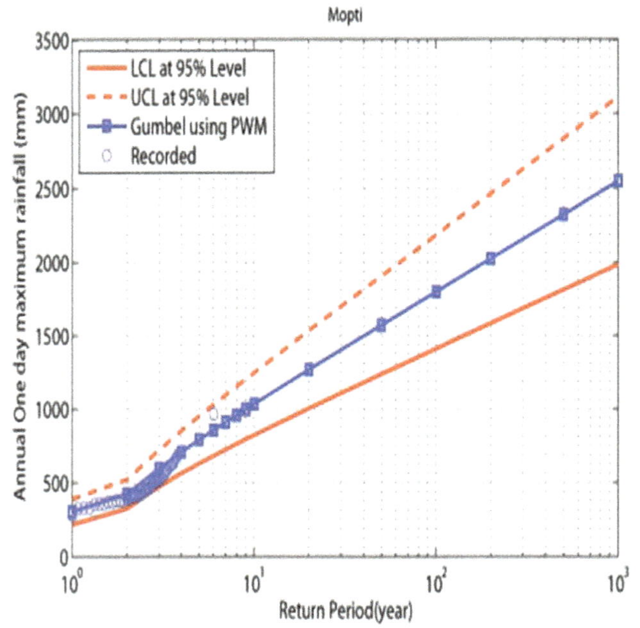

Figure 3. Plot of the recorded and estimated precipitation for the site of Mopti.

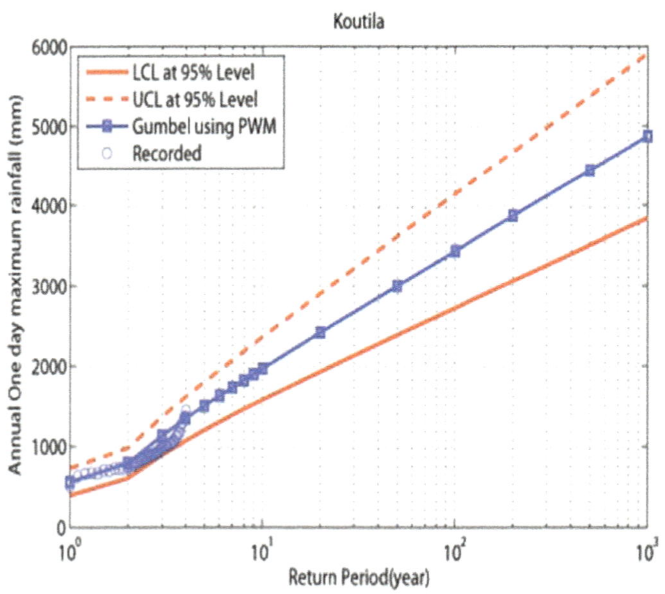

Figure 4. Plot of the recorded and estimated precipitation for the site of Koutiala.

1.27, 1.74, 2.64 and 0.47 for Koutiala, Mopti Hombori (PWM method) and Kayes (MLM method), respectively. So, it may be noted that the percentage of variation of estimate precipitation between the upper and lower limit will be more suitable for small value of the ratio. Finally, the method PWM is considered for modeling the sites of Mopti, Koutiala, and Hombori using PWM and MLM for Kayes (Figure 3, 4, 5 and 6). The plotting of the return period is presented by Figures 3, 4, 5 and 6. The recorded and the estimated precipitation are also represented.

Figures 3, 4, 5 and 6 represent the plotting of the recorded and estimated precipitation using Gumbel Weibull distribution function for the sites of Mopti (Figure 3), Koutiala (Figure 4) and Hombori (Figure 6) using PWM and Kayes (Figure 5 using MLM), with confidence

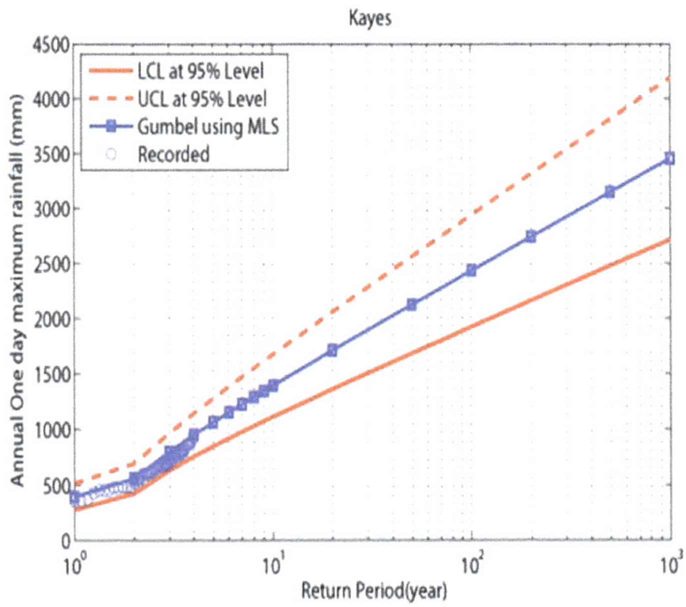

Figure 5. Plot of the recorded and estimated precipitation for the site of Kayes.

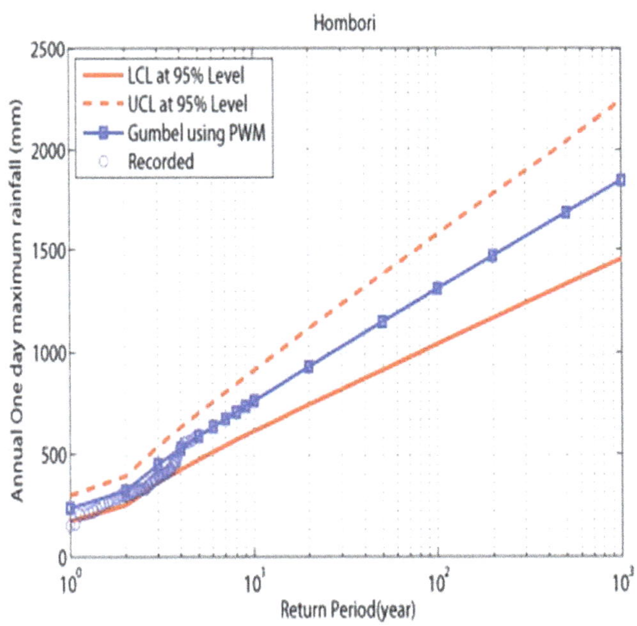

Figure 6. Plot of the recorded and estimated precipitation for the site of Hombori.

limit at 95% level.

The plot showed that the percentages of variations on trends in annual precipitation are strictly inside the confidence limits at 95% (between the UCL and UCL). We can conclude that the selected good fit technique (Anderson-Darling) and methods (PWM, MLM) for predicting are the best estimation.

CONCLUSION

Predicting and forecasting rainfall is very important in risk assessment and decision making. Any change in precipitation affects socio-economic developments and human livelihood. To reduce the risk, a prediction tool is necessary to improve the regional climate modeling.

Based on the limitation of the frequency analysis (Figures 1 and 2) due to the complexity of rainfall analysis, we proposed Gumbel Weibull probability function through the four methods of parameters computation (MLM, MLS, PWM and MOM). The four good fit tests are also used to estimate the efficient method. The results of the good test show the efficiency of Anderson-Darling techniques as compared to the Komorov-smirnov, D-Index and Chi-square proposed techniques. The results of the minimum ratio show the efficiency of PWM method for Koutial, Mopti and Hombori and MLM for Kayes precipitation modeling and predicting. The plotting results of the estimated and the recorded precipitation, show the efficiency of the selected method and test technique (Figures 3, 4, 5 and 6).

Conflict of interests

The authors did not declare any conflict of interest.

REFERENCES

Anderson TW, Darling DA (1954). A test of Goodness of fit. J. Am. Stat. Assoc. 49(268):765-769.

Arthur JG, Maciunas JL, Matalas NC, Wallis JR (1979). Probability Weighted Moments: Definition and Relation to Parameters of several Distributions Expressable in Inverse Form. Water Resour. Res. 15(5):1049-1055.

Bedient P, Huber W (1948). Hydrology and Flooding, Analysis. Addison Wesley Publishing Co.

Bowman KO, Shenton LR (1985). John Wiley and Sons Canada. 5:467-453.

Chausse P (2010). Computing Generalized Method of Moments and Generalized Empirical Likelihood with R." J. Stat. Softw. 34(11):1-35.

Consul PC, Shenton LR (1972). Use of Lagrangian expansion for generating generalized Probability distribution. SIAM J. Appl. Math. 23(2):239-248.

Consul PC, Shenton LR (1973). Some interesting properties of Lagrangian stributions. Commun. Statist. Simul. Computat. 2(3):263-272

Cousineau D, Brown S, Heathcote A (2004). Fitting distributions using maximum likelihood: Methods and packages. Behav. Res. Methods Instrum. Comput. 36 (4):742-756.

Foxa P, Rockstrom J (2003). Supplemental irrigation of dry-spell mitigation of rainfall agriculture in the Sahel. Agric. Water Manage. 61:29-50.

Gumbel EJ (1958). The Statistics of Extremes, New York: Columbia University Press.

Hall AR (2006). Generalized Method of Moments (Advanced Texts in Econometrics). Oxford University Press.

Heathcote A, Brown S, Cousineau D (2004). QMPE: Estimating Lognormal, Wald and Weibull RT distributions with a parameter dependent lower bound. Behav. Res. Methods Instrum. Comput. 36:277-290.

Helmert FR (1876). Ueber fie Wahrscheinlichkeit der Potenzsummen der Potenzsummen der Beobachtungsfehler und uber einige damit im Zusammenhange stehende Frragen. Zeitschrift für Mathematik und Physik 21:S.102-219.

Isaic-Maniu A (1983). Metoda Weibull - aplicatii, Ed. Academiei, Bucharest.

Jones PD, Osborn TJ, Briffa KR, Folland CK, Horton B, Alexander LV, Parker DE, Niasse M (2005). Climate Induced Water Conflict Risks in West Africa: Recognizing and coping with increased climate impacts on shared watercourses. An International Workshop organized by centre for study of Civil War, International Peace research institute, Oslo (PRIO) and Centre for International Environmental and Climate research at University of Oslo (CICERO) for the global environmental change and human security program (GECHS).

Kandji ST, Verchot L, Mackenson J (2006). Climate change and Variability in the Sahel Region: Impacts and Adaptation Strategies in the Agricultural Sector. World Agroforestry Centre/United Nations Environment Programme.

Kolmogorov AN (1941). Confidence limits for an unknown distribution function. Annals of Mathematical Statistics. 12:461-463.

Kolmogorov AN (1933). Sulla Determinazione empirica di une legge di distribuzione. Giornale dell'Istituto Italiano degli Attuari. 4: 83-91.

Landwehr JM , Matalas NC, Wallis JR (1978). Some comparisons of flood statistics in real and log space. Water Resour. Res. 4(5): 902-920.

Landwehr JM, Matalas NC, Wallis JR (1979). Probability weighted moments compared with some traditional techniques in estimating Gumbel parameters and quantiles, Water Resour. Res. 15(5):1055-1106

Manik D, Datta SK (1998). A comparative study of estimation of extreme value. J. River Behav. Control 25(1):41-47.

Mezbahur R, Larry MP, Herbert CH (2006). A modified Anderson-Darling Test for uniformity. Bull. Malays. Math. Sci. Soc. 20(1):11-16.

Phien HN (1987). A review of methods of parameter estimation for the extreme value type–1 distribution. J. Hydraulics 90 (3 & 4): 251-268.

Udual CUI, Ini DE (2012). Analysis of Rainfall Trends in Akwa Idom State, Nigeria. J. Environ. Earth Sci. 2(8):60-70.

United States Water Resources Council (USWRC) (1981). Guidelines for determining flood flow frequency, Bulletin. 17B:50

Vishwa N M, Diwinder K A (2013). Exact Modelling and Analytical Study of Annual Maximum Rainfall with Gumbel and Frechet Distributions Using Parameter Estimation Techniques World of Sciences Journal. (02).

Vivekanandan N (2013). Inter-comparison of parameters of Gumbel Distribution for Rainfall Frequency Analysis. J. Environ. Sci. Comput. Sci. Eng. Technol. 2(4):1336-1346.

Zhang J (2002). Powerful goodness-of-fit tests based on the likelihood ratio. J. R. Stat. Soc. 64(2):281-294.

Geochemistry and hydrogeochemical process of groundwater in the Souf valley of Low Septentrional Sahara, Algeria

CHEBBAH Mohamed and ALLIA Zineb

Centre universitaire de Mila, LSNM, BP 26 RP, 43000, Mila, Algeria.

Ground water is the only hydrous resource exploited in the Souf valley of Algerian Sahara. Several techniques and conventional graphical plots were used to define the geochemistry of the aquifer. Based on the ionic components, hydrochemical facies, the factors controlling groundwater chemistry are assessed and defined. Ground waters are highly saline (total dissolved solids (TDS being 2361-14780 mg/L) and show high contents of major ions that extensively exceed the admitted norms of portability. The relative abundance of major ions was Na^+ > Mg^{2+} > Ca^{2+} > K^+ for cations and Cl^- > SO4 $^{2-}$ > HCO_3^- > NO3$^-$ for anions. Two main waters types are identified; one is sodic facies rich in magnesium, the second is mixed facies. Their spatial distribution is strongly influenced by different parameters linked to the aquifer system in the study area (lithology, structure, conditions and underground circulation directions). Dissolution of gypsum, salts, weathering of carbonate and silicate, ion exchange, mixing and nitrification of the organic matter are the main hydrogeochemical processes of waters salinity. The conjugated action of the different mineralisation processes as well as their combinations had accentuated the salinity of these waters. Their identification allowed a better interrelationship between the different present ions. Indeed, the spatial chemical facies distribution observed is closely linked to the lithologic nature and to the geometry of the aquifer. In this region deprived of treatment means and softening, when choosing captive fields and implantation of drilling of potable water supply, these results would guide decision makers in selecting the right sectors containing less ionised groundwater.

Key words: Groundwater resource, hydrogeochemistry, mineralisation processes, Souf valley, Algerian Sahara.

INTRODUCTION

Chemistry of groundwater is an important factor determining its use. Interaction of groundwater with aquifer minerals through which it flows greatly controls the groundwater chemistry. Hydrogeochemical processes that are responsible for altering the chemical composition of groundwater vary with respect to space and time. The hydrogeochemical processes of the groundwater system help to obtain an insight into the contributions of rock/soil-water interaction and anthropogenic influences on groundwater. These geochemical processes are respon-

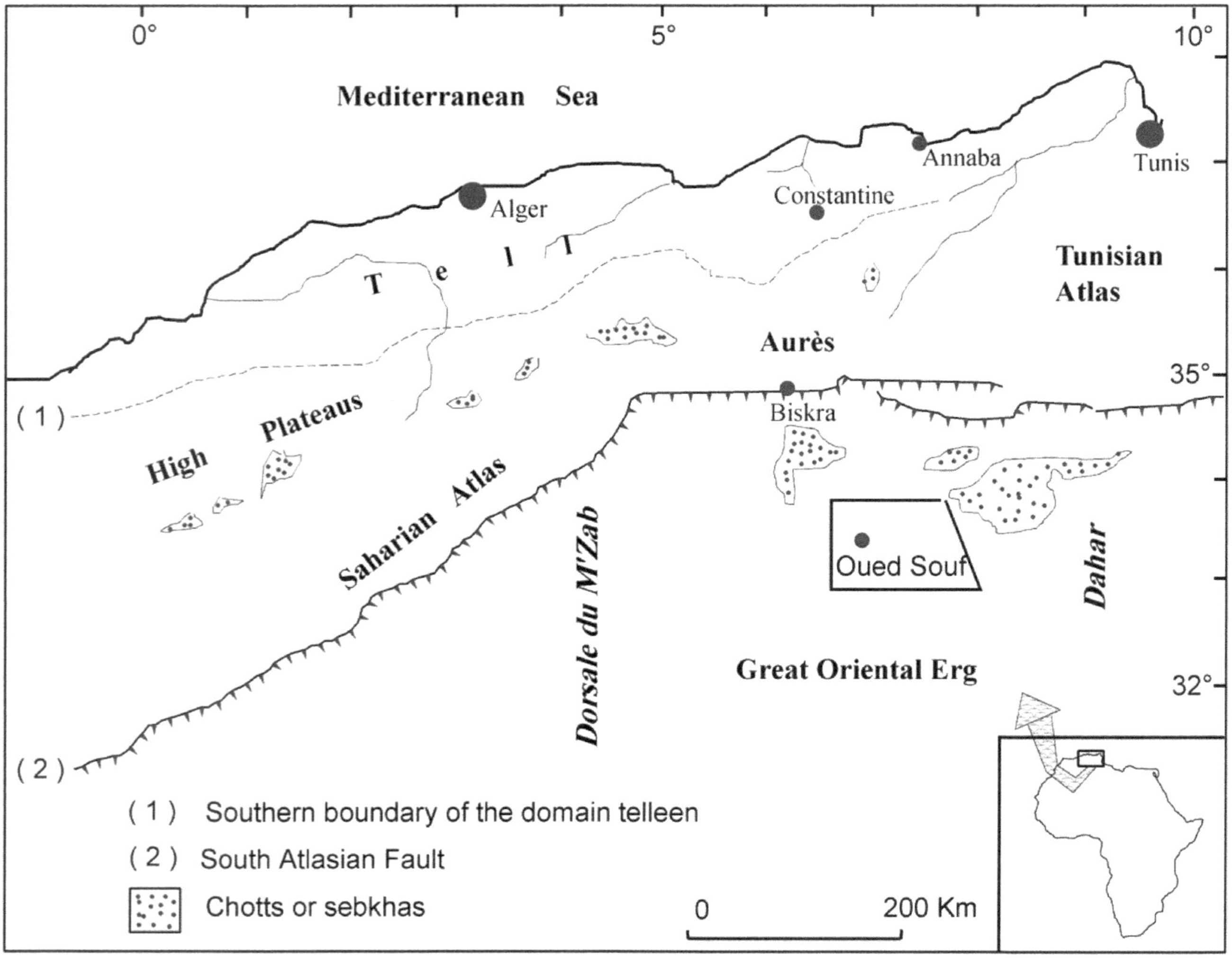

Figure 1. Geomorphologic units of North Africa; in framed: Souf valley.

sible for the spatiotemporal variations in groundwater chemistry (Matthess, 1982; Michard, 1989; Kumar et al., 2006; Swarna Latha and Nageswara Rao, 2011; Kuldip-Singh et al., 2011). Groundwater chemistry, in turn, depends on a number of factors, such as general geology, degree of chemical weathering of the various rock types, quality of recharge water and inputs from sources other than water-rock interaction (Schuh et al., 1997; Kuldip-Singh et al., 2011). Such factors and their interactions result in a various groundwater quality.

In the Souf valley, Low Septentrional Sahara, groundwater is the only source for domestic purposes. Due to rapid urbanization, extension of irrigation, domestic needs of the people has increased tremendously in recent years (D.H.W., 2011). Although some studies have been reported on chemical quality and hydrogeochemical processes operated in other regions of Low Septenrtional Sahara, very little work has been done in the Souf valley. Groundwater is being continuously exploited in the region to meet the growing demand for water supply and irrigation, which is aggravated due to insufficiency of surface water, it is essential to know the hydrogeochemical processes that occur in the aquifer system. Therefore, the main objective of this study is to characterize the aquifer waters and identify the major hydrogeochemical processes that are responsible for groundwater chemistry in the study region.

MATERIALS AND METHODS

The Souf valley is located between the longitudes 6°40' – 6°70' East and the latitudes 33° 41' –33°10' North and covers an area of 80 000 km^2. It is limited in the east by the Tunisian frontier, in the west by the dorsal of the M'Zab, in the south by the extension of Great Oriental Erg and by Choots region (Melrhir and Merouane) in the north (Figure 1). This area is characterised by a hyper arid type climate. The mean annual temperature is 22°C (average of: July 34°C and January 11°C). The diurnal variations temperatures are important in any season and the evapotranspiration potential exceeds 1,700 mm year^{-1}. The region is characterized by low

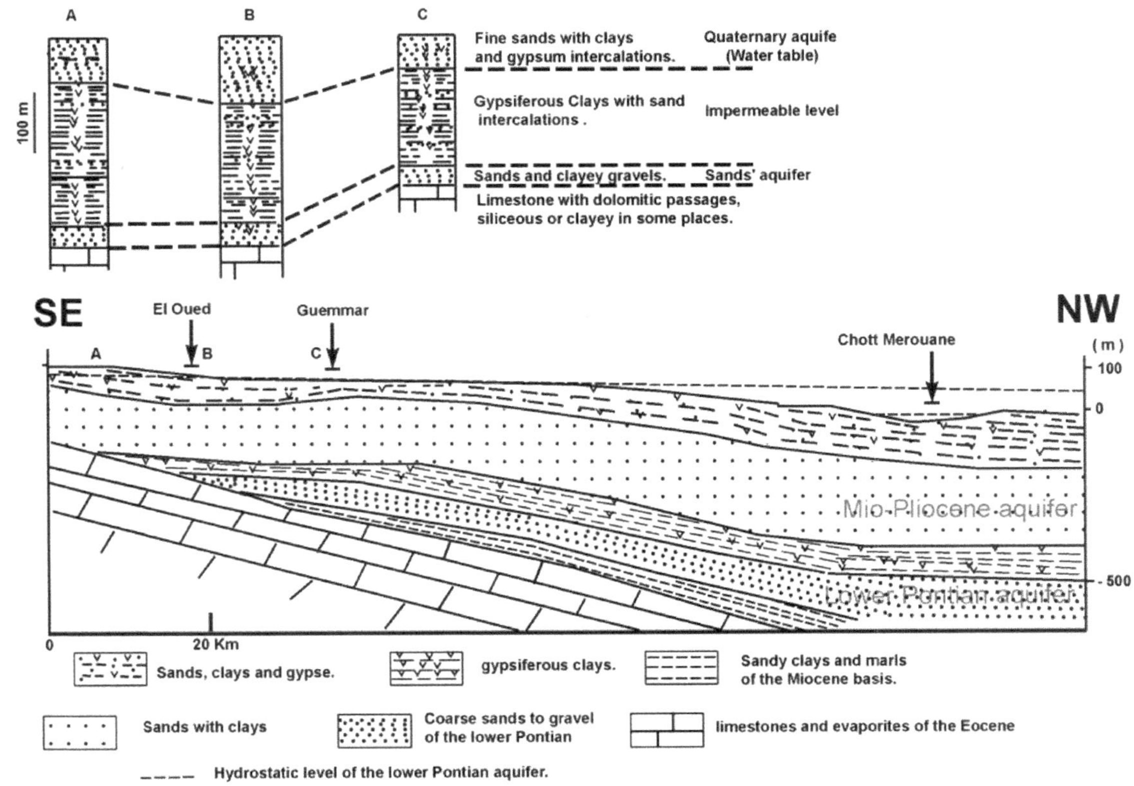

Figure 2. Up: Log correlation of Souf region; down: hydrogeological cross-section of Souf region (After Nelson modified).

rainfall; the average annual rainfall is about 70 mm (located between 60 and 80 mm isohyets). January is the wettest month (approximately 15 mm) and July the driest (less than one mm in average) (D.H.W., 2011). It also forms an independent unit that constitutes one of the transition zones between the Saharan platform and the pre-Saharan through. It is characterized by its extended dunes with a contrasted relief and by its nearly closed local depression (Nesson, 1975).

This valley, presenting a synclinal structure, is filled by thick Meso-cenozoic sedimentary formations, often discordant, essentially composed of detrital, carbonate or evaporates deposits and based on a pre-Cambrian folded basement. Also, the greater part of these sedimentary formations is covered with sand dunes and pebbly or gravely erg thus forming the eastern part of "Great Oriental Erg". This erg, with a tormented relief, rests on a quaternary formation consisting of several tens of meters of compact and homogeneous eolian sand. This formation, which contains the water-table, sees its substratum descend to the north; its waters with those of surface runoff are spreading in the lagoons of salty soil where it evaporates depositing superficial salt crust (Nesson, 1975; Busson, 1970).

The lithostratigraphy of the study region is summarized in Figure 2: from the Cretaceous to the Eocene, successively consists of marly clays, limestone, gypsiferous marly clays and marls with lagunal facies (gypsum, anhydrid, salts and dolomitic passages) on the top. This set is surmounted by the Neogene composed of conglomerates and gravel of the Oligocene and by subhorizontal red sandy clays with gypsiferous sandstones lentils of the Mio-Pliocene. The whole is surmounted by one layer of fine sands with limestone crusts of the Plio-Quaternary (Cornet, 1961; Busson, 1970; Giraud, 1978). These sands, forming the potential aquifer, are constituted of 90% of quartz grains, limestone concretions, gypsum and silicates (Feldspars and silicate accessories: epidote zircon, rutile, tourmaline and garnet) (Dervierux, 1957; Nesson, 1975).

From the hydrogeological viewpoint, in the study area, as throughout the Lower Sahara, geomorphology and disposition of sediments in the basin have favoured the infiltration and subterranean circulation of water precipitated on outcrops with concentric halos of periphery. The alternation of permeable and impermeable deposits assured them charging as they flowed toward the depressed parts. These particularly favorable paleogeographic and geological conditions are the cause of the underground aquifers formation during ancient rainy periods (Nesson, 1975). In this area characterized by a marked aridity and limited surface flows, it is these groundwaters that make its originality, because independence of present of climatic fluctuations, except the saturation level, they form several aquifers regrouped into two major groups. From low to top, one finds: the intercalary continental aquifer (I.C), the terminal continental aquifers (T.C.) including limestones aquifer (Senonian-Eocene), sand aquifer (Mio-Pliocene) and Quaternary aquifer (water table) (Nesson, 1975; Ould, 2005).

Groundwater sampling and analyses

Samples of groundwater from sites in the study area were collected in January-March 2011. *In situ* parameters such as pH, electrical conductivity (EC), temperature and alkalinity (HCO_3) were measured in the field. Samples for major chemical analyses were filtered and collected in plastic bottles. Chemical analyses were done by Hydraulic Direction of Oued Souf's Willaya. The geographical position of analyzed water points is shown in Figure 3.

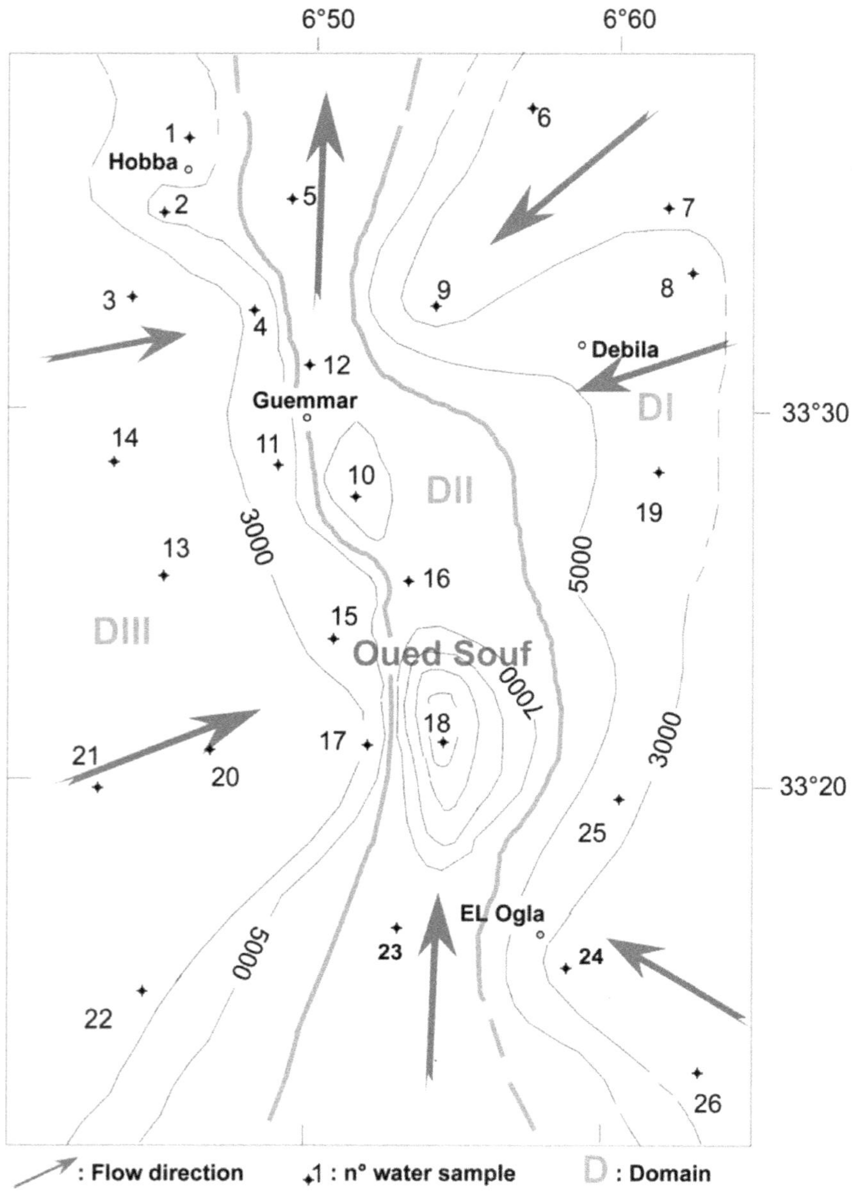

Figure 3. Water sampling locations, iso-concentrations map of total dissolved solids (TDS) in mg/l and domains from Souf valley.

RESULTS AND DISCUSSION

Groundwater chemistry

The statistical parameters (mean, standard deviation, variation coefficient, maximum-minimum) have been calculated for each of the physico-chemical variable and are given in Table 1. Except the carbonates and the pH which show acceptable values, all measured parameters have mean and extreme values that exceed largely the limits of the admitted portability norms.

The total dissolved solids (TDS) ranged from 2361 to 14780 mg/l with a mean of 4277 mg/l. The electrical conductivity (EC) ranged from 3100 to 20000 µs/cm with a mean of 5765 µs/cm. Groundwater was found to be alkaline in nature with an average pH value of 7.6. The relative abundance of major ions was $Na^+ > Mg^{2+} > Ca^{2+} > K^+$ for cations and $Cl^- > SO_4^{2-}$ $HCO_3^- > NO_3^-$ for anions. The variability of elements in solution, expressed by the variation coefficient or by standard deviation, is very important for the majority of the analyzed parameters (Table 1) and indicates that those waters present a large difference of chemical quality from one point to another. However the spatial variability of temperature (m: 22.1, cv: 3.66%) and pH (m: 7.6, CV: 3.80%) remains low. Finally, one notes that the highest variation coefficient, corresponding to potassium variation, cannot be considered because it is essentially due to the high content of

Table 1. Statistical summary of hydrogeochemical parameters of groundwater in Souf valley.

Parameter	Temperature (°C)	pH	Cond (s/cm)	R.S (mg/l)	Ca (mg/l)	Mg (mg/l	Na (mg/l)	K (mg/l)	SO_4 (mg/l)	Cl (mg/l)	HCO_3 (mg/l)	NO_3 (mg/l)
m	22.1	7.6	5765	4277	372	372	948	64	19.65	13.22	140	94
S	0.81	0.29	3889	2825	136	242	1309	194	796	2045	101	45.9
Cv %	3.66	3.80	67.45	66.05	36.6	65.15	138.1	302.1	40.52	154.8	72.3	49
Mini	21.0	7.3	3100	2361	202	153	115	11	1263	216	53	1
Maxi	23.7	8.7	20000	14780	725	1348	5350	909	5000	8582	425	202
Delta	2.7	7.4	16900	12419	523	1195	5235	898	3737	8366	372	199

Table 2. The ionic formula showing water type of samples analysis in the study area.

Ca	Mg	Na	K	SO_4	Cl	HCO_3	NO_3	Water type
25.82	35.68	37.52	0.98	50.31	44.1	1.87	3.72	Na-Mg-SO_4
18.0	34.11	47.14	0.75	51.05	45.31	2.07	1.57	Na-Mg-SO_4
28.13	39.61	31.23	1.03	77.76	15.18	2.45	4.61	Mg -Na- SO_4
40.24	40.38	18.77	0.61	63.31	32.37	1.97	2.35	Mg –Ca- SO_4
11.41	35.17	52.32	1.1	48.03	47.24	2.87	1.86	Na- SO_4-Cl
43.97	39.74	15.38	0.92	79.09	15.81	3.06	2.04	Ca- Mg - SO_4
44.9	42.59	11.45	1.06	78.73	15.73	3.44	2.1	Ca- Mg-SO_4
27.71	46.89	27.79	0.61	64.49	29.46	3.75	2.3	Mg -Na- SO_4
27.07	35.98	35.01	1.94	49.87	38.98	5.98	5.17	Mg -Na- SO_4-Cl
48.36	31.23	19.68	0.73	80.14	15.36	2.53	1.97	Ca- Mg - SO_4
5.5	19.72	74.14	0.64	25.17	72.11	1.66	1.06	Na-Cl
18.62	48.05	32.23	1.1	63.09	30.93	3.33	2.65	Mg -Na- SO_4
12.03	28.2	59.19	0.58	51.66	43.25	2.52	2.57	Na- SO_4
42.05	30.31	26.72	0.92	76.36	18.55	2.45	2.64	Ca- Mg - SO_4
22.34	41.36	35.79	0.51	73.06	20.19	3.33	3.42	Mg –Na-SO_4
28.44	41.72	29.15	0.69	59.57	35.8	3.55	1.08	Mg -Na- SO_4
10.38	30.0	59.17	0.45	49.74	44.0	6.24	0.02	Na-SO_4-Cl
24.32	43.48	31.58	0.62	71.78	21.71	4.03	2.48	Mg -Na- SO_4
8.7	27.51	57.72	5.77	29.43	68.29	1.97	0.31	Na-Cl
27.27	57.67	14.35	0.71	51.74	42.79	2.81	2.65	Mg -SO_4
12.32	45.20	41.17	1.31	58.49	36.9	3.95	0.66	Mg -Na- SO_4
12.89	45.67	39.88	1.56	51.41	40.29	5.25	3.05	Mg -Na- SO_4
10.37	29.98	59.25	0.40	50.80	43.44	5.58	0.19	Na- SO_4
43.53	44.33	11.40	0.74	79.93	15.7	1.86	2.51	Mg -Ca- SO_4
27.19	35.24	35.67	1.89	49.26	38.14	10.82	1.79	Na-Mg -SO_4-Cl
26.88	58.20	14.25	0.67	51.96	42.34	2.93	2.77	Mg -SO_4

the 18 well (k:909 mg/l) whereas for the other wells, the mean is 23.5 mg/l with a variation coefficient of 54%, this can be explained by accentuated saltiness of the 18 well.

Characterisation of hydrogeochemical data

To identify and classify the different waters of the study area, hydrochemical data were processed and projected on the hydrochemical diagrams that aim to facilitate the interpretation of evolutionary trends. They are particularly in groundwater systems, when they are interpreted in conjunction with distribution maps and hydrochemical sections. An overall characterisation of hydrogeochemical data can be possible by knowing the hydrochemical facies of water, generally known as water type, using the ionic composition and various plots.

Ionic formula classification

The ionic formula calculation (Table 2), where the elements relative concentrations are expressed in % with regard to the sum of the same sign elements concentrations, reveals highly variable chemical characteristics. Thus, some of those waters present a magnesian

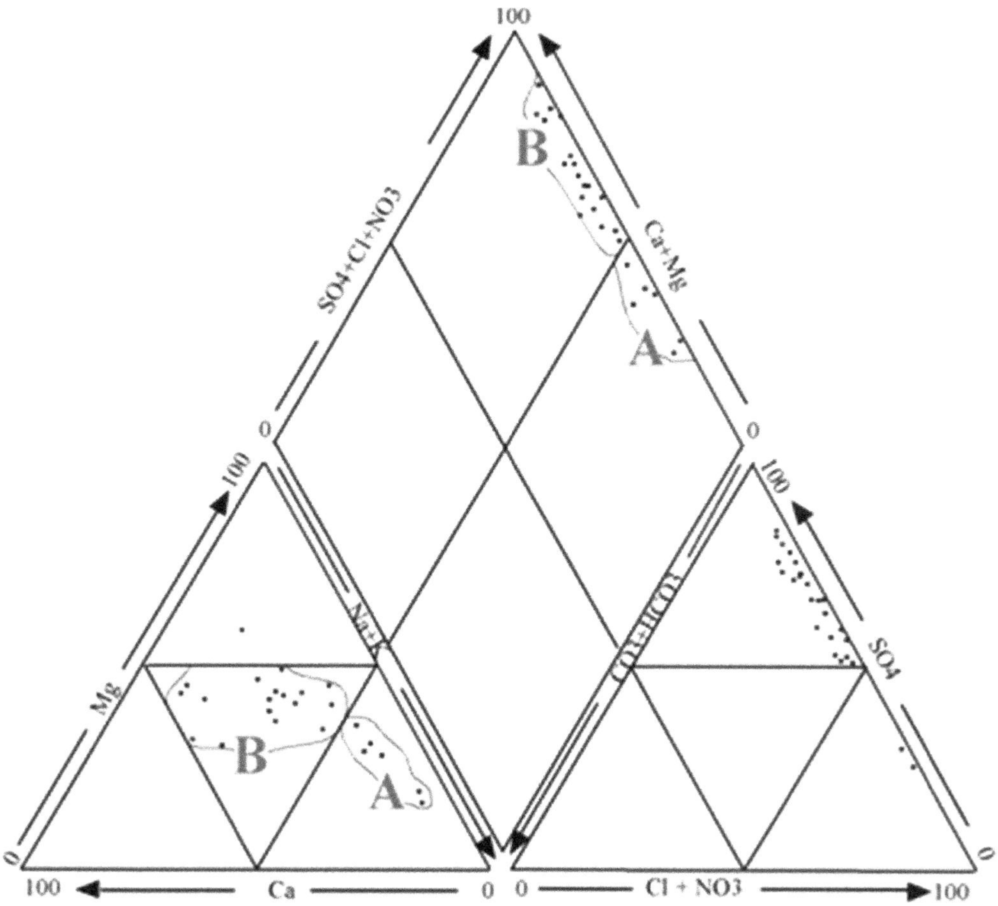

Figure 4. Piper diagram showing the type of water in the study area.

or sulfated calcic facies secondarily chlorided sodic whereas others have a sodic sulfated facies secondarily magnesian sulfated or clearly chlorided sodic.

Piper classification

Results of the projection on the piper diagram (1944) (Figure 4) show that for cations, the waters of different wells are grouped essentially in two main families «A» and «B». The «A» family's waters are sodic with the high concentrations in magnesium, while those of the family «B» present a mixed facies with relatively high magnesium contents except of well 24 that presents a magnesian facies.

For the anions, the distinction between the two families is not clear, in fact, one notes a reversed evolution of SO_4 in comparision with ($Cl+NO_3$) of «A» toward «B» (SO_4 ↑, ($Cl + CO_3$)↓ or inversely). For the whole cations-anions, the distinction between the two families is distinctly clear however; one records an advanced natural evolution of these waters toward a sodic sulfated facies or sodic and potassic chlorided facies (evolution of «B» toward «A»).

Identification of hydrogeochemical process

Reactions between groundwater and aquifer minerals have a significant role on water quality, which are also useful to understand the genesis of groundwater. The hydrogeochemical data are subjected to various conventional graphical plots to identify the hydrogeochemical processes operating in aquifer region of study area. Some of the possible identified processes are explained below.

Weathering and dissolution

Calcium, magnesium and sodium are the dominant cations present in groundwater in this region. Similarly, chloride and sulfate are also present in important amounts. The limestone, gypsum and silicate are the major sources for mineralization in the area. The ions from these sources might have been dissolved and added to the groundwater system by leaching and mixing processes. However, minerals weathering and dissolution can be suggested that a simple plot of Ca^{2+} + Mg^{2+} versus SO_4^{2-} + HCO_3^- could provide information on the relative importance of the main minerals contributing to

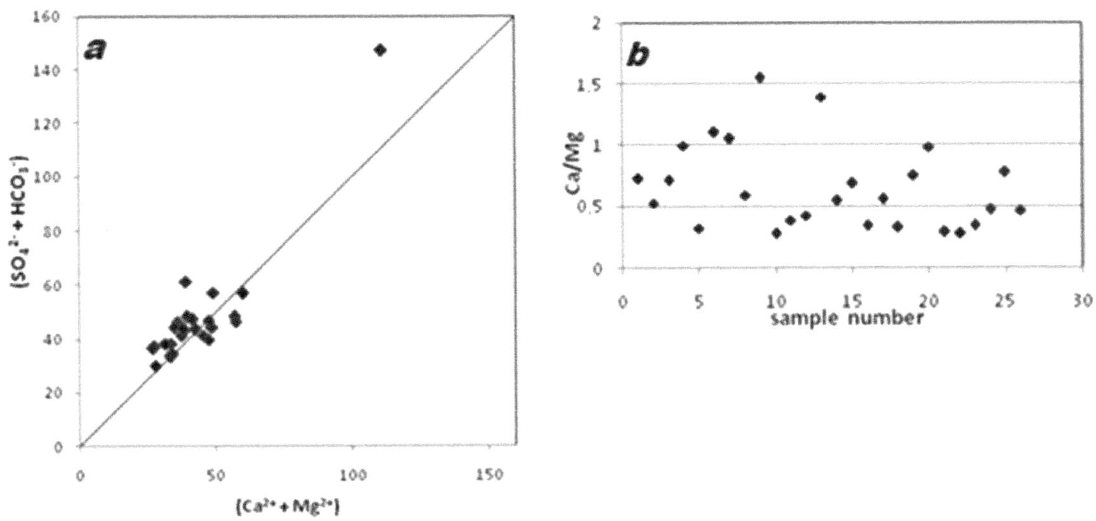

Figure 5. a. (Ca^{2+} + Mg^{2+}) versus (SO_4^{2-} + HCO_3^-) diagram, **b.** Diagram of Ca^{2+} / Mg^{2+} molar ratio for groundwater in the study area.

groundwater mineralization. In Ca^{2+} + Mg^{2+} versus SO_4^{2-} + HCO_3^- (Figure 5a), the points falling along the equiline (Ca^{2+} + Mg^{2+} = SO_4^{2-} + HCO_3^-) suggest that these ions have resulted from weathering of carbonates and silicates (Datta and Tyagi, 1996; Rajmohan and Elango, 2004; Kuldip-Singh et al., 2011). Most of the points in this study fall around the 1:1 line, suggesting that carbonate and silicate weathering are the major hydrogeochemical process operating in this region.

On the other side, if the ratio Ca/Mg = 1, dissolution of dolomite should occur, whereas a higher ratio is indicative of greater calcite contribution (Maya and Loucks, 1995). Molar ratio of Ca/ Mg>2 indicates the dissolution of silicate minerals, which contribute calcium and magnesium to groundwater (Katz et al., 1998). The plot of Ca/Mg ratio of the groundwater samples of the study area reveals that the dominant process is the dissolution of carbonate that present in the calcareous concretions (Figure 5b). The most of the groundwater samples found to be below 1 ratio indicate the calcite weathering prominent process. However, samples lie near the 1 ratio line revealing dolomite dissolutions and the samples lie above the ratio line 2 indicating the effect of silicate minerals. Presence of carbonates and silicates in the sand deposits could favor the weathering process.

In silicate, if the calcium and bicarbonate in groundwater solely originated from calcite, the equivalent ratio of dissolved Ca and HCO_3 in the groundwater is 1:2, whereas from dolomite weathering, it is 1:4 (Garrels and Mackenzie, 1971). Similarly, if the calcium and sulphate in groundwater derived from dissolution of gypsum or anhydrite, then the Ca/SO_4 ratio is almost 1:1 (Das and Kaur, 2001). In Ca versus HCO_3 - scatter diagram (Figure 6a), only two groundwater samples fall on the 1:2 line and most of them lie above the 1:2 equiline, indicating the contribution of calcite weathering on groundwater chemistry. Moreover, in Ca versus SO_4 diagram (Figures 6b and 7a), there are all distributions of the samples above the 1:1 equiline. This shows the dominance of carbonate weathering rather than silicate (Garrels and Mackenzie, 1971). Furthermore, it indicates the presence of dissolution of gypsum in the study area. Therefore, the abundance of Ca + Mg in groundwater can be attributed mainly to gypsum dissolution and carbonate weathering.

Also, silicate weathering can be understood by estimating the ratio between Na^+ + K^+ and total cations (TZ^+). The relationship between Na^+ + K^+ and TZ^+ of the study area revealed that most of the samples are plotted above the Na^+ K^+ = $0.5TZ^+$ line (Figure 7b). This observation indicates the involvement of silicate weathering in the geochemical processes, which contributes mainly sodium and potassium ions to the groundwater (Stallard and Edmond, 1983; Sarin et al., 1989; Kuldip-Singh et al., 2011). Moreover, the (Ca^{2+} + $Mg2^+$)/HCO_3^- ratio of more than 3 of all samples suggests that silicate weathering occurs in this region in addition to the carbonate dissolution.

Ion exchange

The chemical reactions in which ion exchange between the groundwater and aquifer environment occurs during residence or travel can be understood by studying the Chloro Alkaline Indices I and II:

Ratio I: Cl / ($Cl+HCO_3$):

Ratio II: (Na + K) / (Na + K + Ca)

Where, the concentration of ions is in meq/l.

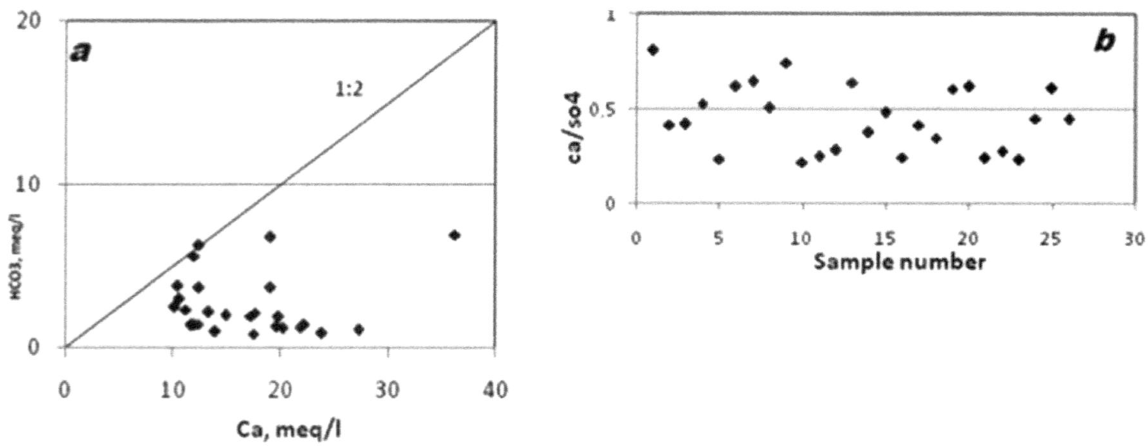

Figure 6. a. Diagram of Ca^{2+} / SO_4^{2+} molar ratio, b. Diagram of Ca^{2+} / SO_4^{2-} molar ratio for groundwater in the study area.

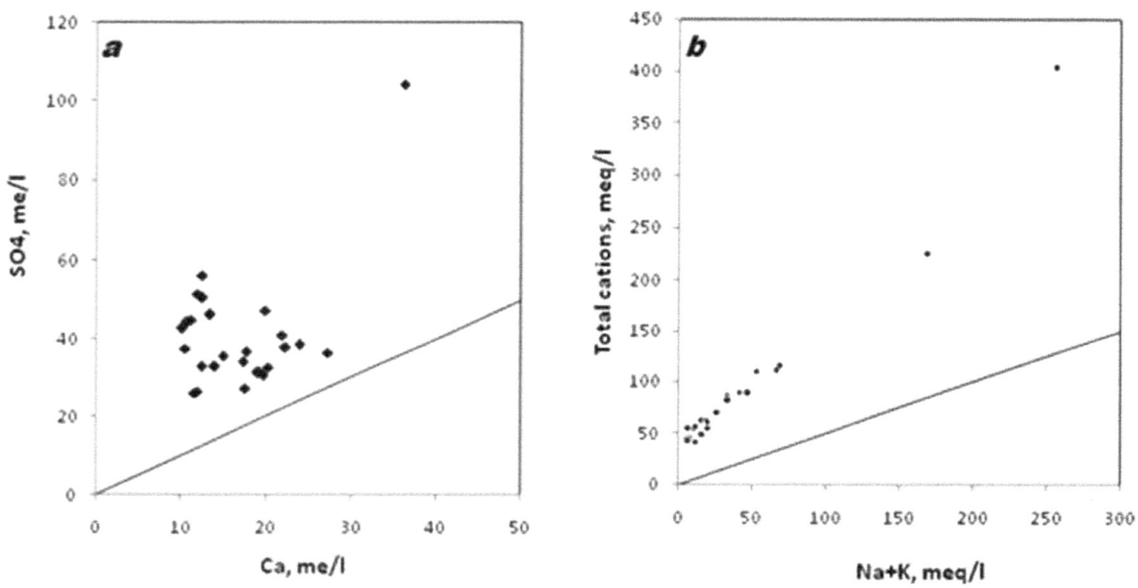

Figure 7. a. Ca^{2+} versus SO_4^{2-} diagram, b. Na^+ + K^+ versus total cations (TZ^+) diagram for groundwater in the study area.

When there is an exchange between calcium or magnesium in the groundwater with sodium and potassium in the aquifer material, both above indices are negative, and if there is a reverse ion exchange, then both indices will be positive (Schoeller, 1965, 1977). The values indicating most of the samples show negative and some wells show positive in the study area (Figure 8). This observation indicates that the normal ion exchange is the dominant process in the groundwater, whereas reverse ion exchange is also observed in some samples.

When ion exchange is a significant geochemical process for controlling the composition of groundwater, the relationship between (Ca^{2+} + Mg^{2+}) versus (SO_4^{2-} + HCO_3^-) diagram illustrates ion exchange should be linear with a slope of 1.0 (Fisher and Mulican, 1997). The plot of (Ca^{2+} + Mg^{2+}) versus (SO_4^{2-} + HCO_3^-) (Figure 5) shows that all the groundwater samples of Souf valley falls left to the 1:1 line (Figure 5a) which indicate that ion exchange tends to be dominant over reverse ion exchange process in the study area.

Evaporation

The study region experiences dry and arid climatic condition, evaporation may also contribute in water

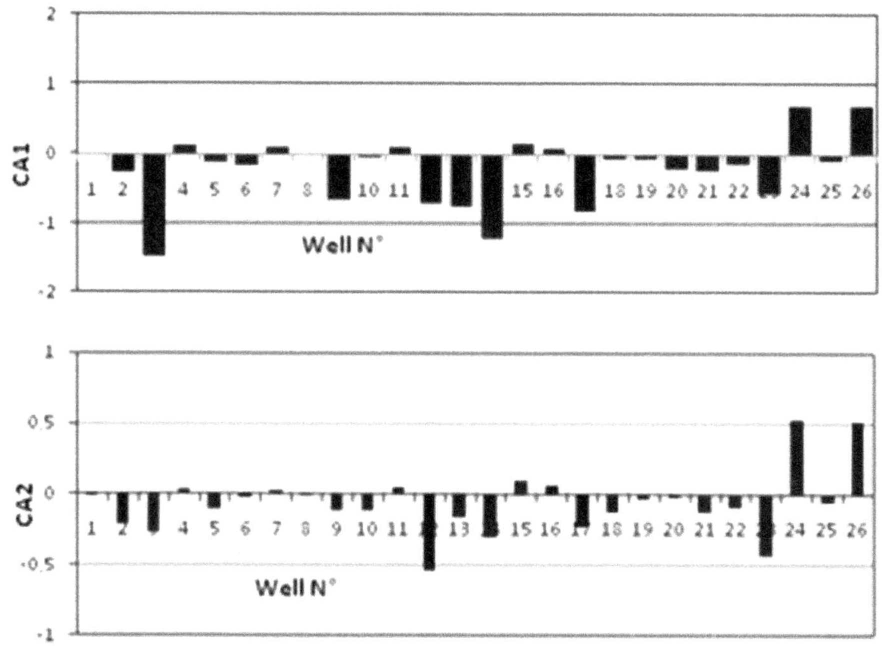

Figure 8. Chloro-alkaline indices 1 and 2 (CAI 1 and CAI 2) for groundwater in the study area.

chemistry. Gibbs (1970) suggested that a simple plot of TDS versus the weight ratio of $Na^+/(Na^+ + Ca^{2+})$ could provide information on the relative importance of the major natural mechanisms controlling groundwater chemistry. From the Gibbs plot (Figure 9a); the dominant process determining the water composition is evaporation and rock/soil-water interaction. Evaporation of surface water and moisture in the unsaturated zone is the main process in the evolvement of groundwater chemical composition. Evaporation concentrates the remaining water and led to precipitation and deposition of evaporates that are eventually leached into the saturated zone. This is expected, as evaporation greatly increases the concentration of ions formed by chemical weathering, leading to high salinity and TDS. This is clearly reflected in the Gibbs plot, which indicates a trend towards evaporations and rock dominated dissolution.

Furthermore, sodium versus chloride ratio can be used to identify the evaporation process in groundwater. Evaporation will increase the concentration of total dissolved solids in groundwater; and the Na/Cl ratio remains the same, and it is one of the good indicative factors of evaporation. If evaporation is the dominant process, Na/Cl ratio should be constant when EC rises. Hence, the plot of Na/Cl versus EC would give a horizontal line, which would then be an effective indicator of concentration by evaporation and evapotranspiration (Jankowski and Acworth, 1997; Kuldip-Singh et al., 2011). The EC versus Na/Cl diagram of the groundwater samples (Figure 9b) shows that one trend line is horizontal indicating that evaporation is the major process. The other trend line is inclined revealing that evaporation may not be the major geochemical process controlling the chemistry of groundwater or dissolution or ion exchange reaction dominating over evaporation.

Spatial repartition of waters

The geographical distribution of the two water types «A» and «B» on the study area is presented in Figure 3. The «A» water type include the well water situated to the center of the structural basin (cuvette), the latter constitutes the main hall of groundwater flow toward the Chotts, by against, the wells of «B» water type located on the basin flanks. According to this spatial distribution, the survey zone has been subdivided in three domains in order to specify the existing relations between the water's chemical quality and the aquifer lithological nature in its natural context.

In every domain, the physico-chemistry has been characterized by the following characteristic ratio: rMg/rCa, rNa+K/rCa, rSO4/rCl, rCl-Na/rCl and the S.A.R (Tables 3 and 4) calculated from means of the different ionic elements in every domain. These characteristic ratios, with others, constitute a good indicator of existing relations between the aquiferous lithology and the underground water chemical composition that circulate there (Back and Hanshaw, 1970; William et al., 1999; Dazy and Razack, 1989; Nkotagu, 1996).

The domain I corresponds to eastern flank of the Souf valley. It is constituted sands with gypsum or sandy clays intercalations. All studied wells belong to the family «B» with a predominance of sulphates on chlorides (rSO4 /

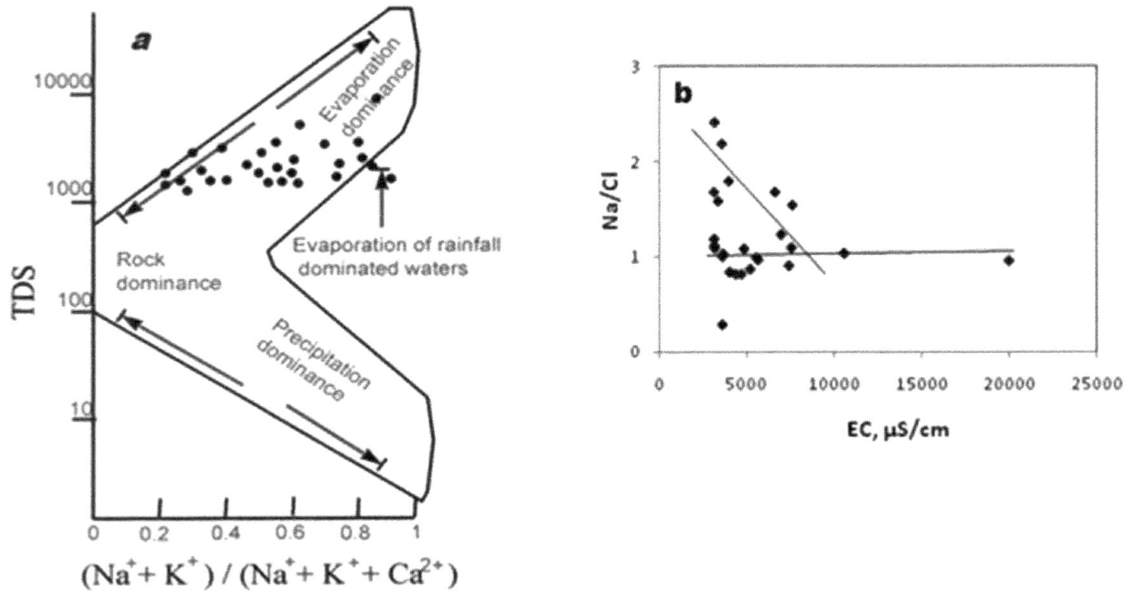

Figure 9. a. Gibbs plot indicating the mechanisms that determine the major ion composition of groundwater. b. EC versus Na/Cl diagram.

Table 3. Mean physico-chemical composition of the different domain waters in the Souf region.

Domain	T (°C)	pH	Cond (s/cm)	R.S (mg/l)	Ca (mg/l)	Mg (mg/l	Na (mg/l)	K (mg/l)	SO_4 (mg/l)	Cl (mg/l)	HCO_3 (mg/l)	NO_3 (mg/l)
D1	22.1	7.6	3984	3024	360	286	287	23	1522	467	118	93
D2	22.0	7.5	10422	7830	338	626	2679	210	2883	3754	267	93
D3	22.3	7.7	4327	2868	384	296	463	20	1750	604	93	94

T°: temperature [°C], Cond.: electric Conductivity [µs/cm], TDS: total dissolved solids [mg/l].

Table 4. Characteristic ratios of waters in every domain of Souf valley.

Domain	rMg / rCa	rNa +K/rCl	rSo4 /rCl	rCl-Na/rCl	S.A.R
D1	1.31	0.73	2.4	0.05	2.74
D2	3.05	7.21	0.57	-0.10	19.92
D3	1.26	1.08	2.14	-0.18	4.3

rCl =2.4), the magnesium on the calcium (rMg / rCa = 1.31) and a ratio (alkaline / calcium) less than 1: (rNa+K / rCa = 0.73).

In general, the mineralization is important and increases grossly from the east to the west, therefore in the flow direction (Figure 3). Means of conductivities and dry residues do not reveal any significant detail and are comparable to those of the other areas of the Low Septentrional Sahara. The pH is nearly constant (7.5 to 7.7).

The importance of SO_4, Cl and HCO_3 in waters can be explained by the respective presence of gypsum, salts and limestone in the sandy matrix crossing of these waters which make these minerals in solution. For the magnesium, it can be only the product of the silicate weathering present in the matrix (garnets and tourmalines) because this phenomenon is very widespread in the (sub) surface conditions (Schoeller, 1967, 1977; Gibbs, 1970; Petalas and Diamantis, 1999). However, if the presence of the calcium is explained by the dissolution of gypsum and limestone and partly by the silicate weathering; its deficit compared to (SO_4^{2+} + HCO_3) suggests an exchange of basis (Ca^{++}---- Mg^{++}) between the liquid phase charged in Mg^{++} and the calcic solids present in the aquifer matrix (Back and Hanshaw, 1970).

Domain II occupies the central part of the basin. It is constituted of calcariferous sand with gypsum and gypseous clays in the basis. All wells' waters belong to the family «A». Waters of this domain, as compared to those of the previous domain, has an anionic abundance characterized by the predominance of chlorides on sulphates (rSO4/rCl = 0.57) with higher average saltiness (the conductivity and the dry residual are 2.5 times more important) and a comparable pH. The ratios (Mg/Ca) and (alkaline/Ca) are highest in the study zone (3.05 and 7.21, respectively). The high salinity in this domain is, presumably, owed to several complementary hydro-geochemical processes that are summarized in the following points:

1. Weathering and dissolution of salts, gypsum, carbonate and silicate, leaching on the basin flanks and the elements concentration in the depressive areas as a consequence of slowdown in flow (increasing mineralisation toward the center on the two flanks of the basin).
2. The dissolution, *in situ* of sulphates, salts and carbonates and the exchange of sodium with potassium contained in clays.
3. The mixture of the saturation level waters with the waters of the underlying aquifers, mainly, those of the Mio-Pliocene; mixture waters confirmed by isotope study (D.H.W, 1997; Guendouz et al., 2003).

In this domain, the reduction of the rSO4/rCl ratio is related to the solubility of salts (NaCl) and to the degree of water saturation in these elements (Kirchmer, 1983). Because waters are first charged with sulphates during their underground trail, then the chlorides (NaCl) take place in the vicinity of saturation, that iss why these waters show their rate of chlorides increasing while going toward the central zone.

The domain III is located in the West flank of the basin. It is constituted, especially, of gypseous sands with clay intercalations. All waters in this area belong to the family «B» with predominance of sulphates on chlorides and the magnesium on the calcium as for the domain I, but with a ratio (alkaline/calcium) nearly equal to 1 (rNa+K/rCa = 1.08). The global mineralisation is the weakest in the region and it increases from west toward East, always in the direction of flow. However, one records an increase of sulphates, chlorides and sodium charges. The pH, here is slightly higher with regards to the two other domains.

The resemblance of chemical facies of these waters and their global context (aquifer lithology and structure) with those of the domain I, induce us to conclude that responsible processes of water salinity in the two domains are comparable if not the same.

Globaly, contents in nitrates are high in the entire region and exceed the admitted norms (in the three domains, means are two times the norm). They are especially, the consequence of the organic matter nitrification contained in worn-out water discharges. In this way, these waters are mixed with those of the water table because of the absence of a purification network in several localities of the region and the absence of an exutory outside of the aquifer system. This origin is confirmed by a bacteriological survey realised par DHW d'El Oued (D.H.W., 1997). We noted that this phenomenon is largely spilled in the Saharan regions and it has been signalled in several countries of Africa (Nesson, 1975; Nkotagu, 1996).

The S.A.R is weak for both I and II domains, whereas it exceeds the value of 10 (19.92) in the domain III. This indicates that waters of this last domain are of very poor quality and then cannot be used even in agriculture.

Finally, ratio rCl-Na/rCl is positive for the domain I, whereas it is negative for the domains II and III. This is closely linked to the sodium big mobility. It indicates that the domain I, probably, constitutes a refill zone of the saturation level, potentially more favorable than the domain III, and that waters would have already done an important course in the aquiferous system in the domain III; which explains the relative importance of SO_4, Cl and Na in domain III with regard to domain I (Dadi et al., 1997; Dazy and Razack, 1989).

Global scenario of waters mineralization

By dissolution of salts, gypsums and limestones or by weathering of carbonate and silicate, waters are charged with SO_4^{2-}, Cl^-, Na^- and Ca^{2+}. At the same time by weathering of silicate accessories, they are charged with Mg^{2+}. While crossing, during their underground circulation, the carbonate concretions or the clayey intercalations are restrained in the sandy matrix forming the aquifer; they can exchange a part of Mg^{++} transported by the calcium contained in the calcic solids (late dolomitization phenomenon). They can also exchange a part of Na^+ by potassium restrained in clays (these basic ion exchanges have consequence on the impoverishment in Ca^{++} and the enrichment in K^+ of these waters). In continuing their progresses in both the western and eastern flank of the basin (domain I and III), they are charged progressively because of the favorable conditions to the ions migration. While arriving at the center (domain II), they are either slowed or braked by the tormented relief of the clayey substratum of the aquifer. They can also be braked by the presence of impermeable limits which encourages, more and more, the concentration of ions especially in the depressive places; added to this, dissolutions or ion exchange that can be produced on site, or possible miscellanies of waters. In this way, saturation level waters observe their saltiness increasing suddenly.

This global scenario of mineralisation waters of the groundwater table in Souf valley is reinforced by distribution of the waters chemistry and their spatio-temporal evolution (Figure 3). This script makes evident the roles of different natural parameters, (intern and external) linked to the aquiferous system, in the acquirement of saltiness. It shows the hydro-geochemical processes in question and can also predict water chemism, probably evolution of the water table in this region.

Conclusions

Ground waters in the Souf valley are highly saline (2361-14780 mg/L TDS) and the hardness is a common characteristic feature of groundwater of this aquifer. Magnesium, calcium and sodium are the dominant cations and sulfates and chlorides are the dominant anions. They present two main waters families with geochemical evolution of facies whose chemistry is strongly influenced by different parameters linked to the aquifer system.

Dissolution, rock weathering, ion exchange, mixing and nitrification of the organic matter are the main hydrogeochemical processes of waters mineralization. The conjugated action of the different mineralisation processes as well as their combinations had accentuated the salinity of these waters. Their identification allowed a better interrelationship between the different present ions. The partial analysis (by domain) of the aquifer in its natural context helped us in the explanation of the global saltiness evolution; since a global scenario of mineralization shows clearly the roles of the different parameters of aquifer system specifying the processes interfering in the acquirement of global waters saltiness and can predict their probable evolution.

In this region deprived of treatment means and softening, when choosing captive fields and during implantation of drilling of potable water supply, these results would guide decision makers in selecting the right sectors containing less ionised groundwater.

Conflict of interests

The authors did not declare any conflict of interest.

ACKNOWLEDGEMENT

The authors wish to thank the staffs of DHW Oued Souf for their help and collaboration.

REFERENCES

Back W and Hanshaw BB (1970). Comparison of the chemical hydrogeology of the carbonate peninsulas of Florida and Yucatan. J. Hydrol. 10:330-368.

Busson G (1970). Le mésozoïque saharien : Essai de synthèse des donnés des sondages Algéro-Tunisiens. C.R.Z.A. Série Géol. N°11, Ed.CNRS, Paris. p. 810.

Cornet A (1961). Géologie de l'Oued Rhir. Terres et eaux, O.C.R.S, 3 et 4 trim. 43:18-24.

D.H.W de l'Oued Souf (1997). Rapport sur l'étude bactériologique des eaux de la région du Souf.

D.H.W de l'Oued Souf (2011). Fiche d'analyses physico-chimiques des eaux de la nappe phréatique.

Dadi S, Boutaleb M, Laziri F, Demassieux L, Dassargues A (1997). Etude hydrochimique d'un aquifère en milieu fissuré; Cas du massif cristallophyllien de l'Oùlmes, Maroc. Proceed.Rabat Synp, Mai 97. IAHS publ. 241:13-20.

Das BK, Kaur P (2001). Major ion chemistry of Renuka lake and weathering processes, Sirmaur district, Himachal Pradesh, India. J. Environ. Geol. 40:908-917.

Datta PS, Tyagi SK (1996). Major ion chemistry of groundwater in Delhi area: chemical weathering processes and groundwater regime. J. Geol. Soc. India 47:179-188.

Dazy J, Razack M (1989). Approche géochimique et structurale des circulations dans un réservoir du domaine alpin (massif d'Allevard, France). Rev. Sci. Eau, Quebec. 2:153-181.

Dervierux F (1957). La nappe phréatique du Souf (région d'El Oued, Algérie) - Etude du renouvellement de la nappe. Terres et eaux, O.C.R.S, 3trim.56 et 1trim.57(29):12-39.

Fisher RS, Mulican WF (1997). Hydrochemical evolution of sodium-sulfate and sodium-chloride groundwater beneath the Northern Chihuahuan desert, Trans-Pecos, Texas, USA. Hydrogeol. J. 10:455-474.

Garrels RM, Mackenzie FT (1971). Evolution of sedimentary rocks. W.W. Norton, New York.
Geol. 46:47-61.

Gibbs RJ (1970). Mechanisms controlling world water chemistry. Sci. J. 170:795-840.

Giraud R (1978). Le continental terminal en Algérie. Ann. Fac. Sci. Dakkar 31:85-87.

Guendouz A, Moulla AS, Edmunds WM, Zouari K, Shand P, Mamou A (2003). Hydrogeochemical and Isotopic Evolution of Water in the Complexe Terminal Aquifer in the Agerian Sahara, Hydrogeol. J. 11(4):483-495.

Jankowski J, Acworth RI (1997). Impact of debris-flow deposits on hydrogeochemical process and the development of dry land salinity in the Yass River catchment, New South Wales, Australia. Hydrogeology J. 5:71-88.

Katz BG, Coplen TB, Bullen TD, Davis JH (1998). Use of chemicalnand isotopic tracers to characterize the interaction between groundwater and surface water in mantled Karst. Groundwater 35: 1014-1028.

Kirchmer JC (1983). Quality control in water analyses. Environ. Sci. Technol. 17 (4):178-181.

Kuldip S, Hundal HS, Dhanwinder S (2011). Geochemistry and assessment of hydrogeochemical processes in groundwater in the southern part of Bathinda district of Punjab, northwest India. Environ. Earth Sci. 64:1823-1833.

Kumar M, Ramanathan AL, Rao MS, Kumar B (2006). Identification and evaluation of hydrogeochemical processes in the groundwater environment of Delhi, India. J. Environ. Geol. 50:1025-1039.

Matthess G (1982). The properties of groundwater. Wiley, New York, p 498.

Maya AL, Loucks MD (1995). Solute and isotopic geochemistry and groundwater flow in the Central Wasatch Range, Utah. J. Hydrol. 172:31-59.

Michard G (1989). Equilibres chimiques dans les eaux naturelles. Ed. Published, 357p

Nesson C (1975). L'évolution des ressources hydrauliques dans les oasis du Bas Sahara Algérien. Mem.Doc, Ed.CNRS 17:96.

Nkotagu H (1996). Origin of high nitrate in groundwater of Tanzania. J. Afr. Sci. 21 (4):471-478.

Ould BSM (2005). Recharge et paléorecharge du système aquifère du Sahara septentrional. Thèse de doctorat, université Tunis El Manar. p 261.

Petalas CP, Diamantis IB (1999). Origin and distribution of saline groundwaters in the upper Miocene aquifer system, coastal Rhodope area, northeastern Creece. Hydrogeol. J. 7: 305-316.

Piper AM (1944). A graphic procedure in geochemical interpretation of water analyses. Trans. Am. Geophys. Union 25:914-923.

Rajmohan N, Elango L (2004). Identification and evolution of hydrogeochemical processes in the groundwater environment in an area of the Palar and Cheyyar River Basins, Southern India. Environ.

Sarin MM, Krishnaswamy S, Dilli K, Somayajulu BLK, Moore WS (1989).

Major ion chemistry of the Ganga-Brahmaputra river system: weathering process and fluxes to the Bay of Bengal. Geochim. Cosmochim. Acta 53:997-1009.

Schoeller H (1965). Qualitative evaluation of groundwater resources. In: Methods and techniques of groundwater investigations and development. UNESCO 54-83.

Schoeller H (1967). Geochemistry of groundwater- an international guide for research and practice, UNESCO.

Schoeller H (1977). Geochemistry of groundwater. In: Brown RH et al (eds) Groundwater studies - an international guide for research and practice. UNESCO, Paris. pp. 1-18.

Schuh WM, Klinekebiel DL, Gardner JC, Meyar RF (1997). Tracer and nitrate movements to groundwater in the Norruem Great Plains. J. Environ. Qual. 26:1335-1347.

Stallard RF, Edmond JM, (1983). Geochemistry of Amazon, the influence of geology and weathering environment on the dissolved load. J. Geophys. Res. 88:9671-9688.

Swarna Latha P, Nageswara Rao K (2011). An integrated approach to assess the quality of groundwater in a coastal aquifer of Andhra Pradesh, India. Earth Sci. 64:1823-1833.

William SL, Miguel PS, Héctor OP (1999). Bicarbonate, sulfate and chloride water in a Shallow, clastic-dominated coastal flow system, Argentina. Groundwater 37 (2):287-294.

Is climate change human induced?

H. N. Misra and Ashutosh Mishra

Geography Department, University of Allahabad, India.

Climate is the most vital element of our planet and its liveability is key concern for every habitat. From Silent Spring till present, debate is on whether humankind has impact on nature. Since its establishment in 1988, the Intergovernmental Panel on Climate Change (IPCC) has been playing pivotal role in raising public concerns on human-induced climate change through its various assessment reports. These reports follow exhaustive review process, and are widely accepted. In 2007, IPCC's 4th assessment report- 'Climate Change 2007 – Impacts, Adaptation and Vulnerability' came into question on Himalayan glacier melt. The climate gate in 2009 further strengthened the confusion on credibility of IPCC`s projections. The present study analyses district level temperature and rainfall patterns of Uttarakhand- a Himalayan state, and examines the validity of IPCC's projection. Uttarakhand is a tourism oriented economy. The state is best known for its religious places and natural sites. Rapid urbanisation in mountainous regions is disturbing regional eco-balance, but increasing vehicular pollution in climate-sensitive areas seems to have greater impact on temperature and precipitation patterns. Result shows a noticeable shift in the variability of temperature and rainfall, and a significant warming especially in mountainous districts. However, human activities do not correlate very well with these changes.

Key words: Climate-sensitive sectors, monsoon, climatic variability, polar caps, vehicular pollution.

INTRODUCTION

India is considered highly vulnerable to climate change, not only because of high physical exposure to climate-related disaster, but also because of the dependency of its economy and majority of population on climate-sensitive sectors (for example, agriculture, forests, tourism, animal husbandry and fisheries). More than 40 million hectares of India land (12%) is prone to floods and river erosion; of the 7,516 km long coastline, close to 5,700 km is prone to cyclones and tsunamis; 68% of the cultivable area is vulnerable to drought and hilly areas are at risk of landslides and avalanches (NDMA, 2007). The country has a unique climate system dominated by the monsoon, and the major physiographic features that drive this monsoon are its location in the globe, the Himalayas, the Central Plateau, the Western and Eastern Ghats and the oceans surrounding the region.

The Himalayas influence the climate of the Indian subcontinent by sheltering it from the cold air mass of Central Asia. The range also exerts a major influence on monsoon and rainfall patterns. They prevent frigid and dry arctic winds from blowing south into the subcontinent keeping South Asia much warmer when compared with regions located between corresponding latitudes throughout the globe.

Himalayan glaciers cover about three million hectares or 17% of the mountain area. They form the largest body of ice outside the polar caps and are the source of water for the innumerable rivers that flow across the Indo-Gangetic

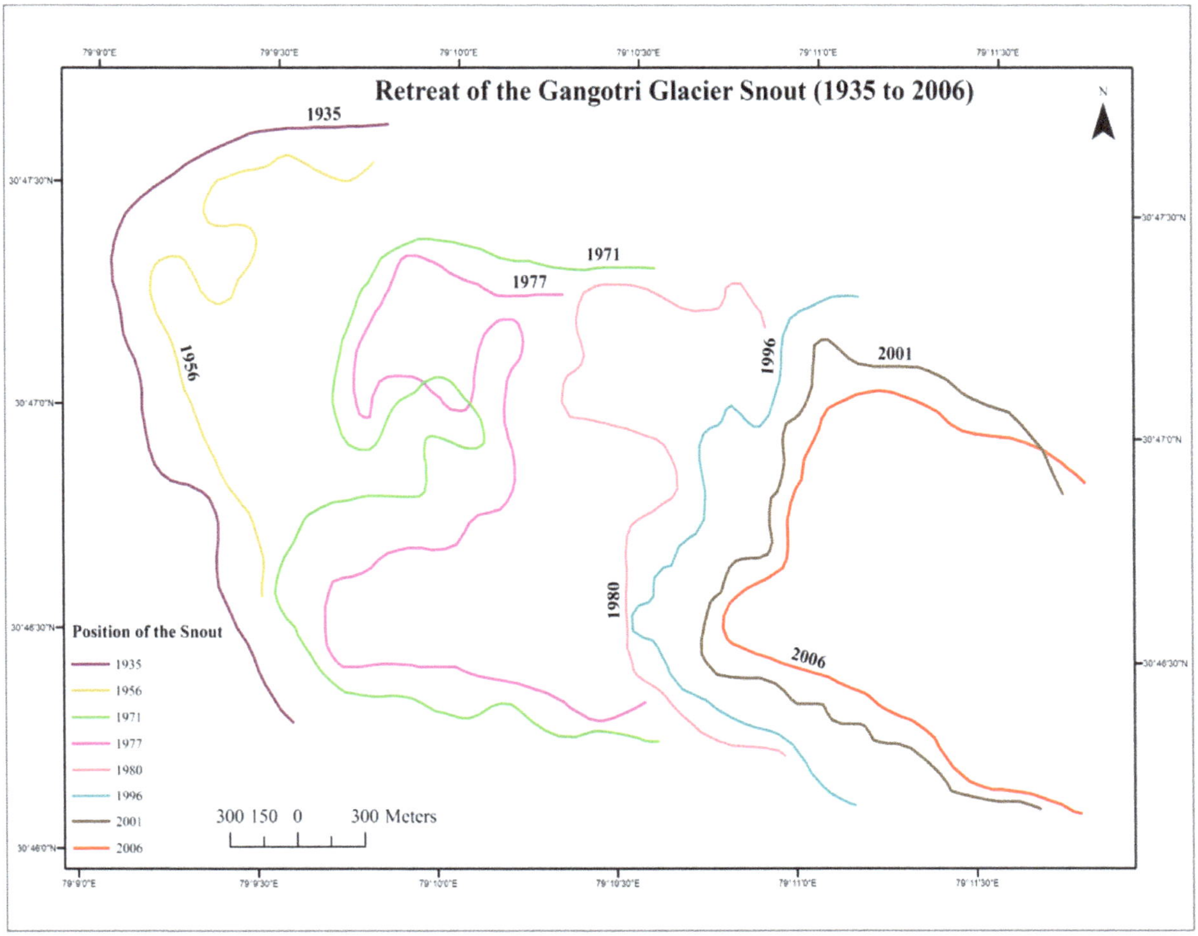

Figure 1. Retreat of the Gangotri glacier snout (1935 to 2006 based on maps made by the Geological Survey of India).

plains. About 15,000 Himalayan glaciers form a unique reservoir which supports perennial rivers such as the Indus, Ganga and Brahmaputra which, in turn, are the lifeline of millions of people in South Asian countries (Pakistan, Nepal, Bhutan, India and Bangladesh). The Gangetic basin alone is home to 500 million people, about 10% of the total human population in the region.

The Himalayan ecosystem is highly vulnerable to the stress caused by increased pressure of population, exploitation of natural resources and other related challenges. Climate change may adversely impact the Himalayan ecosystem through increased temperature, altered precipitation patterns and episodes of drought. According to IPCC`s 4th assessment report "glaciers in the Himalaya are receding faster than in any other part of the world and, if the present rate continues, the likelihood of them disappearing by the year 2035 and perhaps sooner is very high if the Earth keeps warming at the current rate. Its total area will likely shrink from the present 500,000 to 100,000 km^2 by the year 2035" (Cruz et al., 2007 cited in IPCC, 2007).

Syed Iqbal Hasnain, India's well-known Glaciologist, observes that "The Ganga system is about 60 to 70% snow and ice. There are more than 800 glaciers in the Ganga basin. The Gangotri is the big one. It used to cover more than 250 square kilometers, but now it's breaking up in many places. You will see blocks of dead ice that are no longer connected to the main ice body. I'm afraid that if the current trends continue, within 30 or 40 years most of the glaciers will melt out" (Black, 2009).

Contrary to Hasnain`s view, a white paper on the status of Himalayan glaciers and global warming by V.K. Raina, former Deputy Director General of the Geological Survey of India, suggests that "in most cases glaciers have stopped retreating. While the Gangotri glacier stopped receding in the 2007-09 period, glaciers like Pindari in Kumaon continue to record a high annual retreat of almost 10 m annually". He further states that "The glaciers are undergoing natural changes, witnessed periodically" (Raina, 2010). According to assessments made during 1935-2006 by the Geological Survey of India, the Gangotri region has not shown any evidence of major retreat (Figure 1).

A glacier is affected by a range of physical features and a complex interplay of climatic factors. Establishing change in climate of the Himalayan region on the basis of

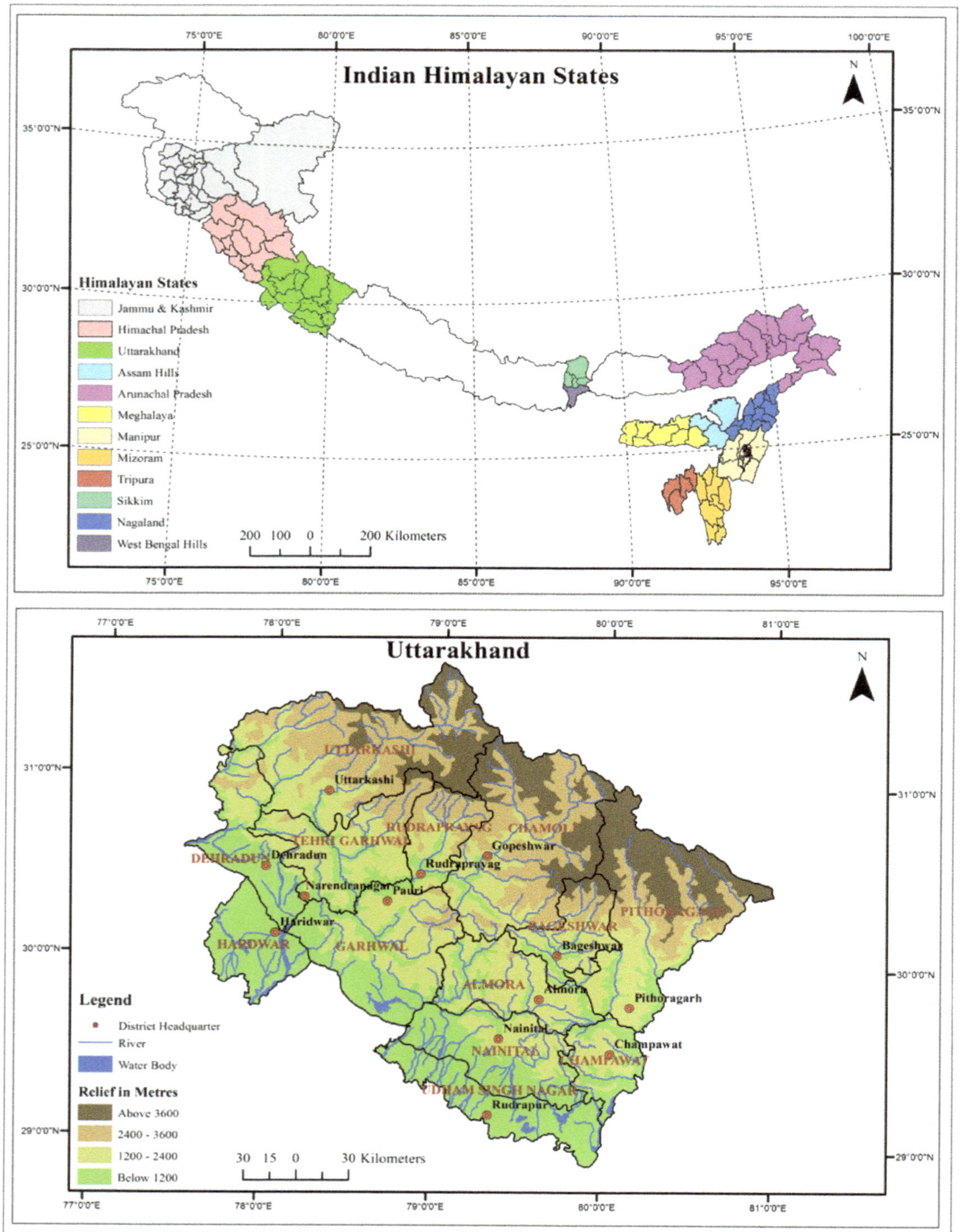

Figure 2. Location map of Uttarakhand.

movement of glaciers, and attributing this change to human activities without analysing the local climate variability, departures and level of human interference does not seems reliable.

The present work is an attempt to identify anthropogenic influence over natural climatic variability of the Himalayan region by considering Uttarakhand as the case of study. The data has been collected from Census of India, India Meteorological Department, Survey of India and Geological Survey of India, and simple correlation and regression techniques have been used for analysis of temperature and rainfall patterns.

Uttarakhand is a part of the Indian Himalayan region (Figure 2). Owing to its immense natural beauty, rich

Table 1. Regression result for temperature and rainfall patterns.

S/N	District	Correlation between temperature and time series (r)	Regression result for temperature series		Correlation between rainfall and time series (r)	Regression result for rainfall series	
			R^2	F		R^2	F
1	Almora	0.266	0.071	7.598**	-0.209	0.044	4.579*
2	Bageshwar	0.299	0.089	9.826**	-0.221	0.049	5.125*
3	Chamoli	0.315	0.099	10.984**	-0.184	0.034	3.515
4	Champawat	0.291	0.085	9.246**	-0.242	0.058	6.211*
5	Dehradun	0.232	0.054	5.694*	-0.035	0.001	0.125
6	Garhwal	0.247	0.061	6.515*	-0.122	0.015	1.503
7	Haridwar	0.221	0.049	5.124*	0.023	0.001	0.051
8	Nainital	0.260	0.068	7.254**	-0.201	0.041	4.229*
9	Pithoragarh	0.342	0.117	13.250**	-0.236	0.056	5.919*
10	Rudraprayag	0.308	0.095	10.514**	-0.153	0.023	2.404
11	Tehri Garhwal	0.263	0.069	7.435**	-0.114	0.013	1.311
12	Udham Singh Nagar	0.260	0.068	7.275**	-0.164	0.027	2.769
13	Uttarkashi	0.308	0.095	10.455**	-0.103	0.011	1.075

**: $p<0.01$, *: $p<0.05$.

biological succession and India's great rivers feeding glaciers- Gangotri, Ponting, Milam, Pindari etc., the region is regarded as Devbhumi- abode of Gods, and Tapobhumi- land of asceticism in Indian scriptures.

The northern region of the state is part of the Great Himalayan Range, covered in snow and glaciers. Two of the Indian subcontinent's most important rivers- the Ganga and the Yamuna- also originate from the glaciers of Uttarakhand. The natural resources of the region provide life supporting, provisioning, regulating and cultural 'eco-system' services to millions of local as well as downstream people. The state lies between the longitudes 77°34′-81°02′E and latitudes 28°43′-31°27′N having a maximum dimension of east-west 310 km and north-south 255 km. It covers an area of 53,484 km^2 with the elevation ranging from 210 to 7817 mt. The state shares border with China (Tibet) in the North and Nepal in the East and inter-state boundaries with Himachal Pradesh in the West, Northwest and Uttar Pradesh in the South. Broadly, the region constitutes of 13 districts falling in two major administrative unit viz., Garhwal (northwest portion) and Kumaon (southeast portion).

The climate of Uttarakhand is temperate, marked by seasonal variations in temperature but also affected by tropical monsoons. January is the coldest month, with daily high temperatures averaging below freezing in the north and near 21°C in the southeast. In the north, July is the hottest month, with temperatures typically rising from 7 to about 21°C daily. In the southeast, May is the warmest month, with daily temperatures normally reaching the highest around 38°C from a low around 27°C. Most of the state's roughly 1,500 mm of annual precipitation is brought by the southwest monsoon, which blows from July through September.

RECENT CLIMATE TREND

Although climate represents a set of factors and determinants showing long term averaged state of the atmosphere over a region but temperature and rainfall are the two most prominent elements among them. The present study examines trends of rainfall and temperature at annual and monthly time scales for the periods of 1911-2012 to understand the climatic variability of the region.

Uttarakhand has two physiographic zones- montane and non-montane. The mountainous regions have recorded more significant warming and declining rainfall trend, while Hardwar, which is almost plain, noticed positive rainfall trend (Table 1). It is evident that the temperature and rainfall departures from centennial average are significantly high in higher altitudes (Figure 3).

Data shows four distinct phases of temperature patterns during the last century- no warming or cooling up to 1950, warming trend during 1950-1080, cooling trend between 1980-2000 and again warming after 2000. On the other hand, the rainfall patterns recorded no significant shift from natural variability at centennial scale (1911-2012) (Figure 4).

Analysis of monthly average temperature data shows very striking results. Colder months- January, February, November and December, have recorded significant warming while the hottest months- June, July, August and September, have shown cooling trend. The study area receives most of its rainfall during the months of July and August but during last century (1911-2011), these months have recorded declining precipitation trend. The months of March and May on the other hand have received more rainfall than normal (Figure 5). Interestingly, the districts of non-montane physiographic

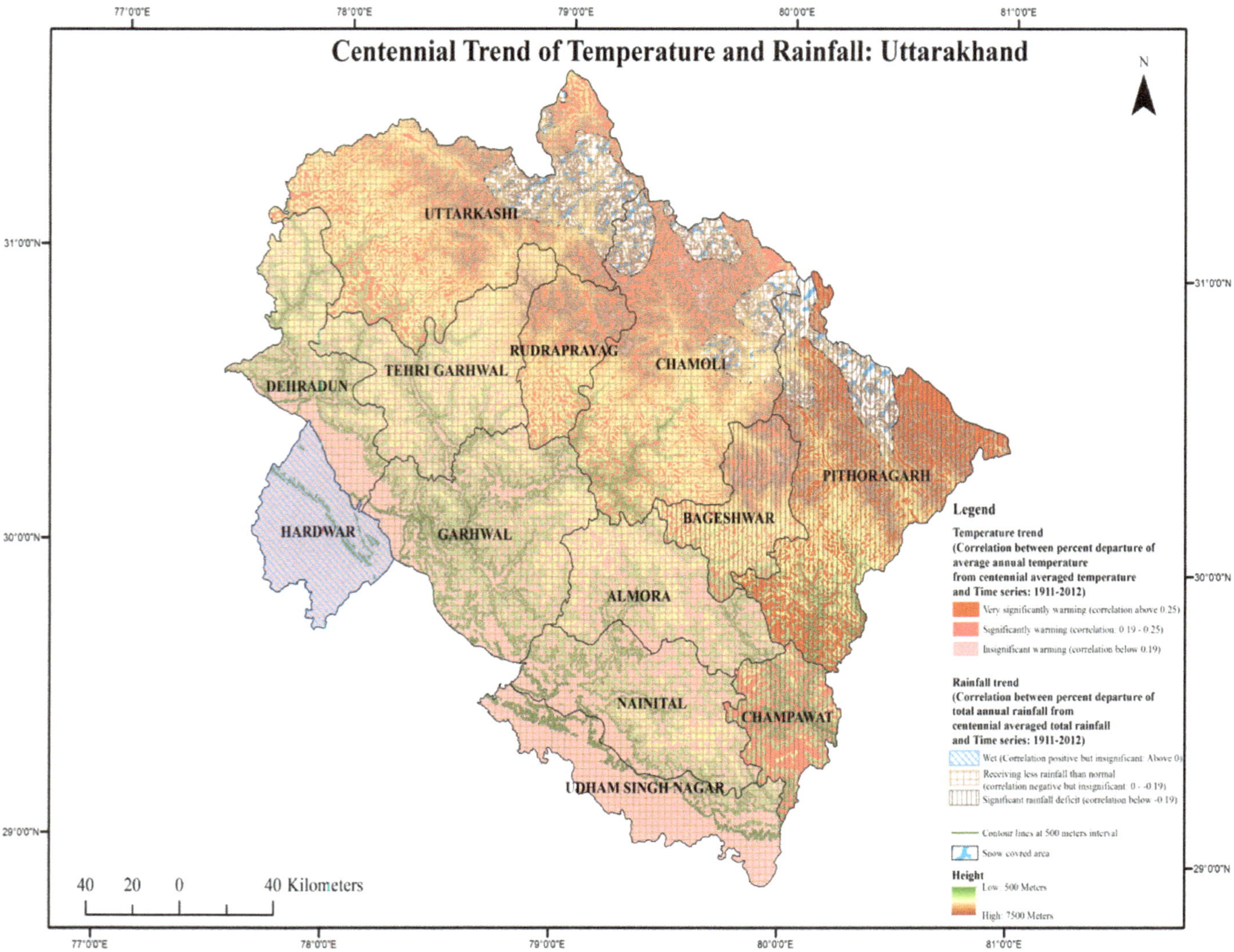

Figure 3. Temperature and rainfall trend of Uttarakhand.

zone have shown lowest departures.

The state has recorded a continuous growth in population during the last century except in 1921-31, however the urban population has grown at faster rate. Most of the population lives in the lower districts where urban share is high (Figure 6).

Hardwar and Dehradun districts record the highest population and percent urbanisation, and have the highest number of census towns (Table 2). These two districts have noticed less warming than others. Contrary to other districts, Hardwar has recorded an increasing rainfall trend although this trend is insignificant. Evidently urbanisation holds no significant association with temperature and rainfall trends.

Heavy forest diversion for basic infrastructures is also being accused for deteriorating local climate's stability. But data suggests that green cover removal is not directly related with the warming. Dehradun and Hardwar which have recorded largest forest diversion are not the warmest districts of the region (Table 3). Forests attract rainfall, but here Hardwar having the noticeable forest diversion, has shown an increasing rainfall trend.

Uttarakhand is famous for religious and adventure tourism. Noticeably the tourist pressure at four major religious centres of the state has been almost doubled during the last twelve years (Table 4).

The three districts – Uttarkashi, Rudraprayag and Chamoli, where these religious centres are located, have recorded very significant warming during past decade. Increasing vehicular pollution seems fuelling temperature rise in these areas. Although growing industrialisation and vehicular density in Dehradun and Hardwar districts have no significant impression on temperature trend on the other hand, it can be said that vehicular pollution is more significantly correlated with temperature patterns in hilly areas while in the plain region, it has less impact on

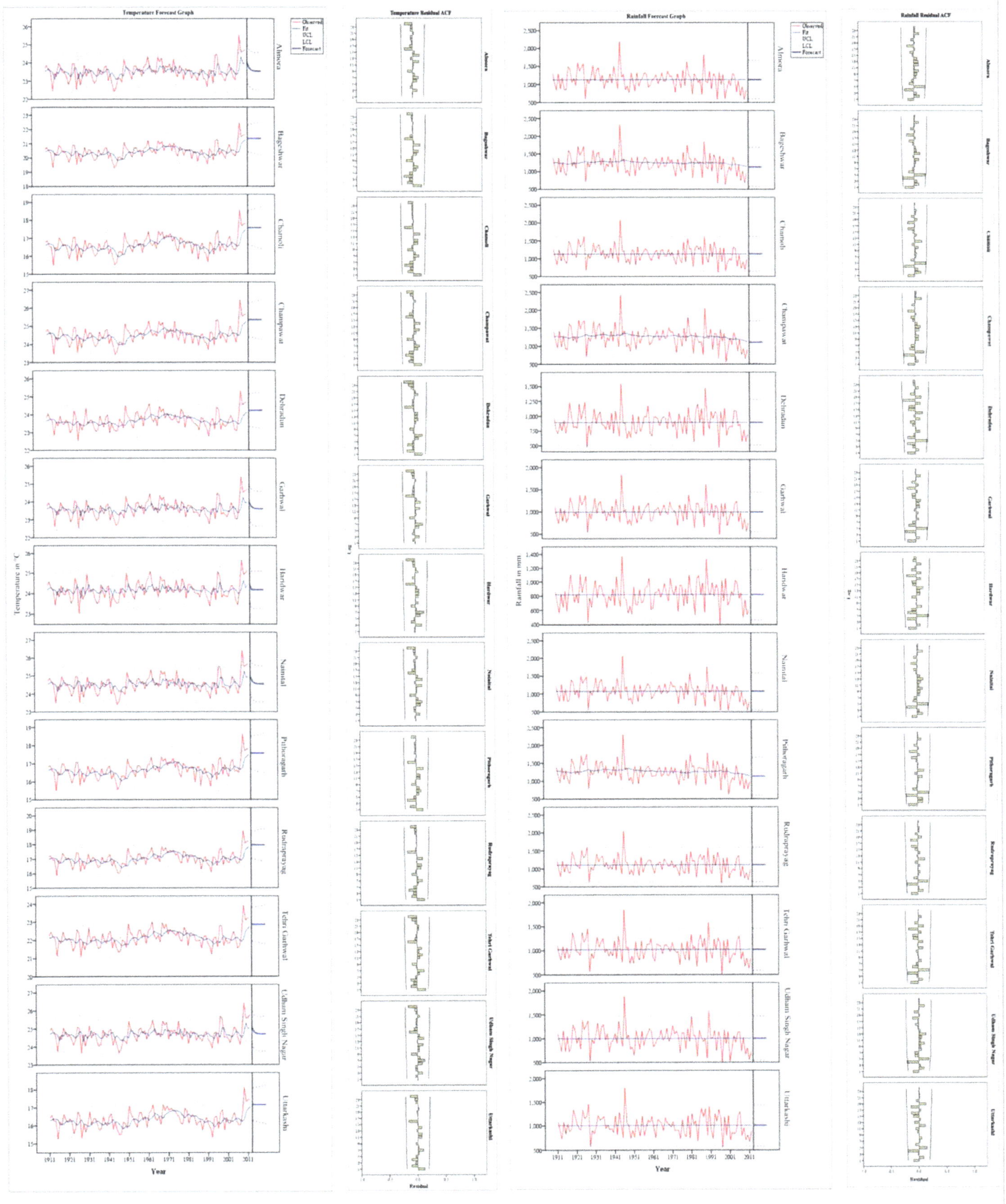

Figure 4. District level temperature and rainfall forecast graph for Uttarakhand.

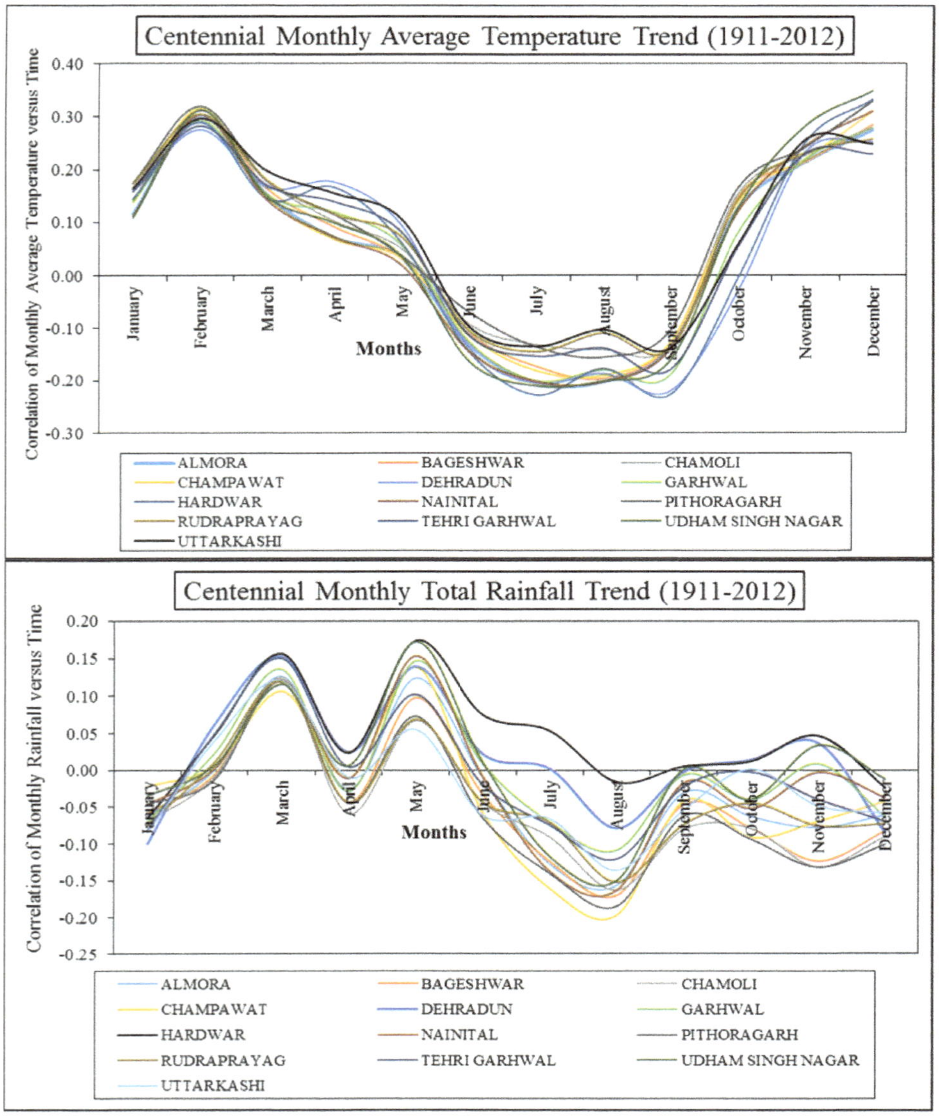

Figure 5. District level centennial monthly temperature and rainfall trend of Uttarakhand (correlation values above 0.19 are significant and above 0.25 are highly significant).

the atmospheric state. In other words, neutralising capacity of plain ecosystem seems greater than mountainous ecosystem.

CONCLUSION

Analysis shows noticeable departures in temperature and rainfall patterns. Months of March and May have recorded more rainfall and significant warming. Temperature of June, July, August and September are at cooler side. Result shows that surface temperatures have risen significantly during the last century, but this may be result of various cooling and warming phases. Besides having significant temperature-time correlations, R^2 values are very weak because of very noisy data. The ARIMA models predicted warming up to 0.3°C till 2035, being maximum for Chamoli district. Results show that this change is almost natural rather than anthropogenic. Warming is unequivocal with decreasing rainfall (except Hardwar), however, temperature and rainfall patterns do not fully support the hypothesis that urbanisation, industrialisation or green cover removal have great bearing on this warming or drying trend. Although increasing vehicular pollution in temperature-sensitive high altitude areas seems to have some impact on these trends, we can say while human interference has fuelled some variations in patterns, natural factors are the major cause behind climatic variability and changes. IPCC's claim that due to human intervention in the Himalayan ecosystem, by 2035 we are going to lose large volume of glaciers, thus, does not seem a real claim.

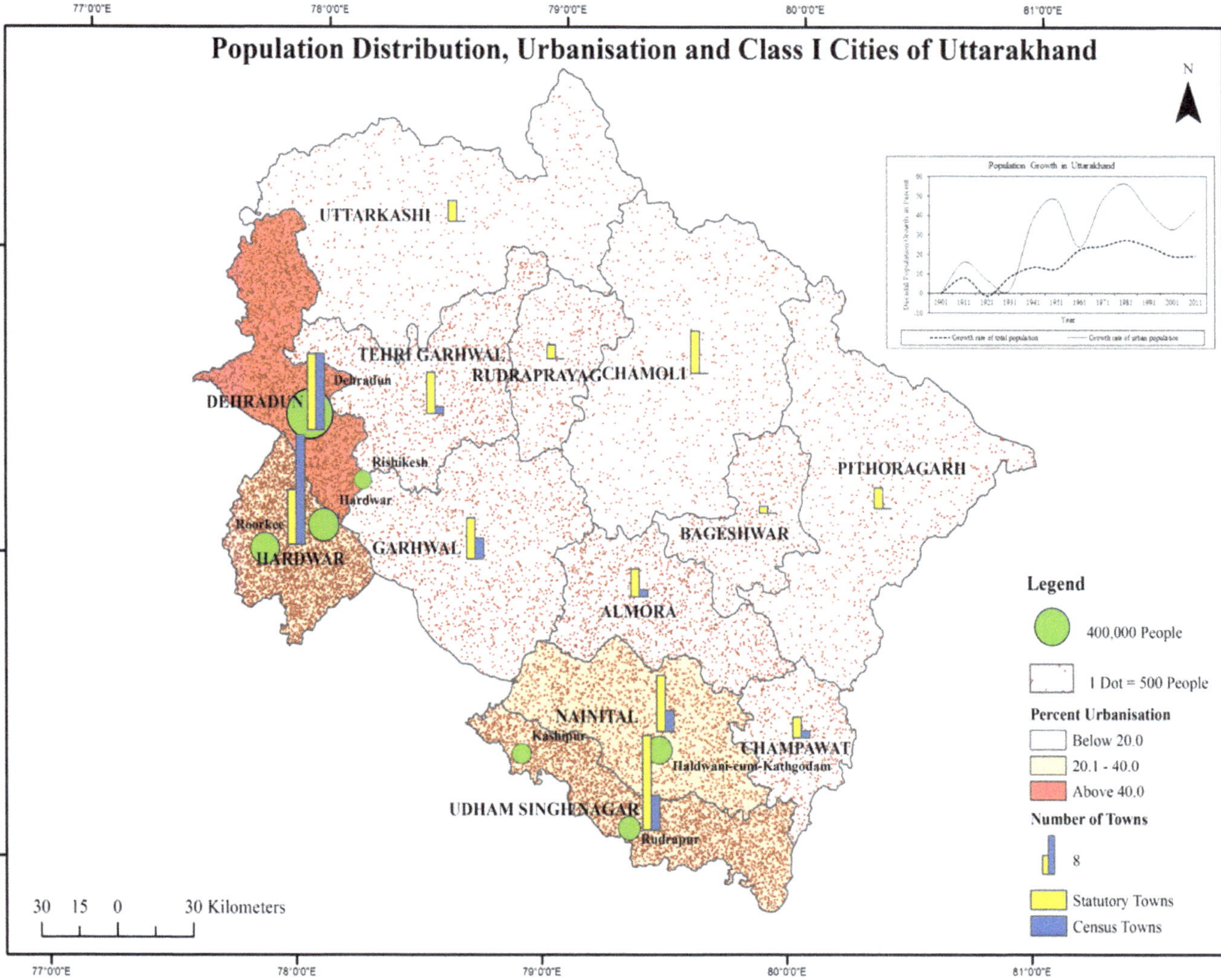

Figure 6. Population distribution in Uttarakhand.

Table 2. Urbanisation in Uttarakhand.

S/N	District	Total population	Percent urbanisation	Statutory towns	Census towns	Village
1	Almora	621927	10.02	4	1	2289
2	Bageshwar	259840	3.5	1	0	947
3	Chamoli	391114	15.11	6	0	1246
4	Champawat	259315	14.79	3	1	717
5	Dehradun	1698560	55.9	11	11	748
6	Garhwal	686527	16.41	6	3	3473
7	Hardwar	1927029	37.77	8	16	612
8	Nainital	955128	38.94	8	3	1141
9	Pithoragarh	485993	14.31	3	0	1675
10	Rudraprayag	236857	4.19	2	0	688
11	Tehri Garhwal	616409	11.37	6	1	1862
12	Udham Singh Nagar	1648367	35.58	14	5	688
13	Uttarkashi	329686	7.35	3	0	707

Source: Census, 2011, GOI and computed.

Table 3. District wise/sector wise details of forest area diverted from 2000 to 2013.

District	Total forest diverted area in hectare	Percent of the total forest diverted area for different purposes						
		Road construction	Managing drinking water	Irrigation	Electricity transmission lines	Hydroelectric power plants	Mining	Other uses
Almora	816.29	88.36	1.7	0.61	2.94	0.01	0	6.39
Bageshwar	559.05	67.73	1.57	0.52	0.99	2.88	20.01	6.3
Chamoli	2097.74	44.85	0.46	0.05	37.45	11.54	0.05	5.59
Champawat	738.57	37.76	0.8	0.28	0.49	0	52.09	8.58
Dehradun	19496.09	2.05	0.05	0.01	0.13	0	8.22	89.54
Garhwal	677.4	43.97	2.95	0.18	24.69	0.43	21.48	6.3
Haridwar	5197.71	1.16	0	0.31	0.16	0	55.79	42.57
Nainital	3165.71	13.71	2.24	0.32	0.61	0	77.14	5.98
Pithoragarh	1667.66	66.42	0.58	0.1	29.6	0.75	0.25	2.3
Rudraprayag	389.56	66.19	1.76	2.36	3.99	16.16	0	9.55
Tehri Garhwal	1591.74	30.35	1.16	0.08	9.88	44.46	0.41	13.67
Udham Singh Nagar	156.23	9.15	0	2.3	3.16	0	0	85.39
Uttarkashi	830.95	52.71	1.1	0.95	13.22	27.59	0	4.44

Source: Uttarakhand forest statistics, Forest Department, Government of Uttarakhand, p. 44.

Table 4. Tourist inflow at selected location in Uttarakhand.

Place	District	Percent increase in tourist Inflow (2001-2012)
Yamunotri	Uttarkashi	240
Gangotri	Uttarkashi	250
Kedarnath	Rudraprayag	378
Badrinath	Chamoli	136

Source: CSE, 2013.

Conflict of interests

The authors did not declare any conflict of interest.

REFERENCES

Black G (2009). India Enlightened. On earth, summer edition, pp. 28-29.

Cruz RV, Harasawa H, Lal M, Wu S, Anokhin Y, Punsalmaa, B, Honda Y, Jafari M, Li C, Huu NN (2007). Asia. In ML Parry, OF Canziani, JP Palutikof, PJ Linden, Hanson CE (eds.), Climate Change 2007: Impacts, Adaptation and Vulnerability, Contribution of Working Group II to the Fourth Assessment Report of the Intergovernmental Panel on Climate Change, Cambridge University Press, Cambridge, UK, pp. 469-506.

NDMA (2007). National Disaster Management Guideline, National Disaster Management Authority, Government of India, p. 1.

Raina VK (2010). Himalayan Glaciers: A State-of-Art Review of Glacial Studies, Glacial Retreat and Climate Change. MOEF Discussion Paper, Ministry of Environment and Forests, Government of India, p. 7.

Challenges of adoption of urine-diversion dry toilets technology as sanitation option by coastal communities of Mkuranga District in Tanzania

Aloyce W. Mayo and Twaha Mubarak

Department of Water Resources Engineering, University of Dar es Salaam, Tanzania.

Mkuranga District council in collaboration with African Medical and Research Foundation (AMREF) introduced ecological sanitation option using urine-diversion dry toilets (UDDT) to the community between 2007 and 2009, which was later declared unsuccessful. This study analyzed factors that hindered the uptake of UDDT by the community by assessing the project implementation strategy. Techniques used for the study were questionnaire, interviews, observation and focus group discussion while respondents were government official and the community at the household level. The study reveals that the literacy level in Mkuranga district is 79.1%, but only 40.6% had some knowledge of ecological sanitation although only 9% proves that. About 52% of the respondents are using conventional pit latrines, but 27.3% have no toilet facilities. There was no evidence of adoption of UDDT technology at household level and only one of the seven UDDTs constructed by the project is operational. There was no sufficient evidence to suggest that Mkuranga District has sufficiently supported the project through supervision, advocacy and addressing community requirements. As a result, the idea came in top-bottom approach which failed because communities were not adequately involved in the project.

Key words: Ecological sanitation, urine-diversion dry toilets, adoption, community participation, Tanzania.

INTRODUCTION

In many towns and rural areas of the world today, people live and raise their children in highly polluted environments (Muench, 2009). Urban and peri-urban areas in developing countries are among the worst polluted and disease ridden habitats of the world. Much of this pollution, which leads to high rates of disease and death, is caused by lack of toilets and inadequate sanitation services (COHRE et al., 2008). The lack of sufficient or adequate services is a result of many factors, including inadequate financial resources, insufficient water and lack of space, difficult soil conditions and limited institutional capabilities. As population increases, the need for safe, sustainable and affordable sanitation systems will be even more critical (UNESCO/IHP, 2006).

In their Joint Monitoring Programme, UNICEF and WHO (2013) have reported that an estimated 2.5 billion people around the world do not have access to improved sanitation, the majority of those (90%) live in rural areas of Asia, Latin America and Africa. It is estimated that 1 billion people worldwide or 15% of world population still practice open defecation (UNICEF and WHO, 2013) including over 5 million people in Tanzania (UNICEF, 2013). In 2011, only 12% of Tanzanians use improved sanitation facilities (UNICEF and WHO, 2013), although others estimated household using improved sanitation are about 33% (Wikipedia, 2013) largely because of low investment in the sector. Although funding in water and sanitation sector has quadrupled since 2002 through multi-donor Water Sector Development Program (WSDP) (WSP and UNICEF, 2013), it is estimated that water supply and sanitation receives annual investment of US$ 175 million only, which is equivalent to US$ 4 per capita (Wikipedia, 2013). Even this little amount is spent largely on water supply only because subsidies for rural household sanitation are not supported by the Tanzanian government policy (WSP and UNICEF, 2013). As a result, individual households are encouraged to invest in their sanitation facilities.

Over the past hundred years, flush-and-discharge has been regarded as the ideal technology, particularly for urban areas (Esrey et al., 1998), although by 2007 only 3% of Tanzanian households use flush toilets (Wikipedia, 2013). For those without access to flush-and-discharge, the conventional alternative is a drop-and-store device, usually a pit toilet, based on containment and indefinite storage of human excreta (Esrey et al., 1998). However, water closets and pit latrines lack some important benefits as compared to ecological sanitation (Esrey et al., 1998). In accordance with Guadagni (2012), the ecological sanitation (EcoSan) approach does not promote a specific sanitation technology, but rather a new philosophy in recycling-oriented resource management, which renders human excreta safe, prevents pollution rather than attempting to control it, and uses the safe products of sanitized human excreta for agricultural purposes.

Ecological sanitation (EcoSan) have been practiced for thousands of years in China (Jurga et al., 2003; Smet and Sugden, 2006) and Japan has introduced the practice of using urine for agriculture about 900 years ago (Abarghaz et al., 2012). Today, application of EcoSan is common in the world over and in places such as in South Asia (Sridevi et al., 2007; WaterAid, 2008; Adhikari et al., 2012), Western Europe (Rhode et al., 2004; Tidaker et al., 2007; Rieck and Muench, 2011), Latin America (Thibodeau and Canaday, 2011), Eastern Europe, the Caucasus and Central Asia (Wendland et al., 2011). In recent years, EcoSan application have been reported in East Africa (Nuwagaba, 2003; Langergraber and Muellegger, 2005; Muellegger, 2011), West Africa (Kiba, 2005), South Africa (Mnkeni et al., 2008; Ingle et al., 2012) and North Africa (Abarghaz et al., 2012).

EcoSan offers a number of advantages including prevention of contamination of groundwater sources, prevention of degradation of soil fertility, provides nutrients to plants and reduces health risks related to sanitation (Werner et al., 2004). They can be constructed on hard rock soils, suitable in areas with high ground water levels and areas prone to flooding. EcoSan is suitable where water is scarce or expensive and hence reduces the burden of the communities on looking for water to use for toilet purposes. In accordance with Langergraber and Muellegger (2005), EcoSan is a holistic sanitation approach that is economically and ecologically sound.

In 2007, Mkuranga District Council in collaboration with AMREF- Tanzania introduced urine-diversion dry toilets (UDDT) technology as a sanitation option with economic benefits. The technology was considered to be a solution to sanitation issues as it is clear that many people in the coastal area tend to practice open defecation, but the technology offers more advantages which go beyond the disposal of faeces (AMREF Mkuranga, 2007). In that project, a total of 7 UDDTs were constructed as a demonstration whereas 2 of them were constructed at a market place and 5 were constructed at the primary schools (Mkuranga District Council, 2008). A total of 560 artisans from 80 villages of Mkuranga were trained on how to construct affordable UDDTs (AMREF Mkuranga, 2009). The objectives of this study are to assess the UDDT project implementation strategy used and to determine setbacks which hindered the adoption of UDDT by the community.

METHODS AND MATERIALS

Description of the project

Mkuranga District has a total surface area of 2,432 km^2 out of which 447 km^2 is part of Indian Ocean, 552 km^2 is forest reserve and 1,433 km^2 of land is suitable for cultivation. It lies between latitude 6° 35' and 7° 30' South of the Equator and between longitudes 38° 45' and 39° 30' to the east. The district boarders with Dar es Salaam Region, Indian Ocean, Rufiji District and Kisarawe District to the North, East, South and West, respectively (Figure 1).

The project started in 2007 by introducing urine diverted dry toilet (UDDT) as an option for ecological sanitation, which will reduce the portion of community using open defecation, a common sanitary method in the area. A three years project was implemented by Mkuranga District council together with their partners, African Medical Research Foundation (AMREF)- Africa. AMREF- Africa were committed to support the project financially by injecting about 100,000 USD and few human resource at the managerial level while Mkuranga District council were committed to provide technical support by giving out the expert to show their skills in both community mobilization, advocacy, technology experts as well as facilitators for capacity building to the community as an agent of sustainability. The expected major outcomes of the project were human behavioral changes, improvement of sanitation status of the community, improvement of agricultural productivity, reduction of communicable diseases and poverty alleviation.

Climate and soil

Mkuranga District is located south of Dar es Salaam along the west coast shores of Indian ocean. The district experiences bimodal rainfall with short rains season between October and December and long rains season covering the month of March to June

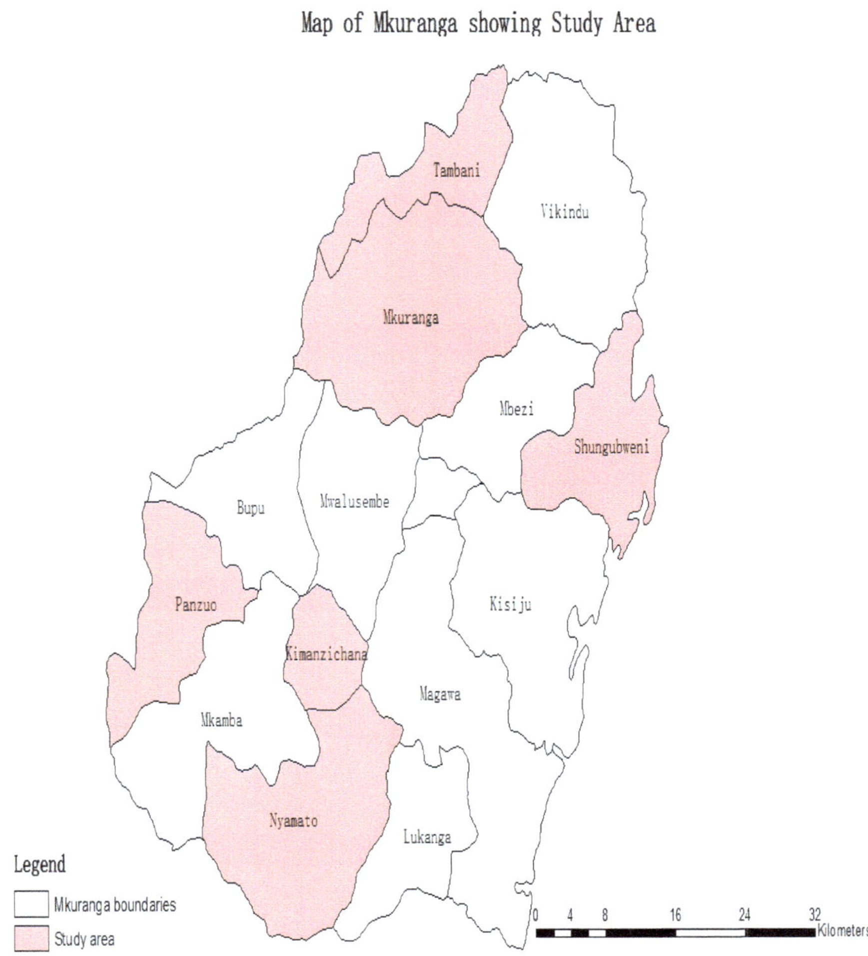

Figure 1. Map of Mkuranga District showing the study area.

(Majule, 2012). The average annual rainfall is about 800 – 1000 mm, but rainfall distribution is not very reliable within the seasons. It is hot throughout the year with average temperature of 28°C (Tanzania Meteorological Agency, 2013). Mkuranga like other parts of the coastal area of Tanzania is largely characterized by sandy soil, which collapses easily when dug, and high water table, which complicates construction of pit latrines (Mubarak, 2013).

Population and economic activities in Mkuranga

In accordance with population and housing census 2012, the district has a total population of 222,921 of which 114,897 (51.5%) are females and 108,024 (48.5%) are males (URT, 2013). The District population growth rate is 3.5% per annum (URT, 2013). The average population density of the District is 95 people per square kilometer, but large concentration of people is found at Kisiju, Magawa, Lukanga and Kitomondo wards due to a number of economic opportunities found in the area like fishing, trading, boat making and port activities. Wards along the main road from Dar es Salaam city to Lindi Region have urban characteristics. The per capital income of Mkuranga District in 2013 was 276.9 USD per annum, which is about 50% of the national per capita income of 550 USD per annum (1 USD = 1600 TZS in year 2013). About 85% of the district inhabitants depend on agriculture.

Sample size and sampling procedure

The sample size of this study was determined by the use of the formulae given by Krejcie and Morgan (1970) as shown by Equation 1.

$$n = \frac{X^{2}.xNxPx\ (1-P)}{[ME^{2}x(N-1)] \div [X^{2}xPx\ (1-P)]} \qquad (1)$$

Where: n = sample size, X^2 = Chi-square for the specified confidence level at 1 degree of freedom (X = 1.96 for confidence interval of 95%), N = population size = 227,990, P = population proportion= 0.5, ME = desired margin of error = 5%. From the input data, the sample size of 360 was obtained.

Wards were selected using purposive sampling method and the respondents were selected using simple random sampling method and lottery technique from three wards located in urban setting (heterogeneous culture) and another three wards from rural settings (homogeneous culture). The number of respondents from each ward was determined proportionally to the population of the ward. Stratified random sampling technique were used to select respondents who were members in project steering committee at district level and ward level whereby groups for focus discussion were formed according to their level they belong (either ward or district level). These key informants participated in the establishment of sanitation promotion project specifically introduction of UDDT within the district.

Data collection tools and processing

A self-administered structured close-ended questionnaire survey, observation (inspection visits) checklist (quantitative), focus group discussions (FGD) and semi-structured in-depth interviews (qualitative) were applied to collect primary data. Secondary data were collected from relevant reports and published research papers. Primary data were collected through administration of questionnaire at the household level; field observation by using observation checklist; interview and focus group discussion with key informant from the government both at the district and ward level as detailed in following sub-chapters. All these tools were pre-tested in Mkuranga ward and revised accordingly before they were used for the study population. The tools were first prepared in English and then translated into Kiswahili. Focus group discussions and semi-structured in-depth interviews were recorded with digital audio recording device. The collected data were entered in the computer and were analyzed using the Statistical Package for Social Science (SPSS) version 11.0 for Windows and Excel sheet. The data presentation forms include averages, frequencies percentages, charts and graphs.

Questionnaires for household

Households used as samples were obtained through a stratified random sampling procedure in accordance with Kothari (2004). Open and closed ended questions, which were prepared and administered to the household members in Swahili language, were designed to capture information on the knowledge, attitudes, perception and practices of the community members at household level with regards to sanitation promotion issues as well as strength, weakness and opportunity of UDDT. The questionnaires sought information on the type of latrine they are currently using, knowledge on ecological sanitation, knowledge on UDDT, their preference between UDDT versus the current latrine; views on UDDT promotion in the area; the possibility of handling dry feacal matter and urine from UDDT if there is any contradiction between their belief or opinion in using UDDT; whether they were involved in the establishment of the project; challenges related to UDDT and their advice for the uptake of the technology. The questionnaires also sought information on demographic data, economic status, household size, occupation, marital status, education level and age of the respondent, duration of their stay in that community and nature of the house tenure.

Key informant interviews

This involved district project steering committee to supplement data from the questionnaires with the help of an interview guide. The government officials involved in the study was District Health Officer; District Water Engineer; District Planning Officer; District Community Development Officer; District Environmental Sanitation Officer; District Education officer; District Agricultural and Livestock Development Officer and District Treasurer. These interviews were aimed at getting an overview of sanitation management in the district by focusing on the coordination of the different stakeholders and challenges faced in the establishment and implementation of sanitation project in the district. It was also a tool to reveal if all necessary steps in the project were taken as well as the participatory approach was fully observed in all stages of the project.

Table 1. Age distribution.

Age	Frequency	Percent
18 - 25 years	42	12.4
26 - 35 years	94	27.6
36 - 55 years	113	33.2
56+	91	26.8
Total	340	100.0

Focus group discussion

Two sets of six people each were formed; one group with officials from three wards with urban setting characteristics and another group includes officials from ward with rural setting characteristics. Each ward was represented by two government officials (Ward Executive Officer and Ward Health Officer). The aim of the focus group discussions was to gather strengths and weaknesses of the project and the response of the people toward the adoption of the technology.

RESULTS AND DISCUSSION

About 340 respondents from the household level, 8 respondents from the district level and 12 respondents from the ward level participated in the study. The demographic data of the respondents from the household indicate that male respondents were 202 (59.4%) while female respondents were 138 (40.6%). The marital status of the respondents were married 72.1% (n=245), single 13.2% (n=45), separated 7.9% (n=27), divorced 3.5% (n=12) and widowed 3.2% (n=11). About 243 (71.5%) of respondents were head of the household and the remaining 97 (28.5%) were just mere members of the household. Table 1 shows the age distribution of the respondents.

Socio economic profile

Source of income

Mkuranga District is fast growing economically because of its close proximity (about 30 to 40 km) with Dar es Salaam City. It has attracted many investors including 9 new industries, which were launched in recent years. This industrial growth within the district has led to the increase of employed people even with its nature of rural setting. Table 2 shows that the majority of the residents of Mkuranga in the surveyed area are peasants (42.7%), although the proportion reported in this study is much lower than 62% reported by Mkuranga District Council (MDC strategic plan, implementation report 2011) and 89.6% reported by National Bureau of Statistics (URT, 2013). This was probably influenced by the large number of respondents who came from the wards which possess urban characteristics. In these areas, most of the people

Table 2. Occupational distribution.

Occupation	Number of respondents	Marginal percentage
Employed	103	35.8
Petty business	38	13.2
Other small industries	24	8.3
Peasant	123	42.7
Total	288	100.0

Table 3. Educational level.

Level	Number of respondents	Percent
Primary school	130	38.2
Secondary school	98	28.8
Tertiary education	41	12.0
Illiterate	71	20.8

are engaged in employment and business. About 62% of those employed are unskilled laborers who earn less than Tshs 150,000/=(about USD 94 in 2013) per month. Self-employment is largely through petty businesses, which may only sustain them in very basic needs such as food and house supplies. Peasants in Mkuranga District own small farms where they cultivate short term crops such as cassava, and perennial crops such as coconuts, mangoes and oranges. Peasants also engage in other economic activities such as fishing whose contribution to their income was not established (MDC report, 2011).

Education level

Table 3 shows that the literacy level is 79.1% (38.2% for primary school, 28.8% for secondary school and 12.0% for tertiary education), which is sufficient for the introduction of new knowledge. Data from the National Bureau of Statistics (TDHS, 2011) indicate that literacy level varies from 78 to 83% depending on the level of urbanization of the area, gender and geographical zones, suggesting that Mkuranga District has the lowest literacy level in Tanzania. In the latest National sensor carried out in 2012, it was reported that only 43% of Mkuranga District residents are literate owing to low rate (51%) of children enrolment rate. It is worth to note that educational level of the community is one of the factors determining the success of a community projects in terms of planning, designing and implementation of new project. Fruman et al. (2012) reported that education was identified as one of the most crucial factors that will lead to acceptance and use of UDDT in rural Georgia. Tumwebaze and Niwabaga (2011) reported that respondents with secondary and tertiary education were 2 to 5 times more likely to adopt to ecological sanitation than respondents with primary education. In another study, Nuwabaga (2003) observed that Ecosan coverage was low among people with primary education level (0.4%) than people with tertiary education (13%).

House tenure

Data from National Bureau from Statistics indicate that only about 20.8% of houses in Mkuranga district are roofed with corrugated iron sheets or tiles, while others (79.2%) are largely thatched roof houses (URT, 2013). Similar findings were reported by Torell and Mmochi (2006) who observed that most residents of Mkuranga District live in poor and simple houses thatched by coconut leaves or grass, poles, and mud walls on earth floors. The houses are also largely comprised of a mixture of huts with walls made of mud and wooden poles (94.7%), as well as conventional bricks and block houses (5.3%), which are traditionally found in wards of urban characteristics like Mkuranga town. Many people in the district are of low income with 89.6% depending on subsistence agriculture for their livelihood. Only 0.6% of the population is connected to electricity. As a result, 98% of the population is depending on charcoal and firewood as a source of energy (URT, 2013). It was observed that about 57.4% of the respondents own houses, whereas 42.7% of the respondents are living as tenants. Most of the tenants (79%) are found in wards, with urban characteristics and the majority of tenants (65%) are employees who normally shift from one working station to another. House tenure is very important aspect in designing the sanitation project as its implementation will need decision, consent and resource investment from the owner of the house either by having a mutual agreement with the tenants or the owner to incur all the "sanitation investments" cost on his structure.

Sanitation (UDDT) project

Project organization

Figure 2 shows the organizational framework of the project. During interview with key informants, it was revealed that the setup of the project meant to be fully participatory as

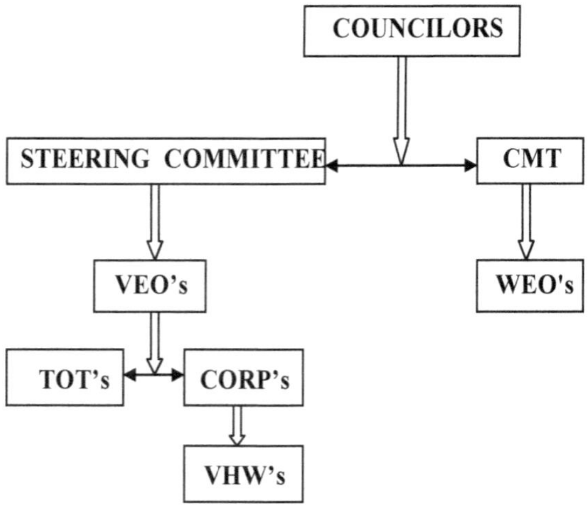

Figure 2. Project hierarchy.

Table 4. Types of latrines used in the study area.

S/N	Type of latrine	Percent of respondents
1	Cistern flush toilet	5.6
2	Conventional pit latrines	52.1
3	Ventilated Improved Pit Latrines	13.2
4	Pour flush toilets	1.8
5	No latrines	27.3

every level at the local government in the district was included. These include councilors, head of departments and steering committee members, which comprises of various professionals at the district level; such as Ward Executive Officers (WEOs), Village Executive officers (VEOs) as well as community member under the title of Trainer of Trainee (TOTs), Community Own Resource Person (CORPs) and the Village Health Workers (VHWs).

The steering committee members at the district level and members of the community (trainers of trainees, community own resource persons and village health workers) were involved in daily activities of the project, but other stakeholders participated only in meetings. In focus group discussions, key informants (District Health Officer, District Planning Officer and District Water Engineer) revealed that the project was mainly supervised by only few members of the steering committee, but there was no evidence suggesting that councilors and heads of departments and ward leaders were involved in the supervision of the project. It was also observed that village executive officers worked as either trainers of trainees or community own resource persons instead of project supervisors at village level in order to earn allowances, which compromised the quality of the work.

Status of latrine use

Table 4 shows that the majority of the respondents (52.1%) are using conventional pit latrines. Other types of latrines used in the study area include ventilated improved pit latrines (13.2%), cistern flush toilet (5.6%) and pour flush toilet (1.8%), but no single UDDT was found at household level. In fact, in the whole study area, the only operational UDDT was found at a primary school, which was one of the seven UDDTs constructed by the project. The remaining six UDDTs constructed during project implementation as demonstration toilets were out of use, which suggests that the technology was completely abandoned instead of being adopted.

It is worth mentioning that the type of sanitation options used by the respondents is related to availability of water in the area and the level of economy and education of the users. For example, water closets were found in Mkuranga town where level of income of community is high and water is available for flushing wastes although tap water in Mkuranga town is unreliable. In wards located in rural areas where water supply is not available, conventional pit latrines are common. In accordance with national population and housing sensors carried out in 2012, only 8.7% of households in Mkuranga District use piped or protected well/spring water as the main source of drinking water (URT, 2013), which is comparable to the proportion of households using cistern and poor flush toilets (7.4%).

The results also show that 27.3% of respondents do not have latrines, suggesting that they are probably using open defecation because community latrines are not available in the area. These findings are similar to those reported by the National Bureau of Statistics (TDHS, 2011) which reported that 22% of household do not have latrine, but differs with those of MDC report (2011) who reported that only 8% of the household do not have latrine. Other reports indicate that over 60% of Mkuranga District households do not have latrines particularly near coastal areas (Towell and Mmochi, 2006). The records from Joint Monitoring Programme conducted by UNICEF and WHO (2013) suggest that in 2011 about 16% of people in rural Tanzania practice open defecation. It is therefore evident that sanitation situation in Mkuranga district is below the average sanitation level of rural Tanzania. Some of the potential reasons are the level of literacy (43%) of Mkuranga District residents, which is among the lowest in the country.

It is worth noting that 69.3% of households, which do not have toilet, are located in a coastal ward of Shungubweni, suggesting that they are using beaches of Indian Ocean for defecation. This factor along with sandy soils in the area, which complicate excavation and construction of the pits, was mentioned by 68% of the steering committee members during interview and focus group discussions. Fortunately, UDDTs do not require deep pits suggesting that soil type is not the cause of failure of the project. It was further observed that about

Table 5. Sources of information on ecological sanitation (sample size = 340).

S/N	Source of information	Population reached	
		Number	Percent
1	Awareness campaigns	188	55.3
2	Community meetings	83	24.4
3	Friends and/or relatives	50	14.6
4	News media	19	5.7

11% of the interviewed head of departments mentioned some culture barriers such as those which restrict in-laws from sharing toilets. Other key informants (23%) mentioned poverty as an obstacle to latrine construction, due to inadequate resources and the habit of the latrine to collapse. A report by Torell and Mmochi (2006) indicated that less than 40% of households in Mkuranga District have toilets partly because of latrine construction difficulties caused by sandy collapsible soil and inadequate enforcement of public health and sanitation regulations and weak by laws. As a result, Mkuranga District Integrated Coastal Management (ICM) Action plan recognized that beach area, which is being used for open defecation and garbage dumping, is a major sanitation issue.

Literature studies have shown that open defecation is influenced by many factors including cultural behavior (Mubarak, 2013; Ashebir et al., 2013), cleanliness of toilet facility (Tumwebaze et al., 2014), poverty, shortage of water supply, lack of housing and illiteracy (Balamurugan et al., 2013). In another study by Pradhan and Heinonen (2010) in Central Nepal, it was observed that hygienic practices are influenced by the level of education largely because of lack of awareness among different socio-economic strata. A research by Ashebir et al. (2013) suggest that it is not enough to provide people with sanitation hardware because they may not be used by the intended users, which correspond to the observation made in this study. Numerous approaches have been suggested in literature to tackle the sanitation software part, which provide behavioral change interventions corresponding to psychological factors to be changed (Mosler, 2012).

Knowledge on ecological sanitation (UDDT)

It was revealed that only 82 of 340 respondents (24.1%) have knowledge of EcoSan. Of the 82 respondents who have knowledge of EcoSan, 46 respondents (55.3%) were informed of ecological sanitation through awareness campaigns and another 20 respondents (24.4%) through community meetings (Table 5). Generally, project initiative disseminated information on ecological sanitation to only 60 of 340 respondents (17.6%), which is contrary to the report given by AMREF (2009) which indicates that more than 63% (143,633 of 227,990 population of Mkuranga) were aware of the sanitation project particularly UDDT technology. It was further observed that an additional 12 of 82 respondents (14.6%) got the information from friends or relatives and the remaining 4 of 82 respondents (5.7%) got the information from news media. Field survey proved that only 7 of 82 responds (9%) knew slightly more than the hearing ecological sanitation.

During the interviews with district officials and focus group discussion with ward officials, it was noted that the steering committee excluded social worker and community development personnel in the project and involved people who were not well informed on ecological sanitation to facilitate the training, sensitization campaigns and advocacy meetings at the community level. During the interview with district officials including steering committee members, it was noted that 4 of them have no idea of new participatory approaches used in sanitation projects, suggesting that participatory was not used during the early stages of the project. As a result, the project failed to reach significant number of people from the target group to stimulate changes within the community. It is therefore evident that there was inadequacy of disseminated and organized advocacy for the technology contributed to its lack of uptake of technology in Mkuranga district by the users.

In accordance with WHO and UNICEF (2009), it is important to allow time for communities to adapt to new technology. Research work in Kabale Uganda indicated that 82% of the respondents in the project area where knowledgeable of EcoSan, which influence 20% of them to use the technology (Tumwebaze and Niwagaba, 2011). In accordance with Tumwebaze and Niwagaba (2011), at least 80% of the community is required to know the new technology in order to influence 20% of them to accept it. Elsewhere, Ashebir et al. (2013) reported that 54.5% of respondents having latrines in Tigray, Ethiopia did not use them at all and only 37.4% of respondents use latrines consistently. Even in more developed countries like European countries, UDDT technology particularly reuse of urine and feces, was initially considered a strange concept by the users (Fruman et al., 2012). Therefore, proper education and training of the communities must be done in order to overcome initial misconceptions that individual have.

Preference of latrine currently used versus UDDT

Table 6 shows the respondents' willingness to convert their existing toilet facility to UDDT. This data was gathered from 81 of 82 respondents who know EcoSan toilets. It was observed that only 1 of 14 respondents (7.1%) who are using flush sanitation (cistern and pour) are willing to convert to UDDT, but 21 of 68 respondents (30.9%) using pit (ventilated and conventional) latrines are willing to use UDDT. Overall, 27.2% of respondents who know EcoSan are willing to convert their current sanitation facility to UDDT, although none have adopted

Table 6. Willingness of converting current latrine into UDDT (n = 81).

Current latrine facility	Number of respondents using this type of latrine	Respondents using this type of toilet	
		Number	Percent
Conventional pit toilet	49	13	26.5
Ventilated improved pit toilet	19	8	42.1
Cistern flush toilet	10	1	10.0
Pour flush toilet	3	0	0
Total	81	22	27.2

Table 7. Views on UDDT promotion.

	Frequency	Percent
Not good idea	138	40.6
Good idea	64	18.8
Don't know	138	40.6
Total	340	100.0

UDDT because of inadequate practical knowledge of EcoSan. In India, Balamurugan et al. (2013) observed that communities using pit sanitation have more adaptability towards EcoSan as it functions far better than what they use in all aspects including health, reuse, hygienic, aesthetics and user-friendliness than pit sanitation. However, flush sanitation has put barriers to changing over to EcoSan as the concept is entirely different.

It was revealed that only 18.8% of the community considered UDDT technology favorably, 40.6% of the community did not support the introduction of UDDT technology and another 40.6% were not committed (Table 7). This was largely because of inadequate practical knowledge of UDDT technology. In accordance with Niwagaba (2003), the community with inadequate knowledge on UDDT, tend to consider it as impure and therefore reject it. However, Nekesa (2007) in his study at Wakiso sub-county in Uganda reported that most of the people in rural area prefer UDDT as an option in sanitation.

Community views on UDDT promotion

Handling and uses of dried faeces and urine

This survey showed that out of 340 respondents, only 70 (20.5%) were willing to handle dried faeces and urine. It was observed that those who rejected handling of excreta did because of various reasons such as nuisance (24%), religious restrictions by Muslims (57.6%) and fear of infectious diseases and lack of knowledge on ecological sanitation (18%). Rosenquist (2005) in analysis of psychosocial of the human-sanitation nexus observed that before dissemination of the knowledge of ecological sanitation, people have a tendency of rejection of the product from the sanitation options. It is therefore not uncommon for people to reject products from sanitation options and handling of excreta, but this is expected to change with improved knowledge on ecological sanitation. It has been documented that usage of decomposed excreta as manure in farmland is difficult to adapt as households fear odour and infectious diseases (Balamurugan et al., 2013). However, a small group within the community could act as a catalyst to changes towards sanitation promotion (Tumwebaze and Niwagaba, 2011). Therefore, the group which shows interest in using UDDT products could easily understand the benefits of the technology hence adopt it.

During focus group discussion, it was revealed that the project employed Participatory Hygiene and Sanitation Transformation (PHAST) approach as a tool to create sense of ownership to the community for sustainability of the project. While it is known that PHAST process is effective at spreading a multitude of health and hygiene messages, it is expensive, relies on donor agencies and non-governmental organizations and has weaknesses of bringing the intended improved and sustained hygiene behaviours (Binamungu, 2007). Various approaches proved to work better depending on environment, society itself and the geographical settings such as urban and rural (MoHSW, 2010). Community led total sanitation (CLTS) which is more suitable in rural setting could be the choice for area with rural setting characteristics while community led urban environmental sanitation (CLUES) which is jointly developed by EAWAG, the Water Supply and Sanitation Collaborative Council (WSSCC) and UN-HABITAT in 2010, could work better in areas with urban setting characteristics.

About 64.5% of the respondents who are willing to handle dry faeces and urine from UDDT said that they will use the product for agricultural purposes while 16% said that they can use the product for some other purposes such as business. It is worth to note that about 20% of the respondents, who refuses to handle dry faeces and urine, are willing to use it. More than 17% are ready to use it in agricultural purpose and more than 2% could use for other purpose. This finding indicates that people are willing to use dry faeces and urine, which could be a motive for them in adopting the technology.

Table 8. Challenges of UDDT.

Challenge	Percent
Technical problems	8.8
Social and religious beliefs	64.7
Financial problems	15.3
Don't know	11.2
TOTAL	100.0

Table 9. Community advice on UDDT uptake.

S/N	Community advice	Percent
1	Technology adjustment	20.6
2	Availability of technical expertise	13.5
3	Affordability of technology	26.8
4	I do not know	39.1
5	Total	100.0

UDDT challenges

About 64.7% (220 of 340) respondents consider social and religious beliefs as a major challenge for adoption of the technology. Other challenges are financial (15.3%), technical (8.8%), but the remaining 11.2% did not have opinions (Table 8). Mkuranga District is mostly dominated by Muslims who considered both urine and faeces as unholy and therefore rejected the idea from the beginning of the project. However, in accordance with Tumwebaze and Niwagaba (2011) more than 30% of the people who adopted UDDT in Kabale, Uganda were Muslims. They further reported that similar challenges were observed among the Muslim communities during the initial stage of the project, but the adoption gradually improved through advocacy. Ecological sanitation has also been used in other Muslim communities such as Pakistan (Nawab et al., 2006). In accordance with Muench (2009), UDDTs are suitable for various cultural settings: they can be designed to suit both sitting and squatting cultures and to cope with the use of water for wet anal cleaning cultures as well. Müllegger (2011) in innovative UDDT designs from East Africa provided various designs of UDDT, some of which have the provision for using water as anal cleansing materials. Similarly, Rosemarin et al. (2007) have suggested ecological sanitation systems that accommodate water as anal cleansing material. Unfortunately, there was no strong evidence to suggest that advocacy and demonstration of UDDT, which allow water to be used as anal cleansing material, was done in Mkuranga District.

About 15.3 % of respondents consider UDDT as more expensive than other toilets. This is contrary to the study done by Nekese (2007) in Wakiso, Uganda where UDDT were reported to be cheaper, affordable and comfortable as it excludes flies and smell. The same finding also were revealed by the study done in Pakistan on preference in designing ecological sanitation systems in north west frontier province, where EcoSan toilets were preferred by the local residents; showing that open defecation was a sign of poverty (Nawab et al., 2006). This discussion shows that there is a chance for UDDT to be adopted by even low income community, but will only be possible if the knowledge were properly disseminated and the community are well exposed to all design of the technology.

Technical problems also were considered to be a challenge by 8.8% of the respondents although during the initial stage of the project, a total of 560 artisans were trained from 80 villages. This is equivalent to 7 artisans per village in almost 70% of the villages in the district, which is a very positive start. However, during focus discussion, it was revealed that currently only 134 artisans are still in the district, but have shifted to towns where they are working as artisans. It was further reported that the selection process of artisans was biased as most of them were handpicked by the village leader and some of the steering committee member for personal rather than technical reasons.

To improve technology uptake by the community, the respondents requested for affordable designs (26.8%), the adjustments of technology to accommodate water as anal cleansing material (20.6%) and availability of technical expertise such as artisans (13.5%). It was noted that 133 of 340 respondents (39.1%) did not have comments on what should be done to influence the adoption of EcoSan technology largely because they do not have practical knowledge of the technology (Table 9). It was noted that although EcoSan was potentially cheaper than other latrines, the issue of affordability was raised by about 26.7% of respondents. It is possible that this group either lack some important information concerning the designs of EcoSan toilets or they do not currently have toilets. In fact, about 11% of the respondent claimed that they have not used or seen UDDT before because demonstration toilet was poorly located.

For new technology to be adopted by the users, it is important that all constraints that may restrict its adaptability such as religious, cultural, legal and financial constraints be identified at the planning stage of the project. This may include involving communities during planning and implementation stages of the project, which is considered as very important because it builds a sense of ownership and commitment among the local people (IRC, 2003; Mayo and Nkiwane, 2013). Evidence of recruitment and training of local masons in the project was available, but the stakeholders were not sufficiently involved in the project. Effort of applying participatory hygiene and sanitation transformation (PHAST) method similar to the one used at Majumbasita in the neighboring district of Ilala (Chaggu, 2002) was not evident in this project. Among the constraints that require information include attitudes of the people to reuse treated human excreta as manure and the responsibilities of various institutions for development of Ecosan with the relevant policies, by-laws, budgets and target levels have to be identified. It is also worth mentioning that financial obstacle, may be one of the constraints for adoptability of Ecosan in Mkuranga District bearing in mind that the district is one of the poorest in the country. In such cases, government subsidies may prove a success provided consumers also contribute to a reasonable extent.

Conclusions

From the study the following conclusions were drawn:

1. District leaders (including councilors) abandoned the project by allowing steering committee to work, supervise and even audit their own transactions. It was further observed that the participatory approach used was not effective. As a result, there was no evidence of the technology adoption in the district.
2. Religious and cultural barriers for adoption of UDDT were observed. Such barriers may only be removed by disseminating the correct information to residents by committed, experienced and knowledgeable personnel. Unfortunately, the personnel from the project were not conversant with ecological sanitation, which was one of the reasons for failure to disseminate the proper contents meant to reach the community.

Conflict of interests

The authors did not declare any conflict of interest.

REFERENCES

Abarghaz Y, Mahi MM, Werner MC, Bendaou N, Fekhaoui M (2012). Ecological sanitation in Morocco Promotion of the Urine-Diversion Dehydration Toilets- A case study of Dayet Ifrah. Am. J. Environ. Sci. 8(3):212-219.

Adhikari RU, Islam M, Tuladhar B, Shakya SK, Thakur KK (2012). Operation and maintenance challenges to promote dry toilet: A case of ENPHO EcoSan implementation area in Nepal. Fourth International Dry Toilet Conference, Tampere, Finland.

AMREF-Mkuranga (2007). UDDT baseline report, Mkuranga Distict council.

AMREF-Mkuranga (2009). Annual report, Mkuranga Distict council.

Ashebir Y, Sharma HR, Alemu K, Kebede G (2013). Latrine use among rural households in northern Ethiopia: A case study in Hawzien distric, Tigray. Int. J. Environ. Stud. 70(4):629-636.

Balamurugan J, Ravichandran M, Nithiya PT (2013). Ecological sanitation and green economy. Int. J. Sci. Res. 2(10):1-4.

Binamungu B (2007). Towards Meeting National and MDG sanitation targets; A review of sanitation status in Tanzania, Water and Sanitation Program, Dar es Salaam, Tanzania.

Chaggu E (2002). Ecological sanitation toilets in Tanzania. Third International conference on Integrated Environmental management in Southern Africa, Johannesburg, South Africa.

COHRE, WaterAid, SDC and UN-HABITAT (2008). Sanitation: A Human Rights. Development Studies and Monograph Series 20. Washington: World Bank DRAFT (2002/3), Dar es Salaam: Economic Research Bureau.

Esrey SA, Gough J, Rapaport D, Sawyer R, Simpson-Hébert M, Vargas J, Winblad U (1998). Ecological Sanitation; Department for Natural Resources and the Environment, Sida, S-105 25 Stockholm, Sweden.

Fruman Y, Wendland C, Samwel A, Deegener S (2012). Acceptance of UDDTs in eastern Europe, Caucasus and Central Asia. Fourth International Dry Toilet Conference, Tampere, Finland.

Guadagni A (2012). Embedding Innovative Sanitation Technology: An Assessment of Urine Diversion Dry Toilets in Bangladesh, Technische Universität – Berlin.

Ingle R, Berdau S, Kleemann F, Arndt P (2012). What does it take to convince decision makers in Omaruru, namibia to scale up urine divesion dyhydration "Otji toilets"? Fourth International Dry Toilet Conference, Tampere, Finland.

IRC (2003). Community water supply management: About community management. Available online at http://www2.irc.nl/manage/whatisit/index.html. Accessed on 25th March 2015.

Jurga I, Gallinat B, Mang H (2003). Implementation of a closed-loop sanitation concept in Yang Song township, China. Proceedings of the 2nd International Symposium in Ecological Sanitation, April 7-11, Lubeck, Germany, 843-848.

Kiba DI (2005). Agronomic valorization of the human excreta: Use of the urine and faeces for the production of aubergine (*Solanus melongema*) and of the corn *(Zea mays)* in the center zone of Burkina Faso, Thesis, Polytechnic University of Bobo Dioulasso.

Kothari CR (2004). Research methodology: Methods and Techniques, Second Edition, New Age International Publishers, New Delhi, India.

Krejcie RV, Morgan DW (1970). Determining Sample Size for Research Activities. Education and Psychological Measurement 30: 607-610.

Langergraber G, Muellergger E (2005). Ecological sanitation – a way to solve global sanitation problems? Environ. Int. 31:433-444.

Majule AE (2012). Implications of ecological and social characteristics to community livelihoods in the coastal areas of Tanzania. Afr. J. Environ. Sci. Technol. 6(1):72-79.

Mayo AW, Nkiwane L (2013). The role of community participation on cost recovery and sustainability of rural water supply projects in Hai District, Tanzania. J. Environ. Sci. Water Resour. 2(11): 388 - 395.

Ministry of Health and Social Welfare (MoHSW) (2010). Annual report on Food, Water and Sanitation.

Mkuranga District Council (2008). Annual report, Health department, Prevention section.

Mkuranga District Council (2011). Annual report, Health department, Prevention section.

Mkuranga District Council (2011). Strategic Plan, Implementation report.

Mnkeni PNS, Kutu FR, Muchaonyerwa P, Austin LM (2008). Evaluation of human urine as a source of nutrients for selected vegetables and maize under tunnel house conditions in the Eastern Cape, South Africa. Waste Manage. Res. 26:132-139.

Mosler HJ (2012). A systematic approach to behavioral change interventions for water and sanitation sector in developing countries: A conceptual model, a review, and a guideline. Int. J. Environ. Health Res. 22(5):431-449.

Mubarak T (2013). Requirements and Challenges of UDDT for Technology Uptake by Coastal Communities: A case of Mkuranga Project- Tanzania, Masters of Sanitation Management Dissertation, Department of Water Resources Engineering, University of Dar es Salaam.

Muench E (2009). Urine-Diversion Dehydration Toilets. Eschborn: German Agency for Technical Cooperation (GTZ) GmbH.

Müllegger E (2011). Innovative urine diverting dry toilet (UDDT) designs from East Africa; EcoSan Club Austria: Vienna, Austria. Sustain. Sanit. Pract. 6:10-14.

National Bureau of Statistics (2011). Tanzania Demographic and Health survey (TDHS); Final report 2010.

Nawab B, Nyborg ILP, Esser KB, Jenssen PD (2006). Cultural preferences in designing ecological sanitation systems in North-west frontier, Pakistan. J. Environ. Psychol. 26: 236-246.

Nekesa J (2007). Community response towards ecological sanitation- Analysis of knowledge based adoption levels among the community

of Kasanje subcount, Wakiso district (Uganda), Project Report, SNV Uganda, West Nile portfolio.
Nuwagaba A (2003). Assessment of factors that influence household's choice of sanitation technologies and excreta re-use (EcoSan), Kampala, Water and Sanitation Program, Makerere Institute of social research.
Pradhan S, Heinonen TH (2010). Knowledge and awareness of eco-sanitation in central Nepal: A questionnaire survey. Environ. Dev. Sustain. 12: 713-726.
Rhode L, Stintzing AR, Steineck S (2004). Ammonia emissions after application of human urine to a clay soil for barley growth. Nutr. Cycl. Agroecosyst., 68:191-198.
Rieck C, Muench E (2011). Technology review of urine diversion dehydration toilets (UDDTs), Design principles, urine and faeces management, Eschborn, Germany; Available at http://www.ecosanres.org/discussion_group.htm. Accessed on 25th March 2015.
Rosemarin A, Kvarnström E, Subbaraman M, Ganapathy V, Dagerskog L, Pasupathiraj K (2007). Ecosan systems that accommodate anal washing. Sustainable Water Management, Issue 2-2007. Available at http://www2.gtz.de/Dokumente/oe44/ecosan/en-analwashing-2007.pdf. Accessed on 25th March 2015.
Rosenquist LED (2005). A psychosocial analysis of the human-sanitation nexus. J. Environ. Psychol. 25(3): 335-346.
Smet J, Sugden S (2006). Ecological Sanitation: Well Factsheet. Available at http://www.lboro.ac.uk/well/resources/fact-sheets/fact-sheets-htm/Ecological%20sanitation.htm. Accessed on 25th March 2015.
Sridevi G, Srinivasamurthy CA, Bhaskar S, Viswanath S (2009). Evaluation of source separated human urine (ALW) as a source of nutrients for banana cultivation and impact on quality parameter. J. Agric. Biol. Sci. 4:44-48.
Tanzania Meteorological Agency (2013). Mean monthly temperature and rainfall variation data.
Thibodeau C, Canaday C (2011). Elegant yet economical urine diverting dry toilets in Equador. Sustain. Sanit. Pract. 6: 23-28.
Tidaker P, Mattsson B, Jonsson H (2007). Environmental impact of wheat production using human urine and mineral fertilizers- a scenario study. J. Clean Prod. 15: 52-62.
Torell E, Mmochi A (2006). Mkuranga Governance Baseline, Coastal Resources Center, University of Rhode Island, p. 18.
Tumwebaze IK, Niwagaba CB (2011). Ecological sanitation uptake knowledge, beliefs and practices in Kabale Municipality, Kabale district; The future of Water, Sanitation and hygiene; Innovation, adaptation and engagement in changing World; 35th WEDC international conference, University of Loughborough, UK.
Tumwebaze IK, Niwagaba CB, Günther I, Mosler HJ (2014). Determinants of households' cleaning intention for shared toilets: Case of 50 slums in Kampala, Uganda. Habitat International, 108-113.
UNESCO/IHP (2006). Capacity building for ecological sanitation; Concepts for ecologically sustainable sanitation in formal and continuing education. International Hydrological Programme (IHP) of the United Nations Educational, Scientific and Cultural Organisation (UNESCO); 1 rue Miollis, 75732 Paris Cedex 15, France and the Deutsche Gesellschaft für Technische Zusammenarbeit (GTZ) GmbH Dag Hammarskjöld Weg 1-5, 65760 Eschborn, Germany.
United Nations Children Education Fund (UNICEF) (2013). Water, Sanitation and Hygiene: WASH - Fast Facts. Available at http://www.unicef.org/tanzania/WASH_factsheet.pdf. Accessed on 25th March 2015.
United Nations Children Education Fund (UNICEF) and World Health Organization (WHO) (2013). Progress on Sanitation and Drinking Water – 2013 Update. WHO and UNICEF Joint Monitoring Programme for Water and Sanitation. Available online at http://www.who.int/iris/bilstream/10665/81245/1/9789241505390_eng.pdf. Accessed on 25th March 2015.
United Republic of Tanzania (URT) (2013). Population and Housing Census 2012. population Distribution by Administrative Areas. national Bureau of Statistics, Dar es Salaam, March 2013.
Water and Sanitation Program (WSP) and United Nations Children Education Fund (UNICEF) (2013). Water Supply and Sanitation in Tanzania: Turning Finance into services for 2015 and beyond. An AMCOW Country Status Overview. Available at http://www.wsp.org/sites/wsp.org/files/publications/CSO-Tanzania.pdf. Accessed on 25th March 2015.
WaterAid (2008). Ecological sanitation latrines: The experience of Nepal. Fieldwork paper. WaterAid in Nepal, Kathmandu, Nepal. Available at WaterAid webpage. Accessed on 25th March 2015.
Wendland C, Deegener S, Jorritsma F (2011). Experiences with urine diverting dry toilets (UDDTs) for households, schools and kindergarten in Eastern Europe, the Caucasus and Central Asia. Sustain. Sanit. Pract. 6:16-22.
Wikipedia (2013). Water supply and sanitation in Tanzania. Available at http://en.wikipedia.org/wiki/Water_supply_and_sanitation_in_Tanzania. Accessed on 25th March 2015.

Accessibility and utilization of mobile phones for governance of water resources in the Lake Victoria Basin: Constraints and opportunities in Tanzania

H. J. Mongi[1] , A. N. Mvuma[2], S. Kucel[3], A. J. Tenge[4] and M. Gabriel[5]

[1]Department of Information Systems, The University of Dodoma, P.O. Box 259, Dodoma, Tanzania.
[2]Department of Telecommunications and Communications Networks, The University of Dodoma, P.O. Box 259, Dodoma, Tanzania.
[3]Department of Mechanical Engineering, Busitema University, P.O. Box 236, Tororo, Uganda.
[4]Department of Development Studies, The University of Dodoma, P.O. Box 259, Dodoma, Tanzania.
[5]Department of Content Development, The University of Dodoma, P.O. Box 259, Dodoma, Tanzania.

Mobile phones are the world's fastest growing technologies in terms of coverage and adoption. This trend makes them a desirable information and communication technology (ICT) platform for addressing the striking challenges of water resources governance at all levels. Some of remarkable challenges in Lake Victoria Basin (LVB) that have negative impact on the water resources are rapid population growth, unfavorable climate change and variability, and non-sustainable urbanization. Mobile phones can contribute to addressing these challenges by addressing three key principles of governance: transparency in a manner that the resources are utilized by increasing population; accountability in both supply and demand sides of water resources for household, industry and agriculture; and participation of key stakeholders in policy formulation and decision making across temporal and spatial scales. This paper analyses the accessibility and utilization of mobile phones among various actors in the governance of water resources within LVB with the aim of identifying opportunities and constraints at community (micro), local government (meso) and national lake/river basin (macro) scale. To achieve this, three data collection sites were identified in Tanzania: Mwanza, Kagera and Mara. Data were gathered from key stakeholders through five methods: key informant interview, actors' panel discussion, observation, literature reviews and questionnaire administration. Qualitative data were subjected to contents analyses while empirical data were analyzed using Statistical Package for Social Sciences (SPSS). The analyses provided information on the constraints in access and utilization as well as opportunities availed by mobile phones. Furthermore, their potential for enhancing transparency, accountability and stakeholders' participation were summarized and discussed. Conclusion drawn from this study is that there is variation in access and utilization of mobile phones across scales of water resources governance in LVB that offer both opportunities and constraints. Integrated and innovative system that bridges the gap by capitalizing on the opportunities and minimizing the constraints could be an option.

Key words: Information and communications technology, mobile phones, Lake Victoria Basin, water resources, governance.

INTRODUCTION

The ecosystem for information and communication technologies (ICTs) has been rapidly changing in the contemporary world. According to Selwhn (2002), it consists of a wide range of technological applications; digital broadcast technologies; telecommunications technologies as well as electronic information processing and resources. ICTs handle both information and telecommunication– including symbols, data, voice, images and video. Common ICT technologies are the internet, radio, television, mobile phones, fixed phones and all computers (mainframes, desktops, laptops, palmtops and tablets). A mobile phone is one of the mobile devices in the group of ICTs. Other devices include PDAs, wireless tablets and mobile computers (Jones and Marsden, 2006; Islam and Masum, 2010).

Mobile technology has been leading with new innovations. Mobile phone research and innovation have brought better brands, better portability and providing various functionalities depending on affordability, usability and interoperability (Ally, 2007; Jones and Marsden, 2006). Mobile phones can provide rigorous data gathering capability using voice-based technique, wireless internet gateway (WIG) or unstructured supplementary service data (USSD). Although when the first products were available in the market, the technology was afforded by wealth people, today it is an important item for every household if not every individual. Ordinary person in the rural area such as a farmer, a fishermen or a livestock keeper can afford owning one. Recent statistics indicate that of the world's seven billion people, six billion have mobile phones (UN News Centre, 2013). According to the International Telecommunication Union (ITU), the mobile phone adoption has far exceeded the total populations in many developing countries. Tanzania has recently recorded 59.2% mobile penetration rate (MDI, 2013). It is estimated that more than 80% of the world's population is covered by a mobile network, while in Africa alone; more than 600 million people have access to mobile phone services (InfoDev, 2013).

Mobile phones are changing the way people perform and interact across economic sectors. Writing in the context of agriculture sector, Loudon (2009) argue that mobile phones can provide data transfer functionalities enabling data sharing between and within the systems. Mobile phones can also be converged with other ICTs like radio and TV to respond to key issues like access to weather forecasts, cropping options as well as market information. Authors in the intersection between ICT and environment have shown that mobile technology can minimize the effect of environmental degradation. Ospina and Heeks (2010) contribute on the redundancy aspect of ICTs; that is, the potential to increase the availability of resources. Redundancy concept has varying implications for computer and human beings. To the computer or mobile phone as a computing device, redundancy leads to inefficient utilization of resources, for example storing the same file in different folders. To human beings, redundancy is a good thing. In this context, Unhelkar (2009) simply define redundancy as "surplus capacity". For example, a mobile phone can enhance achievement of multiple objectives. These include improving access to financial capital which in turn may enhance access to infrastructures for efficient water use, improved systems for water distribution and also markets that are linked to water use efficiency (UNEP, 2010; Schuchardt et al., 2004). Mobile phones also can provide for rapid response through swift access and mobilization of financial assets (Duncombe and Boateng, 2009). Mobile phone services such as m-Banking and m-Finance can improve the speed and efficiency with which local communities and general public at large can be mobilized to respond to water resource challenges. Geographic information systems' enabled mobile phones enhance communication, access to relevant location data, and mobilization of physical, economic and social resources that stakeholders may need to organize themselves while performing their roles in governing the water resources (World Bank, 2013; ITU, 2011).

Governance is often considered a central issue in water resources especially in the developing countries. According to Mbilinyi et al. (2007), governance is a system of administering and exercising power in democratic, transparency, strategic and ethical principles. Under this system, the entire society or concerned group of people are involved in planning, implementing and decision making processes. Rogers and Hall (2003) argue that governance is intensely a political term that refers to effective implementation of socially acceptable allocation and regulation. According to Wong (2009), participation is a core principle of good governance. To achieve sustainability especially within water resources, such participation should be made to happen through meetings, discussions and face-to-face communication at different levels (Mbilinyi et al., 2007). It is through these ways that potential value of the resource to the community and the environment is informed (Medalye and Hubbart, 2008).

Water resources challenges are location specific, although six categories of issues are often mentioned: inadequate water quality (Tebbutt, 1998), competing users, increasing demand amidst declining supply (WRG2030, 2009), climate change, (USDA, 2010; Sehlke, 2008; Frantz, nd), land-use change (Sehlke, 2008) and institutional

Figure 1. The map sketch of LVB showing the study areas.

barriers (Rogers and Hall, 2003; Weston, 2008; Weggoro and Ntumubano, 2010). Governance challenges affecting water resources have continually being addressed by fragmented policies, laws and regulations, weak enforcement of laws, weak institutional coordination and low community participation in decision making across spatial and temporal scales (Weggoro and Ntamubano, 2010, Okurut, 2010). Indicators of these challenges include: budgetary and financial mismanagement, irresponsible public expenditure, lack of transparency, lack of accountability and corruption. Impacts have been degradation of water sources, pollution, declining water quality and quantity and water use conflicts. Integrated approaches will enhance good governance and at the same time promote sustainability of water resources. Mobile phones provide tools that increase awareness, participation, accountability, coordination and communication at local, regional and international levels (Moum, 2006; Fudik et al, nd). These tools can simplify complex management decisions on natural resource management within a trans-boundary ecosystem. Despite the importance of mobile phones, there is inadequate knowledge on their accessibility and utilizations on a sector basis, on rural-urban basis and on various scales of actors in governance of water resources. Furthermore, mobile phones are used alongside other technologies whereby understanding their access and utilization constraints and opportunities would reduce technology duplication and avoid unnecessary wastage of important resources. This paper analyses the accessibility and utilization of mobile phones among various actors in the governance of water resources within Lake Victoria Basin (LVB) with the aim of identifying opportunities and constraints at community (micro), local government (meso) and national lake/river basin (macro) scale. Mobile technology also provides tools that increase awareness, participation, accountability, coordination and communication between stakeholders. These tools can simplify complex management decisions on natural resource management within a trans-boundary ecosystem like LVB. Mobile phones can further be used to predict and timely give early warning for the climate variability and changes.

METHODOLOGY

Brief description of the study area

LVB is a shared ecosystem among five member states of East Africa Community (EAC): Tanzania, Kenya, Uganda, Rwanda and Burundi. These five countries occur between latitudes 5°30"N and 12°S and longitudes 41°50"E and 28°45"E (Figure 1). The basin is blessed with resources such as water, forests, rivers and land for agricultural production, human habitats, wildlife, minerals and

Table 1. Stakeholders at various scales that were consulted

Scale	Region		
	Mwanza	***Kagera***	***Mara***
Micro	Irrigation schemes, Environmental Management and Economic Development (EMEDO), Organisational – Community Livelihood Strategies and Support (COLIS), Beach Management Unit (BMU)	Irrigation schemes, Enviro-comet (ECOT), Misenyi Area Development Project (ADP), Kasambya Gugu Maji, Beach Management Unit (BMU)	Irrigation schemes, Beach Management Unit (BMU)
Meso	Magu District Water Development, District Councils: Ukerewe, Mwanza, Misungwi; Urban Water Supply Authority	Bukoba Municipal Council, Misenyi, Agricultural Office	Musoma Municipal, SCC-VI-Agroforestry Project
Macro	Office of Regional Administrative Secretary (RAS), National Environmental Management Council (NEMC)	Police Force, Tanzania Telecommunications Company Limited (TTCL), Prevention and Combat of Corruption Bureau (PCCB), Misenyi AIDS and Poverty Eradication Crusade (MAPEC), (Association Of Early Childhood Care (AECC), RAS	TTCL, RAS, Lake Victoria Basin sub-office

Source: Field data.

fishery. The study was undertaken in three regions in the Tanzanian part of LVB: Mwanza, Kagera and Mara. Within these regions, three districts/municipal/councils were selected: Nyamagana and Misungwi for Mwanza, Bukoba/Misenyi for Kagera and Musoma for Mara (Figure 1). Criteria for selecting the area were the extent of water resource governance challenges, the availability of key stakeholders and their readiness to participate.

Methods for data collection and analysis

Design of the study involved a series of stakeholders meetings and workshops that were held in each participating region in Tanzania (Mara, Mwanza and Kagera). Discussions during the meetings and workshops included identification of roles in water resource governance, giving information about access to mobile phones and how best they are utilized or could be utilized to improve water resource governance over LVB. The following were the key stakeholders that were involved in different scales:

1. Institutions for water resource beneficiaries (NGOs, CBOs, local communities e.g., water use groups).
2. Institutions for water management (e.g. water use associations, Beach Management Units (BMUs), Lake Basin Water Authority, Local Governments Authorities).
3. Legal and regulatory enforcing organs at national, basin and regional levels (eg. Police, National Environmental Management Council (NEMC)).
4. Policy making institutions (e.g. ministries in-charge of resources management).
5. Research institutions

Other methods of data collection were: key informant interview, observation, document reviews and questionnaire administration. Sample size for questionnaire administration was estimated using a proportion's formulas as follows:

$$n = \frac{t^2 \times p(1-p)}{e^2} \qquad (1)$$

where *n* is the required sample size, *t* is the confidence level at 95% (standard value of 1.96), *p* is the estimated participants in water resource governance in the study area (30%) and *e* is the standard error that tolerated 5% (standard value of 0.05).

The raw data were coded and converted to electronic databases using SPSS software. The data bases were edited and checked for reliability and validity. Then data were analyzed using the SPSS and MS Excel computer programmes. Basic statistics including mean, frequency and percentages were computed and used to compare the proportion of responses in each category. Cross tabulation was used to compare the results within and between country sites. Mean values for data collected by Likert scale questions were computed to obtain the weighted mean (average) for each variable. This computation was done according to the procedures for computing the weighted means when you have different contributions from different groups (Bowerman et al., 2011; Devore and Peak, 1992) in accordance with the equation below:

$$M_w = \frac{\sum w_i x_i}{\sum w_i} \qquad (2)$$

where M_w is the weighted mean (average) for Likert scale data set on variable *i*, W_i is the relative frequency of responses in percentage for variable *i*, and X_i is the value of variable *i* in Likert scale (1, 2, 3, 4). Opportunities and constraints of mobile phones were accessed based on the socio-economic characteristics of the respondents and industry market trends specific to Tanzania.

RESULTS

Characteristics of respondents

Respondents came from various backgrounds and organizations engaged either directly or indirectly in water resource governance as depicted in Table 1. Table 1 further categorizes their organizations with respect to their scales of governance of water resources, that is,

Table 2. Personal characteristics of respondents in the study area.

Characteristics	Description	Frequency	%
Sex	Male	156	71
	Female	64	29
Age (years)	< 18	2	0.9
	18 - 35	86	38.9
	36 – 50	96	43.4
	> 50	37	16.7
Highest education level	Informal	10	4.7
	Primary	133	62.1
	Secondary	71	33.1
	Tertiary		
Length of stay (years)	1-4	31	16.8
	5-10	37	20
	11-15	22	10.8
	> 15	95	51.4

Source: Field data

micro, meso and macro. The table indicates that there was adequate representation of relevant stakeholders in the LVB water governance. Table 2 describes the community respondents with respect to their sex, age, highest level of education and length of stay in a particular area. These variables are important in understanding the opportunities offered and constraints at a very individual point of view.

The results presented in Table 2 show that male were dominant participants. The results do not reflect the actual proportion of female in the population where it is known that the proportion of female is relatively larger. However, the results reflect the situation where majority of head of households are male and they are the ones who frequently appear in meetings and influence decision making. These groups (male and female) are likely to have different perception, knowledge and ability to utilize ICT facilities due to differences in access to resources and information. Normally, women in these countries have limited access due to traditional social barriers although the situation is improving with time. Results in Table 2 further show that about 82% of respondents are aged between 18 and 50 years. This is generally the most active group in water resources management where they participate in building associations; they are the water resource users in production in industries, agriculture, fisheries and at household level. Moreover, about half (51%) of respondents in Tanzania have stayed in their current places of residence for more than 15 years. This suggests that they have accumulated adequate experience and knowledge on issues related to LVB water resources governance challenges.

Respondents' perception on constraints to mobile phone access and utilization

Mobile phone access

Access to mobile phones as measured by individual ownership of the device or accessibility through the household relationship showed that about 30.4% (N=289) of the respondents owned mobile phones as shown in Table 3. This is lower than the average teledensity/penetration of 64% reported by Tanzania Communications Regulatory Authority (TCRA) in June 2014. The discrepancy is attributed to the fact that teledensity reported by TCRA considers number of registered SIM cards without taking into account multiple ownerships while the study considered accessibility to mobile phones. It is not uncommon for users of mobile services to own multiple SIM cards without mobile phones ownership.

However, as compared to other ICT assets, mobile phones rank the highest followed closely by radios while access to computers and modems ranked the lowest. The findings thus suggest that integrating mobile phones with community radios could be the best option for empowering stakeholders at community level for participation in governance of LVB water resources. The same conclusion can be inferred from Figure 2 which shows ownership of ICT facilities by age of respondents.

Figure 2 further shows that the highest percentage of respondents in age groups 18 - 35 and 36 - 50 years own mobile phones (36 and 61%, respectively) followed by radio (33 and 57%, respectively). This implies that respondents in these age groups have more access to mobile

Table 3. Community access to mobile phones against other ICT facilities by communities.

Type of ICT Facilities		Own ICT asset			
		Yes		No	
	Sex	Frequency	%	Frequency	%
Radio	Male	64.0	32.0	136.0	68.0
	Female	22.0	24.7	67.0	75.3
	Total	86.0	29.8	203	70.2
Mobile phones	Male	63	31.5	137	71.9
	Female	25	68.1	64	71.9
	Total	88	30.4	201	69.6
Television	Male	18	9	182	91
	Female	9	10.1	80	89.9
	Total	27	9.3	262	90.7
Computer	Male	11	5.5	183	92
	Female	4	4.5	85	95.5
	Total	15	5.2	268	93.1
Internet modem	Male	10	5	190	95
	Female	5	5.6	84	94.4
	Total	15	5.2	274	94.8

Source: Field data.

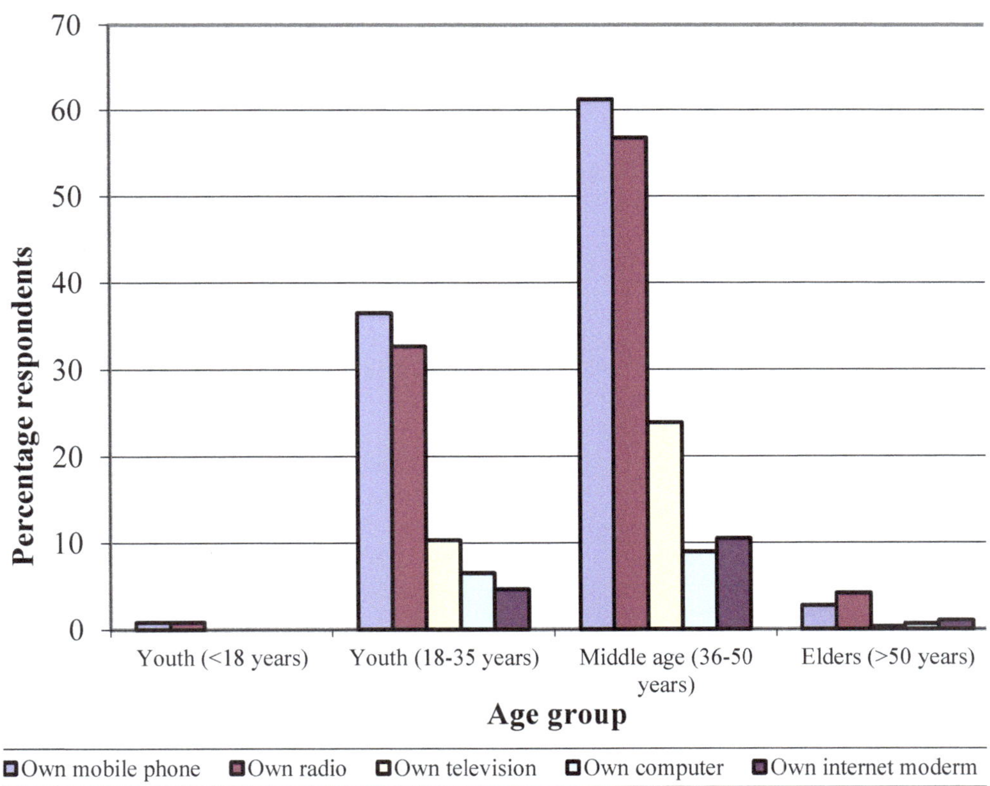

Figure 2. Ownership of ICT facilities by age of respondents of community's members (Source: Field data).

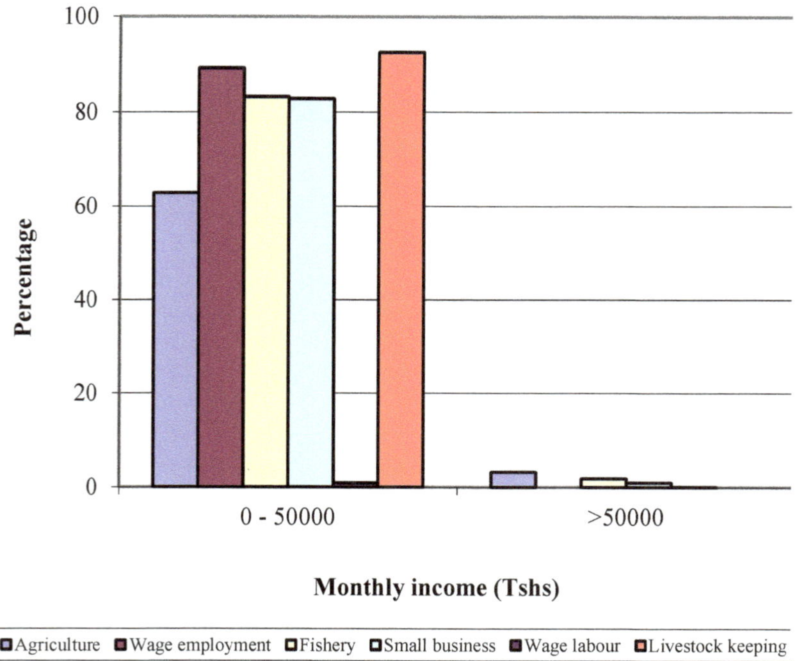

Figure 3. Monthly income levels of community members without mobile phones (Source: Field data).

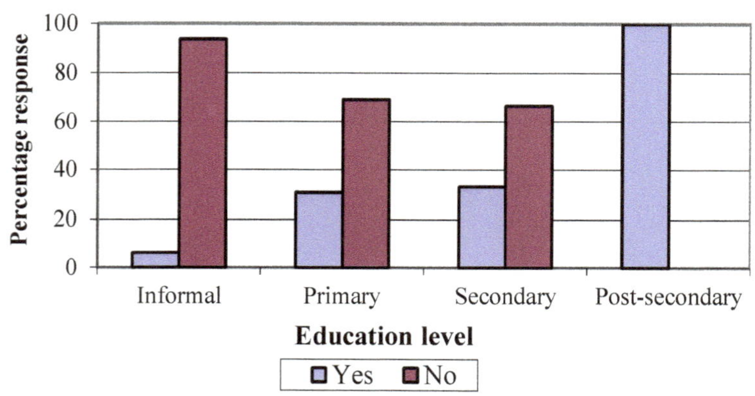

Figure 4. Community members with and without mobile phones by education level (Source: Field data).

phones than other ICT facilities.

Figure 3 summarizes the respondents who do not own mobile phones by their income categories across different economic activities. Results show that between 65 and 90% of those who do not own mobile phones (69.8% of all respondents) were earning monthly income equal or less than Tshs 50,000/= or close to US$ 32. Although it is sometimes difficult to ascertain income of an individual, this reflects the major obstacle facing stakeholders at the lower scale of governance of water resources.

Figure 4 shows the distribution of respondents with and without mobile phones across different education levels. The results indicate a declining gap in ownership as someone advances into higher education. For example, while for respondents with informal education the gap was 87%, for post-secondary education the gap dropped to 0%. This trend can be an indication of a correlation between education level, likelihood of employment and income.

Awareness on capability of mobile phone services

The study indicated that there is generally low awareness about potential services that mobile phones can offer. However, the difference was more apparent in rural-based community members (Kagera and Mara) as

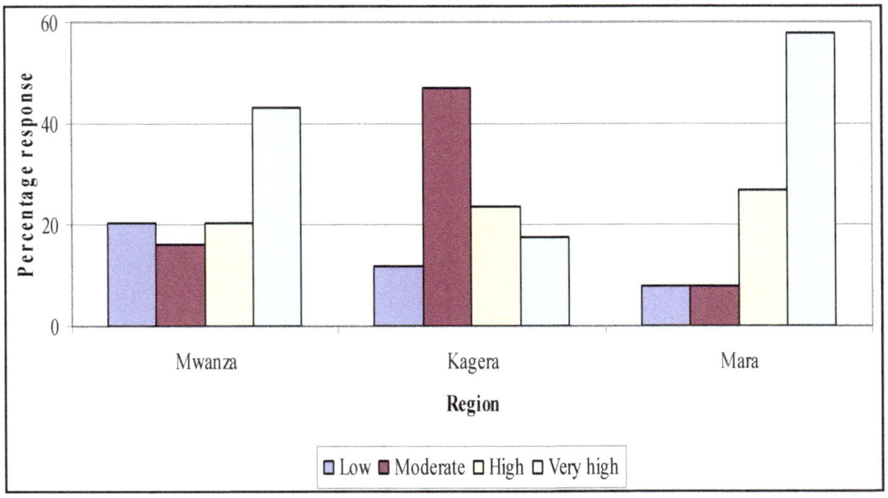

Figure 5. Awareness on the potential of mobile phones to provide reliability of information (Source: Field data).

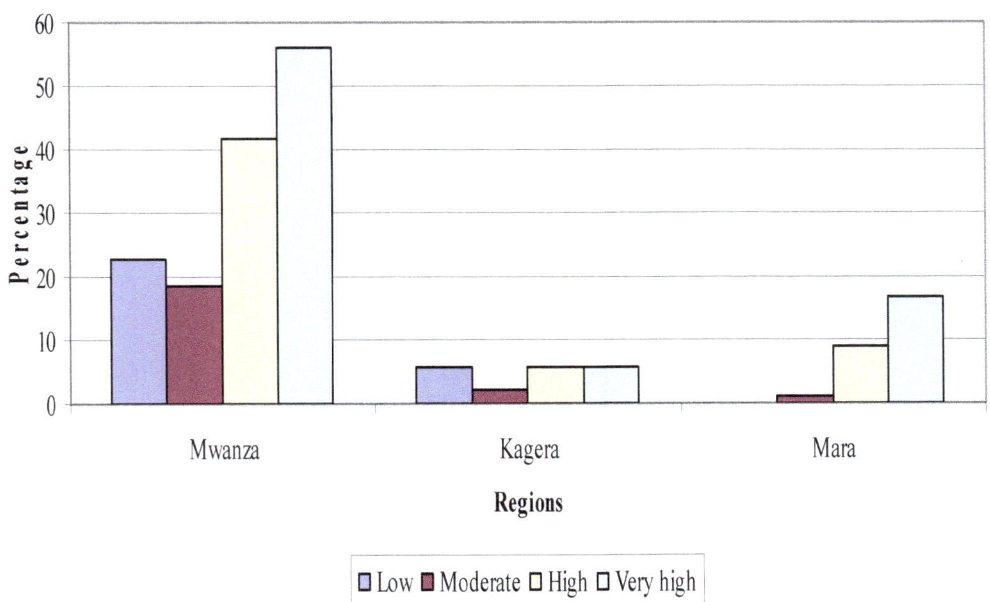

Figure 6. Awareness on the potential of mobile phones to provide for rapidity (Source: Field data).

compared to their counterparts in urban and peri-urban areas (Mwanza).

Figure 5 summarizes the respondents' perception on reliability of mobile phone services such as M-Payment. Most respondents seem to have high to very high trust on mobile phone money services but others especially those from rural areas (for example, Bukoba) have shown little hesitance.

Rapidity is an advantage that a mobile phone can provide in mobilizing resources and responding to urgent needs for water resource challenges. Respondents' perception on rapidity is shown in Figure 6. Majority of respondents were aware of the capacity of mobile phone to offer rapid service delivery.

Perceived cost of mobile phone and network coverage

There was general perception that the cost of acquiring and running the mobile phone is high. Majority of respondents attached the low access and usability to costs of acquiring and operating the mobile phone. Respondents had a varying perception on mobile network coverage (Figure 7b). Those in mostly urban and peri-urban areas reported good to very good network coverage. However, many rural dwellers in all the study areas reported moderate to poor coverage.

DISCUSSION

Constraints and possible halting strategies

There was generally acceptable level of awareness on the quality of service that a mobile phone can provide among respondents who owned the device. However, issues related to acquisition cost and coverage emerged among the same group. In addition, there was a huge group of respondents (69.6%) who did not own a mobile device. In fact, this was relatively smaller when compared with those who did not own other ICT assets such as radio (70.2%), television (90.7%), computer and internet modem (>93%) as detailed in Table 3. To this, very large group awareness is crucial along with empowerment through enabling them to possess a handset. This can be achieved through concerted efforts by the government and other development partners by embracing supportive policies to underprivileged members of the community. Ownership by age is favorable as the best majority owners belonged to age groups between 18 and 50 years (Figure 2). These are groups actively engaged in water resource governance and share most of its challenges. Majority were small irrigation water users, fishermen and water users at household level.

Studies in Nigeria indicated that despite good penetration rate at the national level, many rural dwellers, approximately 71% had no cell phones, thus excluded from the benefits of the mobile phone revolutions (Hultman, 2013). The government facilitated development of mobile based system to curb corruption activities in agricultural input distributions, identify subsidies recipient's name, address and phone numbers; therefore keep proper records for monitoring and evaluation. The government envisages further benefit from the system including using the records for policy and decision making since government policy must be based on evidence and well analyzed data. In order to make the project more inclusive, the government of Nigeria planned to distribute over 10 million mobile phones to the neediest famers (*ibid*). This is an example of government driven initiative that would increase community empowerment in governing their water resources.

Experience has indicated that income poverty and low education have greater chances of isolating the vulnerable communities from access to public services and hence hinders participation in water resource governance activities. Results from this study indicated that there was a close relationship between access to mobile phones and levels of education and income. Over 90% of respondents interviewed through questionnaires (N=289) were at the micro scale. They had qualification of varying levels (Table 2). The higher the educations level, the more the chance of an individual to own a mobile phone. This was so for the income levels (Figures 3 and 4). The best explanations could be that as one goes up the education ladder, has a greater chance of earning a higher income and eventually own a mobile phone. Lower education and income levels could also reflect in the ways in which respondents perceived quality of service of mobile networks (Figures 5 and 6). It was to be expected that most respondents without mobile phone could not express their perceptions. While the industry figures show improvement in network coverage and total cost of ownership, respondents' perception indicated this was still a constraint especially in rural areas (Figures 7a and b). This is also a matter of policy and regulation enforcement at lower levels of governance.

Vertical and horizontal approaches have to be considered in order to improve governance of water resources through mobilization of active participation, high levels of accountability and transparency. The structure of water resource governance may be considered as a triangle with a wide base. This bottom level or the low scale (micro) consists of many participants with a lot of transactional information among them. Addressing access and utilization constraints at this level is mandatory.

The opportunities from mobile phones and how to harness them

The mobile industry in Africa consists of a wide range of solutions which most of them can be adopted in governing water resources in Tanzania and beyond. Vodafone and Accenture identified 12 opportunities for mobile phone technology to increase income and productivity (Vodafone/Accenture, 2011). Some of these platforms are already widely used in Africa, while others are still in early stages of implementation. The list include mobile payment systems, micro-lending platforms, mobile information platforms, help lines, small logistics, traceability and tracking systems, mobile management of supplier networks, mobile management of distribution networks, trading networks, tendering platforms and bartering platforms. Although most of these examples have been drawn from agricultural sector, they can serve other sectors as well with or without slight modification.

To take full advantage of the opportunities availed by these systems, other pre-conditions have to be fulfilled. These include the availability of mobile network in the areas where water resources are threatened by governance challenges, affordable acquisition costs of handsets and their day to day use, supportive institutional arrangement that would enhance the users' ability to benefit from the technologies.

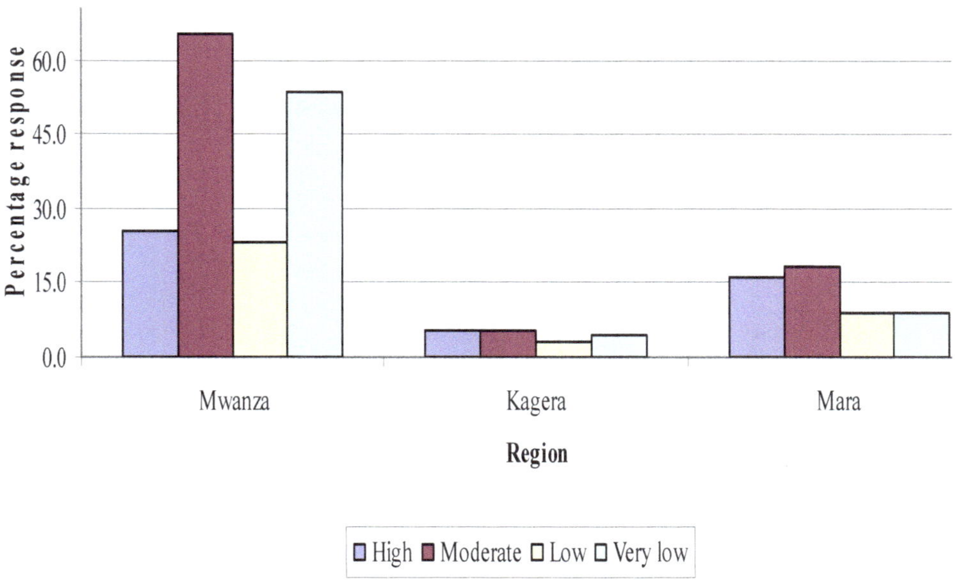

Figure 7a. Perceived cost of ownership of a mobile phone (Source: Field data).

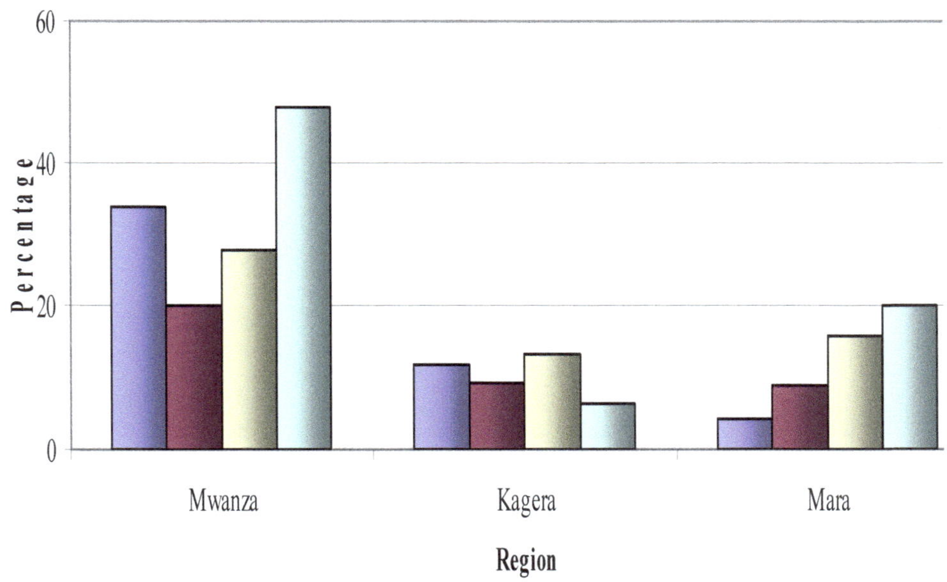

Figure 7b. Perceived mobile network coverage (Source: field data).

Results in Figure 8 shows that penetration rate has been increasing over the past seven years with a very good trend (R^2> 0.9). Penetration is measured as ratio of total mobile connections to the total population. Therefore, the results indicate that about 59% of population is already reached by the mobile connection. Further to this, Figure 9 shows that the users are responding to the increasing coverage. The number of subscribers to mobile phone services (voice, SMS and data) has been increasing in the same trend as the penetration.

Although fixed phone services have been there for long, their performance in terms of subscription has remained in the lower tail. Both penetration and subscription trends are opportunities to increase use of the mobile phones both in urban and rural areas.

There was slight variation between respondents' perception on total cost of ownership (TCO) of mobile phones and the industry position. While most respondents perceived the costs to range from high to moderate (Figure 7a), the industry data has shown that TCO had been declining sharply (R^2> 0.89) over the last four years (Figure 10). It is therefore a matter of policy to

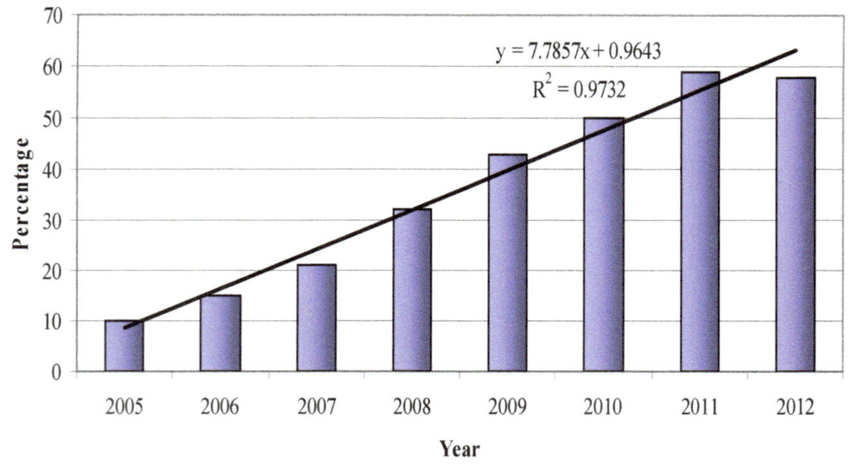

Figure 8. Mobile network market penetration (Source: TCRA (2012)).

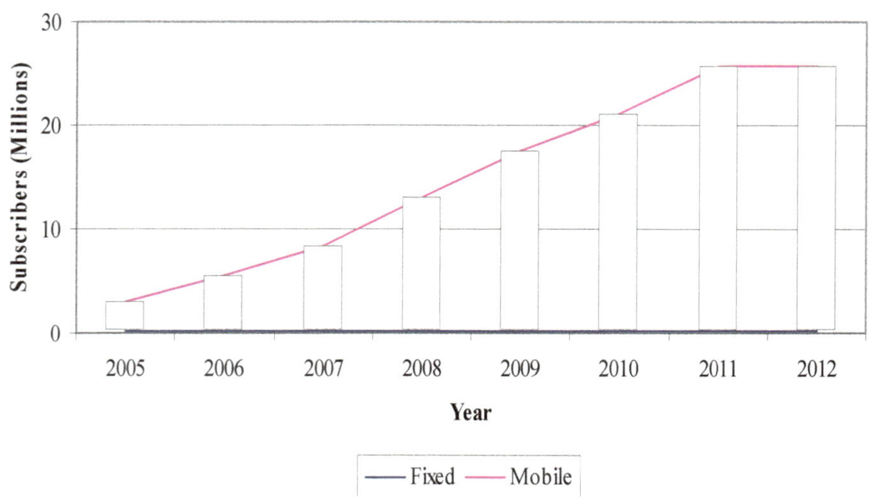

Figure 9. Subscription trends for mobile networks (Source: TCRA (2012)).

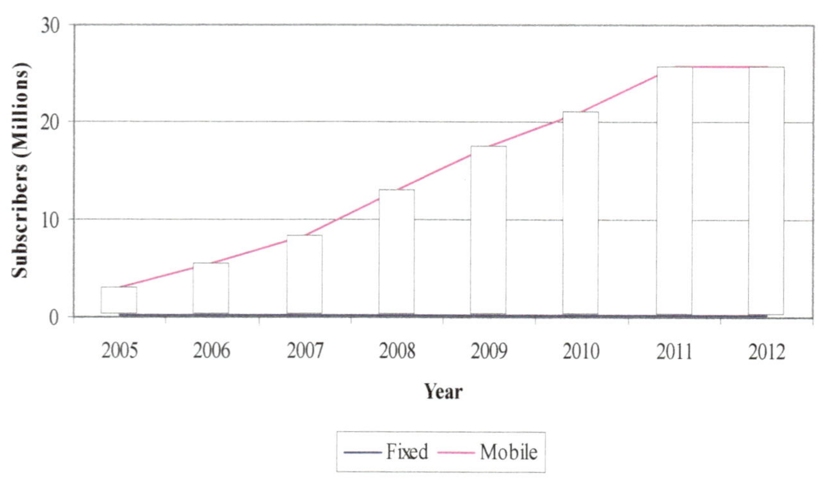

Figure 10. Total cost of ownership of a mobile phone in Tanzania (Source: MDI (2013)).

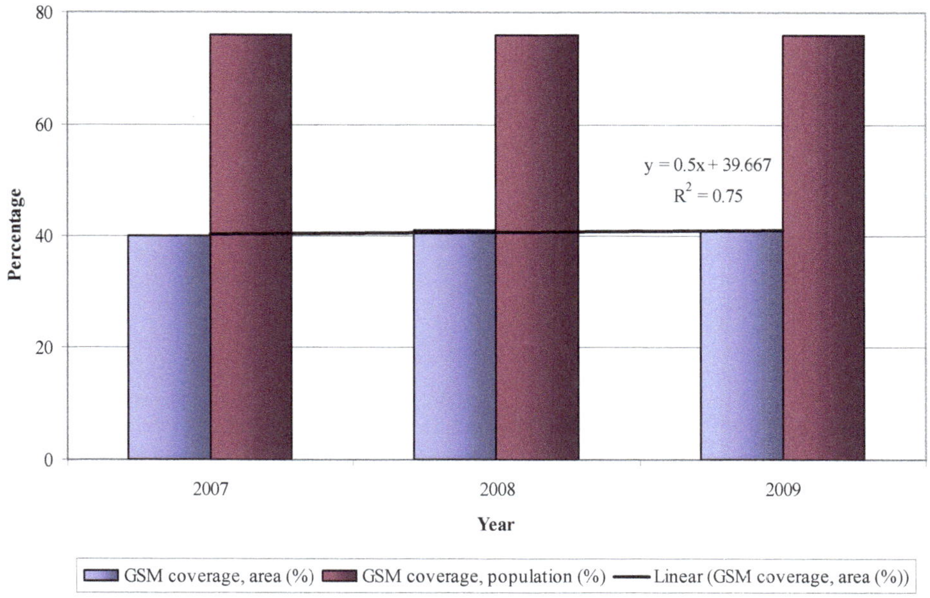

Figure 11. GSM technology coverage (Source: MDI (2013)).

ensure that the changes in TCO is reflected in the consumers' end and becomes potential for the industry's contribution to develop other sectors specially those sensitive to governance challenges like the water sector.

The industry's data also shed some light on coverage of mobile networks. Results of data analysis shown above indicate that GSM technology coverage by population has remained constant over years. However, coverage by area has had a promising increment over the same period (Figure 11).

The later development is important as it may reflect the size of rural areas being reached by this leading mobile technology.

Conclusions

In order to overcome traditional and emerging challenges on governance of water resources in LVB, new integrated and innovative tools and techniques are needed. The effectiveness of the tools and techniques depends to a large extent on their accessibility and utilization by stakeholders at each governance scale. Moreover, development and implementation of such tools and techniques should consider various technological and institutional aspects at various governance scales in order to avoid frustrating failure and wastage of invaluable resources. This study has identified main constraints affecting accessibility and utilization of mobile phones for addressing water resources governance in LVB to include income poverty and low education prevailing among stakeholders. The study serves as part of situation analysis for further research work. It partly builds the foundation upon which other activities of the research depend. Although, there are a number of constraints in accessing and utilizing mobile phones, many opportunities are availed by the industry for more inclusive technology. The ever increasing penetration rate and coverage of mobile networks combined with the wide range of solutions and platforms have potential to make mobile phones a choice technology in addressing water resources governance challenges in Tanzania. However, the role of public policy to top up as the inclusive strategies is paramount. One of them is through empowering lower scales with mobile phone access and playing a more supportive role in the regulations of mobile phone service provision.

ACKNOWLEDGEMENTS

Research team acknowledges Lake Victoria Research Initiative (VicRes), Inter-University Council for East Africa (IUCEA) and Swedish International Development Agency (SIDA) for supporting the research of which this paper is one of the outputs. The team also acknowledges the kind and material support received from stakeholders at different levels in the Lake Victoria Basin during the entire period of the research. The team wishes to thank the management and staff of Busitema University, Kigali Institute of Science and Technology and the University of Dodoma for their administrative support.

REFERENCES

Ally M (2007). Mobile Learning. Int. Rev. Res. Open Distance Learn. 8 (2):1492-3831.

Devore JL, Peck R (1993). Statistics: The exploration and analyisis of data 2nd edition Duxbury Press (Belmont,Califonia)

Duncombe R, Boateng R (2009). Mobile Phones and Financial Services in Developing Countries: A Review of Concepts, Methods, Issues, Evidence and Future Research Directions. Third World Q. 30(7):1237-1258.

East African Community (EAC) (2010). EAC Facts and Figures – 2009, Arusha: EAC Secretariat.

Frantz D (nd). Global Warming as a Water Resource Challenge in Southern Asia, http://www.globalministries.org/get-involved/justice-and-advocacy/water/global-warming-as-a-water-resour.html, [Accessed on 12 March 2014]

Fudík, K., Sabol T, Dulinová V (nd). Policy Modelling Supported by e-Participation ICT Tools, OCOPOMO Project, http://web.tuke.sk/fei-cit/furdik/publik/metteg10_OCOPOMO_paper.pdf

Hultman T (2013). Nigeria: Cell Phones for Farmers to Cut Corruption, Deliver Services, All Africa, http://allafrica.com/stories/printable/201301150001.html, [Accessed on 10 March 2014]

InfoDev (2013). The Mobile Innovation Roadshow: the 'next frontier' for economic growth in Southern Africa. http://www.infodev.org/en/Article.985.html [Accessed on 10th March 2014]

Islam MM, Masum AKM (2010). Mobile Commerce and Mobile Payment: A Study on Islamic Perspective, IIUC Studies Vol. 7.

ITU (2011). The Role of ICT in Advancing Growth in Developing Countries, International Telecommunication Union, http://www.itu.int/ITU-D/ldc/turkey/docs/The_Role_of_ICT_in_Advancing_Growth_in_LDCs_Trends_Challenges_and_Opportunities.pdf [Accessed on 10 March 2014]

Jones M, Marsden G (2006). Mobile Interaction Design, West Sussex, John Wiley and Sons.

Loudon M (2009). Mobile Phones for Data Collection, MobileActive.com,http://www.mobileactive.org/howtos/mobile-phones-data-collection [Accessed on 10 March 2014]

Mbilinyi DAS, Masanja A, Sakara A, Mahundi B (2007). Guidelines for Establishing Community Based Collaborative Fisheries Management In Marine Waters Of Tanzania, Ministry of Livestock and Fisheries Development, Fisheries Development Division & WWF Tanzania Programme Office,

MDI (2013). Mobile Industry Market Statistics, Mobile Development Intelligence (MDI), http://www.mobiledevelopmentintelligence.com/statistics#, [Accessed on 10th March 2014]

Medalye J, Hubbart JA (2008). Water resources. In: Encyclopedia of Earth. Eds. Cutler J.

Moum A (2006). A Framework for Exploring the ICT Impact on the Architectural Design Process, ITcon Vol. 11 (2006), Moum, pg. 410, http://www.itcon.org/data/works/att/2006_30.content.07890.pdf

Okurut, T. (2010) Key Note Address: Integrated Environmental Protection Approaches: Strengthening the role of Water Supply Operators in Resource Conservation, 15th International African Water Congress and Exhibition Commonwealth Resort, Munyonyo, Kampala, Uganda, 16th March 2010

Ospina AV, Heeks R (2010). Unveiling the Links between ICTs & Climate Change in Developing Countries: A Scoping Study, IDRC/CRDI

Rogers P, Hall A (2003). Effective Water Governance, Global Water Partnership Technical Committee (TEC), Background paper No. 7, Global Water Partnership, http://www.gwp.org/Global/ToolBox/Publications/Background%20papers/07%20Effective%20Water%20Governance%20%282003%29%20English.pdf [Accessed on 15 January 2015]

Schuchardt L, Steinmetz H, Ehret J, Ebert A, Schmitt TG (2004). Scalable mobile information system to support the treatment. Water Sci. Technol. 50:11(199-206).

Sehlke G (2008). Water Resources Challenges in the Next Decade and Beyond: Global Change, American Water Resources Association, http://awramedia.org/mainblog/2008/04/23/water-resources-challenges-in-the-next-decade-and-beyond-global-change/, [Accessed on 08 March 2014]

Selwhn N (2002). Telling tales on technology. Quantitative studies of Technology and education. Burlington, VT: Ashgate

TCRA (2013). Quarterly Telecommunications Statistics, Tanzania Communications Regulatory Authority, http://www.tcra.go.tz/publications/telecom.html, [Accessed on 10th March 2014]

Tebbutt THY (1998). Principles of Water Quality Control, 5th Edition, Elsevier Ltd.

UN News Centre (2013). Deputy UN chief calls for urgent action to tackle global sanitation crisis: http://www.un.org/apps/news/story.asp?NewsID=44452&Cr=sanitation&Cr1=#.UU7Y2oaCmji [Accessed on 10 March 2014]

UNEP (2010). Africa Water Atlas. Division of Early Warning and Assessment (DEWA). United Nations Environment Programme (UNEP). Nairobi, Kenya, https://na.unep.net/atlas/africaWater/downloads/chapters/africa_water_atlas_123-174.pdf [Accessed on 10 March 2014]

Unhelkar B (2009). Handbook of Research in Mobile Business: Technical, Methodological, and Social Perspectives, New York: Information Science Reference

Vodafone/Accenture (2011). Mobile Phone Technologies for Improving Agricultural Income and Productivity http://www.vodafone.com/content/dam/vodafone/about/sustainability/2011/pdf/connected_agriculture.pdf [Accessed on 10 March 2014]

Weggoro NC, Ntamubano W (2010). EAC governance structure with regard to watershade management. Paper presented at the "Symposium on science and policy linkages for trans boundary water management" http://www.eac.int/environment/index.php [Accessed: 25 July 2014]

Weston TR (2008). Development of the Marcellus Shale–Water Resource Challenge, Kirkpatrick & Lockhart Preston Gates Ellis LLP, http://www.wvsoro.org/resources/marcellus/Weston.pdf, [Accessed on 12 March 2014]

Wong S (2009). Lesson from a Participatory Trans-boundary water governance project in West Africa, In: IIED (2009) Participatory Learning and Action: community-based Adaptation to climate change, IIED, London

World Bank (2013). Philippines: Geo-tagging for Efficient, Cost-Effective Project Management http://www.worldbank.org/en/news/feature/2013/02/19/geo-tagging-for-efficient-cost-effective-project-management?cid=EXT_FBWB_D_EXT, [Accessed on 10 March 2014]

WRG2030 (2009). Charting Our Water Future: Economic frameworks to inform decision-making, Water Resource Group 2030, The Barilla Group, The Coca-Cola Company, The International Finance Corporation, McKinsey & Company, Nestlé S.A., New Holland Agriculture, SABMiller Plc, Standard Chartered Bank, and Syngenta AG., http://www.mckinsey.com/client_service/sustainability/latest_thinking/charting_our_water_future [Accessed on 13 March 2014].

Permissions

All chapters in this book were first published in AJEST, by Academic Journals; hereby published with permission under the Creative Commons Attribution License or equivalent. Every chapter published in this book has been scrutinized by our experts. Their significance has been extensively debated. The topics covered herein carry significant findings which will fuel the growth of the discipline. They may even be implemented as practical applications or may be referred to as a beginning point for another development.

The contributors of this book come from diverse backgrounds, making this book a truly international effort. This book will bring forth new frontiers with its revolutionizing research information and detailed analysis of the nascent developments around the world.

We would like to thank all the contributing authors for lending their expertise to make the book truly unique. They have played a crucial role in the development of this book. Without their invaluable contributions this book wouldn't have been possible. They have made vital efforts to compile up to date information on the varied aspects of this subject to make this book a valuable addition to the collection of many professionals and students.

This book was conceptualized with the vision of imparting up-to-date information and advanced data in this field. To ensure the same, a matchless editorial board was set up. Every individual on the board went through rigorous rounds of assessment to prove their worth. After which they invested a large part of their time researching and compiling the most relevant data for our readers.

The editorial board has been involved in producing this book since its inception. They have spent rigorous hours researching and exploring the diverse topics which have resulted in the successful publishing of this book. They have passed on their knowledge of decades through this book. To expedite this challenging task, the publisher supported the team at every step. A small team of assistant editors was also appointed to further simplify the editing procedure and attain best results for the readers.

Apart from the editorial board, the designing team has also invested a significant amount of their time in understanding the subject and creating the most relevant covers. They scrutinized every image to scout for the most suitable representation of the subject and create an appropriate cover for the book.

The publishing team has been an ardent support to the editorial, designing and production team. Their endless efforts to recruit the best for this project, has resulted in the accomplishment of this book. They are a veteran in the field of academics and their pool of knowledge is as vast as their experience in printing. Their expertise and guidance has proved useful at every step. Their uncompromising quality standards have made this book an exceptional effort. Their encouragement from time to time has been an inspiration for everyone.

The publisher and the editorial board hope that this book will prove to be a valuable piece of knowledge for researchers, students, practitioners and scholars across the globe.

List of Contributors

ELEGBEDE Bernadin
Direction Départementale de l'Hydraulique, Ouémé-Plateau, 04 BP 1412, Bénin
Centre Interfacultaire de Formation et de Recherche en Environnement pour le Développement Durable (CIFRED), Université d'Abomey-Calavi (UAC), 03 BP 1463, Jéricho, Cotonou, Bénin

EDORH A. Patrick
Département de Biochimie et de Biologie Cellulaire, Université d'Abomey-Calavi (UAC), 01 BP 526 Cotonou, Bénin
Centre Interfacultaire de Formation et de Recherche en Environnement pour le Développement Durable (CIFRED), Université d'Abomey-Calavi (UAC), 03 BP 1463, Jéricho, Cotonou, Bénin

CLEDJO Placide
Laboratoire Pierre Pagney : Climat, Eau, Ecosystème et Développement (LACEEDE), Faculté des Lettres, Arts et Sciences Humaines, Université d'Abomey-Calavi, 03 BP 1122 Cotonou, Bénin

KOUMOLOU Luc
Centre Interfacultaire de Formation et de Recherche en Environnement pour le Développement Durable (CIFRED), Université d'Abomey-Calavi (UAC), 03 BP 1463, Jéricho, Cotonou, Bénin

BOKO Michel
Laboratoire Pierre Pagney : Climat, Eau, Ecosystème et Développement (LACEEDE), Faculté des Lettres, Arts et Sciences Humaines, Université d'Abomey-Calavi, 03 BP 1122 Cotonou, Bénin

Messanvi GBEASSOR
Département de Physiologie/Pharmacologie, Faculté des Sciences, Université de Lomé (UL), BP 1515 Lomé, Togo

E. E. CREPPY
Laboratory of Toxicology and Applied Hygiene/UFR of Pharmaceutical Sciences, 146, Street Léo Saignat, 33076 Bordeaux Cedex France

Aaron Suh Tening
Department of Chemistry, University of Buea, P.O Box 63, Buea, Cameroon

George Bindeh Chuyong
Department of Botany and Plant Physiology, University of Buea, P.O Box 63, Buea, Cameroon

Godswill Azinwie Asongwe
Department of Geology and Environmental Science, University of Buea, P.O Box 63, Buea, Cameroon

Beatrice Ambo Fonge
Department of Botany and Plant Physiology, University of Buea, P.O Box 63, Buea, Cameroon

Lydia Likowo Lifongo
Department of Chemistry, University of Buea, P.O Box 63, Buea, Cameroon

Antoine David Mvondo-Ze
Department of Soil Science, Faculty of Agronomy and Agricultural Sciences, University of Dschang, BP 67 Dschang, Cameroon

Vivian Bih Che
Department of Geology and Environmental Science, University of Buea, P.O Box 63, Buea, Cameroon

Cheo Emmanuel Suh
Department of Geology and Environmental Science, University of Buea, P.O Box 63, Buea, Cameroon

Elizabeth O. Oloruntoba
Department of Environmental Health Sciences, Faculty of Public Health, College of Medicine, University of Ibadan, Ibadan, Nigeria

Mynepalli K. C. Sridhar
Department of Environmental Health Sciences, Faculty of Public Health, College of Medicine, University of Ibadan, Ibadan, Nigeria

Ramon T. Alabi
International Institute of Tropical Agriculture, Ibadan, Nigeria

Stephen A. Adebowale
Department of Epidemiology and Medical Statistics, Faculty of Public Health, College of Medicine, University of Ibadan, Ibadan, Nigeria

O. M. Nweke
Department of Geology and Exploration Geophysics, Ebonyi State University, P.M.B. 053, Abakaliki, Ebonyi State, Nigeria

O. P. Aghamelu
Department of Geology and Exploration Geophysics, Ebonyi State University, P.M.B. 053, Abakaliki, Ebonyi State, Nigeria

I. A. Obasi
Department of Geology and Exploration Geophysics, Ebonyi State University, P.M.B. 053, Abakaliki, Ebonyi State, Nigeria

Linus K. Munishi
School of Life Sciences and Bio-Engineering, Nelson Mandela African Institute of Science and Technology, P O Box 447, Arusha, Tanzania

Pudensiana C. Sawere
Pangani Basin Water Office, P O Box 3020, Arusha, Tanzania

Luqman K. Abidoye
Department of Chemical Engineering, Loughborough University, LE11 3TU, Loughborough, Leicestershire, England

Maina R. Wairagu
Department of Chemical Engineering, Loughborough University, LE11 3TU, Loughborough, Leicestershire, England

Virendra Kumar Soni
Department of Zoology, Bahauddin Science College, Junagadh-362001, Gujarat, India

Manish Visavadia
Department of Zoology, Bahauddin Science College, Junagadh-362001, Gujarat, India

Chirag Gosai
Department of Zoology, Bahauddin Science College, Junagadh-362001, Gujarat, India

Mohammad Dilshad Hussain
School of Studies in Zoology & Biotechnology, Vikram University, Ujjain-456010, M.P, India

Mukesh Singh Mewada
Department of Zoology, Indira Gandhi National Tribal University, Amarkantak-484886, M.P, India

Suresh Gor
Department of Botany, Government Science College, Gandhinagar-382010, Gujarat, India

Khwaja Salahuddin
Department of Botany, Bahauddin Science College, Junagadh-362001, Gujarat, India

Mariam Zameer
College of Earth and Environmental Sciences, University of Punjab, Lahore

Maryum shuja
College of Earth and Environmental Sciences, University of Punjab, Lahore

Arooj Ashraf
College of Earth and Environmental Sciences, University of Punjab, Lahore

Nadia Mukhtar
College of Earth and Environmental Sciences, University of Punjab, Lahore

Bilal Munir Ahmad
Microbiology Diagnostic and Research Laboratory, Mayo Hospital/Kemu, Lahore, Pakistan

Aziz-ur-Rahman
Department of Energy and Environment, Hamdard University, Karachi, Pakistan

Muhammad Owais Chughtai
Department of Energy and Environment, Hamdard University, Karachi, Pakistan

BOUFEKANE Abdelmadjid
Water and Environment Engineering Laboratory, Higher National School of Hydraulics, BP 31, 09000, Blida, Algeria

SAIGHI Omar
Laboratory of Geo-Environment, Faculty of Earth Sciences and Country Planning, University of Sciences and Technology, Houari Boumediene (U.S.T.H.B), El Alia, BP 32, Bab Ezzouar, 16111 Algiers, Algeria

K. S. Agrawal
Department of Chemical Engineering, Faculty of Technology and Engineering, The M. S. University of Baroda, Vadodara, Gujarat, India

Elsayed M. Ali. Nafea
Aquatic Environment Department Faculty of Fish Resources, Suez University, Egypt

Mohamed A. Zyada
Environmental Sciences Department Faculty of Science Dameitta University, Egypt

Harold W. T. Mapoma
School of Environmental Studies, China University of Geosciences (Wuhan), Lumo Road 388, Wuhan 430074, Hubei, China
Department of Physics and Biochemical Sciences, University of Malawi, The Polytechnic, P/ Bag 303, Blantyre 3, Malawi

Xianjun Xie
School of Environmental Studies, China University of Geosciences (Wuhan), Lumo Road 388, Wuhan 430074, Hubei, China

BOUALLA Nabila
Department of Civil Engineering, Laboratory Materials, Soil and Thermal, 31000 Oran, Algeria

BENZIANE Ahmed
Department of Civil Engineering, Laboratory Materials, Soil and Thermal, 31000 Oran, Algeria

MOUSSA Kacem
Department of Earth Sciences, Laboratory Geodynamics Basins and Sediment Budget, 31000 Oran, Algeria

DERRICH Zoubir
Department of Chemistry, Laboratory of Physico-Chemical Material/Catalyst/Environment, 31000 Oran, Algeria

Hassana I. Mustapha
Department of Agricultural and Bioresources Engineering, School of Engineering and Engineering Technology, Minna, Nigeria

Omotayo B. Adeboye
Department of Agricultural and Environment Engineering, Obafemi Awolowo University, Ile-Ife, Osun State, Nigeria

I. B. M. Kosamu
The Polytechnic, University of Malawi, Private Bag 303, Blantyre 3, Malawi

S. C. Gama
COM/MLWTB Laboratory, College of Medicine, Private Bag 360, Blantyre 3, Malawi

M. Tsakama
The Polytechnic, University of Malawi, Private Bag 303, Blantyre 3, Malawi

B. U. G. Mughogho
The Polytechnic, University of Malawi, Private Bag 303, Blantyre 3, Malawi

C. Tenthani
The Polytechnic, University of Malawi, Private Bag 303, Blantyre 3, Malawi

Jacob Atser
Department of Urban and Regional Planning Faculty of Environmental Studies, University of Uyo, PMB 1017, Uyo, Nigeria

Peter U. Udoh
Department of Architecture, Faculty of Environmental Studies, University of Uyo, PMB 1017, Uyo, Nigeria

Luke Amadi
Department of Political Science and Administrative Studies, University of Port Harcourt, Nigeria

C. U. Mac Ogonor
Department of Political Science and Administrative Studies, University of Port Harcourt, Nigeria

A. A. Adekunle
Department of Botany and Microbiology, University of Lagos, Yaba, Lagos, Nigeria

A. O. Adeniyi
Department of Botany and Microbiology, University of Lagos, Yaba, Lagos, Nigeria

Musa Dalil
Department of Urban and Regional Planning, Federal University of Technology, Minna, Niger State, Nigeria

Nda Hauwa Mohammad
Government Reserve Area, Minna, Niger State, Nigeria

Usman Mohammad Yamman
Department of Urban and Regional Planning, The Federal Polytechnic, Bida, Niger State, Nigeria

Abdul Husaini
Niger State Geographic Information Systems, Minna, Nigeria

Sanni Lekan Mohammed
Department of Urban and Regional Planning, Federal University of Technology, Minna, Niger State, Nigeria

Zahid Mahmood
Minerals Processing Research Centre, PCSIR Laboratories Complex, Lahore-54600, Pakistan

Saqib Nasir
Minerals Processing Research Centre, PCSIR Laboratories Complex, Lahore-54600, Pakistan

Nadia Jamil
College of Earth and Environmental Sciences, University of the Punjab, Lahore-54590, Pakistan

Asma Sheikh
Minerals Processing Research Centre, PCSIR Laboratories Complex, Lahore-54600, Pakistan

Adnan Akram
Minerals Processing Research Centre, PCSIR Laboratories Complex, Lahore-54600, Pakistan

N. Asuoha, Adaku
Safety Health and Environmental Department, Mobil Producing Nigeria, Nigeria

I. Osu Charles
Department of Pure and Industrial Chemistry, University of Port Harcourt, P.M.B 5323 Choba, Port Harcourt, Rivers State, Nigeria

I. Yakubu
University of Mines and Technology, Tarkwa, Ghana

D. Mireku-Gyimah
University of Mines and Technology, Tarkwa, Ghana

A. A. Duker
Kwame Nkrumah University of Science and Technology, Kumasi, Ghana

Lemma Bekele
Geography Department, Kotebe Univesity College, Addis Ababa, Ethiopia

Kya Abraham Berthe
Optical Spectroscopy and Sciences of Atmospheres Laboratory (LOSSA)

B Abdramane
Optical Spectroscopy and Sciences of Atmospheres Laboratory (LOSSA)

Setephan Reichenbach
Department of Computer Science and Engineering, University of Nebraska-Lincoln, Nebraska, United States

CHEBBAH Mohamed
Centre universitaire de Mila, LSNM, BP 26 RP, 43000, Mila, Algeria

ALLIA Zineb
Centre universitaire de Mila, LSNM, BP 26 RP, 43000, Mila, Algeria

H. N. Misra
Geography Department, University of Allahabad, India

Ashutosh Mishra
Geography Department, University of Allahabad, India

Aloyce W. Mayo
Department of Water Resources Engineering, University of Dar es Salaam, Tanzania

Twaha Mubarak
Department of Water Resources Engineering, University of Dar es Salaam, Tanzania

H. J. Mongi
Department of Information Systems, The University of Dodoma, P.O. Box 259, Dodoma, Tanzania

A. N. Mvuma
Department of Telecommunications and Communications Networks, The University of Dodoma, P.O. Box 259, Dodoma, Tanzania

S. Kucel
Department of Mechanical Engineering, Busitema University, P.O. Box 236, Tororo, Uganda

A. J. Tenge
Department of Development Studies, The University of Dodoma, P.O. Box 259, Dodoma, Tanzania

M. Gabriel
Department of Content Development, The University of Dodoma, P.O. Box 259, Dodoma, Tanzania